Advances in Industrial Control

Advances in Industrial Control is a series of monographs and contributed titles focussing on the applications of advanced and novel control methods within applied settings. This series has worldwide distribution to engineers, researchers and libraries.

The series promotes the exchange of information between academia and industry, to which end the books all demonstrate some theoretical aspect of an advanced or new control method and show how it can be applied either in a pilot plant or in some real industrial situation. The books are distinguished by the combination of the type of theory used and the type of application exemplified. Note that "industrial" here has a very broad interpretation; it applies not merely to the processes employed in industrial plants but to systems such as avionics and automotive brakes and drivetrain. This series complements the theoretical and more mathematical approach of *Communications and Control Engineering*.

More information about this series at http://www.springer.com/series/1412

Sandro Corsi

Voltage Control and Protection in Electrical Power Systems

From System Components to Wide-Area Control

Springer

Sandro Corsi
Consultant
Via N. Sauro 10, 21053 Castellanza
Italy

ISSN 1430-9491 ISSN 2193-1577 (electronic)
Advances in Industrial Control
ISBN 978-1-4471-7165-2 ISBN 978-1-4471-6636-8 (eBook)
DOI 10.1007/978-1-4471-6636-8

Springer London Heidelberg New York Dordrecht

Softcover reprint of the hardcover 1st edition 2015

Printed on acid-free paper

Springer is part of Springer Science+Business Media (www.springer.com)

In memory of my mother, Lavinia, and my father, Elio.
To my wife, Daniela, and my children, Lucia and Marco, for their love and support over the years

Series Editors' Foreword

The series *Advances in Industrial Control* aims to report and encourage technology transfer in control engineering. The rapid development of control technology has an impact on all areas of the control discipline. New theory, new controllers, actuators, sensors, new industrial processes, computer methods, new applications, new philosophies…, new challenges. Much of this development work resides in industrial reports, feasibility study papers and the reports of advanced collaborative projects. The series offers an opportunity for researchers to present an extended exposition of such new work in all aspects of industrial control for wider and rapid dissemination.

Electric power systems are an essential enabler in any country's infrastructure and there is much ongoing technological change in this field. Developing nations are constructing and commissioning new power systems all the time to advance the standard of living of their citizens. Meanwhile, mature industrial nations seem to be working to a rather different agenda. In these countries there is a "tug of war" between political ideals arising from climate-change concerns and the engineering community concerned with maintaining the viability of a working and reliable electric-power-system infrastructure. Climate-change concerns have driven the increasing use of renewable-energy power-generating resources such as wind-turbine farms, solar-power systems, and micro-generating systems like small-scale community hydro-power plants and individual domestic-scale power-generating systems. The growth in the use of these weather-dependent systems has been accompanied by moves to decommission or substantially reduce the use of coal-fired power stations, along with an increased use of natural gas for power generation and in some countries, since the Fukushima disaster in Japan, the abandonment of nuclear-fuelled power stations. All this change and the introduction of intermittent and often small-scale electric-power suppliers poses a substantial engineering challenge for the control, stability and operation of the electric-power transmission and distribution system. These challenging times for electric-power-system technology provide a very suitable context for the *Advances in Industrial Control* monograph series to publish Sandro Corsi's monograph: *Voltage Control and Protection in Electrical Power Systems: from System Components to Wide Area Control.*

Sandro Corsi was formerly a Manager at Italy's ENEL Research Department. From 2000, when significant reorganization took place, he became a senior scientist and research manager at CESI S.p.A. Milano, Italy. His main interests are in studies, design and applications of grid-voltage controls, generator controls, power electronics, HVDC systems, automation systems, security and protection systems, and advanced control and communication technologies in real power systems. For renewable-energy systems, he has long experience in studies and field applications of special control systems in photovoltaic, wind and fuel cells, generators and power stations.

His substantial industrial experience has been distilled to produce this very welcome monograph contribution for the *Advances in Industrial Control* series. The reader will find the monograph falls into two parts, Part I: *Voltage Control Resources* and Part II: *Wide Area Voltage Control*. There are three chapters in Part I covering the basis for the relationships between active and reactive power, and voltage; the equipment employed in reactive power control of voltage; and a concluding chapter on reactive power control of grid voltage.

Part II has some eight chapters covering topics such as the hierarchical control of voltage (developed through a full understanding of the hierarchical structure of transmission-grid systems), analyses for secondary and tertiary voltage regulation (SVR and TVR), power-system voltage stability, real-time indicators of voltage stability, the economic justification for voltage ancillary services, wide-area voltage protection and smart grids.

Also in Part II is a chapter (Chap. 5) containing international examples of hierarchical voltage-control systems. These are the French, the Italian, the Brazilian, the Romanian and the Chinese systems, all discussed in different levels of detail. Some of the other chapters in the monograph contain examples from other national networks. A brief instructive presentation of theory and electrical-system models is provided in an appendix.

The monograph provides some lessons on how important it is fully to understand the workings of an industrial system in order to generate control solutions that find acceptance from both process operators and company accountants. Hierarchical control is a tool that often provides a physically justifiable framework for a large-scale-system control solution, and the hierarchical voltage-control system described in this monograph is an excellent example of such a control solution.

The monograph will be of interest to a wide readership in both the industrial and academic power system and control communities. Engineers in electrical-power-system companies, manufacturers' research centres and utilities will find the monograph essential reading. Academics, final-year undergraduate students, postgraduate students and academic researchers in the disciplines of power engineering, and control engineering will also find the monograph of considerable interest.

Industrial Control Centre, Glasgow, Scotland, UK M.J. Grimble
M.A. Johnson

Preface

Nihil difficile volenti

Two fundamental control functionalities are required for any electrical power system to operate:

- The equilibrium existing between the real power delivered by generators and that absorbed by loads and losses must be continuously maintained. This equilibrium, characterised by constant frequency of a system's AC variables, is achieved by controlling the generated active power in order to compensate for variations in load;
- Grid voltages must be maintained around nominal values with power transfer taking place at low current values (i.e., operation is carried out far below that which would cause line overload) and at low losses, guaranteeing safe and reliable operation of system components (far from over- or under-voltage, which would compromise the normal working of components). Voltage management is generally achieved by controlling the available on-field reactive powers as well as transformer tap positions through on-load tap changers.

According to this schematic subdivision, the two main controls of a power system are:

- Independent of each other:
  - Constant frequency is maintained as much as possible by controlling generated active powers;
  - High, constant voltage is maintained as much as possible by controlling the system reactive powers and transformer tap positions.
- Achievable in practice by clear control solutions:
  - Generator active powers have to be modified in real time to maintain an unchanged system frequency;
  - On-field reactive powers provided by compensating equipment and generators have to be modified in real time to maintain an unchanged proper voltage in the grid.

When we consider the complexity of a multivariable, nonlinear, real power system, the above-mentioned simplified subdivision of the two control functionalities remains valid even if changes in active power also impact system voltages and, conversely, variations in voltage also change power transfer and, therefore, frequency. That notwithstanding, the main contribution to frequency change is still given by generated active powers, whereas the most relevant contribution to grid voltage change is still determined by reactive power flows.

Furthermore, the main objective of generated active power is satisfaction of load demands in accordance with contractual requirements. Any generator production required by a dispatcher for controlling system frequency is only a small amount of total power. Therefore, controlling the active power flow by changing system voltages is theoretically possible, but this method is not used for practical problems, in part because of the difficulty operators would encounter in changing grid system voltages to their correct values at any instant.

On the other hand, the main objective of system reactive power control is grid voltage sustenance. Controlling voltage by changing the active production of the generator is theoretically possible, but, again, this method is never used in practice except under extreme operating conditions where there are high system security risks.

Therefore, separating voltage control in a power system from aspects of generator speed and grid frequency control is fully justified: distinguishing between the two is not only technically possible (interactions that exist are easily managed by the main controls); it is also the common and practical way a power system operates.

This book provides a general overview and detailed descriptions of the principal voltage control aspects of a power system, distinguishing between continuously operating real-time, stabilising controls and discontinuous stepping controls, which are always ready to operate but which are active only when system voltage protection is needed. Moreover, among continuous solutions, the book distinguishes wide area transmission network control from distribution grid with renewable-energy generator control.

Introductory to an analysis of grid/wide area voltage is an in-depth survey of power system component voltage control solutions. In fact, generators, compensating equipment, power electronic equipment and transformers with on-load tap changers basically support grid voltages. Therefore, any proper analysis of multiple and overlapped grid voltage control loops asks for an all-inclusive view of the complexity of different but simultaneous control actions, as well as a deeper understanding of control functions and of each solution's performance. With this aim attention is given to:

- Differences that exist among available voltage control resources, their peculiarities and limits;
- Relevant aspects of each control system that aid an understanding of their functionalities and dynamic performance;
- Hierarchical differences among the control systems considered and coordination needed for each to realize its proper contribution;
- Benefits related to each control and the working conditions required for their achievement.

Only at the end of this thorough and complex preliminary analysis can we see clear evidence of the true benefits and limitations of the more traditional voltage control solutions and gain a better understanding and appreciation of the innovative grid voltage control and protection solutions proposed here. Such solutions aim to improve the security, efficiency and quality of electrical power system operation.

This is not a traditional academic book: it does not give a wide overview of the contributions of major experts to each considered topic, nor does it dedicate equal space to each. On the contrary, it mainly relates the author's experience and belief in each aspect's importance, its usefulness in practice and its effectiveness, giving more space to those contributions he deems most important. Other contributions are therefore mentioned when needed for comparison or to help readers see differences and/or to clear up possible misunderstanding or incorrect beliefs, some of which are widespread.

Moreover, the book does not dedicate much space to those aspects of voltage control and protection already widely addressed and gathered in classic books on power system control. The presentation of these basic topics is limited to their essential points, serving only as introductory. In keeping with this approach:

- References herein cannot cover exhaustively the available contributions to each topic; the papers most often cited are my own.
- The book is not for beginners but rather for those who are versed in electrical power systems and possess basic competencies in automatic control of dynamic/multivariable processes.

Finally, those basic competencies in electrical power systems which are assumed and therefore not assisted by the book include:

- Electrical technology and principles; electrical generators, electronic converters and electrical grids;
- Dynamic modelling of power systems in accordance with process physics, related automatic control objectives and applicable simplification of models to aid analysis/understanding of the results presented;
- Automatic control theory applied to dynamic processes and related design/analysis aspects.

Milano, Italy Sandro Corsi
February 2015

Acknowledgements

This book is primarily a collection of industrial applied research results gained mostly during the author's working years at Enel Automation Research Centre (CRA). There, under the guidance of CRA director G. Quazza and power system dynamics experts including E. Ferrari, F. Saccomanno, V. Arcidiacono and R. Marconato, my understanding of power system modelling and control greatly benefited. The work allowed me to make innovative control proposals, which have been internationally considered and appreciated.

CRA no longer exists; much of the skills its researchers gained and the research approach itself are now largely dispersed. Hence, this book has two objectives: to preserve a record of the type of applied research done there, to clearly demonstrate the relevance of dynamic analysis to electrical power system studies; and to propose innovative technology and advanced automatic control solutions.

This book would not have been written if Prof. M. J. Grimble and Prof. M. A. Johnson, of Strathclyde University, Glasgow, had not asked me to initiate such a monograph, convincing me to travel this arduous path. I sincerely thank them for their kind encouragement and support.

I also extend sincere gratitude and appreciation to the large number of collaborators whose dedication, commitment and professionalism attended these studies and for assistance given me in the laboratory development of innovative solutions and in tests on real power systems and control centres with technologically advanced prototypes. Significant, concrete experience was gained during the years of very intense applied research to which this book is largely linked.

A tribute is also due to the growing efforts of numerous international investigators in the area of power system voltage control, whose scientific contributions are directly responsible for motivating this book. Over the years a large group of international friends provided me with the opportunity to exchange competent opinions, debate the proposed results and direct my research towards solving widely recognised and still pending problems. Some are named in this book's references; others are among past and present members of IEEE and CIGRE international committees on voltage stability and control.

I am very grateful to Prof. G. N. Taranto, Federal University of Rio de Janeiro/ COPPE, Brazil, for joint collaboration on voltage instability indicator studies in the year previous to the writing of the book and whose results this book cites. I thank Prof. M. Eremia, University Politehnica of Bucharest, for recent joint collaboration on the subject of voltage and reactive power control. Technical aspects to which this book refers can be found in Chap. 7, entitled "Voltage and Reactive Power Control", of IEEE's Handbook of Electrical Power System Dynamics—Modeling, Stability, and Control (Wiley & Sons, 2013).

Thanks are also due to Springer UK editorial staff, especially to Engineering Editor, Oliver Jackson, Senior Editorial Assistant, Charlotte Cross, and to Kathy McKenzie, who served as copy editor, paying close attention to all aspects of the book's presentation.

Finally, I thank my wife, Daniela, for her profound patience, sacrifice and consistent encouragement, enduring the many long evenings and weekends I was immersed in the writing and editing of this book.

Contents

Abbreviations and Acronyms

AVR	Automatic voltage regulator
B	Susceptance
C	Capacitance, condenser
DMS	Distribution management system
DSA	Dynamic security analysis
DSO	Distribution system operator
DSTATCOM	Distribution STATCOM
ECS	Excitation control system
EHV	Extra high voltage
EMS	Energy management system
FACTS	Flexible AC transmission system
FC–TCR	Fixed capacitor and thyristor controlled reactor
GTO	Gate turn-off thyristor
HSVC	High side voltage control
HV	High voltage
I	Current
IGBT	Insulated-gate bipolar transistor
IPRT	In-phase regulating transformer
ISO	Independent system operator
IT	Information technology
LF	Load flow
LMC	Loss minimisation control
LV	Low voltage
MOSFET	Metal oxide semiconductor field effect transistor
MSC	Mechanically switched capacitors
MSR	Mechanical switched reactors
MV	Medium voltage
N	Transformer turns ratio
NVR	Network voltage regulation
OEL	over-excitation limit
OLTC	On-load tap changer
OPF	Optimal power flow

P	Active power
PAR	Phase angle regulator
PMU	Phasor measurement unit
PST	Phase shifting transformer
PVR	Primary voltage regulation
PWM	Pulse width modulation
Q	Reactive powerIn-phase and in-quadrature regulating transformers
R	Resistance
RCS	Remedial control scheme
RTU	Remote terminal unit
RVR	Regional voltage regulator
SC	Synchronous compensator
SCADA	Supervisory, control and data acquisition
SE	State estimation
SG	Synchronous generator
SGs	Smart grids
SPS	Special protection scheme
SQR	Power station secondary reactive power regulator
SSG	Static synchronous generator
SSSC	Static synchronous series compensator
STATCOM	Static compensator
SVC	Static VAR compensator
SVG	Static VAR generator
SVR	Secondary voltage regulation
TCR	Thyristor controlled reactor
TSC	Thyristor switched capacitor
TSO	Transmission system operator
TVR	Tertiary voltage regulation
UEL	Generator under-excitation limit
UPFC	Unified power flow controller
V	Voltage
VAR	Volt-ampere reactive (unit of power)
VSI	Voltage source inverter
V-WAP	Wide area voltage protection
V-WAR	Wide area voltage regulation
WAR	Wide area regulation
WAP	Wide area protection
X	Reactance
Z	Impedance

Introduction

Frequency control and automatic voltage control in electrical power systems have always been considered the two fundamental regulating functionalities [1–12]. Frequency regulation through active power control was considered, worked out and settled first and foremost because it relates more to the power system energy trade, the physical running speed of generators and the cost of energy to consumers. Voltage control problems were evident mainly to system operators ever since the time of the first grids, but the push to solve them not been adequate, and so a clear and standard solution has not yet been achieved. A full understanding of voltage problems and what are thought to be their proper solutions varies widely among cultures and countries. In addition, there are differences in the practical ways voltage is controlled in the field by individual utilities, methods that are generally inadequate and ineffective for meeting real voltage needs. There are many reasons for this deficiency:

- From a theoretical point of view, static analyses of grid voltage have not been up to the task of linking study results to real system performance. In fact, only recently have engineers come to a general consensus on what constitutes a proper dynamic analysis of voltage instability. Up until now, such uncertainty has made it difficult for system operators to trust theoretical analyses or static simulation result;
- Lack of reliable dynamic simulation tools in the past, despite their more recent availability for large power systems, with tested dynamic models including operating on-field controls and protections. Nowadays, modern simulation tools allow a system operator to better understand and reconstruct the links between voltage and reactive power of real power system phenomena;
- The complexity of the subject of voltage control, which requires, in principle, voltage regulation of all grid buses—as compared to the simplicity of single-variable frequency regulation;
- Practical difficulties system operators have in properly defining overall grid bus voltage values as well as in making decisions on the proper choice for determining the grid operating state and on tracking the system state dynamics. Moreover, the difficulty of operating on-field control of available reactive power resources and of correctly fixing their values by avoiding useless reactive power flow or

circulation between generators and transformers. Lastly, the difficulty of properly changing transformer taps with satisfactory results or facing possible stability problems related to on-load tap changer closed-loop operation;

- Inadequate voltage control proposals by manufacturers who consider local problems only, at a given bus. Such proposals often include expensive compensating equipment controls. Conversely, voltage problems often call for overall network management and coordinate control of available resources in specific areas;
- The unavailability at dispatcher control centres of system voltage continuous control, which should replace the less effective manual control/dispatching, always inadequate to meeting on-time real voltage problems. In fact, only large voltage variations can be recognised and managed in practice by manual recovery controls. From this perspective the distinction between continuous regulating control and stepping protection fades. In fact often the system operator's understanding of "voltage control" is so confused that he assumes it to be only voltage protection control;
- Previous unavailability of adequate information technology (IT), phasor measurement units (PMU) and SCADA/EMS control systems, which nowadays conversely support most ISO/TSO control centres, therefore providing in real time most of the information coming from the field, including bus voltages and available reactive power reserves;
- Unavailability of an existing SCADA/EMS system with adequate control functionality for real-time continuous, reliable and fast control of grid voltages.

The above list does not exhaust all the possible reasons voltage control problems are not adequately met today. The main reason is surely linked to the unified tradition system operators follow of manual control; for this reason *they are often far from recognising the importance of innovative automatic solutions and promoting them in practice,* in spite of their great concern about voltage problems. Often, their manual intervention comes too late, that is, in extreme voltage conditions, when uncoordinated control risks failing. *In spite of this, general dispatcher scepticism of grid voltage control improvement through technology innovation is widespread.*

This is an impasse which the present book seeks to overcome by showing in detail the practicality of the modern, conceptually new, wide area voltage control. Evidence is given of its great advantages (with respect to traditional control methods) as well what can be gained by new control functionalities which modern technologies that are now available can provide. In addition, we present the distinction between solutions of wide area voltage regulation (V-WAR) and wide area voltage protection (V-WAP), demonstrating the due synergy between them when they operate on the same power system as well as the simplicity and effectiveness of the protection solution in this case.

The new trend in electricity marketing is characterised by open access and restructuring of the industry into generation, transmission and distribution companies. This trend is accompanied by a growing demand on energy, which places power systems in higher-risk operational states. For this reason, the need to sustain grid voltages by controlling all available reactive powers is more urgent; generators with larger reactive power reserves are the prime candidates for sustaining grid voltages this way.

The problem of effective and automatic voltage and reactive power control in large and complex electrical power systems has been seriously considered since 1980; its solution demands the definition and realisation of sophisticated control schemes able to increase system security and operational efficiency. Utility and transmission system operators (TSO) are, on the one hand, certainly interested in enlarging the reliability, security and quality of supply with an effective solution that has a minimum impact on investment costs. However, due to the novelty of this area, utility companies and TSOs are monitoring one another to gain an advantage, each learning from its competitors' on-field experiences before making its own investment. Unfortunately, this approach is too prudent and often stalls decisions and delays the application of advanced and currently available wide area control solutions.

Voltage-reactive power control is indispensable in power systems that operate under normal or emergency conditions. During normal operation power/voltage control ensures the transmission of electrical energy at the required voltage quality and in conditions that are most convenient for suppliers and users. In emergencies the role of voltage control is to increase system security by enlarging the margin with respect to system voltage instability limits, thus ensuring continuity in system operation and proper operating conditions for the largest number of consumers.

Voltage regulation and reactive power compensation problems generally require a different approach whether we consider transmission or distribution level. At transmission level, the high voltage (HV) network can benefit from voltage-reactive power support that is provided by the largest generators, through which the overall grid is controlled. At the distribution level, voltage control generally concerns independent, individual distribution areas: each area represents a small, separate part of the overall distribution system.

Lastly, transmission and distribution levels are controlled by different dispatching centres and operators, and while extra high voltage (EHV) system operation strongly impacts distribution area voltages, distribution voltage variations only lightly impact on the EHV voltages. Moreover, whereas transmission networks are characterised by large generators and very low resistance lines with respect to reactance values, distribution networks have a high load density, radial structure with a higher R/X ratio, and they host few and small generators. Because of these differences, the objective and modality of voltage-reactive power (V-Q) control can vary in these overlapping networks, even when we consider future distribution grids with renewable energy applications, which have an increased number of distributed generators and an increased need of "smartness".

On a transmission grid the main voltage control objectives are:

- Continuous maintenance of a high voltage profile;
- Minimisation of power system losses;
- Increase in a system's voltage stability margin.

To achieve these objectives, there must be present on the transmission level:

- Sufficient controllable reactive power reserves to face contingencies;
- An effective and automatic wide area voltage control system.

In a distribution network the primary voltage control objectives are:

- Maintenance of voltage at consumer terminals in an acceptable range;
- Minimisation of system local losses;
- Increase in voltage stability margin in the distribution area.
- Increased distribution area ability to local load feeding with continuity, even in the presence of area switching to a stand-alone operating condition.

Achieving these objectives in each distribution area requires:

- Strong voltage support by the local transmission network (high voltages, maintained as much as possible at constant value);
- Areas with high voltage controllability by on-load tap changers (OLTC);
- Adequate compensating equipment, well located to face extreme load conditions;
- An effective, automatic distribution area medium voltage (MV) regulation system, coordinating distributed generators (when available) and OLTC controls, and operating on local compensating equipment only when needed, i.e., at the moment when switching provides real voltage support, and minimising the number of manoeuvres;
- Distribution area voltage regulation that ensures local stability ahead of large perturbations and grid separation/connection switching.

From the above considerations, transmission network voltage control and distribution network voltage control appear to be distinct solutions, to be achieved separately. Evidence must be given to the great advantage the distribution area achieves by an effective transmission network voltage control, due to the opportunity it affords for minimising the MV level control effort in terms of the number of manoeuvres made by OLTCs and compensating equipment.

Furthermore, distribution losses minimisation as well as distribution area voltage stability increase can easily be achieved under the support of a transmission grid automatic voltage control. Nevertheless, the trend in future power system development is towards distributed generators in MV grids, mainly due to reasons of security and smaller investment. This trend offers evidence that the complexity of such small grid controls must increase in order to ensure the safe operation of interconnected HV/MV grids, as well as in stand-alone MV conditions, guaranteeing load feeding with continuity, even during heavy voltage transient. Accordingly, future distribution active grids/microgrids are also referred to as “smart grids” (SGs). Nowadays, the distribution management system (DMS) does not cover the above described smart control functionalities.

Generally speaking the voltage control problem is strongly influenced by the actual operating conditions of power systems that continuously change in a way that prevents the dispatching operator from manually tracking them, as was previously mentioned. In fact, the operator recovers voltage lowering with some delay and with clear difficulty caused by uncoordinated and in some cases discretionary manual controls. Moreover, the tendency to exploit the electrical lines near the loadability limit determines a system’s voltage vulnerability increase.

An automatic voltage control system, one that is able to coordinate all control variables and available reactive power resources in the amount and at the moment they are needed, is therefore the route to important improvements at both the transmission and distribution levels.

All considerations regarding power system control addressed in this Introduction so far assume a power system's main objective is load feeding under all possible operating conditions.Therefore, load shedding for voltage control during normal operation has not been considered, except for heavy contingency use, to protect and save part of a system in the event of a real voltage instability risk. This obvious consideration does not find, in practice, coherent generalised examples, again for the reason that the shortcut route to voltage control via load shedding is often proposed as, or justified to be, the only available solution. The author is against this unnatural practice unless it is done for protection.

Load shedding around the world is currently practiced under the impetus of energy market liberalisation, which often entails a system operator's uncritical adaptation to rules of the energy market. Thus, optimisation of voltage control systems which seeks to minimise the customer's vulnerability to power interruption does not occur, an objective which is again overlooked in most monopolistic electrical energy regimes, where the absence of innovative voltage control is also relevant.

Any voltage control strategy is obviously strongly influenced by established rules of operation and a power system's available control structure, and by the commercial relationship between supplier and consumer. As such, several factors contribute to increasing the vulnerability of a system's voltage plan, i.e., energy interruptions to consumers as well as inability of a system to meet power quality requirements. These factors are

- TSO/DSO tendency to exploit electrical lines near their loadability limits;
- Frequently insufficient interconnecting lines between neighbouring power systems;
- Increasing power quality requirements of customers.

With this introduction we have sought to provide a clear, preliminary awareness of why many significant improvements in power system voltage control are still pursued. We hope this book will help further the understanding of already practicable innovative voltage control solutions in real power systems.

References

1. Concordia C (1951) Synchronous machines: theory and performance. Wiley, New York
2. Kimbark EW (1956) Power system stability, vol 3. Wiley, New York
3. Quazza G (1966) Non-interacting controls of interconnected electric power systems. IEEE Trans Power App Syst PAS-85(7):727–741
4. De Mello FP, Concordia C (1969) Concepts of synchronous machine stability as affected by excitation control. IEEE Trans Power App Syst PAS-88:316–329
5. Elgerd OI (1971) Electric energy systems theory: an introduction. McGraw-Hill, New York

6. Weedy BM (1979) Electric power systems, 3rd edn. Wiley, New York
7. Taylor CW (1994) Power system voltage stability. McGraw-Hill, New York
8. Kundur P (1994) Power system stability and control. McGraw-Hill, New York
9. Miller TJE (1982) Reactive power control in electric systems. Wiley, New York
10. Saccomanno F (1992–2003) Electric power systems: analysis and control. Wiley, New York (English version)
11. Arcidiacono V, Ferrari E, Saccomanno F (1975) Studies on damping of electromechanical oscillations in multi-machine system with longitudinal structure. IEEE/PES 1975 Summer Meeting, Paper, F75, pp 450–460
12. Quazza G (1977) Large scale control problems in electric power systems. Automatica 13:579–593

Part I
Voltage Control Resources

Chapter 1
Relationship Between Voltage and Active and Reactive Powers

The analytical link between voltage (V) and reactive power (Q) in electrical lines and loads is presented. Acceptable simplifications of equations are introduced to highlight dominant aspects of the V-Q link, which strongly impacts our understanding of grid voltage phenomena, voltage control, as well as performance required of protection solutions and design characteristics. An essential presentation of the main V-Q relationships in electrical lines often referred to by this book is provided.

1.1 Grid Short Lines

In order to establish the relationship between active and reactive power flow and voltage, one-line (Fig. 1.1a) and phasor (Fig. 1.1b) diagrams of a short line are analysed.

$\bar{v}_1$ and $\bar{v}_2$ are the phase voltages, and $\bar{\imath}_1$ and $\bar{\imath}_2$ are the currents at the line extremities. The supposed absence of the shunt admittance leads to $\bar{\imath} = \bar{\imath}_1 = \bar{\imath}_2$ in all points along the line. Denoting by φ the angle between $\bar{\imath}$ and $\bar{v}_2$, the components of the current $\bar{\imath}$ are $I_i = I\cos\varphi$ and $I_r = I\sin\varphi$. Assuming that voltage $\bar{v}_1$ is constant and $\bar{v}_2$ is the phase origin, the complex voltage drop $\Delta\bar{v} = \overline{Z}\,\bar{\imath}$ has two components:

$$\Delta u = RI_i + XI_r, \quad \delta u = XI_i - RI_r, \tag{1.1}$$

where $\bar{\imath} = I_i - jI_r$ for inductive loads, and Δu, δu are the longitudinal and transversal components of the voltage drop.

Let us denote by $\overline{S} = 3\overline{S}_2$ the three-phase complex power and by $\overline{S}_2 = V_2\left(I_i + jI_r\right) = P_2 + jQ_2$ the single-phase complex power. Introducing the active power P_2 and the reactive power Q_2, Eq. (1.1) become

$$\Delta u = \frac{RP_2 + XQ_2}{V_2}, \; \delta u = \frac{XP_2 - RQ_2}{V_2}. \tag{1.2}$$

S. Corsi, *Voltage Control and Protection in Electrical Power Systems*,
Advances in Industrial Control, DOI 10.1007/978-1-4471-6636-8_1

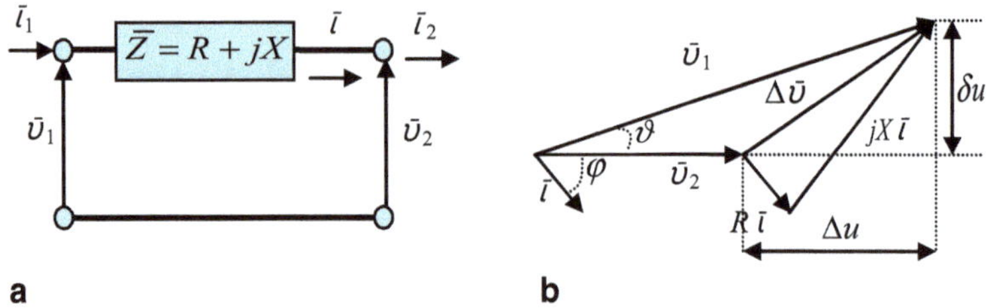

Fig. 1.1 Short line model: **a** one-line diagram; **b** phasor diagram

Because generally $R << X$ for transmission lines, it turns out that

$$\Delta u \approx \frac{XQ_2}{V_2}, \ \delta u \approx \frac{XP_2}{V_2}. \tag{1.3}$$

Therefore, the voltage drop Δu is mainly determined by the reactive power flow along the line. That is, the magnitude difference between $\bar{\upsilon}_1$ and $\bar{\upsilon}_2$ depends mainly on the reactive power transits. Instead, the active power substantially affects the phase difference between $\bar{\upsilon}_1$ and $\bar{\upsilon}_2$. According to this, the flow of reactive power has to be reduced first in order to contain the voltage drop. In practice, this is possible when the reactive power generation occurs near the consumption area.

Starting from relation (1.3) and taking into account the fact that V_1/V_2 is close to 1 per unit (p.u.), it follows that

$$\frac{\Delta u}{V_2} \approx \frac{\Delta u}{V_1} \approx \frac{XQ_2}{V_1^2} \approx \frac{Q_2}{S_{2cc}}, \tag{1.4}$$

where $S_{2cc} = V_1^2/X$ is the short-circuit power at node 2.

Characteristic *V*-Q of the system is therefore given by expression

$$V_2 \approx V_1\left(1 - \frac{Q_2}{S_{2cc}}\right). \tag{1.5}$$

This means the voltage at node 2 will depend on the amount of reactive power injected into it by node 1 as well as on the weakness of node 2.

From the above considerations, the following points concerning receiving bus and sending bus are clear:

- Voltage magnitudes are largely determined by reactive power flows;
- Reactive power flow on the line increases the voltage difference between sending and receiving buses;
- The stronger the short-circuit power in a given bus, the less reactive power flow from the line reduces its voltage with respect to the sending bus value.

1.1.1 Reactive Power Transfer

Because of the relevance of reactive power on voltage value, it is important to highlight the basic equation describing reactive power transfer along a line. Referring again to Fig. 1.1 and considering a pure inductive line, the complex power at bus 2 is given by

$$\overline{S}_2 = \overline{V}_2 \overline{I}_2^{\,*}, \tag{1.6}$$

with

$$\overline{I}_2 = \frac{\overline{V}_2 - \overline{V}_2}{jX}.$$

From Fig. 1.1, considering the angle ϑ between the two voltage phasors $(\overline{V}_1 \angle \overline{V}_2)$, we find that current I_2 becomes

$$\overline{I}_2 = \frac{V_1 \cos\vartheta + jV_1 \sin\vartheta - V_2}{jX}.$$

and

$$\overline{I}_2^{\,*} = \frac{V_1 \sin\vartheta}{X} + \frac{j(V_2 - V_1 \cos\vartheta)}{X}.$$

Therefore,

$$\overline{S}_2 = \overline{V}_2 \overline{I}_2^{\,*} = V_2 \left[\frac{V_2 \sin\vartheta}{X} - \frac{j(V_2 - V_2 \cos\vartheta)}{X} \right].$$

Writing as

$$\overline{S}_2 = P_2 + jQ_2 = \frac{V_2 V_2 \sin\vartheta}{X} + \frac{j\left(V_2 V_2 \cos\vartheta - V_2^2\right)}{X}.$$

with

$$P_2 = \frac{V_2 V_1 \sin\vartheta}{X}. \tag{1.7}$$

and

$$Q_2 = \frac{V_1 V_2 \cos\vartheta - V_2^2}{X}. \tag{1.8}$$

Analogously, the reactive power at the sending bus V_1 is given by

$$Q_2 = \frac{V_1^2 - V_1 V_2 \cos\vartheta}{X}. \tag{1.9}$$

From the above equations the following observations about the receiving bus are clear:

- The reactive powers at extreme edges of a line are inversely proportional to line reactance value. The same law of dependence applies to the active powers;
- Line reactive power flow increases if voltage at sending bus increases or if the voltage at the receiving buses decreases;
- Small magnitude variations of extreme line edge voltages (values around 1 p.u.) and cosϑ (values around 1) combined significantly impact the amount and direction of line reactive power flow. This point illustrates the critical effect that changing voltages and/or voltage vector angles has on reactive power flow control, especially in manual operation;
- The active power flow is, more than reactive power, influenced by load angle variation due to the high value of the slope of sinϑ around ϑ=0.0.

1.1.2 Losses

Minimisation of reactive power transport is also motivated by the reduction of the Joule losses on the line. In fact, these losses are expressed by the equation

$$\Delta P_L = 3RI^2 = 3R\frac{S_2^2}{\left(\sqrt{3}V_2\right)^2} = R\frac{P_2^2 + Q_2^2}{V_2^2}, \tag{1.10}$$

where not only the active power but also the reactive power flow contributes to line losses.

Because the thermal limit is defined by the admissible current for any network element, the reactive power transfer also reduces the amount of active power transmission flow. Loss minimisation does, therefore, require reactive power compensation as well as system operation at the highest voltage values.

From the above considerations, the following statement concerning reactive power line transfer is apparent:

- Reactive power flow on the line increases losses.

1.2 Reactive Loads

We further consider load voltage. Figure 1.2 shows two basic reactive loads: inductive type ($+jX$), shown in Fig. 1.2a, and capacitive type ($-jX$), shown in Fig. 1.2b.

Considering the inductive load (a) and because the current is $\pi/2$-delayed with respect to the voltage:

$$\bar{V} = jX(-jI) = XI = V.$$

Since $I = Q/V$, the right side becomes $XQ/V = V$, so

$$V^2 = XQ.$$

At small variations (i.e., differentiating),

$$2V\Delta V = X\Delta Q,$$

so

$$\Delta V = X\frac{\Delta Q}{2V}.$$

Or

$$\frac{\Delta V}{V} = X\frac{\Delta Q}{2V^2}. \tag{1.11}$$

- According to (1.11), the injection of reactive power into an inductive load determines the load's voltage increase. Obviously, for a given amount of reactive power absorbed by the load, the higher the short circuit power of the load bus the less the voltage increase effect.

Analogously, considering the capacitive load (Fig. 1.2b) with the reactive power delivered by the load:

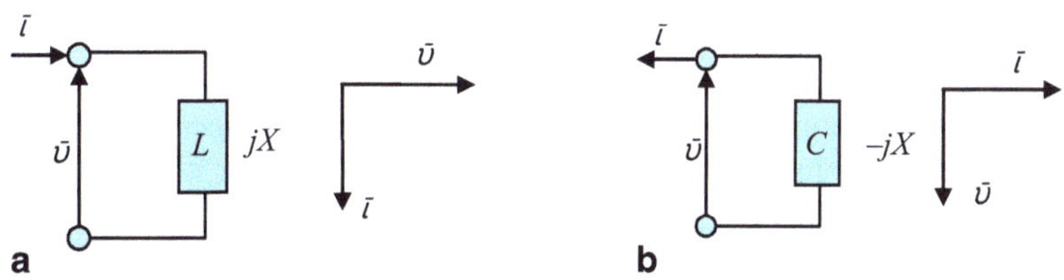

Fig. 1.2 One-line diagram: **a** inductive load ($+jX$); **b** capacitive load ($-jX$)

$$\bar{V} = -jX(+jI) = XI = V, \text{ so again, } V^2 = XQ.$$

As before,

$$2V\Delta V = X\Delta Q, \text{ giving } \Delta V = X\frac{\Delta Q}{2V},$$

and, again,

$$\frac{\Delta V}{V} = X\frac{\Delta Q}{2V^2}. \tag{1.12}$$

- According to equation (1.12) and Fig. 1.2b, it is clear that the voltage of a capacitive load increases when the load injects reactive power into the grid.

In extreme synthesis, reactive power injection into a load bus increases or reduces the bus voltage depending on the inductive or capacitive nature of the load seen by the bus. In a predominant inductive grid, as a real power system is, the evaluation of the positive or negative effects on a given bus voltage by the reactive power injection on that load bus would require a comparison between the line voltage drop (due to the reactive power flow) and the load bus voltage increase (due to the reactive power injection).

Because transmission line reactance is very small in comparison with the loads seen by the transmission buses and due to the fact that real loads are of reactive nature basically, the obvious conclusions follow:

- Transmission network voltage increase is properly operated by injecting reactive power into the load buses (inductive loads) from nearest resources. The opposite control determines the voltage lowering.
- The higher the voltage in the transmission grid the lower the control effort to sustain voltages; this is because of capacitive effect of the lines.

1.3 Grid Medium-Long Length Lines

Examining the medium/long line case, admittance can no longer be neglected as was the assumption in the short line case (Fig. 1.1). We refer now to the line Γ scheme (see Fig. 1.3), where the shunt parameters are modelled by a concentrated capacity C.

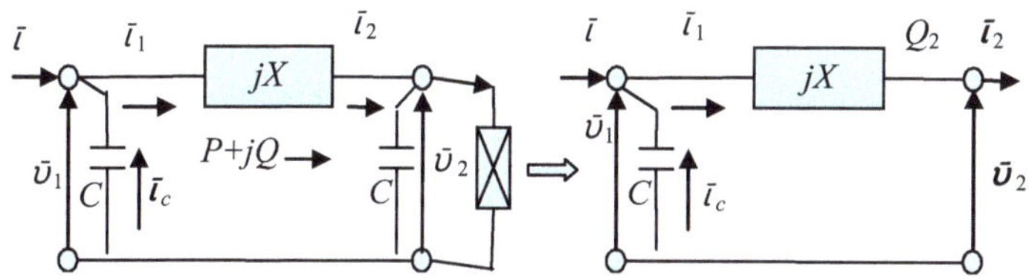

Fig. 1.3 One-line diagram Γ scheme

Equations (1.3) and (1.4) also describe the scheme in Fig. 1.3b, assuming $\bar{I}_1 = \bar{I} + \bar{I}_c$. So, the same conclusions as before can be reached when shunt admittance is considered. Obviously, when including shunt admittance, Q_2 is lower due to the contribution $\omega C V_2^2$ to the load reactive power. On the other hand, Q_2 will also depend on the additional contribution of shunt admittance at the V_1 side:

$$Q_2 = Q_1 + \omega C V_1^2 - \frac{\omega L S_{02}^2}{V_2^2},$$

where $\bar{S}_2 = 3S_{02}$ indicates the complex power at a bus of voltage V_2 and current $I_2 = S_{02} / V_2$; Q_1 is the reactive power input due to $\bar{I}$.

In this case it is interesting to note that the reactive power balance between the amounts produced by the line and absorbed by the line reactance is

$$Q_L = \omega C V_1^2 - \frac{\omega L S_2^2}{V_2^2}.$$

If S_{02} assumes the value of the line "natural power",

$$S_{02} = P_N = V_2^2 \sqrt{\frac{C}{L}} = \frac{V_2^2}{Z_C},$$

then

$$Q_L = \omega C V_1^2 - \omega C V_2^2.$$

If $V_1 = V_2$ then $Q_L = 0.0$; but $Q_2 = 0.0$ as well. Therefore, from the above conditions there exists an active power P_N injected into the load that makes the reactive balance zero.

Under these conditions in fact, the power transmitted on the line is at constant voltage magnitude and unitary power factor. If transmitted power S is higher than natural power P_N, as it is for high loaded overhead lines, then the line absorbs the reactive power.

For cables, the term ωCV^2 is predominant and the reactive power generation overcomes the absorption. For cable lines, the admissible maximum thermal power is always lower than natural power.

In conclusion:

- Shunt admittance gives, in general, a significant contribution to voltage support, partially compensating local loads, except at particular operating conditions.
- The higher the line voltage, the larger the reactive power production by the shunt admittance.

Without a doubt, the shunt admittance reactive power contribution must be considered in a power system voltage analysis, but only in terms of the reactive power resources determining the operating point; shunt admittance is not to be used for real-time voltage control. In other words, voltage control generally would not be operated by the continuous switching of operator lines on and off.

1.4 Grid as a Combination of Loads and Lines

In large real electrical grids, usually characterised by the prevailing inductive nature of the loads, the effects of the active (*P*) and reactive (*Q*) power injections into the system buses are generally seen in terms of voltage (*V*) and frequency (*f*) variations.

Voltage variations in system buses (vector *dV*) are usually described in terms of differential equations through linearised models making use of matrices $\partial V/\partial P$ and $\partial V/\partial Q$, otherwise called *sensitivity matrices*, denoted by S_{VP} and S_{VQ}, respectively. The voltage variation in a given network bus corresponds, as seen before, to active, but it mostly corresponds to reactive power flow changes on the concurrence lines in that bus. Therefore, considering overall grid buses, the matrix equation describing the dependence of vector *dV* on injection vectors *dP* and *dQ* is here after shown as

$$dV = \frac{\partial V}{\partial P} dP + \frac{\partial V}{\partial Q} dQ = S_{VP}\, dP + S_{VQ}\, dQ. \tag{1.13}$$

Coefficients of the sensitivity matrices obviously depend on load and line characteristics and show, at each bus and for a given local injection, the resultant effect of local and remote changes either in voltages, reactive power flows or line losses.

Numerically speaking and from a voltage control perspective, the most relevant matrix is $\partial V/\partial Q$, whose coefficients at a given column indicate the voltage variation contribution at each grid bus corresponding to the injection into the bus linked to the selected column of a unitary amount of reactive power.

With respect to grid voltage controllability, it is important one recognise the dependence of bus voltage on reactive power injections. In fact, active power variations are basically managed in connection with the production plan, which in turn follows contractual agreements. Therefore, it is not usually required that active power production deviate toward contributing to voltage support, unless an emergency condition accompanied by extreme security risk justifies it.

From this perspective, changes in active power production or load shedding finalised to grid voltage control is a not economical, and it is a far from ethical grid operation praxis during grid normal operating conditions. In fact, the truer and more effective variable to be used, giving the best outcome for power system voltage control, is the reactive power. This confirms the power system operator's need for adequate controllable reactive power reserves and for an adequate control system, one that is aimed at achieving a proper and safe power system voltage control.

Further information on the power system linearised model and its use in the design of grid voltage control systems is provided in Chap. 4.

References

1. Elgerd OI (1971) Electric energy systems theory: an introduction. McGraw-Hill, New York
2. Weedy BM (1979) Electric power systems, 3rd edn. Wiley, New York
3. Kundur P (1994) Power system stability and control. McGraw-Hill, New York
4. Miller TJE (1982) Reactive power control in electric systems. Wiley, New York
5. Eremia M, Shahidehpour M (2013) Handbook of electrical power system dynamics—modeling, stability, and control. Wiley, New York

Chapter 2
Equipment for Voltage and Reactive Power Control

Chapter 1 explained how voltage support requires reactive power control. In this chapter, we describe in detail the main equipment in power systems that are able to deliver or absorb the reactive power through particular aspects of control as they relate to voltage and reactive power.

Reactive power switchable compensating equipment is discussed first, then voltage and reactive power continuous control devices are described, with a distinction made between rotating electrical machines and static power electronic converters (i.e., static VAR compensator (SVC), static compensator (STATCOM) and unified power flow controller (UPFC)). A detailed description of the features of voltage-reactive power control schemes and dynamic performances is provided. Lastly, we present the on-load tap changing transformer (OLTC), a voltage discrete control device, explaining in detail its operation and applications.

2.1 Introduction

We recall from previous considerations that the practical way to perform voltage regulation in a power system requires, in large part, control of generated and consumed reactive power and its flow at different voltage levels (i.e., in transmission or distribution grids).

The main equipment in a power system is the synchronous generator, which is able to deliver or absorb a significant amount of reactive power. The automatic voltage regulator (AVR) controls the generator's excitation in order to maintain stator edge voltage at set-point value. Because this local priority control is mainly concerned with generator voltage at its MV/LV bus level, it does not use the generator's available reactive power resources to the fullest to cover the needs of real voltage control at the HV load buses.

Compensating equipment, which is generally installed in the substation, also contributes to system voltage support. This equipment can be categorised as:

S. Corsi, *Voltage Control and Protection in Electrical Power Systems*,
Advances in Industrial Control, DOI 10.1007/978-1-4471-6636-8_2

- Reactive power sources or loads; includes: shunt capacitors, shunt reactors, synchronous compensators and static compensators;
- Equipment providing compensation of line inductive reactance; includes: fixed or switched series capacitors;
- Equipment providing variable ratio on transformer windings; includes: tap-changing transformers.

Shunt reactors and capacitors as well as series capacitors are passive compensation devices: they can be permanently connected or they can be switchable. In the first case, these devices are designed as part of the basic grid, the one to be controlled; in the second, they are part of control resources that support basic grid voltages by recovering voltage variations. From here, the discussion mainly concerns "switchable" and therefore controllable reactive power resources. Stepping control of these devices is usually of a manual, local or remote type.

Synchronous and static compensators are continuous, closed-loop units. The reactive power they absorb or generate is automatically adjusted, so the voltage level of the buses to which they are connected remains constant. This equipment, similar to generators, maintains the controlled bus voltage at a set-point value. In terms of voltage control they do not differ from real generators.

The above mentioned devices can be used alone or in any combination. Some are only suitable for constant or slow-varying compensation, whereas others allow for fast variation of reactive power or shunt susceptance.

2.2 Reactive Power Compensation Devices

Shunt capacitors are used to increase a lagging power factor contribution, whereas shunt reactors are employed when leading power factor corrections are required, as in the case of lightly loaded cables. In both cases the device supplies/absorbs reactive power to recover voltage values around the nominal value.

When voltage is lowered, there is a decrease in VARs produced by shunt capacitors or absorbed by reactors; thus, when the need is greatest, capacitor effectiveness generally decreases, unless it is controlled before a significant decrease in voltage occurs. On the other hand, when loads are light, voltages are high, and the reactive power produced by capacitors or absorbed by reactors is larger than the nominal values, so their contribution increases if they are not properly controlled.

2.2.1 Shunt Capacitors

Capacitors are connected either directly to a bus bar or to the tertiary winding of a main transformer and are disposed along the route to minimise losses and voltage drops. They compensate locally the reactive power used by consumers and are distributed throughout the system. The main advantages of shunt capacitors are low cost and flexibility of installation and operation.

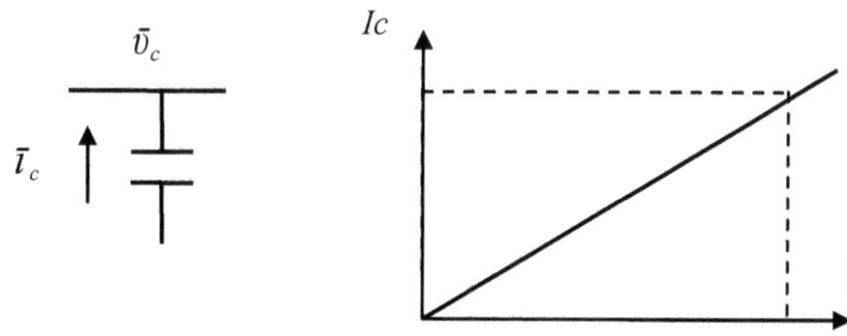

Fig. 2.1 Current-voltage characteristic of a capacitor

The shunt capacitor's principal disadvantage is that its reactive power output reduction at low voltages is proportional to the voltage squared. Moreover, switching reduces capacitor lifetime. The primary application of this compensation device is generally in distribution grids to supply the reactive power as close as possible to the point where it is consumed, i.e., at load buses.

The output characteristic (*I*–*V*) is linear, defined by rated values of voltage and current, as shown in Fig. 2.1. From Chap. 1, § 1.2:

$$\bar{V}_C = -jX_C(jI_C) = X_C I_C = V_C.$$

Therefore,

$$I_C = \frac{V_C}{X_C} = \omega C V_C \quad \text{and} \quad Q_C = \omega C V_C^2.$$

Compensation schemes include both fixed and switchable capacitor banks. In the case of transmission systems, shunt capacitors are used to compensate for inductive ($\omega L I^2$) losses and to ensure satisfactory voltage levels during heavy load conditions. Capacitor banks are switched either manually or automatically by voltage relays. Their location in the field is determined after completion of detailed power flows, contingency analysis and studies of dynamic transients.

On/off switching of capacitor banks provides a conventional means of controlling system voltages to recover large voltage deviation, typically due to the load difference from night to day or after a large contingency. It cannot contribute to real-time voltage continuous control because the number of switching manoeuvres possible is limited.

Shunt capacitors are sensitive to over-voltages and over-currents, which are limited by appropriate protections.

2.2.2 Mechanically Switched Capacitors (MSC)

The basic scheme of a mechanically switched capacitor (MSC) typically consists of a single capacitor unit or a bank of capacitor units connected to the power system either directly by a circuit breaker or via a transformer. Pre-strike- and re-strike-free circuit breakers are needed to avoid system over-voltages due to capacitor switching transients, possibly damped by series small reactors, which also reduce harmonics.

Response time is equal to the switching time dictated by the circuit breaker arrangement, which is on the order of 100 ms following initiation of an operating instruction. Frequent switching is not possible unless discharge devices are provided. Normal switching frequency is 2–4 times/day with the capacitors connected under heavy system load and disconnected under light system load conditions.

Harmonics from a power system may provide additional load (current and voltage stress) to the capacitor. Losses are very low, typically 0.02–0.05 % of the nominal MVA rating. Shunt capacitors in use range in size from a few KVARs at low voltage (LV) in a single unit to hundreds of MVARs in a bank of units at EHV applications.

Because of the linear voltage versus current characteristic, the output of a shunt capacitor during system disturbances is most unfavourable as its reactive output is proportional to the square of the voltage, thus giving a much reduced reactive power output at a reduced voltage.

2.2.3 *Shunt Reactors*

Generally, shunt reactors are used to compensate line capacitance effects by limiting voltage rise when a circuit is open or when a load is light. They are often used for EHV overhead lines longer than 150–200 km, where capacitive line-charging current flowing through high-value inductive reactance causes a voltage rise, with the highest values present at the sending end of the line.

The output characteristic (*V–I*) is linear in the operating range and deviates from linearity for iron-core or shrouded iron reactors due to saturation, as shown in Fig. 2.2.

From Chap. 1, § 1.2 and during linear performance,

$$\bar{V}_L = jX_L(-jI_L) = X_L I_L = V_L.$$

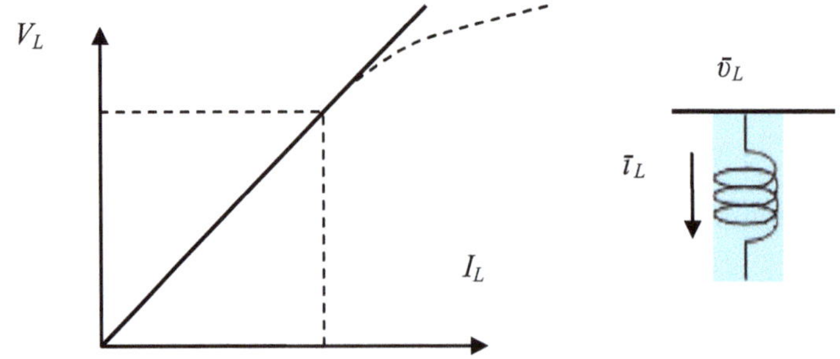

Fig. 2.2 Voltage-current characteristic of a shunt reactor

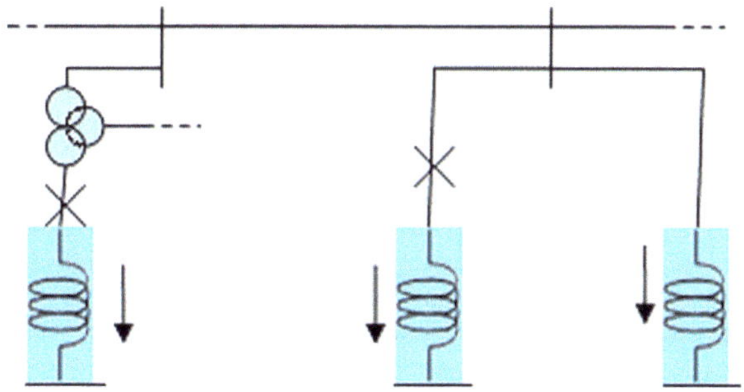

Fig. 2.3 Connection configurations of shunt reactors: switchable and permanent reactors

Therefore,

$$I_L = \frac{V_L}{X_L} = \frac{V_L}{\omega L}; \quad Q_L = \frac{V_L^{\;2}}{\omega L}.$$

Shunt reactors can be connected directly to an electric line or through a transformer installed in the terminal station (Fig. 2.3). In the case of robust systems, shunt reactors are permanently connected to the long electrical lines to limit temporary (lasting less than 1 s) or switching over-voltages up to 1.5 p.u. Additional shunt reactors can be also used on electrical lines to limit over-voltages due to lightening.

Response time is equal to the switching time given by the circuit breaker arrangement, which is on the order of 100 ms following initiation of an operating instruction. Frequent switching is not possible. Normal switching frequency is 2–4 times/day, with reactors connected under a light system load and disconnected under heavy system load conditions.

Harmonics are produced by reactor current distortion in a saturation range at higher than nominal voltages. Losses are low, typically about 0.2–0.4 % of nominal MVA rating. Shunt reactors, in use, range from a few MVARs to hundreds of MVARs at HV-EHV applications.

2.2.4 Mechanically Switched Reactors (MSR)

During heavy load conditions, shunt reactors must be disconnected; for this reason they are equipped with switching devices. Mechanically switched reactors (MSR) are used only on short lines supplied by weak systems. Shunt reactors cannot contribute to real-time voltage continuous control due to limits on the number of switching manoeuvres.

The basic scheme of the MSR typically consists of a shunt reactor connected by a circuit breaker or a disconnect switch to a transmission line bus bar or a transformer tertiary winding.

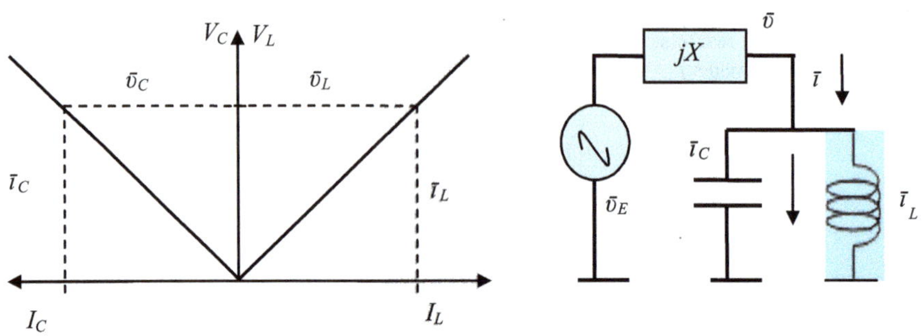

Fig. 2.4 *V–I* characteristics of shunt reactor and shunt capacitor in parallel

2.2.5 *Multiple Compensation Device Operating Point*

Representing the linear characteristics of capacitor and reactor on the same *V–I* plane, where it is assumed the current enters at the positive sign, Fig. 2.4 shows the case of the contemporary operation of a shunt capacitor in parallel with a shunt reactor.

The figure demonstrates two facts:

- The reactor absorbs current while the capacitor delivers current. According to the link between voltage and current discussed in § 2.2.1 and § 2.2.3:

$$I = I_C + I_L, \qquad V = V_C = V_L,$$

$$I_L = \frac{V_L}{X_L} = \frac{V_L}{\omega L}, \qquad Q_L = \frac{V_L^2}{\omega L},$$

$$I_C = -\omega C V = V\frac{1}{X_C}, \qquad Q_C = -\omega C V_C^2.$$

- When the reactance values of the two passive components $\bar{v}_C$ and $\bar{v}_L$ are equal in absolute value, their algebraic sum is zero; then the operating point is fixed by the external voltage, with no impact of the shunts on the grid voltage ($I = 0.0$); that is, the voltage axis also represents the resultant characteristic of the two shunts. In this case the full recirculation of reactive power between the two compensating devices is active, in the amount

$$Q_L = \frac{V^2}{\omega L} = -Q_C = \omega C V^2,$$

with $V = V_E$ (see Fig. 2.4).

This obvious result confirms that simultaneous use of the two types of permanently connected compensating equipment does not make sense. On the other hand,

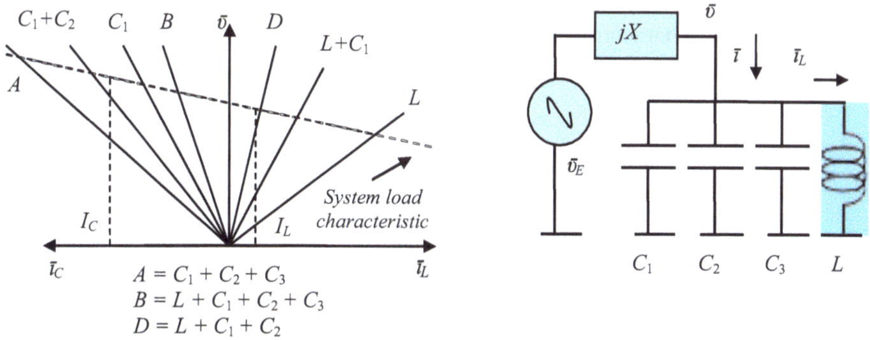

Fig. 2.5 *V–I* characteristics of MSC and MSR in parallel, fixing different operating points with system load characteristic, under a hypothesis of constant V_E and X

their switchable use by means of MSC and MSR in parallel is a possible solution for buses with a wide range of voltage variation.

It can be seen in Fig. 2.5 that the operating point is defined by the intersection of the system load characteristic with the combined characteristic of the shunt compensating equipment. The I value can be of capacitive or inductive type, depending on the combined operating conditions of the MSR and the MSC.

To better recognise the impact of compensating equipment on bus voltage it is necessary we eliminate the wrong hypothesis: namely, that the equivalent system seen by the local bus does not take into account the shunt commutations. In fact, any switching of reactor or capacitor impacts the equivalent values of V_E and X, thus changing the shape of the system load characteristic, as Fig. 2.6 shows.

Two different shunt resultant characteristics can be seen in Fig. 2.6: Case A, where the capacitive effect is dominant, and Case B, where the inductive effect prevails. Starting from A and switching off a capacitor shunt, a new resultant shunt characteristic B is determined, with current I changing from delivery I_C to absorption I_L. This produces not only a change in V but also in the equivalent V_E and X values being the grid voltages less sustained by a change in the reactive power from injection into the grid to absorption from the grid.

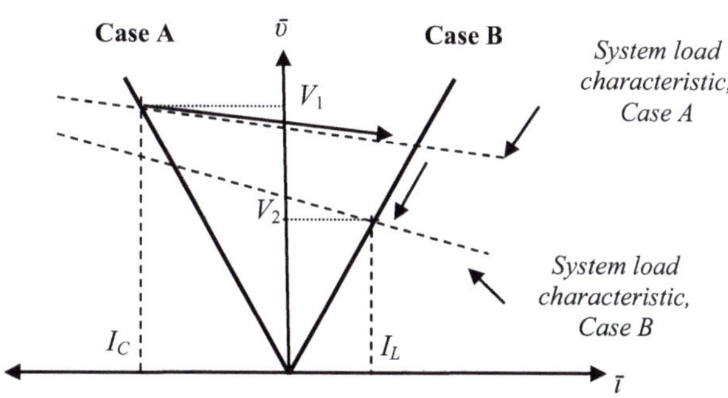

Fig. 2.6 *V–I* shunt characteristics of capacitive and inductive dominant effects and trajectories following commutation from operating point *A* to *B*, taking into account consequent change in equivalent system load characteristic

The result would be in a presumably lower V_E value and/or a different slope of the system load characteristic, thereby determining a different operating point at lower voltage V (from V_1 to V_2 in the figure).

2.3 Voltage and Reactive Power Continuous Control Devices

VAR generators are distinguished as either rotating electrical machines or static power electronics converters. The unique feature of VAR generators is their ability to deliver or absorb reactive power with continuity and in a repetitive way, without significant equipment fatigue on building materials or without internal losses. This happens until the generator's working point is maintained inside an operation within a field of continuity, bounded by capability curves that fix the maximum reactive power generation or absorption to be compatible with allowed thermal stresses, available cooling and/or design rating.

Among rotating electrical machines, the synchronous generator is not simply a megawatt generator but also a VAR generator: it allows the functional separation between the active and reactive power controls and the delivery or absorption of VARs up to limits without appreciable impact on the active power produced. Accordingly, as it is not able to deliver MW of power, a synchronous compensator is a pure VAR generator.

Considering power electronic converters, the main VAR generators are the so-called:

- Static VAR compensator (SVC);
- Static compensator (STATCOM);
- Unified power flow controller (UPFC).

Obviously, any VAR generator can, in principle, support voltages at its terminal edges or in the local grid buses.

2.3.1 Synchronous Generators

Synchronous generators are primary voltage control devices and they are primary sources of a spinning reactive power reserve. Through excitation controllability they allow continuous fast control of their stator voltages and of reactive power delivered to or absorbed by the grid. A closed-loop control scheme with an automatic voltage regulator (AVR), such as the basic one pictured in Fig. 2.7, is generally used for this purpose.

Excitation control systems (ECS) of synchronous generators can be classified as either "rotating" or "static". The first category comprises rotating machines such as DC power amplifiers that feed the synchronous generator field. Rotating types include:

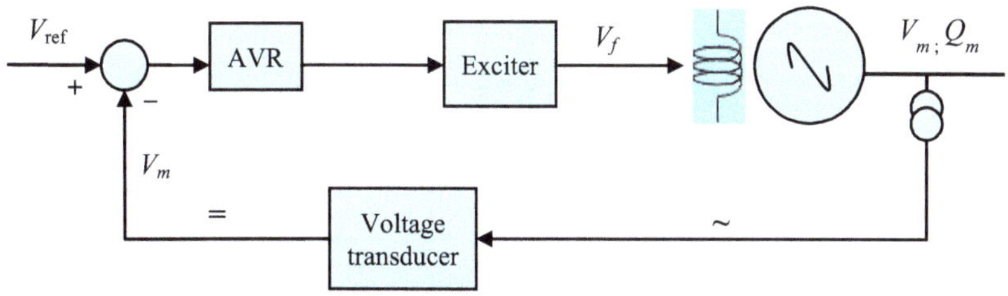

Fig. 2.7 Basic voltage control scheme of a synchronous generator

- ECS with exciting dynamo and electromechanical voltage regulator;
- ECS with exciting dynamo and electronic\microprocessor-based voltage regulator;
- ECS with alternator and rotating diodes, with electromechanical voltage regulator.

The second category considers as a DC power amplifier that feeds the synchronous generator field, a power electronic converter, typically thyristor-based. Static types include:

- ECS with static exciter and electronic/microprocessor-based voltage regulator.

ECS with Exciting Dynamo

We refer to the exciting dynamo represented in Fig. 2.8, where symbols have the obvious meaning. At high values of flux linkage ϕ, magnetic saturation $S_{at}[V_f]$ modifies the linear dependence between control current I_c and output voltage V_f, thus determining the represented static nonlinear characteristic to be taken into account in the exciter control scheme.

The exciting dynamo in the ECS is coaxial to the synchronous generator and achieves generator stator edge voltage regulation by controlling the dynamo excitation, usually through a thyristor bridge fed by an auxiliary, coaxial, permanent magnet alternator.

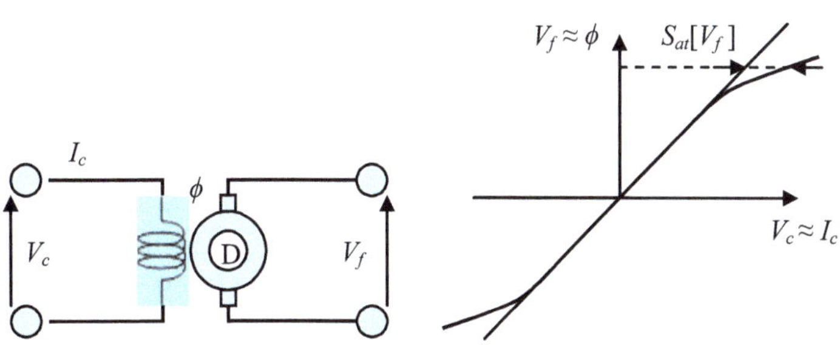

Fig. 2.8 Basic scheme of a dynamo and its magnetic characteristic

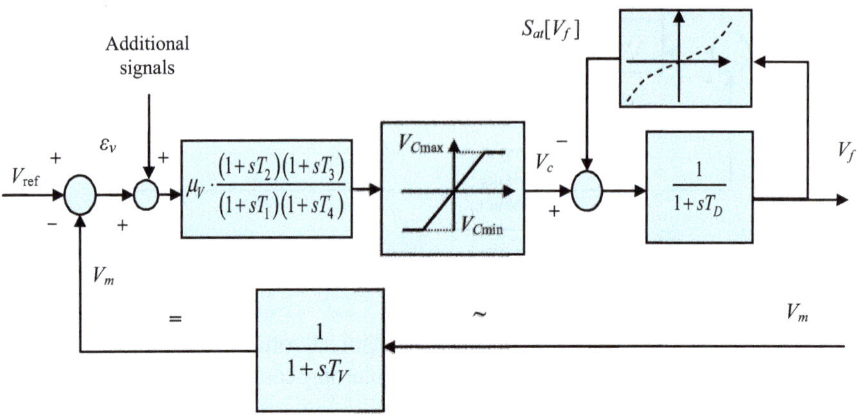

Fig. 2.9 Block diagram of ECS with exciting dynamo and AVR of electronic type

The block diagram of an ECS with dynamo is represented in Fig. 2.9, where the dynamo is characterised by a first order linear model (time constant T_D), which has a feedback that takes into account the magnetic saturation effect $S_{at}[V_f]$ on the control voltage, V_f. The control amplifier, which is of the second order when it is of an electronic type, has a linear operating field between its saturations represented by V_{Cmax} and V_{Cmin} values. A first order measurement filter (with small time constant T_V) of the generator voltage recloses the main feedback on the entering summing junction that compares the voltage set-point V_{ref} with the generator voltage measurement V_m. The voltage regulator is designed and tuned to achieve an adequate stability of the synchronous generator's voltage control loop up until the time it operates inside its saturation field.

In the case of an electromechanical voltage regulator of the first order, loop stability requires an additional negative transient feedback $sKT/(1+sT)$ from V_f to the second summing junction, where the voltage error ε_v is combined with other signals.

ECS with Alternator and Rotating Diodes

An alternator coaxial to the main synchronous generator feeds a diode bridge that provides the field excitation. Synchronous generator voltage regulation is achieved by controlling the field voltage of the exciting alternator. This solution, employing as it does rotating diodes, offers many advantages because slip rings and brushes are absent.

The block diagram of the ECS with alternator and rotating diodes seen in Fig. 2.10 is the same as that of Fig. 2.9, the case of an electronic regulator. Time T_D is the dominant time constant of the linear model representing the alternator feeding excitation windings through the bridge. $S_{at}[V_f]$ is the magnetic saturation effect on the control voltage, V_f, due to the alternator field winding.

With an electromechanical voltage regulator of the first order, outer voltage control loop stability would require an additional negative transient feedback

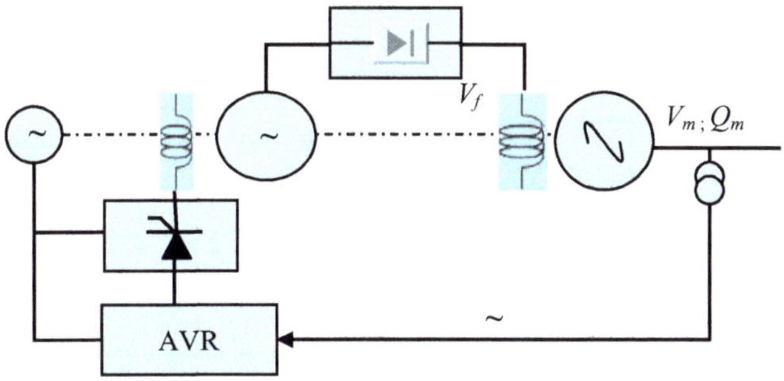

Fig. 2.10 Principal scheme of ECS with alternator and rotating diodes

$sKT/(1+sT)(1+sT_a)$ from the regulator output V_C instead of V_f (due to the absence of brushes). An additional transient feedback time constant T_a is required to compensate the delay introduced by the exciting alternator.-

ECS with Static Exciter

The thyristor-based static exciter with electronic voltage regulator is nowadays widely used throughout the world, often substituting for obsolete solutions that make use of dynamo and electromechanical amplifiers. Apart from design and maintenance advantages, the static exciter has a high speed dynamic response allowing for very fast voltage regulation.

A half-controlled—or, better, a full-controlled—thyristor bridge fed by an excitation transformer is connected to power station auxiliary bus bars, which in turn are fed by a synchronous generator through an auxiliary services transformer.

In the case of a reverse exciting current, which is needed in extreme conditions of high capacitive loads (e.g., the case of long line voltage launch during a black start-up), a scheme with two antiparallel bridges is required. Bridge output voltage control is achieved by changing the phase (with respect to the AC voltage applied to the thyristor anodes) of the current pulses (firing pulses) sent to the thyristor gates.

Regarding the dynamic performance of a static exciter, its functional characteristics allow synchronous generator excitation voltage to change from the minimum to the maximum value (the ceiling) with a delay on the order of 20 ms (the 50-Hz case). This time delay is very low with respect to the time constant of the dynamo or alternator field winding.

The static exciter should therefore improve the stability of the voltage control loop with respect to the rotating exciters. In fact, this is true only when the voltage feeding the exciter is independent from the synchronous generator voltage. This objective can be achieved by feeding the exciter transformer by an independent alternator coaxial to the synchronous generator. In turn, the alternator excitation field is fed by an auxiliary, coaxial, permanent magnet alternator through a diode bridge.

The ECS block diagram for a static exciter and electronic/microprocessor-based voltage regulator is shown in Fig. 2.11, with the exciter fed by the synchronous

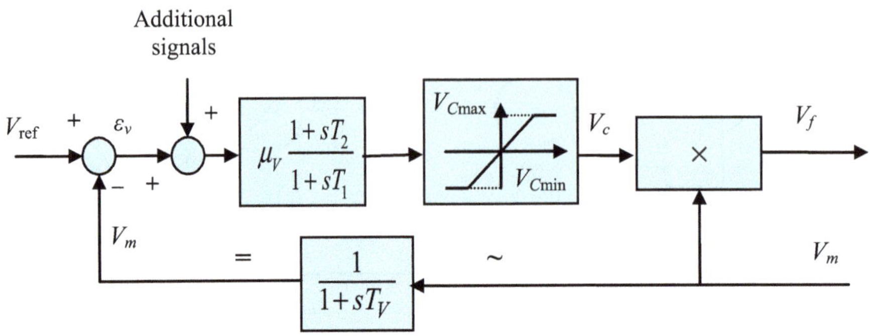

Fig. 2.11 Block diagram of static exciter fed by synchronous generator voltage output

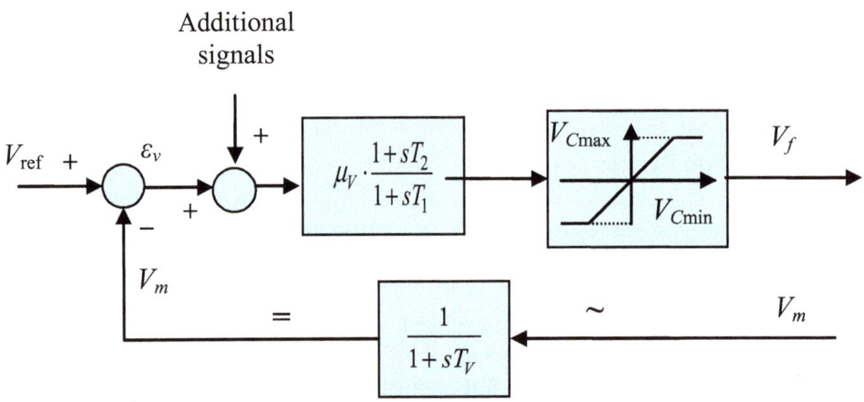

Fig. 2.12 Block diagram of static exciter fed by independent AC voltage source

generator's voltage output; Fig. 2.12 shows the exciter transformer fed by an independent AC voltage source. Figure 2.11 shows the dependence of bridge voltage output on synchronous generator stator voltage clearly is given by the multiplier. This dependence is avoided in the case shown in Fig. 2.12.

The voltage regulator with $T_1 > T_2$ allows for a voltage control loop static precision increase, simply increasing the static gain μ_V.

ECS Model Parameters

Table 2.1 shows typical values of the parameters used in the ECS block diagrams in Figs. 2.9, 2.11 and 2.12.

Synchronous Generator as Reactive Power Source

Referring to the basic scheme of a generator connected to an inductive load (see Fig. 2.13), the P_2+jQ_2 power is delivered to the load through the transformer, and the line here is represented as pure reactances jX_T and jX_L.

Table 2.1 ECS model parameters

ECS model parameters	ECS with dynamo		ECS with alternator and rotating diodes		Static ECS with thyristor bridge	
	AVR with magnetic amplifier or amplidyne	AVR with analogical electronic amplifiers	AVR with magnetic amplifier	AVR with analogical electronic amplifiers	Half-controlled bridge	Full-controlled bridge
μ_V	400	400	400	400	400	400
T_1	0.1	20	0.1	20	20	20
T_2	0.0	1.5	0.0	1.5	1.5	1.5
T_3	0.0	1.0	0.0	0.7	0.0	0.0
T_4	0.0	0.1	0.0	0.1	0.0	0.0
K	0.1	0.0	0.1	0.0	0.0	0.0
T	2.0	–	2.0	–	–	–
T_a	0.0	–	1.0	–	–	–
T_V	0.05	0.03	0.05	0.03	0.03	0.03
T_D	1.0	1.0	0.7	0.7	0.0	0.0
$V_C$max	7	7	7	7	5	5
$V_C$min	−7	−7	−7	−7	0.0	−4

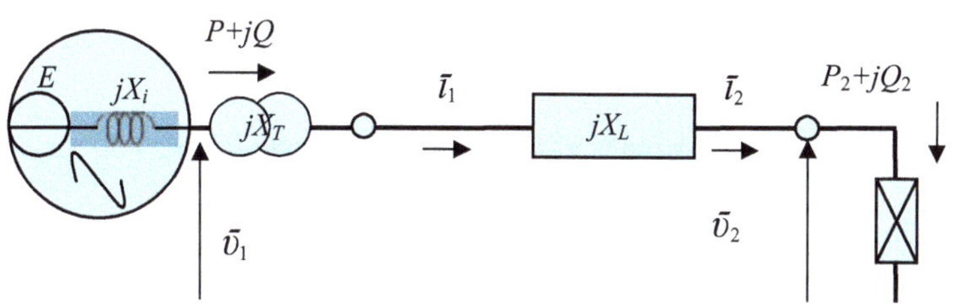

Fig. 2.13 Principal scheme of a synchronous generator connected to a load

As already introduced by Eqs. (1.7) and (1.8),

$$P_2 = \frac{V_2 V_1 \sin \vartheta}{X}, \quad Q_2 = \frac{V_1 V_2 \cos \vartheta - V_2^2}{X},$$

where $X = X_T + X_L$. The generator produces power $P + jQ = P + j(Q_2 + XI^2)$ under the hypothesis of a line and transformer without losses. The term XI^2 represents the reactive power absorbed by the line.

It is seen that, even for long lines, angle ϑ is small enough to allow the simplification: cos $\vartheta \approx 1$. Moreover, in normal operating conditions, $P \leq 1.0$ p.u.; $V_1 \approx V_2 \approx 1.0$ p.u. and sin $\vartheta \leq 0.3$, even for high values of X: $X \leq 0.3$ p.u.

Accordingly, the Q_2 equation can be simplified as follows:

$$Q_2 = \frac{V_1 V_2 - V_2^2}{X}. \tag{2.1}$$

Starting from this result, it is now easy to see that changes in Q_2 modify the amplitude V_2, under the assumption of constant V_1 (the generator is subject to AVR voltage regulation). At small variations, Eq. (2.1) becomes

$$\Delta Q_2 = \frac{V_1^0 \Delta V_2 - 2V_2^0 \Delta V_2}{X} = \left(\frac{V_1^0 - 2V_2^0}{X} \right) \Delta V_2,$$

and so

$$\Delta V_2 = \left(\frac{X}{V_1^0 - 2V_2^0} \right) \Delta Q_2 \approx -X \Delta Q_2.$$

Therefore, at constant V_1, the Q_2 load increase (ΔQ_2) determines a reduction of the local voltage V_2 proportional to the total line reactance X. In turn, the grid voltage reduction (negative ΔV_2) determines an increase in the generator-delivered reactive power $\Delta Q = -\Delta V_2 / X$.

- At constant generator voltage, generator reactive power decreases if grid voltage increases, and, conversely, it increases when grid voltage decreases. This has a stabilising effect on grid voltage. The increased or decreased reactive power output acts in a way to partially counteract grid voltage variation.

When, conversely, V_1 variation is allowed and controlled, that is,

$$\Delta Q_2 = \frac{V_1^0 \Delta V_2 + V_2^0 \Delta V_1 - 2V_2^0 \Delta V_2}{X} \approx \frac{\Delta V_1 - \Delta V_2}{X}.$$

so that

$$\Delta V_2 \approx \Delta V_1 - X \Delta Q_2,$$

then the same load increase considered before (ΔQ_2) determines a lower reduction of the local voltage V_2 due to the ΔV_1 contribution (that is, to the higher ΔQ).

In fact, in the previous case the AVR automatically controls excitation voltage to maintain constant V_1 while changing the $P+jQ$ delivered by the generator. Conversely, in the second case, V_1 changes, and in the first approximation the generator's internal e.m.f. ($\overline{E}$) is constant.

This operating condition can be described with Eq. (2.1), substituting V_1 with E, and X with $X+X_i$ (X_i being the generator's internal reactance, usually higher than X; for a turbo alternator, $X_i \approx 1.9$ p.u., whereas $X \approx 0.3$ p.u.). With so high an equivalent line reactance, generator voltage control becomes quite critical in the absence of AVR control.

- With controlled voltage set-point variation, a generator's increased or decreased reactive power output can be larger than would be possible with constant voltage, to better counteract grid voltage variation.

To meet reactive power demand, generators can be controlled inside their over- and under-excitation operating limits. Due to the slow dynamics associated with generator rotor and stator thermal stresses, a short time overload capability is allowed and can be well utilised through transient over-voltage and over-excitation limiting circuits, to better handle contingencies.

On a continuous basis, a generator is successfully operated within both its voltage limits (usually between + 5 % and − 5 %) and its over- and under-excitation limits that define the reactive power control field.

Generator capability curves are voltage-dependent; therefore, over- and under-excitation limits change dynamically. More precisely, if generator voltage increases, the over-excitation limit reduces deliverable reactive power, whereas the under-excitation limit increases absorbable reactive power.

Over- and Under-Excitation Limits

To each value of a generator's active power there corresponds a pair of reactive powers (opposite in sign) that define, at operating generator voltage, the maximum reactive power values that the generator can deliver and absorb. These two limiting values are required for thermal reasons, as is now clarified.

From no-load up to turbine nominal power, the different pairs of maximum and minimum reactive powers define a generator's limiting curves: the constant voltage family of curves usually represent a generator's operating field (between the over-excitation and under-excitation limits) (see Fig. 2.14). The maximum performance

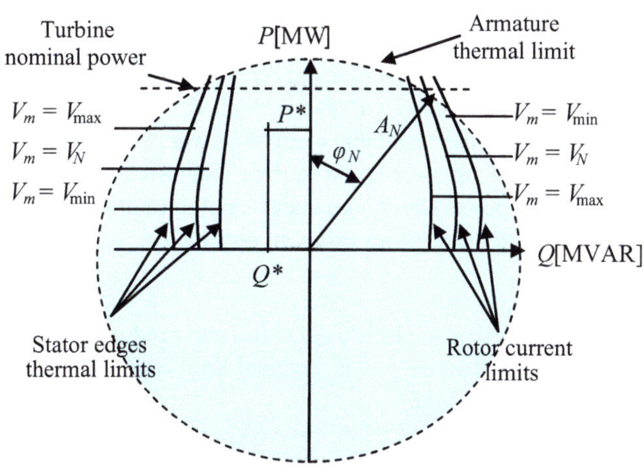

Fig. 2.14 Over- and under-excitation limiting curves of a synchronous generator

curves, also called "capability curves", are usually represented on the (*P*, *Q*)-plane with the assumption of positive delivered reactive power (the generator sees an inductive load), which, conversely, becomes negative when absorbed by the generator (capacitive load).

There are basically three thermal limits for a synchronous generator:

a. Armature thermal limit;
b. Field current limit;
c. Stator edge thermal limit.

The over-excitation field is defined by the (a) and (b) limits. The lower between the two fixes the over-excitation field; in turn, the lower of the (a) and (c) limits fixes the under-excitation field.

In addition to thermal limits, a synchronous generator has to avoid operating outside stability limits (there is a static stability limit at small perturbations), which could further reduce the allowed operating field in the (*P*, *Q*)-plane (see Chap. 8, Voltage Stability). With a proper AVR control the static stability limit is usually larger than the thermal limit.

The turbine nominal power limit fixes the maximum active power that a generator can deliver, which is more restrictive than the armature current limit, never reached unless near its points of intersection with the rotor current limits.

Inside the field bounded by the working limits considered, the synchronous generator's operating point (P^*, Q^*) can be modified to achieve the desired *P* delivery and *Q* delivery/absorption. When a surplus of reactive power reserve exists in a power system, generators can absorb reactive power operating in the under-excitation domain up to the under-excitation limits.

Figure 2.14 also shows that in an over-excitation operation, when terminal voltage decreases, the generator's available reactive power reserve increases (even instantaneously), whereas in an under-excitation operation it decreases, and vice-versa, as illustrated by the field current limit and stator edge thermal limit curves. Moreover, by decreasing the active power generated by the machine, a small increase of reactive power reserve is created during both over- and under-excitation due to the shape of the limiting curves. Therefore, the possibility of increasing the operating power distance from the reactive power limit by reducing active power production exists, but this contribution is not relevant, and its use becomes reasonable only when critical grid voltage values are reached, that is, only after the system operator has activated all other available reactive power resources in the field

Generators provide voltage regulation in their admissible voltages ranges (usually ±5 %) and reactive power (over- and under-excitation limits): the more restrictive power should dictate the limiting value. In conclusion:

- The reactive power that an electric generator can produce or absorb mainly depends on delivered active power, thermal limit capability and the terminal voltage's allowed range.

The above described generator's capability curves refer to constant cooling operating conditions. The thermal limiting curves become visibly more or less restrictive according to the higher or lower performance of the generator cooling system (determining the generator operating class).

Therefore, different families of capability curves are linkable to a given generator. Those applicable should be in accordance with generator cooling operating conditions (i.e., generator class).

All generator limits mentioned are imposed through the generator's AVR with closed-loop controls (see Chap. 3), to be tuned according to manufacturer's prescription, generator cooling characteristics, stability restrictions (if any), generator internal failures (if any, and when they do not compromise an operation) and, in practice, by exercising greater or lesser caution in the tuning of AVR limiting controls.

The generator's contribution to the grid voltage ancillary service, according to the contractual obligation with the system operator, could also interfere with the selection and tuning of a generator's allowed capability field.

Armature Thermal Limit

This generator thermal limit is linked to the stator winding resistance and stator iron losses. Copper losses (P_{Cu}) are proportional to the square of the stator current:

$$P_{Cu} = K_{Cu} I^2 = K_{Cu} \frac{P^2 + Q^2}{V_m^2}.$$

Iron losses (P_{Fe}) do not significantly depend on P and Q but grow with stator flux and therefore with stator voltage (at constant rotor speed):

$$P_{Fe} = f(V_m).$$

The stator thermal limit represents the maximum allowed total losses $P_{Cu}+P_{Fe}$ at constant voltage V_m. Because P_{Fe} is constant, the limit is represented by a circle centred on the (P, Q)-plane with radius $A(V_m)$:

$$\begin{aligned} P_{Cu} + P_{Fe} &= K_{Cu} \frac{P^2 + Q^2}{V_m^2} + f(V_m) = K, \\ P^2 + Q^2 &= \frac{V_m^2}{K_{Cu}} [K - f(V_m)] = [A(V_m)]^2. \end{aligned} \tag{2.2}$$

$[A(V_m)]^2$ in Eq. (2.2) is roughly constant in a given field due to the compensation effect of voltage dependence laws on iron and copper losses. Around nominal voltage V_{mN}, the term $A(V_m)$ is constant and indicates apparent nominal power A_N of the generator. Usually, the allowed voltage field is $V_{mN} \pm 5\,\%$. To summarise:

- Stator thermal limit is usually represented in the allowed voltage field $V_{mN} \pm 5\,\%$ by a single circle with radius indicating alternator apparent nominal power A_N.

Rotor Current Limit

Rotor current limit is linked simply to losses due to field windings resistance. As such, it is an excitation current limit. The rotor winding is dimensioned to pass the maximum excitation current that allows the generator to operate at the nominal power factor with apparent nominal power A_N in the field $V_{mN} \pm 5\,\%$.

Due to magnetic saturation, the worst operating condition occurs at $V_{m\max}$. As seen before, in Fig. 2.14, this limit is characterised by a family of voltage-dependence static curves, which, due to magnetic saturation effect, increase the reactive power field when V_m is decreased.

Stator Edge Thermal Limit

A smooth-rotor synchronous generator operating in under-excitation can reach operating conditions with very hot extreme edges, which can be dangerous and provoke failures. Warming is due to dispersion flux at the extreme rotor edges, cutting the stator's metallic structural parts at synchronous speed. Such flux cutting determines the parasite currents induced in the stator's metallic parts crossed by high density flux. The salient pole's synchronous generator, due to high reluctance in the dispersion flux course, has fewer thermal problems at stator edges.

Synchronous Generator Thermal Limits Overloading

Due to the thermal nature of the considered limits characterised by slow dynamics, generator thermal limits can be transiently overcome without provoking machine failure. Transient limit overloading allows the generator an important transient support to grid voltage in the face of large contingencies, followed by a high-risk voltage reduction.

When authorised, transient overloading is pursued by a dedicated AVR control, designed and tuned by taking into account the dynamic performance of the generator's thermal phenomena.

2.3.2 *Synchronous Compensators*

A synchronous compensator/condenser is a synchronous machine that runs without a mechanical load; depending on the value of the excitation, it can absorb or generate reactive power in the same way a synchronous generator does.

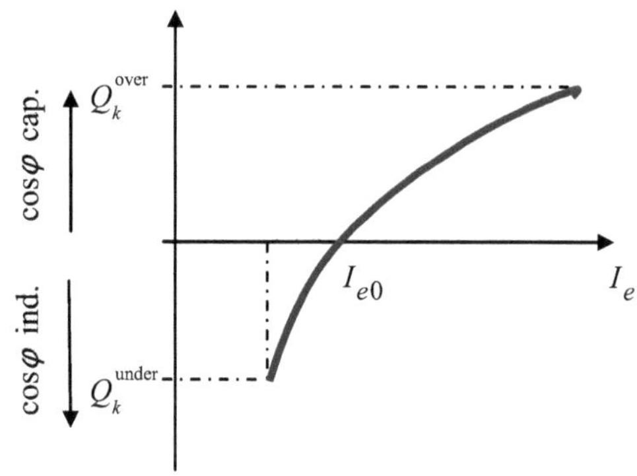

Fig. 2.15 Variation of reactive power Q_k of a synchronous compensator in terms of field current I_e

Even if synchronous compensator losses are considerable compared with static capacitors, its power factor is near zero: that is, it can operate up to the maximum delivery or absorption of reactive power in accordance with the operating limits (Fig. 2.15).

When the compensator operates in the over-excitation domain, it injects reactive power into the network but absorbs a field current that leads the voltage at the connection bus bar by 90°. When the compensator operates in the under-excitation domain it absorbs reactive power from the network and absorbs a field current that again leads the voltage by 90°. Acting on the machine excitation, the reactive power injected into/absorbed from the network by the synchronous compensator can be controlled.

When used with a voltage regulator, the compensator can automatically run over-excited at high loads and under-excited at light loads, but the amount of reactive power it supplies/absorbs will strongly depend on operating voltage set-point value.

The ratio between the maximum absorbed reactive power Q_k^{under} and the maximum produced reactive power Q_k^{over} is an important characteristic of a synchronous compensator. Due to the fact that in under-excitation the compensator becomes unstable at low field currents, synchronous compensators are usually designed with a ratio

$$\frac{Q_k^{\mathrm{under}}}{Q_k^{\mathrm{over}}} = \frac{0.5}{0.65},$$

which can be increased up to the value 1 by enlarging the compensator air gap. This has the disadvantage of increased cost.

A great advantage of the synchronous compensator is its flexible operation at all load conditions. Even if the cost of such installations is high, it is justified in some circumstances, such as at the bus bar receiving end of a long, high voltage line, where transmission at less than a unity power factor cannot be tolerated. Maintenance costs of this machine are usually high. The synchronous compensator is run up as an induction motor for a few minutes and then synchronised.

2.3.3 SVG: Static VAR Generators

The static VAR generator is a static electrical device/equipment/system that draws controlled capacitive or inductive current from the electrical power system and thereby generates or absorbs reactive power. Permanently connected and switchable reactor and capacitor banks have been always used in AC power systems to ensure desirable voltages along transmission lines and to minimise voltage variations in the face of changing daily power demand.

By definition, capacitors generate and reactors absorb reactive power when connected to an AC power source and have been used with mechanical switches for coarsely controlled VAR generation or absorption since the early days of AC power transmission. To summarise:

- VAR generators can provide stepping variable shunt impedance by synchronously switching shunt capacitors/reactors "on" and "off" the grid, from maximum capacitive to the maximum inductive output at a given grid voltage.

Continuous variable reactive power for dynamic system compensation can be achieved by different semiconductor switching devices and circuits. Thyristor valves are used almost exclusively in conjunction with capacitor and reactor banks in practical compensators. Other techniques using gate turn-off (GTO) thyristors can generate VAR without the use of capacitors and reactors. The GTO technological advance of controlled on/off capability at high power level offered new possibilities for power electronic equipment, permitting better management of transmission grids through rapid, continuous and flexible control of reactive and active power flows.

The high switching frequency of the insulated gate bipolar transistor (IGBT) allows extremely fast control, which can be used in areas such as mitigation of voltage flicker caused by electric arc furnaces. Although circuit configurations and operating principles significantly differ, all controlled static VAR generators have fast response, freely repeatable VAR output control and virtually unlimited life. To summarise:

- Most continuous VAR control can be achieved by VAR generators based on semiconductor switching devices that provide a continuous variable shunt impedance from maximum capacitive to maximum inductive output at a given grid voltage.

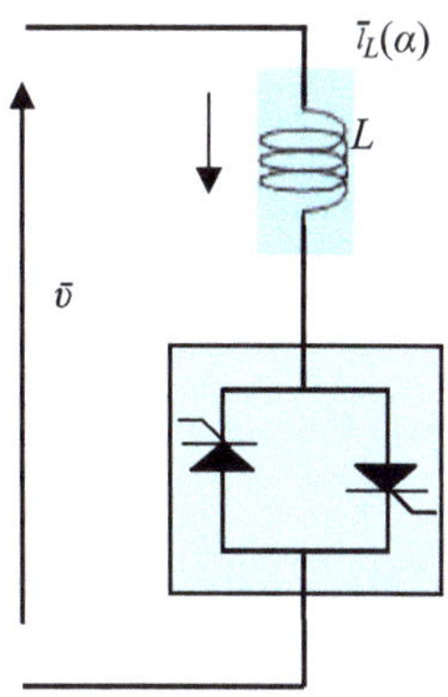

Fig. 2.16 *TCR*: Basic thyristor-controlled reactor with bidirectional thyristor valve

Thyristor-Controlled Reactor (TCR)

An elementary single-phase thyristor-controlled reactor (TCR) is shown in Fig. 2.16. It consists of a fixed reactor of inductance L and a bidirectional thyristor valve (switch). A thyristor valve can be brought into conduction by the simultaneous application of a gate pulse to all thyristors of the same polarity. The valve automatically blocks the current circulation immediately after the current crosses the zero value, unless the gate signal is reapplied.

The nonsinusoidal reactor current waveform can be controlled from maximum (thyristor valve closed) to zero (thyristor valve open) by the method of firing delay angle control (α). That is, the closure of the thyristor valve is delayed with respect to the peak of the applied voltage in each half-cycle, and therefore current conduction interval duration is controlled.

The magnitude of the current in the reactor can therefore be varied continuously from the maximum to zero by increasing delay angle α from 0 to 90 electrical degrees (which reduces the conduction angle from 180 to 0 electrical degrees). An adjustment in the reactor current can take place only once during each half cycle, in the 0-to-90-degree interval (the firing angle).

The amplitude $I_L(\alpha)$ of the fundamental reactor current $i_L(\alpha)$ can be expressed as a function of angle α:

$$I_L(\alpha) = \frac{V[1-(2/\pi)\alpha-(1/\pi)\sin(2\alpha)]}{\omega L} \tag{2.3}$$

where:

V = amplitude of applied AC voltage,
L = inductance of thyristor controlled reactor,
ω = angular frequency of applied voltage.

Figure 2.17, which shows the equivalent grid seen by the TCR, points out the effects of reducing the current. Here, a reduction in current can also be interpreted as an increase in inductor reactance. Therefore, a delay angle α increase from 0 to 90

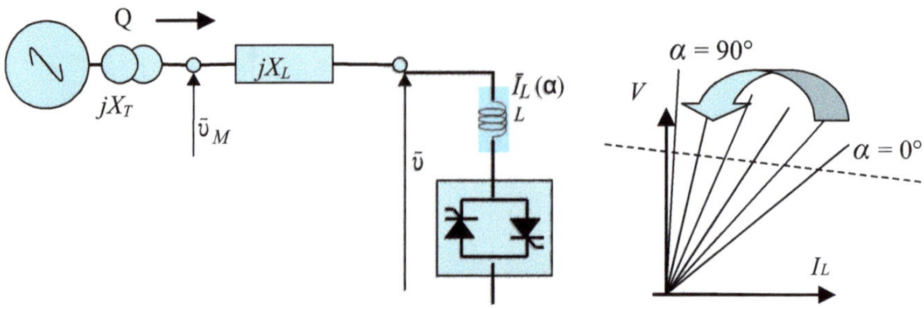

Fig. 2.17 TCR family of V–I_L characteristics at different delay angle α values and corresponding operating point given by the equivalent system load characteristics

electrical degrees represents an increase in the slope of the voltage-current linear characteristic (V–I_L) up to ∞ at the open loop condition.

The TCR generates harmonics. For identical positive and negative current half-cycles, only odd harmonics are generated. In a three-phase system, three single-phase thyristor-controlled reactors are used, usually in a delta connection. Under balanced conditions, the triplet harmonic currents (3rd, 9th, 15th, etc.) circulate within the delta-connected TCRs and do not enter the power system. The magnitudes of the other harmonics can be reduced by various methods (for example, by a 12-pulse TCR arrangement) or by filters.

Thyristor-Switched Capacitor (TSC)

The single-phase thyristor switched capacitor (TSC) consists of a capacitor, a bidirectional thyristor valve and a relatively small surge current limiting reactor (Fig. 2.18).

The purpose of the reactor is to limit switching transients and to damp inrush currents (such as what is caused by a control malfunction resulting in the capacitor switching at a "wrong time"). It may also be used to avoid resonances with AC system impedance at particular frequencies.

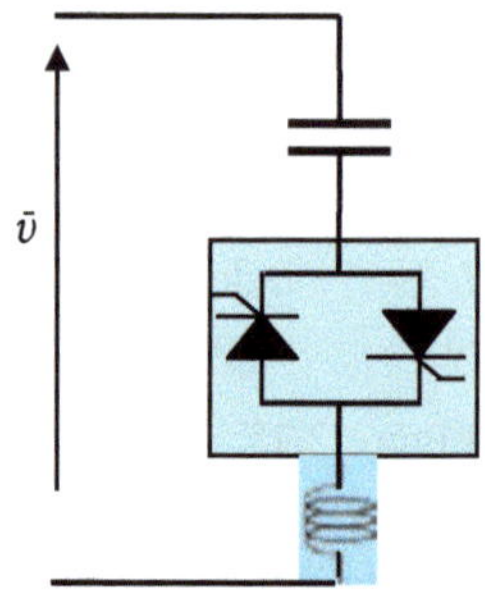

Fig. 2.18 *TSC*: Basic thyristor-switched capacitor bank with bidirectional thyristor valve

Under a steady-state condition, when the thyristor valve is closed and the TSC branch is connected to a sinusoidal AC voltage source $\bar{v} = V\sin(\omega t)$, the steady state current in the branch is given by

$$I = V\frac{X_C}{X_C - X_L}\omega C\cos(\omega t) = \omega C V_C \cos(\omega t). \tag{2.4}$$

The instant of TSC switching determines the switching-in transients.

The minimum transient in the current will occur if the TSC branch is disconnected at any point where the current is zero by a previous removal of the gate drive to the thyristor valve. At zero-current crossing, the capacitor voltage is at peak value V_C. The disconnected capacitor stays charged at this voltage value; therefore, the voltage across the nonconducting valve varies between zero and the peak-to-peak value of the applied AC voltage (see Fig. 2.18). In this case the TSC bank could be switched-in again, without any transient at the appropriate peak of the applied AC voltage.

Usually, the capacitor bank is allowed to discharge after disconnection. Thus, reconnection of the capacitor may have to be executed at residual capacitor voltage. This can be accomplished with minimum possible transient disturbance if the thyristor valve is turned on at the instant the capacitor residual voltage is equal to the applied AC voltage; that is, when the voltage across the thyristor valve is zero. Large transients in the current will occur if the firing required at zero system voltage or maximum voltage exists across the thyristor.

In practice, capacitor switching must take place at that specific instant in each cycle at which the condition for minimum transients is satisfied (zero or minimum voltage across the thyristor valve). In practical TSC circuits, there must always be sufficient series inductance to keep the slope $di(t)/dt$ of the worst switch-in transient within the limits given by the capability of the thyristor.

From the above it follows that firing delay angle control is not applicable to capacitors: capacitor switching must take place at that specific instant in each cycle at which conditions for minimum transients are satisfied. For this reason, a TSC branch can provide only a step-like change in the reactive current it draws (maximum or zero). Therefore, in order to approximate continuous current variation, several TSC branches in parallel have to be employed. The basic TSC scheme is represented in Fig. 2.19 and consists of a number of parallel TSC elements and the thyristor-firing synchronising controller.

The output characteristic of the TSC compensator is therefore discontinuous and determined by the rating and the number of parallel connected units. Unlike what occurs for the TCR, the voltage support provided is discontinuous, as shown in Fig. 2.19. The TSC, as a reactive power source, is controllable in discrete steps only. Voltage V is thus controlled in the range: $V_{ref}+\Delta V/2$, where ΔV is the deadband.

The dynamic response of a TSC is fast and typically around 0.5–1 cycle of supply frequency, but delay in the measurement and control circuits may impose settings that give a slower response for control stability of typically around 3–10 cycles of supply frequency. Harmonic generation of the TSC is zero, but there is a danger of

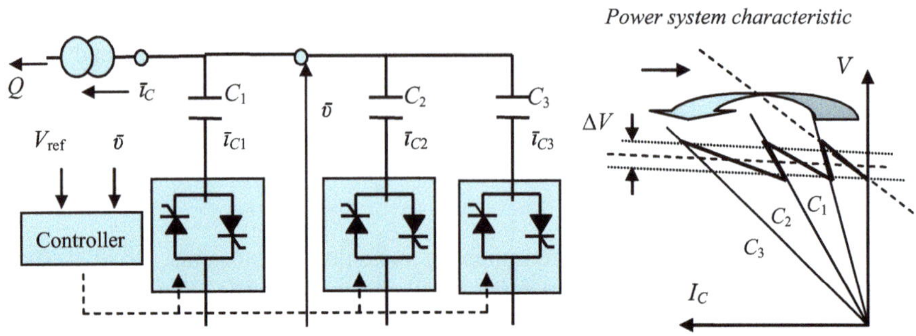

Fig. 2.19 Thyristor-switched capacitor scheme (*TSC*) and output characteristics under voltage control with deadband ΔV

series resonance when the power system experiences harmonic frequencies. Thus, careful coordination with the series reactor impedance is required.

Fixed Capacitor and Thyristor Controlled Reactor (FC-TCR)

A basic VAR generator that uses a permanently connected capacitor with a thyristor controlled reactor (FC-TCR) is shown in Fig. 2.20.

Reactor current is varied according to the TCR shown previously, which is continuously linked to the firing control of the delay angle. The fixed capacitor is, in practice, usually fully or partially substituted by a filter network that has the necessary capacitive impedance at the fundamental frequency to generate the required VARs. Furthermore, the filter provides low impedance at selected frequencies to shunt the dominant harmonics produced by the TCR.

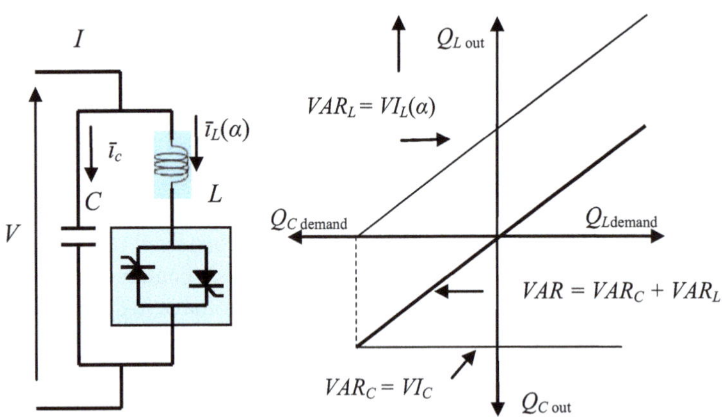

Fig. 2.20 Basic FC-TCR-type static *VAR* generator and its *VAR* demand versus *VAR* output characteristics

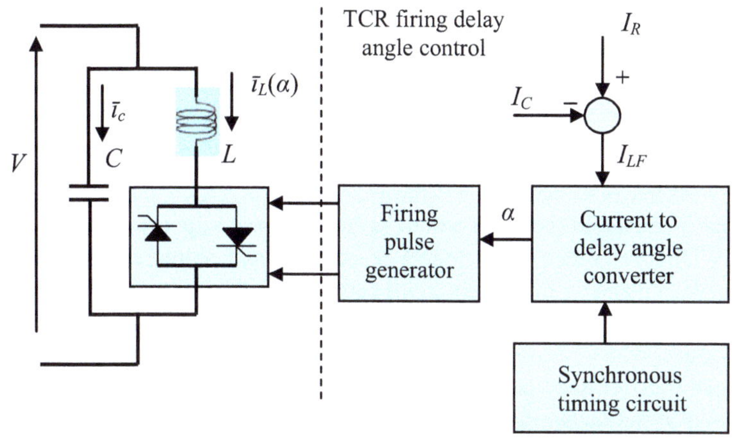

Fig. 2.21 Functional control scheme for the FC-TCR-type static VAR generator

An FC-TCR consists of a variable reactor and a fixed capacitor, with an overall VAR demand versus VAR output characteristic as shown in the Fig. 2.20: the constant capacitive VAR generation VAR_C of the fixed capacitor is opposed by the variable VAR absorption VAR_L of the TCR to yield the total output:

$$VAR = VAR_C + VAR_L.$$

At the maximum capacitive VAR output, the TCR is off ($\alpha=90°$). To decrease capacitive output, the current in the reactor is raised by decreasing delay angle α. At zero VAR output, the capacitive and inductive currents become equal and perfectly cancel the VAR output. With a further decrease in the angle α (assuming that the rating of the reactor is greater than that of the capacitor) the inductive current becomes larger than the capacitive one, resulting in a net inductive VAR output. At zero delay angle the TCR conducts current over the full 180-degree interval, resulting in maximum inductive VAR output equal to the difference between the VAR generated by the capacitor and that absorbed by the fully conducting reactor.

Control of the TCR in the FC-TCR VAR generator must fulfil four basic functions, as shown in Fig. 2.21:

1. *Synchronous timing function*: Usually provided by a phase-locked loop circuit that runs in precise synchronism with the AC system voltage and generates appropriate timing pulses with respect to the voltage peak.
2. *Reactive current conversion to firing angle*: Usually provided by a real-time function, implementing the mathematical relationship between the amplitude of the TCR fundamental current ($I_L(\alpha)$) and the delay angle α, given by Eq. (2.3).
3. *Computation of the required fundamental reactor current* $I_L(\alpha)$: This derives from the required generator VAR output current that is provided as the amplitude reference input to the VAR generator control. This is achieved by simply subtracting the scaled amplitude of the capacitor current I_C from the reference I_R.

Positive polarity for I_R indicates inductive output current, and negative polarity indicates capacitive output current.

4. *Thyristor firing pulse generation function*: This is accomplished by the firing pulse generator circuit, which produces a relatively large gate current pulse for the thyristors in response to firing angle converter output.

Taking a "black box" perspective, the FC-TCR-type VAR generator can be considered a controllable source of reactive power whose output faithfully follows an arbitrary input reference signal (either reactive current or susceptance) in a given frequency band and within the specified capacitive and inductive ratings.

The dynamic performance of the VAR generator is limited by the firing angle delay control, which results in a time lag with respect to the input reference signal. The actual transfer function of the FC–TCR-type VAR generator can be expressed with a transportation lag in the form

$$G(s) = ke^{-T_d s},$$

where s is the Laplace transform operator and T_d the transportation lag corresponding to firing delay angle α. For a single-phase TCR, the maximum transportation lag is $1/2f$, where f is the frequency of the applied voltage. For a three-phase six-pulse balanced TCR, the maximum transportation lag is $1/6f$.

Thyristor-Switched Capacitor, Thyristor-Controlled Reactor (TSC-TCR)

The thyristor-switched capacitor, thyristor controlled reactor (TSC-TCR)-type compensator was developed primarily for dynamic compensation of power transmission systems with the intention of minimising standby losses and providing increased operating flexibility.

A basic single-phase TSC-TCR is shown in Fig. 2.22. It consists of n TCS branches and m TCR branches. Often, $m=1$, as in the figure, and the VAR reactor rating fully compensates each VAR capacitor rating. In this case of relatively small

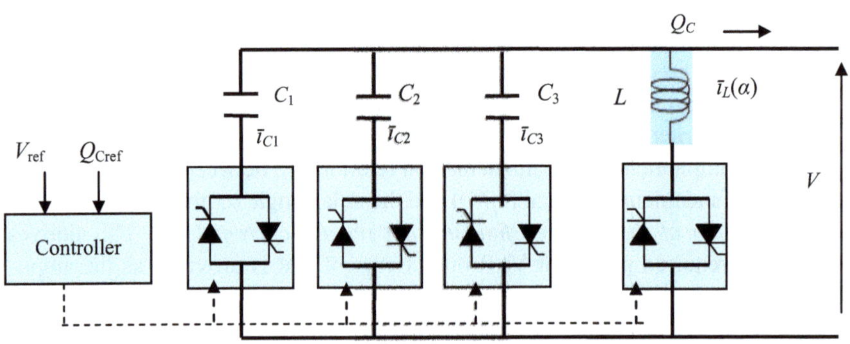

Fig. 2.22 Basic TSC-TCR-type static VAR generator

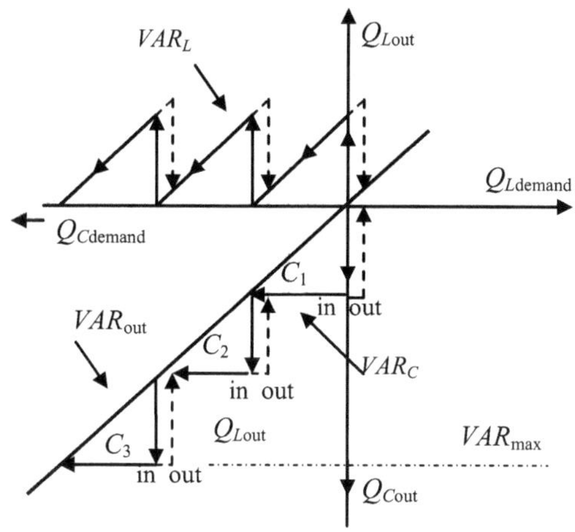

Fig. 2.23 Basic TSC-TCR-type static VAR generator: VAR demand versus VAR output characteristic

inductive VAR output, the TCR is used to cancel the surplus of each step-capacitive VAR. During the first interval, one capacitor bank is switched in (by firing, for example, the thyristor valve linked to C_1, and, simultaneously, the current in the TCR is set by the appropriate firing delay angle so that the sum of the TSC VAR output (negative in Fig. 2.23) and that of the TCR (positive) equals the capacitive output required.

The operation can be described as follows. Under the assumption of a TCR with the same VAR rating of each TSC, total capacitive output range is divided into n intervals. In the first interval, the output of the VAR generator is controllable by TCR in the zero to VAR_{max}/n range, where VAR_{max} is the total VAR rating provided by all TSC branches.

In the subsequent intervals, achievable by switching on available capacitor banks, the output is controllable inside the single TSC interval by the TCR that absorbs surplus capacitive VARs inside a single TSC interval only. This is not an operating limit, due to the possibility of the entire capacitor bank switching within one cycle of the applied AC voltage. However, to avoid an indeterminate switching condition, the TCR VAR rating has to be somewhat larger in practice than that of the TSC in order to provide enough overlap (hysteresis) between the switching "on" and "off" VAR levels. The resulting VAR_{demand} versus VAR_{output} static characteristic (VAR total) of the TSC-TCR-type VAR generator is therefore linear, covering the field from VAR_{max} capacitive to TRC inductive (Fig. 2.23).

A functional control scheme of the TSC-TCR-type VAR generator is shown in Fig. 2.24.

The TSC-TCR fulfils three major functions:

1. *Needed TSC branches*: Determines the number of TSC branches to be switched in, to approximate the required capacitive output current (with a positive surplus) and computes the amplitude of the inductive current needed to cancel the surplus capacitive current. The input current reference I_{ref} representing the magnitude of the requested output current is divided by the amplitude I_C of the current that a

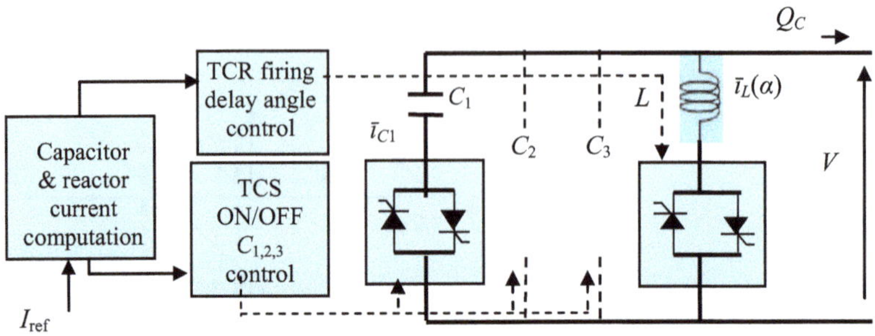

Fig. 2.24 Functional control scheme for the TSC-TCR-type static VAR generator

TSC branch would draw at the given amplitude V of the AC voltage. The result, rounded to the next integer, gives the number of capacitor banks needed. The difference in magnitude between the sum of the capacitor currents and I_{ref} gives the amplitude I_{LF} of the fundamental reactor current required.

2. *TSC switching*: Controls of the switching of the TSC branches in a "transient-free" manner. It switches the capacitor bank either when the voltage across the thyristor valve becomes zero or when the thyristor voltage is at a minimum.
3. *TCR current control*: Varies the current in the TCR by firing delay angle control in the identical way seen for the FC-TCR.

From the black box perspective the TSC-TCR-type VAR generator, in a manner similar to its FC-TCR counterpart, can be considered a controllable source of reactive power whose output follows an arbitrary input reference signal (either reactive current or reactive power or voltage or susceptance). Therefore, an external observer, by monitoring the output current, generally would not be able to detect internal capacitor switching nor to tell whether the VAR generator employs fixed or thyristor-switched capacitors.

Note that the maximum switching out delay for both the TSC and TCR is a half cycle. The transfer function of the TSC-TCR-type VAR generator is the same as that of its FC-TCR counterpart, except that the maximum transportation lag T_d, encountered when the capacitive output is to be increased, is theoretically twice as large; that is, $1/f$ or $1/3f$, respectively.

In the literature we can find the term "mechanically switched capacitor, thyristor controlled reactor (MSC-TCR)"; it is not to be confused with TSC-TCR. The MSC-TCR arrangement does not have the response or the repeatability of operation generally needed for dynamic compensation of power systems. In the final analysis it is the response of mechanical breakers that mostly determines the elapsed time between capacity VAR demand and actual capacitive VAR output. Since precise control of the mechanical switch closure is not possible, the capacitor bank must be switched without any appreciable residual charge to avoid high and possibly damaging transients.

Considering a practical discharge time of about 3–4 cycles (a typical breaker closing time is about 3–7 cycles), under worst-case conditions the MSC delay time may be 6–11 times higher than the TSC's. Moreover, switched capacitors in a compensator are occasionally subjected to repeated switching operations. Considering a typical life of 2000–5000 operations for mechanical breakers or switches, repeated switching of capacitor banks in practice would be prohibited, and therefore the actual VAR output would have to be allowed to settle above or below the amount needed for proper compensation.

2.3.4 Static VAR Compensators (SVCs)

A static VAR compensator (SVC) is a static VAR generator (SVG) capable of drawing capacitive and/or inductive current from an electrical power system. Among flexible AC transmission systems (FACTS), the SVC is dedicated basically to voltage support. An SVC operates the electronic soft switching of its own shunt reactors and/or capacitors, achieving continuous reactive power variation. It is ideally suited to the control of varying reactive power demand of large fluctuating loads and overvoltage dynamics due to load rejection. It is also used in HVDC converter stations and where fast control of voltage and reactive power flow is required.

The term "static" is used to indicate that the SVC, unlike the synchronous compensator, has no moving or rotating components. The static VAR compensator is defined as a shunt-connected static VAR generator or absorber whose output is adjusted to exchange capacitive or inductive current so as to maintain or control specific parameters of the electrical power system (typically bus voltage). Thyristor-switched or thyristor-controlled capacitors/inductors and combinations of such equipment with fixed capacitors and inductors come under the SVC category.

According to TSC-TCR control philosophy (see Fig. 2.22) but applied to a system with larger inductive resources, SVCs consist of *n* TCS and *m* TCR branches, separately controllable. In this case of relatively large inductive VAR output, the TCR is not used to cancel the surplus of each step-capacitive VAR (see Chap. 2, § 2.3.3.4, on TSC-TCR control), but TCS and TCR are controlled to move with continuity from maximum capacitive contribution (all TCS "on" and all TCR "off") to maximum inductive absorption (all TCS "off" and all TCR "on" at the maximum inductive VAR output), along a linear characteristic with positive slope, fixed by an SVC control law.

The primary SVC objective is the increase of power transmission capability of a local grid. That is, the reactive power output (capacitive or inductive) of the SVC is varied to control the voltage at the transmission network local bus so as to maintain the desired active power flow under possible system disturbances and contingencies.

The output of the static VAR generator is modified to stabilise specific parameters of the power system in the face of network contingencies such as load changes, generator and line outages, and disturbances such as faults and load rejections. These parameters usually fall into one of two categories:

a. Direct voltage support, to increase power system capability and prevent voltage instability;
b. Transient and dynamic stability improvements, to increase the first swing stability margin and provide damping for power oscillations. This second point is analysed in Chap. 8, on power system stability.

Compensation along the electrical line requires a midpoint dynamic shunt. With a dynamic compensator at the midpoint, symmetrical line behaviour is achieved. Midpoint voltage will vary with load, and an adjustable midpoint susceptance serves to maintain constant voltage magnitude at the load, so the advantage of SVC use is evident. With rapidly varying loads, reactive power demand can be speedily corrected by SVC, with small overshoots and voltage rise.

Power system oscillation damping can be also obtained by rapidly changing the output of the SVC from capacitive to inductive so as to counteract the acceleration or deceleration of interconnected machines.

SVC Voltage Control Requirements

In order to meet the general compensation requirements of a power system, the output of the static VAR generator is controlled to either maintain or vary the voltage at selected buses in the transmission grid. A general control scheme converting SVG into SVC is shown in Fig. 2.25.

The static VAR generator includes TSC and TCR banks that are controlled so that the amplitude I_{comp} of the reactive compensating current i_{comp} drawn from the power system follows the current reference I_{ref}.

The power system as seen by the SVC is represented with an equivalent dynamic generator including an electromechanical loop and AVR control loop, together with

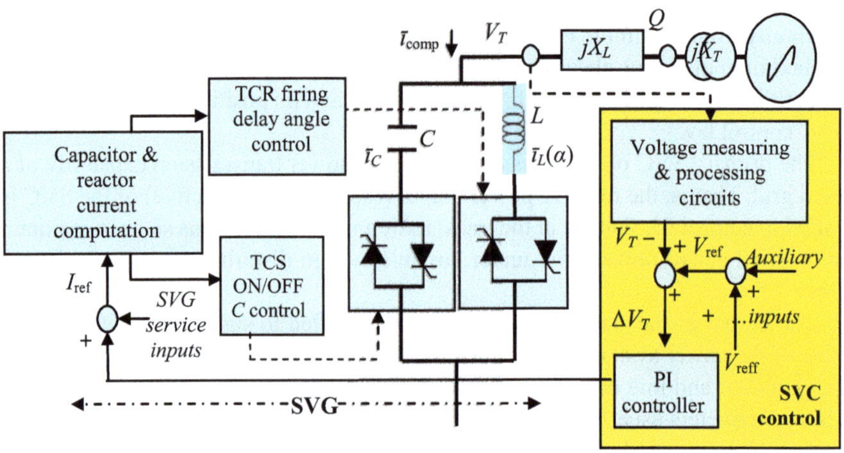

Fig. 2.25 General control scheme of a SVC

a source equivalent impedance Z (including generator transformer X_T and transmission line impedances and loads X_L). The terminal voltage of the considered HV bus is therefore characterised by a generally varying amplitude V_T and angular frequency ω_T.

Basic SVC control operates an SVG as a perfect terminal voltage regulator in the following way:

- Amplitude V_T of terminal voltage v_T is measured and compared with voltage reference V_{ref}.
- Error ΔV_T is processed and amplified by a PI controller to provide the current reference I_{ref} for the SVG. Therefore, I_{comp} is closed-loop controlled via input I_{ref}, so the V_T is maintained with continuity and precisely at the level of the V_{ref} in the event of power system and load changes.

If a proper compensation of an AC power system requires some specific variation in the amplitude of the terminal voltage against time or some other variable (for example, reactive power output control), then an appropriate correcting signal, derived from the auxiliary inputs, can be added to the fixed reference V_{reff} in order to obtain the desired actual and variable V_{ref} that controls the terminal voltage in a closed-loop manner: this describes the SVC case in which SVG control imposes a compound with positive slope on V_{ref} in a way that easily intersects the load characteristic at a V_{ref} value that provides a control margin (see below).

SVC Regulation Slope

Usually the SVC is not used as an integral terminal voltage regulator, but rather the terminal voltage is allowed to vary in proportion to the compensating current. There are several reasons for this so-called "compounding" (see § 3.3.3):

- Allowing a regulation "drop" extends the linear operating range of a compensator within available maximum capacitive and inductive ratings;
- Terminal voltage is allowed to be smaller than nominal no-load value at full capacitive compensation and to be higher than the nominal value at full inductive compensation;
- Perfect regulation (zero drop or slope) could result in poorly defined operating point with a tendency toward oscillation if the system impedance were to exhibit a "flat" region (low impedance) in the operating frequency range of interest;
- A regulation slope tends to enforce automatic load sharing between SVCs as well as between other voltage-regulating devices located in the same grid area.

The desired terminal voltage versus output current characteristic of the SVC can be established by an external control loop on I_{comp} by use of the previously defined auxiliary input, as shown in § 3.3.6 on SVC control.

A typical SVC characteristic (terminal voltage V_T versus output current $\overline{i}_C$, with a given slope) is shown in Fig. 2.26, together with a particular load characteristic of

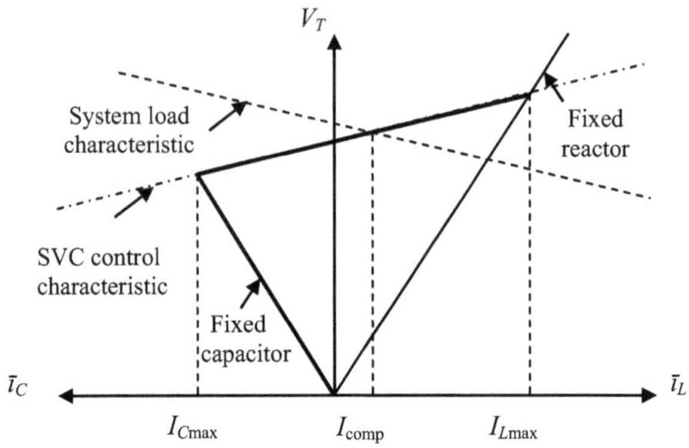

Fig. 2.26 Voltage-current characteristic of a static VAR compensator

an AC system (the voltage versus a reactive current's linear characteristic, assuming constant voltage at the equivalent generator and Z as a reactance, where $X=X_T+X_L$).

The system load characteristic, due for example to load rejection, intersects the SVC V–I characteristic at a point that asks for the inductive compensating current I_{comp}. The intersection point of the load lines with the voltage vertical axis defines terminal voltage variation without compensation. The terminal voltage variation with compensation is entirely determined, in Fig. 2.26, by the SVC regulation slope. More detail on SVC control is given in § 3.3.6 on SVC control.

2.3.5 *Static Compensators (STATCOMs)*

The STATCOM is a static synchronous generator (SSG) having characteristics similar to the synchronous compensator (capable of delivering lagging/leading VARs to a load or to a bus in the power system). As an electronic device it has no inertia, which brings with many associated advantages: better dynamics, lower investment cost and lower operating and maintenance costs.

Generally speaking, an SSG is a self-commutated switching power converter supplied from an appropriate electric energy source and operated to produce a set of adjustable multiphase voltages, which may be coupled to an AC power system for the purpose of independent exchange of controllable real and reactive powers. When the active energy source is replaced by a DC capacitor, which cannot absorb or deliver real power except for short durations, the SSG becomes a static synchronous compensator (see Fig. 2.27). Therefore, the STATCOM possesses no long-term energy support on the DC side and cannot exchange real power with the AC system.

A STATCOM handles only fundamental reactive power exchange with the AC system and provides voltage support to buses. In transmission systems it is also used to modulate bus voltage during transient and dynamic disturbances in order to improve transient stability margins and to damp dynamic oscillations.

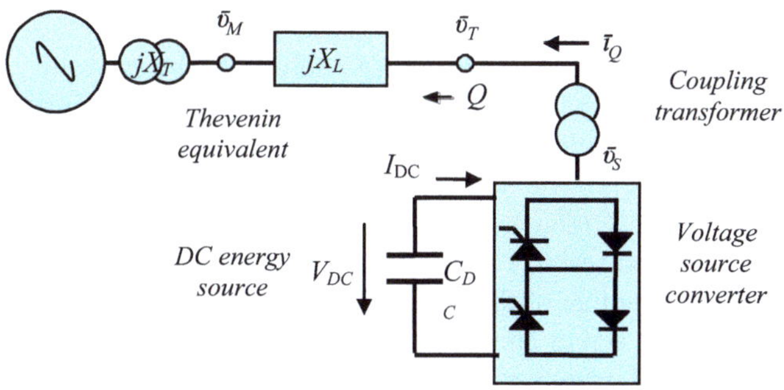

Fig. 2.27 Static synchronous compensator

High power STATCOMs essentially consist of a three-phase pulse width modulation (PWM) voltage source inverter (VSI) using:

- GTOs (high losses), thyristors (high losses) or insulated gate bipolar transistors (IGBTs);
- A DC side capacitor, which provides the DC voltage required by the inverter;
- Filter components to filter out the high frequency harmonics of inverter output voltage;
- A link inductor, which links inverter output to AC supply side (provided by the coupling transformer);
- A control system.

PWM control techniques allow the development of more general models that can readily be adapted to represent voltage but also other control techniques such as phase angle control.

The VSI generates a three-phase voltage, which is synchronised with the AC supply. The DC side is linked to the AC source by the capacitor and the link inductance. The current drawn by the inverter from the AC supply is controlled to be mainly reactive (leading or lagging as per requirement) with a small active component, needed to supply the losses in the inverter and link inductor (and in the magnetics, if any).

A STATCOM provides reactive power generation as well as absorption purely by means of electronic processing of voltage and current waveforms in the VSI. This means capacitor banks and shunt reactors are not needed for generation and absorption of reactive power (as in the SVC case), facilitating a compact design and size. In summary, STATCOM definition is given:

- Static compensator (STATCOM): A static synchronous generator operated as a shunt-connected static VAR compensator whose capacitive or inductive output current can be controlled independent of AC system voltage.

The basic voltage source converter scheme is shown in Fig. 2.27.

The charged capacitor C_{DC} provides a DC voltage to the converter, which produces a set of controllable three-phase output voltages with the frequency of the AC power system. By varying the amplitude of output voltage (V_S), reactive power exchange between the converter and the AC system can be controlled. If the amplitude of output voltage (V_S) is increased above the AC system voltage (V_T), a leading current is produced and reactive power is generated. Decreasing the amplitude of output voltage to below that of the AC system produces a lagging current, and the STATCOM is then viewed as an inductor. In this case, reactive power is absorbed. If amplitudes are equal no power exchange takes place.

A practical converter is not lossless. Moreover, in the STATCOM the required voltage source output is generated by inverting the DC voltage, which is assumed available across the capacitor on the DC side. Obviously, if the active power going into the inverter from the grid is held at zero, the initially charged capacitor will soon discharge down to zero due to active power losses in the inverter that must be supplied by the DC side. DC side voltage will remain constant (or at least controlled) only if the power drawn from AC mains is just enough to supply all the losses that take place everywhere due to the flow of the reactive current demanded.

The mechanism of the V_S phase angle ϑ adjustment can therefore be used to compensate losses but also to control the reactive power generated or absorbed by increasing or decreasing the capacitor voltage (V_{DC}) and thereby the output voltage (V_S). From these observations the different ways a STATCOM can be controlled on the AC and DC sides can is clear.

STATCOM operation in "current regulated mode" is employed to force the inverter to deliver set values of current to the AC system. However, high speed current control schemes would require high frequency switching in the inverter, so high that low frequency devices like the thyristor and GTO are ruled out. Hence, these schemes must be based on MOSFET or IGBT. In distribution grids, a DSTATCOM uses IGBT to achieve high speed switching levels, simple inverter structures and high frequency PWM or hysteresis control to function as current regulated sources.

High power STATCOMs in a transmission system are usually made with devices of low (under − 3 kHz) switching frequency capability (GTOs) and, hence, needing special PWM patterns to optimise switching behaviour. Such STATCOMs use the synchronous link principle in control blocks (discussed later in Chap. 2, § 2.3.5.1).

With the advent of the STATCOM, still better performance can be reached in areas such as:

- Dynamic and steady-state voltage control in transmission and distribution systems;
- Simultaneous control for both reactive and active[1] power.

[1] The exception with respect to the above considered case of low active power is the use of voltage source converters connected in "back-to-back configuration" between two AC bus bars, which enables active power transfer between two AC grids (synchronous or asynchronous or even with different frequencies) by modulating the two voltage phasor angles, simultaneously providing reactive power support to the AC networks.

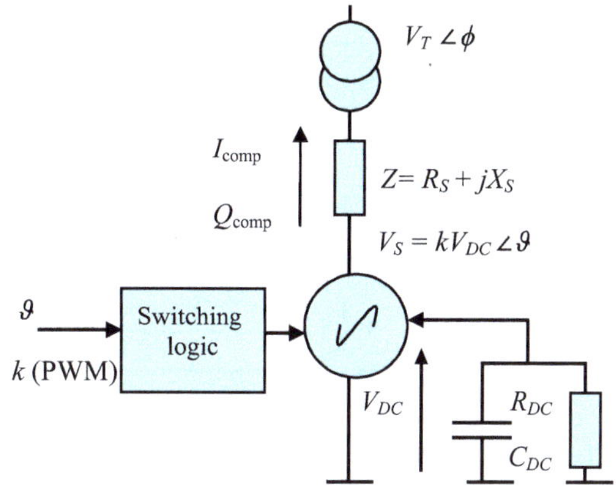

Fig. 2.28 STATCOM model with PWM voltage control

STATCOM Voltage Control Requirements

The STATCOM (Fig. 2.28) can be seen as a voltage source behind the transformer impedance (R_S+jX_S) having an amplitude proportional to the DC side voltage V_{DC}. Switching inertia can be neglected, due to the high frequency, therefore, V_{DC} modulation can be simply represented by a proportional term (k) provided by the PWM control. The angle ϑ does include the phase difference to be imposed between V_S and V_T and depends on the operating control on V_{DC} in conjunction with the main V_T or I_{comp} or Q_{comp} control.

Generally speaking, the DC side capacitor voltage is maintained constant (or allowed to vary with a definite relationship maintained between its value and the reactive power to be delivered by the inverter) by controlling a small active current component. AC currents are controlled indirectly by controlling the phase angle ϑ of the inverter output voltage with respect to the AC side source voltage in a so-called "synchronous link-based control scheme", whereas they are controlled directly by current feedback in the case of a "current-controlled scheme". In the latter case the inverter is current-regulated: its switches are controlled in such a way that the inverter delivers a commanded current at its output, rather than a commanded voltage (the voltage required to see the commanded current flowing out of the inverter will be synthesised automatically by the inverter). The current control scheme produces a very fast STATCOM, which can adjust its reactive output within a period of tens of microseconds after a sudden change in reactive demand.

However, as control at the transmission/subtransmission level is our concern here, such extreme speed is not required; the STATCOM should address more directly the voltage support at HV buses by delivering/taking as much reactive power to/from the HV bus as required to maintain the voltage V_T within pre-decided limits. This task is achieved by the STATCOM with fewer limitations than on an SVC because in this situation, without the physical components of the SVC reactor and a capacitor that links voltage and current delivery, the current I_{comp} can reach the maximum capacitive/inductive values in a large V_S field. Accordingly, we notice the

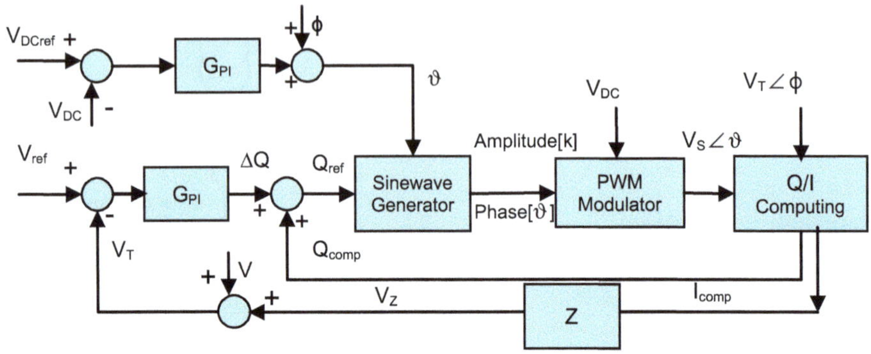

Fig. 2.29 STATCOM: AC HV bus bar and DC voltage control loops

relevant differences of the STATCOM static characteristics in Fig. 2.30 with respect to SVC. The vertical trend at the allowed maximum capacitive and inductive currents is due to the STATCOM's wider controllability.

The STATCOM therefore maintains bus voltage by injecting leading or lagging VARs into the system, up to the maximum capacitive/inductive current, even at very low HV bus bar voltages.

The outer STATCOM control loop senses bus voltage, compares it with a set value and processes the error in a PI controller that sets the reactive reference *I* or *Q* for the inner control loop. Direct control of V_S is possible but not recommended, due to its low sensitivity (small voltage amplitude control band) with respect to current and reactive power, which have larger control ranges around a given operating condition.

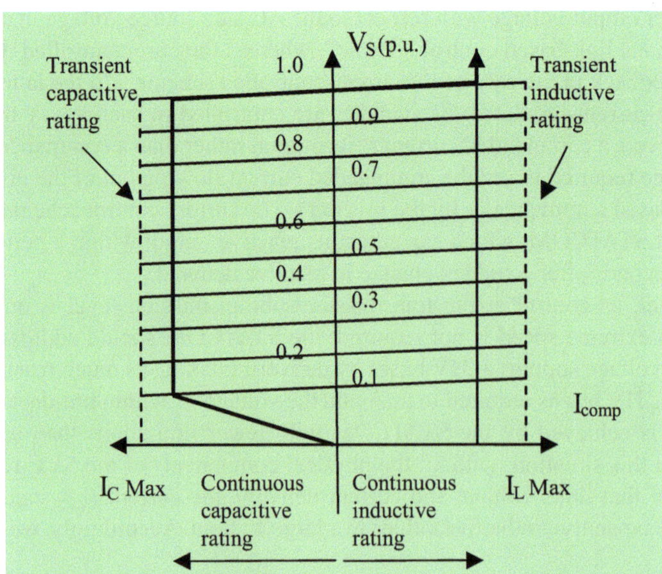

Fig. 2.30 STATCOM *V*–I characteristics

The effectiveness of a given STATCOM with a specified rating in providing voltage support at a particular bus will depend on the short circuit capacity at that bus. A low value of short circuit capacity implies that the Thevenin impedance behind the bus voltage is large. In addition, the STATCOM at that location will be more effective than a similarly rated STATCOM at another bus with a higher short circuit capacity. Therefore, a bus with a higher value of short circuit capacity will have better immunity against sags/flickers/swells caused by problems elsewhere. And hence it probably needs no STATCOM at all, unless placing one would strongly influence surrounding bus voltages and thus be useful for regulating voltage there.

Figure 2.29 synthesises the above comments on STATCOM control, showing the two DC and AC side controls and the dependence of voltage output V_S on V_{DC} and ϑ.

STATCOM Regulation Slope

In many transmission applications the STATCOM does not function as a perfect voltage regulator; rather, terminal voltage is allowed to vary in proportion to the compensating current. The useful lifetime of equipment is extended if this kind of droop regulation is used. Regarding this kind of compounding, considerations that were already brought up for the SVC can be fully extended to the STATCOM, including the fact that "droop" regulation allows automatic load sharing between various local static compensators.

A STATCOM's desired terminal voltage-versus-output current characteristic can be established by an external control loop by reclosing I_{comp} with a drop gain γ on the input summing junction of the V_T control loop (see Chap. 3, § 3.3.7, on STATCOM voltage and reactive power control).

The static characteristics of a STATCOM as shown in Fig. 2.30 are a clear illustration of its ability to support very low system voltage, down to about 0.15 p.u., the value associated with coupling transformer reactance. This is in strong contrast with an SVC, which at full capacitive output becomes an uncontrolled capacitor bank at low voltage.

A STATCOM can support system voltage at extremely low voltage conditions as long as the DC capacitor is able to retain enough energy to supply losses. It is interesting to note that a STATCOM's *V–Q* static characteristics from here are very similar to the SVC's *V–I* static characteristics (compare Fig. 2.31 with Fig. 2.26). This points to the STATCOM's extremely poor reactive power control margins at very low voltages.

2.3.6 *Unified Power Flow Control (UPFC)*

The unified power flow controller (UPFC), which incorporates a static synchronous series compensator (SSSC) as the series part and a STATCOM as the shunt element, represents an alternative approach to transmission angle control; it can be operated

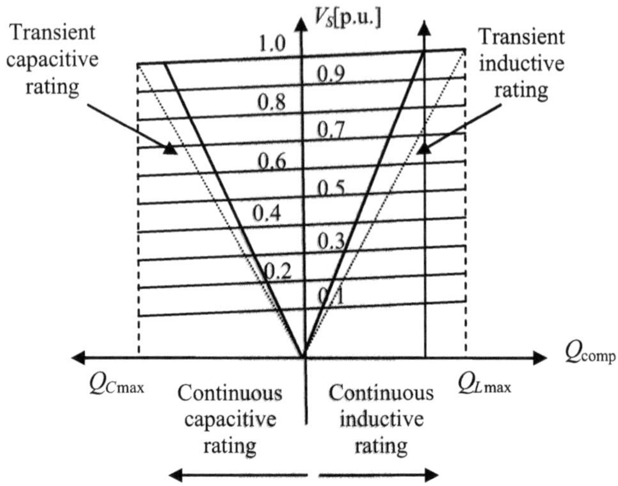

Fig. 2.31 *V–Q* characteristic comparison between STATCOM and SVC

as an ideal phase shifter (a specialised transformer used to control the flow of real power through the mechanical rotation of windings).

The most important characteristic of the UPFC is its ability to provide an output voltage that is fully controllable in amplitude and phase, which can be vectorially added to the network voltage. The added voltage on a line allows independent control of both the reactive and active power flow in the line. Alternately, it can independently control local bus voltage, line impedance and phase angle between voltage vectors at the buses on the edges of the line where the UPFC is installed. The UPFC can also contribute to carrying out, through appropriate control, the following functions:

- Transient stability improvement;
- Power swing damping;
- Voltage stability improvement.

Accordingly, the complexity and cost of the UPFC is clearly justified for real power transfer control in place of the obsolete electromechanical phase shifter; however, it is less justified for voltage and reactive power control. Nevertheless, taking advantage of its on-field presence, the UPFC also can be used for voltage or reactive power flux control to support grid voltage by contributing to improved voltage stability.

The general structure that provides a UPFC with its control functions consists of two voltage-sourced AC/DC converters, both connected by a common DC link: one in parallel and the other in series with the AC line (see Fig. 2.32). As mentioned, the UPFC incorporates as its series part the static synchronous series compensator, while its shunt element is a static compensator.

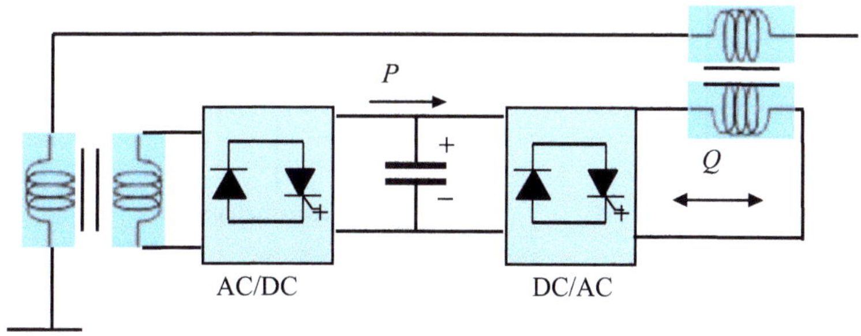

Fig. 2.32 Schematic configuration of the UPFC with two GTO-based voltage-sourced converters

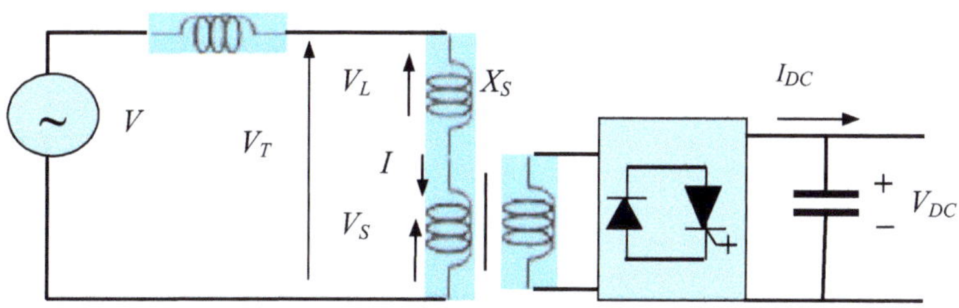

Fig. 2.33 Scheme of shunt converter connections

Fundamentals of the Shunt Voltage Source Converter

The shunt-connected converter in a UPFC structure, shown in Fig. 2.33, is the already-seen STATCOM alternating synchronous voltage source V_S behind a coupling reactance X_S, provided by the per-phase leakage inductance of the coupling transformer. Like a STATCOM, it provides reactive shunt compensation and covers the active power demand of the series-connected converter through the DC link with a battery/capacitor. In this book, reactive shunt compensation functionality is relevant, linked as it is with converter output voltage V_S, kept in phase with AC system voltages and modulated in amplitude to control the reactive power exchange between the converter and the AC system.

That is, if the amplitude of the output voltage is increased to above that of the AC system, then the corresponding current flows through the tie reactance from the converter to the AC system, and the converter generates reactive (capacitive) power to sustain the local AC system voltage V_T. If the amplitude of the output voltage is decreased below that of the AC system, then the reactive current flows from the AC system to the converter, which absorbs reactive (inductive) power. Obviously, changing the phase of the converter output voltage V_S by imposing a lead (lag) angular position with respect to the AC system voltage V_T determines the converter delivery (absorption) of active power.

Generally, the shunt converter is controlled to provide the active power demand of the series converter and to cover system losses by regulating the DC link voltage.

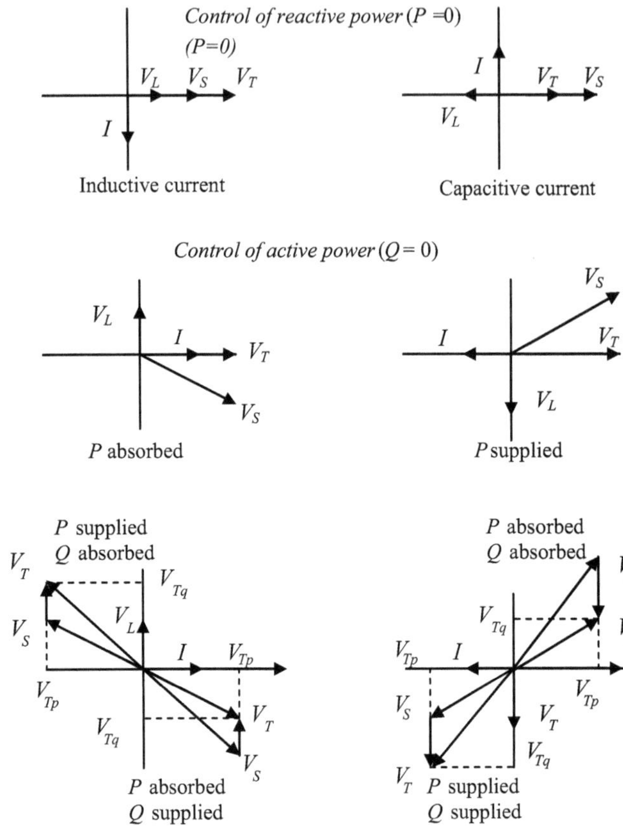

Fig. 2.34 Phasor diagrams showing the condition for reactive and active power exchange between AC system and voltage source converter

The phasor diagram shown in Fig. 2.34 shows the magnitude and phase requirements for converter output voltage to determine positive or negative reactive and active power exchanges.

Fundamentals of the Series Voltage Source Converter

The generalised series synchronous compensator, implemented by a DC-to-AC converter with an energy storage device on the DC side, is shown in Fig. 2.35.

Assuming the injected voltage V_{pq} in series with the line can be controlled without restrictions (DC energy storage with infinity capacity), the amplitude and phase angle of phasor V_{pq} can be chosen, independent of the line current, between 0 and 2π, with a variable module value between zero and a defined maximum value V_{pqmax}. This implies the synchronous voltage source must be able to generate or absorb both active and reactive power. The reactive power is therefore internally generated or

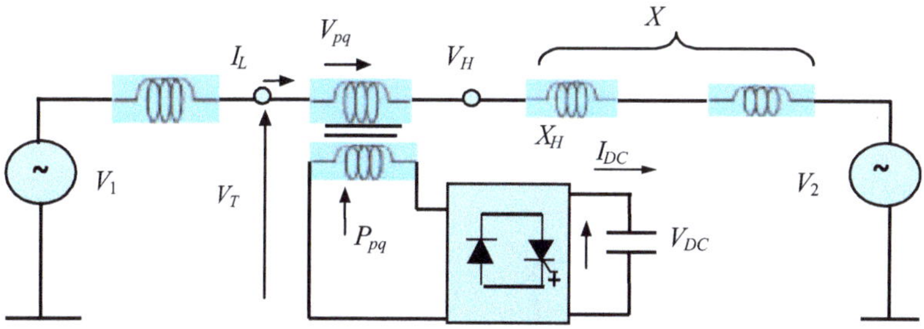

Fig. 2.35 Scheme of shunt converter connections

absorbed by the converter. However, the active power is supplied from, or absorbed by, the DC energy storage side, as illustrated in Fig. 2.36.

The generalised series synchronous compensator can achieve all basic power flow control functions by adding an appropriate voltage phasor V_{pq} to terminal voltage V_T, as shown in Fig. 2.37. By appropriate control of V_{pq} the following basic power flow controls can be accomplished:

a. Terminal voltage regulation ($|V_H|$ regulation): $V_H = V_T + V_{pq}$;
b. Series impedance compensation combined with terminal voltage regulation;
c. Phase shifting regulation combined with terminal voltage regulation;
d. Terminal voltage regulation combined with series impedance compensation combined with phase shifting regulation.

Fig. 2.36 Generalised series-connected synchronous voltage source with energy storage. Dependence of *P* and *Q* at AC side on V_{pq} phase; dependence of *P* at DC side on I_{DC} current direction

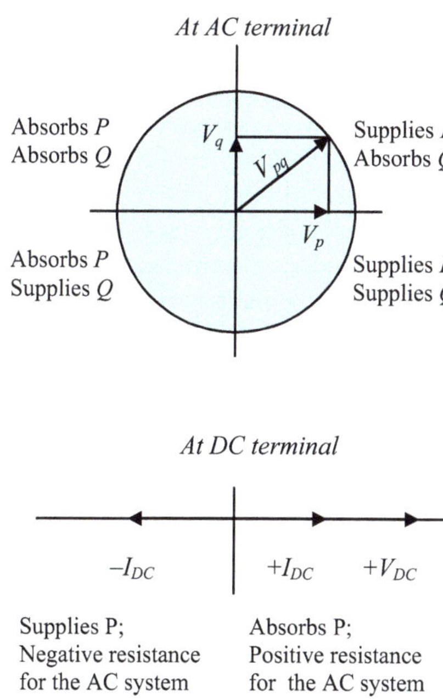

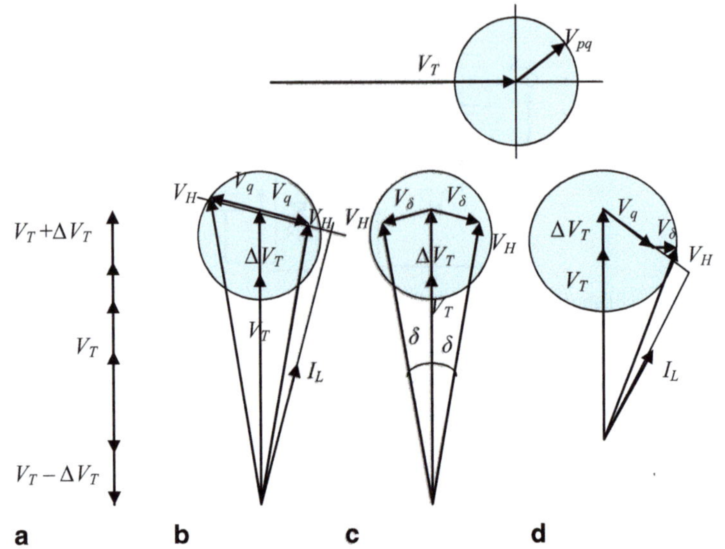

Fig. 2.37 Phasor diagrams illustrating general concept of series voltage injection and power flow control functions

- In Fig. 2.37 terminal voltage regulation (a) is achieved by imposing V_{pq} in phase with V_T and controlling the amplitude of the injected voltage:

$$V_H = V_T + \Delta V_T$$

- Line impedance compensation (b) imposes the squareness of V_{pq} with respect to I_L: ($V_{pq} = V_q$), whereas the injected voltage versus defines inductive or capacitive compensation.
- Phase shifting regulation (c) basically imposes the squareness of V_{pq} with respect to V_T ($V_{pq} = V_\delta$), with angle variation depending on amplitude and versus of injected voltage. In this way, the UPFC operates as a phase angle regulator able to internally generate the required reactive power.
- Multifunction power flow control (d) with $V_{pq} = \Delta V_T + V_q + V_\delta$ is an exclusive peculiarity of the UPFC.

Generalised voltage injection control, allowing for variation of the angle of injected voltage through a full 360° simultaneous with its magnitude, makes it possible to control both the magnitude and the angle of a line current. Therefore, it is possible to independently control both the active and reactive power flow in a transmission line, as explained in the following section.

Fundamentals of the UPFC

The generalised series compensator with an infinite energy source can be implemented by various power converter arrangements. One proposed by Gyugyi employs two AC/DC converters operated from a common DC link capacitor, as shown schematically in Figs. 2.32 and 2.35. The left converter in Fig. 2.35 is in shunt with the transmission line, while the other converter, on the right, is in series with the line and generates the voltage $v_{pq}(t) = V_{pq}\sin(\omega t + \rho)$ at the fundamental frequency ω with variable amplitude $0 \leq V_{pq} \leq V_{pq\max}$ and phase angle $0 \leq \rho \leq 2\pi$, which is added to the AC system terminal voltage $v_T(t)$ by the series-connected coupling transformer.

The transmission line current flowing through the injected voltage source determines a VA injection or absorption at the maximum rate given by the maximum injectable voltage and maximum line current. A converter connected in shunt with the AC power system is primarily used to provide active power demanded by the other converter through the DC link. The series-connected converter itself generates reactive power demanded by series voltage injection. Therefore, the transmission system is not burdened by reactive power flow from a remote source due to the UPFC operation.

It is worth noting that since the shunt converter can also generate or absorb reactive power at its AC terminals, it can also fulfil, with proper control, the function of an independent STATCOM, providing reactive power compensation for the transmission line and thus executing an indirect voltage regulation at the V_T input terminal of the UPFC.

The two-machine power system of Fig. 2.38, with sending-end voltage V_1, receiving-end voltage V_2 and line or tie impedance X (assumed for the sake of simplicity to be inductive), is introduced to establish the capability of the UPFC to control transmitted active power P and reactive power flow Q_1 and Q_2 at the sending and receiving ends of the line.

The system voltages in the phasor diagram show a transmission angle δ. The active (P) and reactive (Q) powers depend on the modules of V_1 and V_2 and δ values for a given X value. Therefore, P and Q are not mutually independent:

$$P = \frac{V_1 V_2 \sin\delta}{X}, \quad Q_1 = \frac{V_1^2 - V_1 V_2 \cos\delta}{X}, \quad Q_2 = \frac{V_2^2 - V_1 V_2 \cos\delta}{X}.$$

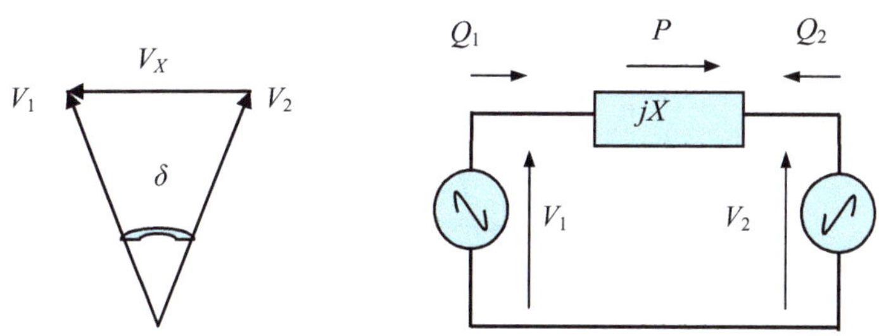

Fig. 2.38 Simple two-machine system with active and reactive power flows

With $V_1 = V_2 = V$, transmitted power P and reactive power Q supplied at the line ends are shown in the figure. For stable values of δ: $0 \leq \delta \leq \pi/2$, the reactive versus active power equation confirms that any δ angle variation determines changes in P and in Q_1, Q_2:

$$Q = P\frac{1-\cos\delta}{\sin\delta}, \tag{2.5}$$

An analogous result is produced when the voltage modules or the tie reactance are changed.

In this scenario, Fig. 2.39 includes the UPFC via a controllable voltage source in series with the line, which, as previously explained, can generate or absorb the reactive power it exchanges with the line. On the contrary, the active power exchanged must be supplied to or absorbed from it through the sending-end bus via the shunt converter, which also controls local bus reactive power injection/absorption or local HV bus voltage.

Line current I_L flows through series voltage V_{pq} and results in both a reactive and an active power exchange. The transferred active and reactive powers P_2 and Q_2 to the receiving-line end are formulated as follows, assuming V_2 lies on the real axis:

$$\bar{V}_T = V_T e^{j\delta}, \quad \bar{V}_{pq} = V_{pq} e^{j(\delta+\sigma)}, \quad \bar{V}_H = \bar{V}_T + \bar{V}_{pq}.$$

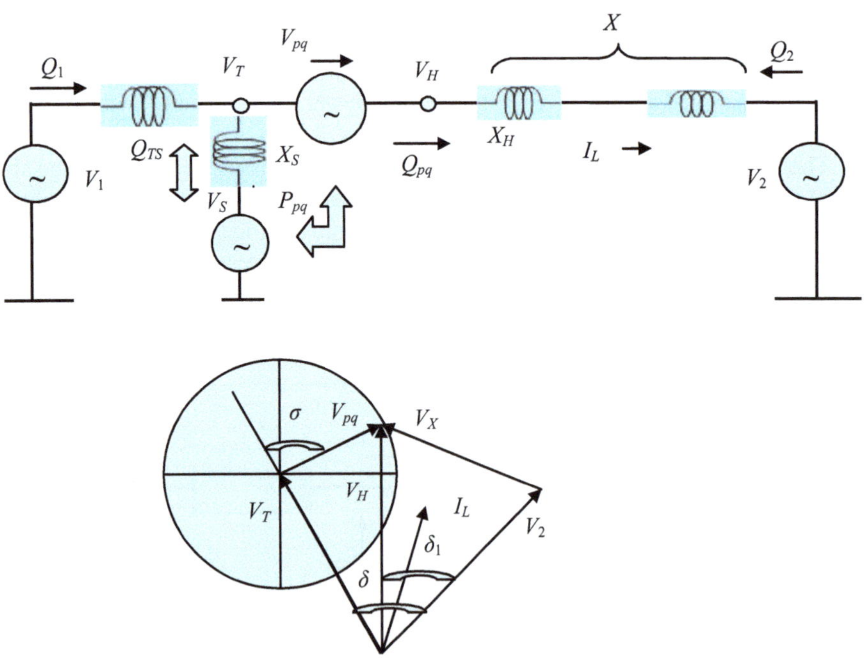

Fig. 2.39 Simple two-machine system with unified power flow controller

with

$$0.0 \le \delta \le \pi, 0.0 \le \sigma \le 2\pi.$$

The entering current to the receiving end corresponds to

$$\overline{i_L} = \frac{\overline{V_X}}{jX} = \frac{\overline{V_T} + \overline{V_{pq}} - \overline{V_2}}{jX} = \frac{\overline{V_T} - \overline{V_2}}{jX} + \frac{\overline{V_{pq}}}{jX}.$$

Therefore, the entering powers to the receiving line end are

$$P_2 + jQ_2 = \overline{V_2 i_L}^*$$
$$= V_2 \left[\frac{\overline{V_T} - \overline{V_2}}{jX} + \frac{\overline{V_{pq}}}{jX} \right]^*$$

$$= \left[\frac{V_2 V_T \sin(\delta)}{X} + \frac{V_2 V_{pq} \sin(\delta + \sigma)}{X} \right] - j \left[\frac{V_2^2 - V_2 V_T \cos(\delta)}{X} - \frac{V_2 V_{pq} \cos(\delta + \sigma)}{X} \right]$$

$$= P(\delta) + P_{pq}(\delta,\sigma) - jQ(\delta) + jQ_{pq}(\delta,\sigma) = P(\delta,\sigma) + jQ(\delta,\sigma). \tag{2.6}$$

This equation shows the following:

- According to the assumed current direction, both active and reactive powers enter the receiving bus;
- The additional contributions $P_{pq}(\delta,\sigma)$ and $Q_{pq}(\delta,\sigma)$ can be controlled ($0.0 \le \sigma \le 2\pi$) independently from the δ value ($0.0 \le \delta \le \pi$) between $\pm V_2 V_{pq} / X$.

Accordingly, the transferred active power can be regulated between the values as follows:

$$P_2(\delta) - \frac{V_2 V_{pq\max}}{X} \le P_2 \le P_2(\delta) + \frac{V_2 V_{pq\max}}{X},$$
$$Q_2(\delta) - \frac{V_2 V_{pq\max}}{X} \le Q_2 \le Q_2(\delta) + \frac{V_2 V_{pq\max}}{X}.$$

To show the regions of P_2 and Q_2 controllability, the attainable reactive and active powers at the receiving-end bus are now plotted under the assumption of V_T, V_2 and X at values of 1 p.u.

With $\delta=0$ and $V_{pq}=0$, P_2, Q_2 are both zero (the origin of the coordinates). The Q_2 supplied by the receiving-end generator, plotted against transmitted power P_2 is represented by Eq. (2.5) for stable values of δ: $0 \le \delta \le \pi/2$ and $V_{pq}=0$. Obviously, if Q_2 is delivered by the receiving-end bus, the curve is reflected across the P-axis.

At a given δ, Eq. (2.5) gives the origin of the circle $Q_{pq}(\delta, \sigma)$, which shows the variation of P/Q as the voltage phasor V_{pq} with its maximum magnitude $V_{pq\max}$ rotated a full revolution. The area within this circle defines all P_2, Q_2 values obtainable by controlling the magnitude and angle σ of the phasor V_{pq}. We observe that the given V_{pq} voltage rating can establish power flow in either direction without imposing any reactive power demand on receiving-end/sending-end generators.

In general, at any angle δ transmitted powers P_2 and Q_2 can be freely controlled by the UPFC within the boundaries obtained by rotating the injected voltage phasor V_{pq} with its maximum magnitude a full revolution ($0 \le \sigma \le 2\pi$). The boundary in each plane is centred on the basic power transmission defined by δ at $V_{pq}=0$.

With $V_{pq} \neq 0$, the previous $P(\delta, \rho)$, $Q(\delta, \rho)$ equations show that P_2 and Q_2 change with respect to values of $P(\delta)$, $Q(\delta)$ without series compensation. The control region edge of P_2 and Q_2 is described by a circle with centre at $P_2(\delta)$, $Q_2(\delta)$ and radius of $V_{pq}V_2/X$. The limiting circle is described by the equation:

$$\left[P_{pq}(\delta,\sigma)-P_2(\delta)\right]^2+\left[Q_{pq}(\delta,\sigma)-Q_2(\delta)\right]^2=\frac{V_2 V_{pq\max}}{X}.$$

The circular regions described by this equation, under the assumption of $V_2=1.0$, $V_{pq\max}=0.5$ and $X=1.0$ (p.u. values), are represented in Fig. 2.40 for transmission angles $\delta=0°, 30°, 90°$.

Considering the case with $\delta=0°$, the values of active and reactive powers are all zero with $V_{pq}=0$, i.e., the system is at a standstill at the origins of the P_2 and Q_2 coordinates. The circle around the origin shows the variation of P_2 and Q_2 as the voltage phasor V_{pq}, with its maximum amplitude $V_{pq\max}$, is rotated a full revolution ($0 \le \rho \le 2\pi$). The area within this circle defines all P_2 and Q_2 values obtainable by controlling the magnitude and angle of the phasor V_{pq}. We observe for the assumed data values the UPFC is able to establish 0.5 p.u. of power flow ($X=1.0$) in either direction without imposing any reactive power demand on receiving- or sending-end generators. Of course, the UPFC can also force the equivalent generator at one end to supply reactive power to the generator at the other end.

In general, at any given transmission angle δ the transmitted power P and the reactive power demand of the transmission line at the receiving end can be freely controlled within the corresponding boundaries in the (P_2, Q_2)-plane by rotating the injected phasor V_{pq}. The boundary in each plane is centred around the point defined by the transmission angle on the Q_2 versus P_2 curve that characterises the basic power transmission at $V_{pq}=0$.

Considering the case of $\delta=30°$, again a circular control region is defined for the receiving-end bus that allows injection of more active and reactive powers. The control regional boundary for P_2 and Q_2 remains a circle at all transmission angles. This is true for the $\delta=90°$ case also. In this case, the UPFC contributes to the increase of the already high value of the active power transferred to bus 2 as well as the reactive power up to values greater than a maximum of 1 p.u., owing to contribution of 0.5 p.u. already provided by the sending bus T.

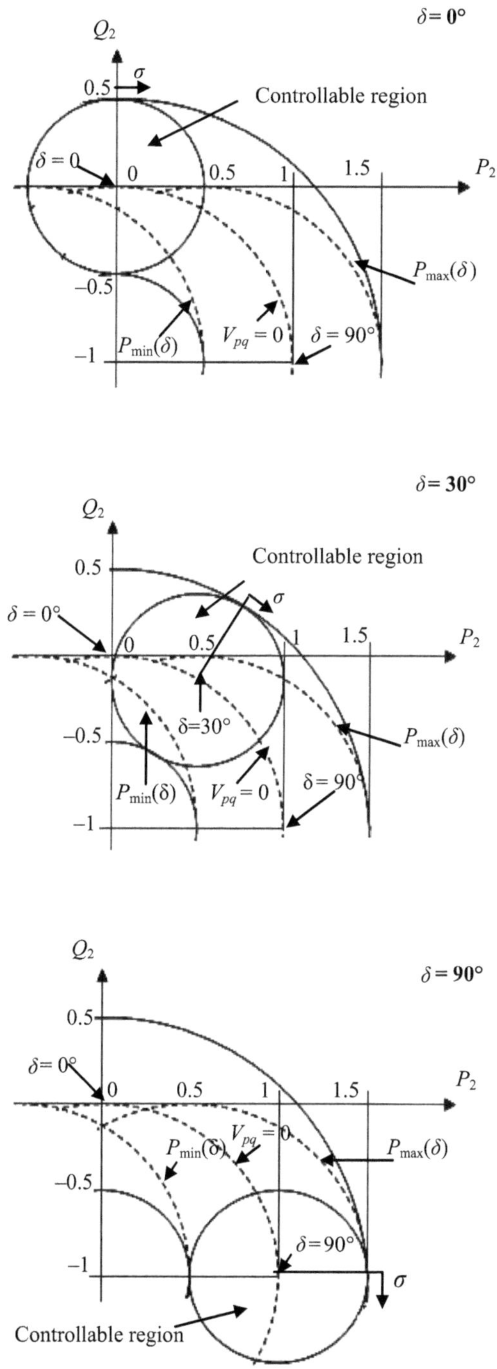

Fig. 2.40 Attainable receiving-end reactive power versus transmitted power with UPFC at $\delta=0°$, 30°, 90°

UPFC Voltage Control Requirements

In addition to active and reactive power flow control, the UPFC can also operate as a series impedance compensator or as a static VAR source or as local HV bus voltage control. These applications are all subsets of active and reactive power flow control and depend upon the structure and characteristics of the control system used, which should also provide real-time dynamic compensation of the AC transmission system (lending stability improvement).

Power system control therefore involves execution of a UPFC's simple functions, such as active and reactive power and voltage control, or of its more complex functions of transient stability control and oscillation damping. The equivalent circuit of a UPFC system is represented in Fig. 2.39. The UPFC is located at the sending end of the transmission line that follows changes in the control system's reference values of the active and reactive power in the line and the reactive power of the shunt branch. More precisely, referring to Fig. 2.32:

- The AC/DC shunt converter connected to the HV sending-end bus via a coupling transformer is primarily used to provide active power demand by the series DC/AC converter via the common DC link. It can also generate or absorb reactive power at its AC terminal, independent of the active power transferred via the DC link; therefore, with proper controls it can also fulfil the functions of an independent STATCOM providing reactive power compensation to the local HV bus or executing voltage regulation at the UPFC input terminal. The control also maintains the necessary DC link voltage and ensures smooth active power transfer between the two converters.
- The DC/AC control is structured to accept external reference values whose order of priority can be preselected for the desired reactive shunt compensation, series compensation, transmission angle and output voltage power transfer.
- These closed loop control reference signals force the converters to impose AC voltage at the UPFC input and output terminals, thereby establishing the desired transmission parameters. If the UPFC is operated only with phase angle reference input, it automatically becomes a perfect phase shifter.

Figure 2.41 shows a synthesis of the large number of automatic control loops with which a UPFC can be provided. A proper combination of them is sure to make different control objectives possible. These ideas are summed up precisely:

- The UPFC shunt converter exchanges reactive power with the grid at bus V_T. Two overlapped control loops are represented. The inner loop refers to the reactive power injected or absorbed by the grid and allows this variable to remain fixed when it operates alone. When, conversely, it is overlapped by an outer voltage control loop, voltage V_T is automatically controlled at the set-point value externally fixed by the operator or, if necessary, required by the V_H control operating on the UPFC output.
- The UPFC series converter basically injects an AC voltage (V_{pq}), derived from the DC voltage, in series with the transmission line. This converter control has to regulate the magnitude and phase of V_{pq} to achieve two main control objectives:

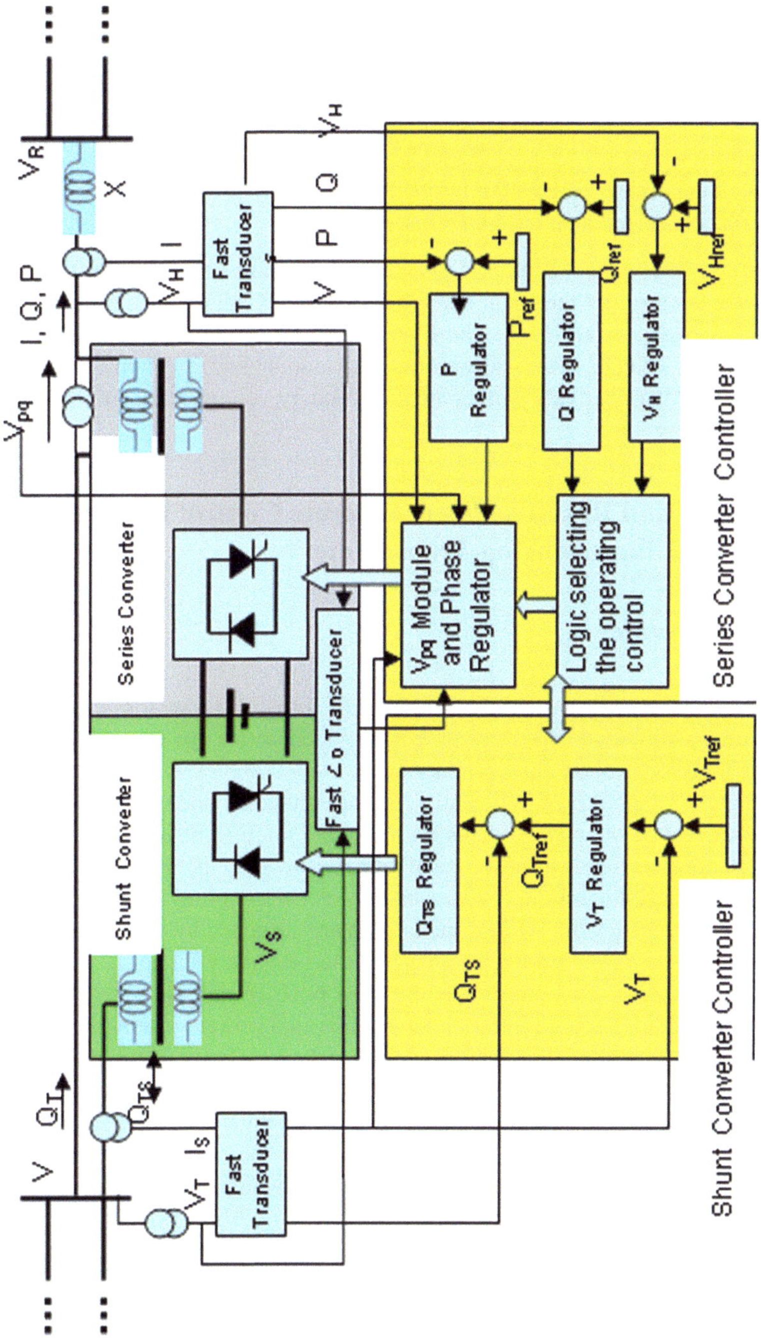

Fig. 2.41 UPFC control loops, focusing on its extreme power and versatility for voltage and reactive power flow control

- Output P and Q control: P by $\angle V_{pq}$ and Q by $|V_{pq}|$;
- Output P and V_H control: P by $\angle V_{pq}$ and V_H by $|V_{pq}|$.

The three control loops are represented in the figure together with a logic selecting the operating control and managing the expected interchanges between the two converter controls.

Another, simpler, control operation could be as follows:

- The shunt converter operates the V_T control;
- The series converter operates the P control.

The series converter cannot control P, Q and V_H at the same time, whereas the combination of V_T control by the shunt converter and the Q control by the series converter is manageable up to $V_{pq\max}$. The V_T control corresponds to high side voltage control (HSVC). The Q_{TS} control as well as the Q amount and direction controls the candidate UPFC to effectively participate in secondary voltage regulation (SVR).

2.4 Voltage and Reactive Power Discrete Control Devices: On-load Tap-changing Transformers

2.4.1 Generalities

A transformer tap changer selects the number of turns of a transformer winding. A transformer with variable turns ratio N is provided with a tap changer, enabling stepping voltage control on the secondary side.

Tap-changing transformers efficiently control voltage on one winding side only (*if it is sustained by the voltage value on the other side*). In other words, a transformer with tap changing is not a source of reactive power in the way the equipment previously introduced was (§ 2.3); it serves only as a decoupling interface between two different voltage levels—where one or both are sustained by voltage sources.

By changing the turns ratio, the voltage in the secondary circuit is varied and its control is obtained. This constitutes the most popular and widespread form of voltage control among the different voltage levels in a power system, the success of which simply depends, at each level, on the voltage solidity of the next higher level.

Figure 2.42 gives a schematic diagram of a tap changer placed in the primary winding

Figure 2.43 shows the equivalent Γ circuit of a two-winding transformer with the tap changer located in the secondary winding

If the connections of the primary and secondary windings are identical (e.g., star-star or delta-delta), the following equations describe the model:

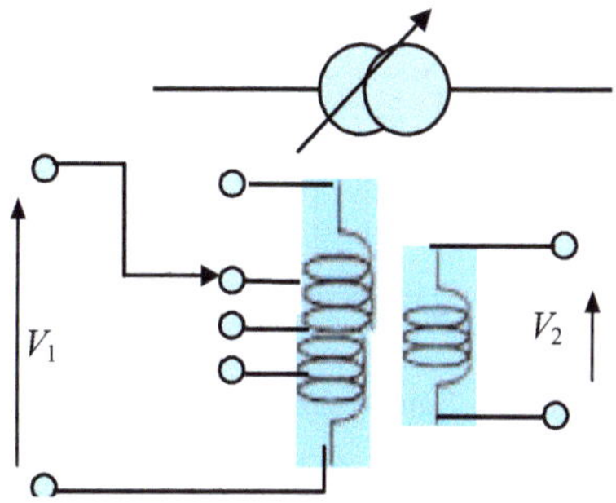

Fig. 2.42 Single-phase OLTC with tap changer in the primary winding

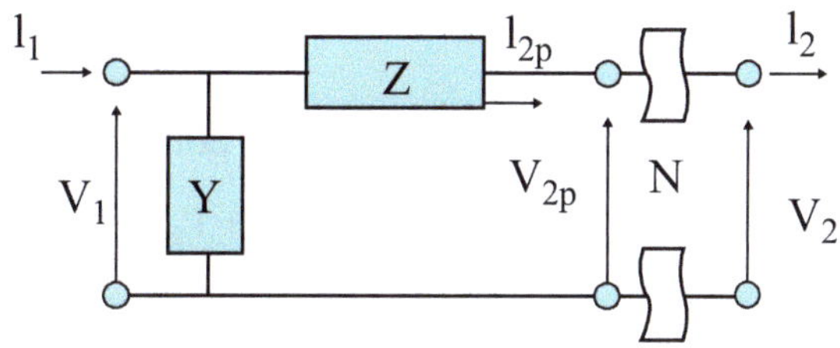

Fig. 2.43 Single-phase Γ equivalent circuit of two-winding transformer with secondary winding effect seen at the primary level

$$V_{2p} = NV_2, \quad I_{2p} = \frac{1}{N} I_2,$$

$$N_{nom} = \frac{V_{n1}}{V_{n2}} = \frac{I_{n2}}{I_{n1}}, \quad \frac{V_{2p}}{V_{n1}} = N \frac{V_{n2}}{V_{n1}} \frac{V_2}{V_{n2}}, \quad \frac{I_{2p}}{I_{n1}} = \frac{1}{N} \frac{I_{n2}}{I_{n1}} \frac{I_2}{I_{n2}}.$$

N is the actual turns ratio, which is usually different from the nominal value N_{nom}. The nominal voltages of the primary and secondary windings are represented by V_{n1} and V_{n2}, respectively; the nominal phase currents of the primary and secondary windings are given by I_{n1} and I_{n2},

In p.u., turns ratio m is shown, along with v_{2p} and i_{2p}:

$$m = \frac{N}{N_{nom}}, \quad v_{2p} = mv_2, \quad i_{2p} = \frac{1}{m} i_2.$$

2.4.2 Output Voltage Dependence on Current Turns Ratio

In order to determine the operating turns ratio of a transformer providing a given output voltage value, we consider the simple configuration of a load supplied from a source through a transmission line and a step-down transformer.

The circuit in Fig. 2.44 is obtained by neglecting the no-load losses of the transformer and including its series impedance into the line equivalent impedance Z of the source line.

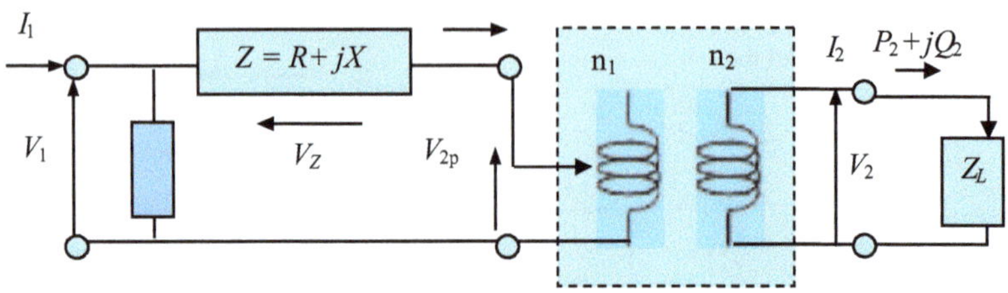

Fig. 2.44 Simple circuit source-line transformer with tap-feeding load

For a given operating point, the values of V_1, P_2 and Q_2 are known. Moreover,

$$N = \frac{V_{2p}}{V_2}.$$

The equation relating voltages on the primary side is

$$V_1 = V_{2p} + V_Z.$$

Neglecting the transversal component of the voltage drop and considering the voltage at the load as the phase reference, the above relations become (see also Eq. (1.2)):

$$V_1 = V_{2p} + \frac{RP_2 + XQ_2}{V_{2p}}, \quad V_{2p} = \frac{V_1 + \sqrt{V_1^2 - 4(RP_2 + XQ_2)}}{2},$$

$$N = \frac{V_{2p}}{V_2} = \frac{V_1 + \sqrt{V_1^2 - 4(RP_2 + XQ_2)}}{2V_2}.$$

Therefore, the actual turns ratio at no-load (i.e., $I_2 = 0.0$) is given by

$$N_0 = V_1 / V_2,$$

while in all other operating conditions the ratio is reduced with a load increase. In fact, for a desired V_2 output, the tap position changes according to the input voltage V_1, but it also depends on the operating load: tap stepping goes up when V_1 goes down or when P_2 and Q_2 increase.

Voltage output V_2 clearly depends on N and V_1, as is now described in the analysis of the transformer from a secondary winding.

The circuit in Fig. 2.45 is obtained by neglecting no-load losses of the transformer and including its series impedance in line equivalent impedance Z of the source line, transferred to the secondary winding.

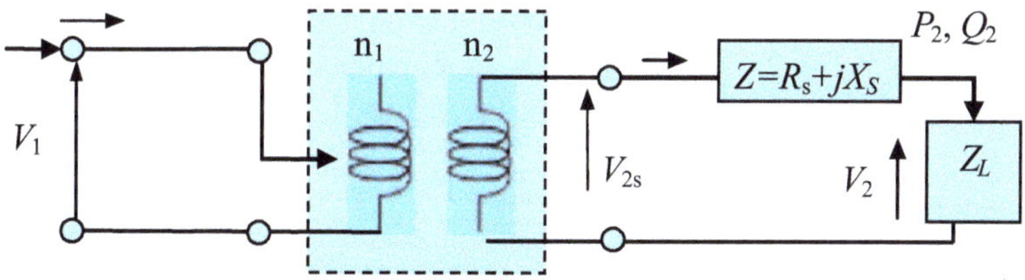

Fig. 2.45 Simple circuit from secondary winding. Transformer with tap load

$$R_S = \frac{R}{N^2},\quad X_S = \frac{X}{N^2}$$

$$\frac{V_1}{N} = V_{2S} = V_2 + \frac{\frac{R}{N^2}P_2 + \frac{X}{N^2}Q_2}{V_2},$$

$$P_2 + jQ_2 = \frac{V_2^2\left(R_L + jX_L\right)}{R_L^2 + X_L^2},\quad V_1 = V_2\left[N + \frac{\frac{R}{N}R_L + \frac{X}{N}X_L}{R_L^2 + X_L^2}\right].$$

From the last equation it is clear that voltage V_2 is a function of V_1 and N. From the linearised model:

$$\Delta V_1 = \left[V_2 - \frac{V_2}{N^2}\left(\frac{RR_L + XX_L}{R_L^2 + X_L^2}\right)\right]_0 \Delta N + \left[N + \frac{\left(\frac{R}{N}R_L + \frac{X}{N}X_L\right)}{R_L^2 + X_L^2}\right]_0 \Delta V_2,$$

$$\Delta V_2 = -\left[\frac{V_2}{N}\frac{N^2\left(R_L^2 + X_L^2\right) - \left(RR_L + XX_L\right)}{N^2\left(R_L^2 + X_L^2\right) + \left(RR_L + XX_L\right)}\right]_0 \Delta N$$
$$+\left[\frac{N\left(R_L^2 + X_L^2\right) + \frac{RR_L + XX_L}{N}}{R_L^2 + X_L^2}\right]_0 \Delta V_1. \tag{2.7}$$

Equation (2.7) confirms that variation of V_2 increases with V_1 and decreases when N increases.

2.4.3 *Static Characteristic of the Transformer*

On-load tap changing is used to keep voltage V_2 at the secondary winding of the transformer close to a reference value V_{2sch} by modifying the turns ratio N. In order to analyse the influence of on-load tap changing on the primary winding, we refer to the example in Fig. 2.44, where impedance Z_L represents the load. The following V–I relationships accompany tap changing model used:

$$\overline{V}_1 = \overline{V}_{2p} + j\overline{Z}\overline{I}_{2p}, \quad \overline{V}_2 = \overline{Z}_L \overline{I}_2.$$

To simplify the calculation, the resistances of both line and transformer are neglected, i.e., $\overline{Z} \approx jX$, and the load is modelled by a resistance, i.e., $\overline{Z_L} = R_L$, absorbing the active power $P_2 = V_2^2 / R_L$.

Choosing voltage at the load terminals to be the phase origin, i.e., $\overline{V}_2 = V_2\angle 0^\circ$, and taking into account the expression of the turns ratio N:

$$\overline{V}_1 = \overline{V}_{2p} + jX\overline{I}_2, \quad \overline{V}_2 = R_L \overline{I}_2.$$

Taking into account the dependence on N,

$$\overline{V}_1 = NV_2 + j\frac{X}{N}\frac{V_2}{R_L} = NV_2\left(1 + j\frac{X}{N^2}\frac{1}{R_L}\right),$$

and absolute value,

$$V_1 = NV_2\sqrt{1^2 + \left(\frac{X}{R_L}\right)^2 \frac{1}{N^4}},$$

gives

$$V_2 = \frac{V_1}{N\sqrt{\cdot}} = \frac{V_1}{\frac{X}{NR_L}\sqrt{1+\left(\frac{R_L N^2}{X}\right)^2}}$$

or

$$V_2 = f(N) = \frac{NR_L}{X}\frac{V_1}{\sqrt{1+(R_L / X)^2 N^4}}. \tag{2.8}$$

Equation (2.8) defines the dependency of the voltage at the load terminals (with pure active load) in terms of the turns ratio. (Equation parameters are source voltage V_1 and resistance R_L.) This function defines the static characteristic of the on-load tap changing transformer (Fig. 2.46).

It can be seen that the function $V_2 = f(N)$ has a maximum, which can be determined by setting the derivative equal to zero: $dV_2/dN = 0$. Denoting

$$a = \frac{V_1 R_L}{X}, \quad b = \left(\frac{R_L}{X}\right)^2,$$

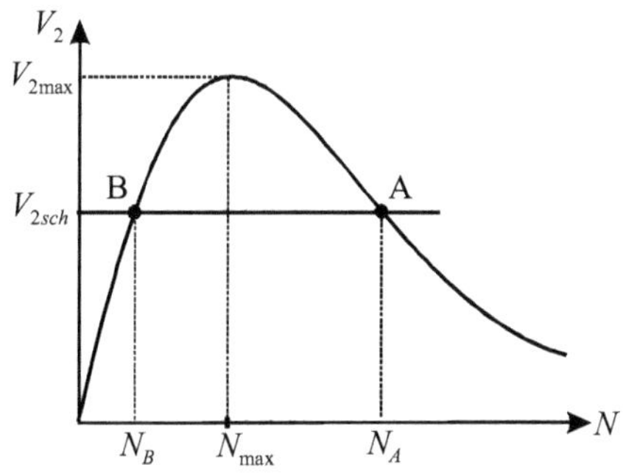

Fig. 2.46 Static characteristic of the on-load tap changing transformer $V_2=f(N)$

Equation (2.8) becomes

$$V_2 = \frac{Na}{\sqrt{1+bN^4}}.$$

Therefore,

$$\frac{dV_2}{dN} = \frac{a\sqrt{1+bN^4} - aN\left(4bN^3 / 2\sqrt{1+bN^4}\right)}{\left(\sqrt{1+bN^4}\right)^2} = \frac{a\left(1+bN^4\right)-2abN^4}{\left(\sqrt{1+bN^4}\right)^3}.$$

Imposing

$$\frac{dV_2}{dN} = \frac{a\left(1-bN_{\max}^4\right)}{\left(\sqrt{1+bN_{\max}^4}\right)^3} = 0$$

results in

$$1-bN_{\max}^4 = 0,$$

that is,

$$N_{\max} = \sqrt[4]{\frac{1}{b}} = \sqrt{\frac{X}{R_L}}.$$

The maximum value of the voltage at the load terminals is obtained from (2.8) at $N=N_{\max}$ by replacing $\frac{R_L}{X} = \frac{1}{N_{\max}^2}$, which gives:

$$V_{2\max} = N_{\max}\frac{R_L}{X}\frac{V_1}{\sqrt{2}} = \frac{V_1}{\sqrt{2}}\sqrt{\frac{R_L}{X}}.$$

Therefore, V_{2max} increases when R_L increases or X decreases, that is, when N_{max} decreases.

Figure 2.46 shows that for a scheduled voltage $V_{2sch} < V_{2max}$, two operating points A and B exist, while for $V_{2sch} > V_{2max}$ no operating point can be found. This demonstrates the fact that no value V_{2sch} can be obtained at the load by changing the turns ratio. Moreover, around point B the OLTC is unstable because by increasing N voltage V_2 is increased.

In the case where the load is modelled by a sole reactance, i.e., $\overline{Z}_L = jX_L$, the absorbing reactive power $Q_2 = V_2^2 / X_L$ results in the voltage equations

$$\overline{V_2} = jX_L \overline{I_2},$$

$$\overline{V_1} = \overline{V}_{2p} + jX\overline{I}_{2p} = NV_2\left(1 + \frac{X}{X_L N^2}\right).$$

Therefore,

$$V_2 = \frac{V_1}{N\left(1 + \frac{X}{X_L N^2}\right)} = \frac{V_1 N X_L}{X + X_L N^2}.$$

Imposing

$$\frac{dV_2}{dN} = 0 \Rightarrow V_1 X + N_{max}^2 (X_L V_1 - 2X_L V_1) = 0$$

Gives

$$N_{max} = \sqrt{\frac{X}{X_L}}, \quad V_{2max} = \frac{V_1\sqrt{X_L}}{2\sqrt{X}}.$$

Therefore, N_{max} decreases when X_L increases or when X decreases. Moreover, the static characteristic is similar to that in Fig. 2.46, with V_2 increasing up to N_{max}. The same can be said in this case (see also Chap. 8 on power system stability).

An OLTC turns ratio change is usually ordered by the operator through local or remote manual control. Open loop control is often preferred by system operators concerned that a system will fall into voltage instability when OLTCs operate in closed loop control. Another reason for this custom is the possible lack of coordination among OLTCs working in the same grid area, with the risk that some controls will drift in opposite directions.

Figure 2.47 Illustrates schematically the on-load tap changing operating mechanism. The tap changer decreases the turns ratio when $V_2 < V_{2sch} - \varepsilon$ and increases it when $V_2 > V_{2sch} + \varepsilon$; or, it takes no action when voltage V_2 assumes values inside the admissible limits $[V_{2sch} - \varepsilon, V_{2sch} + \varepsilon]$.

Figure 2.48 shows a simplified block diagram of turns ratio closed loop control characterised by a nonlinear block followed by an integrator. The integration time T_v (equal to about 10 s at the least) corresponds to the motor, which acts on the

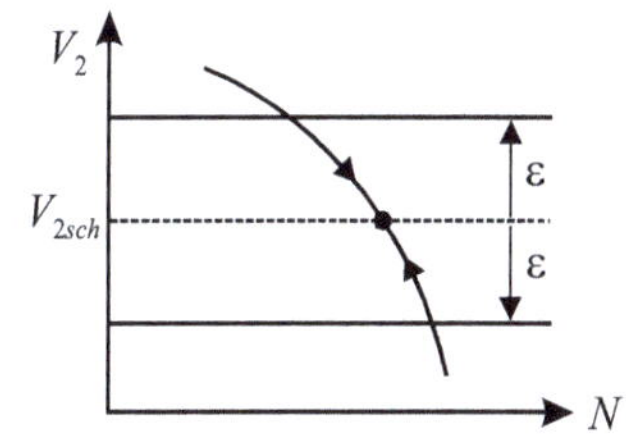

Fig. 2.47 Commutation logic of the on-load tap changer

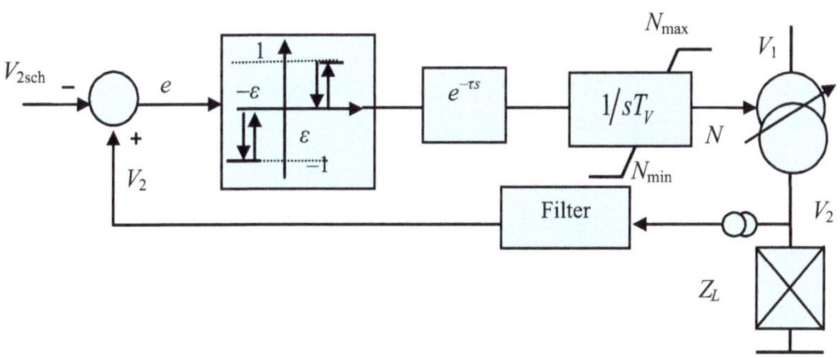

Fig. 2.48 Block diagram of closed loop load voltage control by tap changer steps

changer; V_2 is the voltage to be regulated, V_{2sch} is the desired voltage and N_{tap} is the required windings ratio inside the allowed range: $N_{min} \leq N \leq N_{max}$.

The action of the tap changer control is further characterised by delays between subsequent turns ratio steps represented by the block with time delay τ on the order of tens of seconds. More precisely: in order to avoid employing the tap changer operation in the case of voltage transient variations, the on-load tap changer is blocked during a time delay before the first commutation and between two successive commutations. Usually, if the voltage exceeds admissible limits, the first commutation has a greater delay compared to the subsequent commutation. Furthermore, for cascading located transformers the greater the voltage level the smaller the delay of the first commutation.

Under steady-state conditions, the integrator ensures a zero voltage error if the required change of N lies within the specified range ($N_{min} \leq N \leq N_{max}$). This discontinuous regulating control is slow compared to other types of voltage continuous regulation, and it ensures the desired bus voltage value at steady-state operating conditions only, unless voltage should fall into instability by a reduction in N (point B in Fig. 2.46).

Transformers must be able to provide regulation under normal conditions, such as a load variation according to a daily load curve, and under emergency situations, when voltage assumes values outside admissible operating limits. This leads to equipment damage due to over-voltages or system collapse because of the voltage instability phenomenon.

From the above, we understand it is desirable that the number of OLTC on-load commutations be as small as possible, a practice that also reduces tap-changer life degradation linked to the number of manoeuvres.

An operator will be able to stop the tap changer operation when a voltage instability risk is encountered only if the OLTC loop control opening is structured in a way that blocks tapping during heavy transients.

2.4.4 *Link of Voltage, Reactive Power and Turns Ratio in OLTC Transformer Applications*

Combined Use of OLTC and Reactive Power Injections in Transmission Networks

First Scheme A common practice for reactive flow control in transmission networks is the use of the tertiary of a three-winding transmission transformer for reactive power injection via synchronous compensators or capacitor/reactor banks, as shown in Fig. 2.49.

For a given load condition, the transformer tap ratio setting allows the required reactive generation/consumption on the tertiary bus. Representing the three-winding transformer with its equivalent star (or Y) connection and neglecting winding resistances and transformer shunt losses, the impedance diagram of the system under consideration is shown in Fig. 2.50.

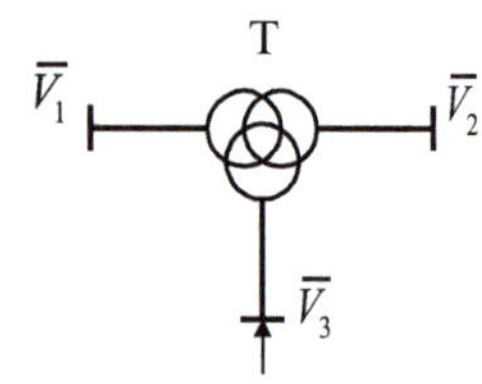

Fig. 2.49 Three-winding OLTC transformer with reactive power source at its tertiary winding

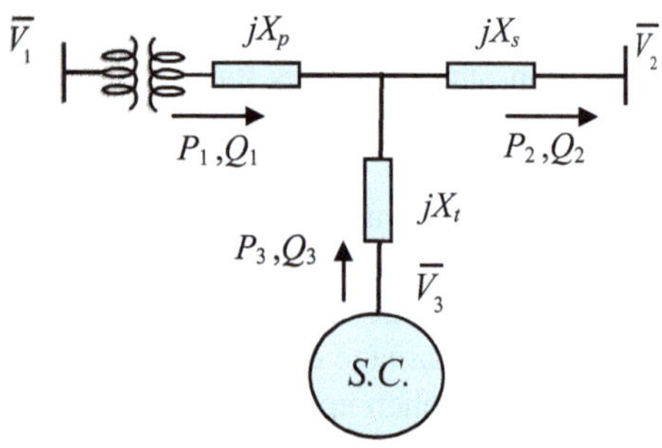

Fig. 2.50 Equivalent impedance diagram of a three-winding OLTC transformer with synchronous condenser connected to its tertiary winding

For a given secondary load P_2, Q_2, assuming $P_3 \approx 0.0$, voltage drop between buses 1 and 2 when the transversal component is neglected is given by

$$\left|\overline{\Delta V}\right| \approx \Delta V = \frac{V_1}{N_{12}} - V_2 = \frac{X_P Q_1 + X_S Q_2}{V_2} = \frac{X_P (Q_2 - Q_3) + X_S Q_2}{V_2}$$

or

$$-V_2 V_1 + N_{12} V_2^2 + N_{12}[(X_P + X_S) Q_2 - X_P Q_3] = 0.$$

The last equation gives the relationship between V_1, V_2 and Q_3. Then, for V_1 and V_2 known and for specific Q_3, the required tap ratio is given by

$$N_{12} = \frac{V_2 V_1}{V_2^2 + [(X_P + X_S) Q_2 - X_P Q_3]}.$$

Or, from another perspective, the value of the reactive power injection Q_3 for a specified tap ratio is determined by the relation

$$Q_3 = \frac{-V_2 V_1 + N_{12}(X_P + X_S) Q_2 + N_{12} V_2^2}{N_{12} X_P}.$$

The arrangement of this operation usually skips computation of Q_3 and simply applies the manual control of the OLTC transformer tap ratio and the automatic control of the synchronous condenser excitation to sustain V_2 under the assumption of a robust V_1.

Second Scheme Voltage at the load terminals cannot be maintained at the scheduled value by tap changing; an additional reactive power is required.

In order to dimension the minimum required rating of a reactive power compensation device, tap changing flexibility is considered at its extreme operating limits. Therefore, the turns ratio is imposed at the minimum value (N_{min}) for maximum load as well as at the maximum value (N_{max}) for minimum load.

If no reactive power compensation device is used, from Fig. 2.44, considering voltage V_1 constant and neglecting the transversal component of voltage drop, the following is obtained:

- For maximum load,

$$V_1 = V_{2P\max} + \frac{RP_2^{\max} + XQ_2^{\max}}{V_{2P\max}};$$

- For minimum load,

$$V_1 = V_{2P\min} + \frac{RP_2^{\min} + XQ_2^{\min}}{V_{2P\min}}$$

- subject to the condition

$$V_{2P\min} \le V_{2P\text{sch}} \le V_{2P\max}.$$

In order to achieve the scheduled voltage $V_{2P\text{sch}}$ for both operating conditions, the synchronous compensator has to provide a reactive power $Q_{\text{comp}}^{\text{gen}}$ for the peak load condition and to absorb a reactive power $Q_{\text{comp}}^{\text{abs}}$ for the minimum load case.

Introducing the synchronous compensator as in Fig. 2.50, we need to modify the equations of voltage:

- For maximum load:

$$V_1 = V_{2P\text{sch}} + \frac{RP_2^{\max} + X(Q_2^{\max} - Q_{\text{comp}}^{\text{gen}})}{V_{2P\text{sch}}}.$$

- For minimum load:

$$V_1 = V_{2P\text{sch}} + \frac{RP_2^{\min} + X(Q_2^{\min} - Q_{\text{comp}}^{\text{abs}})}{V_{2P\text{sch}}}.$$

Equating the V_1 equations with and without the single-phase reactive power generated by the compensator in an over-excited condition, we obtain

$$Q_{\text{comp}}^{\text{gen}} = \frac{V_{2P\text{sch}}}{X}\left(V_{2P\text{sch}} - V_{2P\max} - \frac{P_2^{\max}R + Q_2^{\max}X}{V_{2P\max}} + \frac{P_2^{\max}R + Q_2^{\max}X}{V_{2P\text{sch}}}\right).$$

Similarly, equating the V_1 equations with and without the single-phase reactive power generated by the compensator in an under-excited condition, we obtain

$$Q_{\text{comp}}^{\text{abs}} = \frac{V_{2P\text{sch}}}{X}\left(V_{2P\min} - V_{2P\text{sch}} + \frac{P_2^{\min}R + Q_2^{\min}X}{V_{2P\min}} - \frac{P_2^{\min}R + Q_2^{\min}X}{V_{2P\text{sch}}}\right).$$

Ignoring the difference of the latter terms in the last two expressions, the simplified equations of the reactive power generated/absorbed by the synchronous compensator are obtained:

$$Q_{\text{comp}}^{\text{gen}} = \frac{V_{2P\text{sch}}}{X}(V_{2P\text{sch}} - V_{2P\max}) = \frac{N_{\min}^2 V_{2\text{sch}}(V_{2\text{sch}} - V_{2\max})}{X},$$

$$Q_{\text{comp}}^{\text{abs}} = \frac{V_{2P\text{sch}}}{X}(V_{2P\min} - V_{2P\text{sch}}) = \frac{N_{\max}^2 V_{2\text{sch}}(V_{2\min} - V_{2\text{sch}})}{X}.$$

Taking into account the typical proportion

$$Q_{\text{comp}}^{\text{abs}} \leq (0.5\ldots0.65)Q_{\text{comp}}^{\text{gen}},$$

the rated power Q_n of a synchronous compensator results from the more restrictive of the two conditions:

$$Q_n \geq Q_{\text{comp}}^{\text{gen}} \geq \frac{Q_{\text{comp}}^{\text{abs}}}{(0.5\ldots0.65)}.$$

The general problem of placing and sizing reactive power compensation sources in electrical networks is mostly solved by performing technical-economical optimisation calculations. With turns ratio being constant, the values $N_{\max}=N_{\min}=N$ are introduced in the above expressions and the turns ratio N is determined once Q_n is defined, by imposing the restriction $N_{\max}<N_{\min}<N$.

In order to use the entire regulation range of the compensator (for instance, at maximum load to provide $Q_{\text{comp}}^{\text{gen}}$, and during minimum load, ($Q_{\text{comp}}^{\text{abs}} = 0.6Q_{\text{comp}}^{\text{gen}}$), under the assumption of constant V_1 for both regimes, the maximum load can be written:

$$V_{2P\text{sch}} + \frac{RP_2^{\max} + X(Q_2^{\max} - Q_{\text{comp}}^{\text{gen}})}{V_{2P\text{sch}}} = V_{2P\text{sch}} + \frac{RP_2^{\min} + X(Q_2^{\min} - 0.6Q_{\text{comp}}^{\text{gen}})}{V_{2P\text{sch}}}$$

giving

$$Q_{\text{comp}}^{\text{gen}} = \frac{R(P_2^{\max} - P_2^{\min}) + X(Q_2^{\max} - Q_2^{\min})}{1.6X}.$$

Reactive Power Flow Control Between Two High-Voltage Networks Connected Through an OLTC Transformer

Transmission levels of different voltages are usually interconnected through OLTC transformers (Fig. 2.51). In the case of infinite-bus networks at both sides, OLTC transformers are used as a means for reactive flow control between two connected networks.

By using the equivalent impedance diagram shown in Fig. 2.52 with tap winding assumed as placed on the transformer primary windings, and by neglecting transformer resistances and shunt losses, the following equation can be written:

$$\left|\overline{\Delta V}\right| \cong \Delta V = V_1 - NV_2 = \frac{X_T Q_1}{NV_2}$$

or

$$V_2^2 - \frac{V_1 V_2}{N} + \frac{X_T Q}{N^2} = 0.0.$$

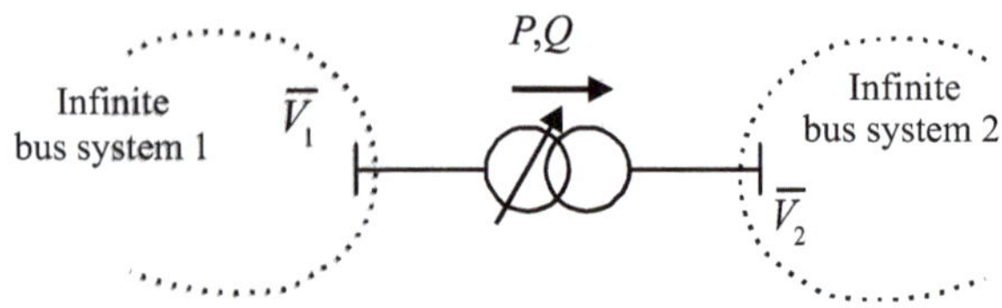

Fig. 2.51 Interconnection of two powerful networks through an OLTC

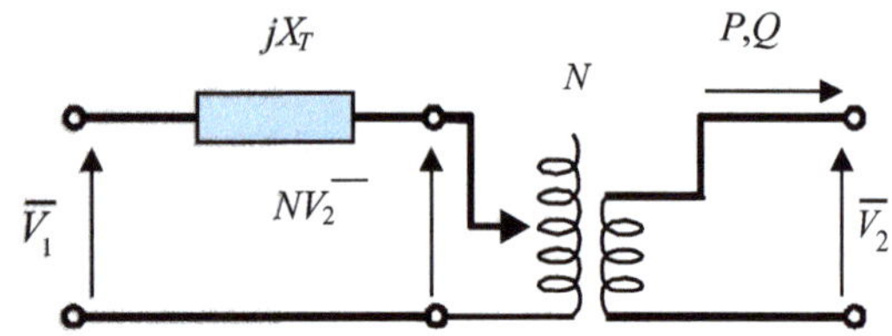

Fig. 2.52 Interconnection of two powerful networks through an OLTC: equivalent impedance diagram

The last equation gives the relationship among V_1, V_2, Q and N for a known X_T. If, for example, the zero reactive flow between two networks is required for specific values of voltage magnitudes V_1 and V_2, the transformer tap ratio has to be

$$N = V_1 / V_2.$$

That is, it must match precisely the actual voltages of the two powerful connected networks.

Radial Transmission/Distribution System with Two Cascaded OLTC Transformers

Very often, OLTC transformers are connected in series with the feeder line, from the transmission to the distribution network, as shown in Fig. 2.53.

Let N_S and N_R be the tap ratios of OLTC transformers on sending and receiving ends, respectively. It is of interest to determine, at steady state, the specified relationship of the tap ratios N_S and N_R with voltage magnitudes V_1 and V_2. The following relations relate to the impedance diagram in Fig. 2.53 (bottom),

$$\overline{V_{1s}} = \frac{\overline{V}_1}{N_S}, \quad \overline{V_{2P}} = N_R \overline{V}_2, \quad \overline{Z} = R + jX = \overline{Z_{TS}} + \overline{Z_L} + \overline{Z_{TR}}.$$

The voltage throughout the system is

$$\overline{\Delta V} = \frac{\overline{V}_1}{N_S} - N_R \overline{V}_2 = \overline{Z}\overline{I}_R = (R + jX)\overline{I}_R = \frac{P + jQ}{N_R \overline{V}_2^*}.$$

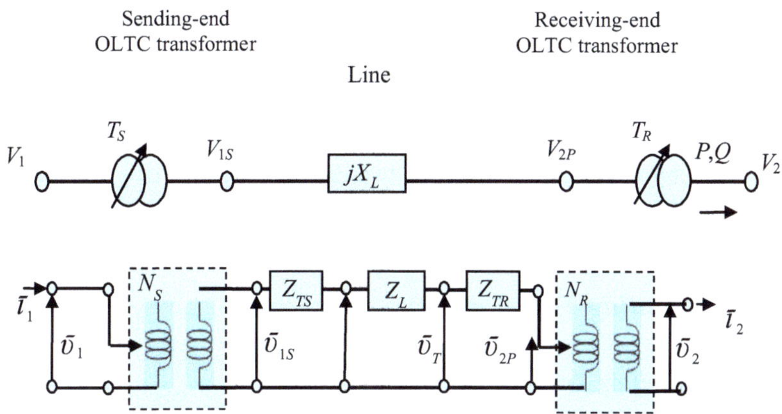

Fig. 2.53 Radial transmission/distribution system with two cascaded OLTC transformers: (*top*) one-line diagram; (*bottom*) *impedance* diagram

Assuming $\overline{V_2} = V_2\angle 0^\circ$ and neglecting the transversal component of the voltage drop, the above becomes

$$\left|\overline{\Delta V}\right| \approx \Delta V = \frac{V_1}{N_S} - N_R V_2 = \frac{RP + XQ}{N_R V_2}.$$

Rearranging the above equation gives

$$V_2^2 - \frac{V_1 V_2}{N_S N_R} + \frac{RP + XQ}{N_R^2} = 0.0,$$

whose solution is

$$V_2 = \frac{1}{2N_R N_S}\left(V_1 + \sqrt{V_1^2 - 4(RP + XQ)N_S^2}\right).$$

This equation is characterised by four variables (V_1, V_2, N_S, N_R) for a known load (P, Q) and a value of system impedance ($R+jX$). Its application requires additional specifications on some of the variables. For example, if equal magnitude of voltages V_1, V_2 is required (i.e., complete compensation of the voltage drop in the system is required), the equation becomes ($V_1 = V_2 = V$):

$$2N_R N_S V = V + \sqrt{V^2 - 4(RP + XQ)N_S^2}$$

or

$$\frac{N_R}{N_S}(1 - N_R N_S) = \frac{RP + XQ}{V^2}.$$

Then, for known R, X, P, Q and V, the relationship between tap ratios N_S and N_R is easily found.

This notwithstanding, we clearly see the increased difficulty of manually operating cascaded OLTC tap ratios when regulating grid voltages are aimed in front of a continuous load change. Instead, when automatic closed loop control is considered, a time hierarchy is required between the two cascaded OLTCs, one which imposes a slower tap control speed on the OLTC operating at the lower voltage level.

2.4.5 *Regulating Transformers*

In-Phase Regulating Transformer (IPRT)

The in-phase regulating transformer, or "booster", is used to control voltage amplitude in the line to which the device is connected (Fig. 2.54).

Each phase has one winding that is series-connected to the bus whose voltage is to be controlled, while the other winding is fed by the same bus via a variable-ratio auxiliary transformer. Varying an auxiliary transformer's turns ratio brings about a change in the amplitude(ΔV) of the series-connected winding, leaving the phase unaltered. This is the case of the OLTC.

Phase-Shifting Transformers (PSTs)

There are two types of phase shifters, corresponding to the schemes of Figs. 2.55 and 2.56 (only phase R is shown, for convenience).

i) In-quadrature regulating transformers

In the first type, the turns ratio variation of the auxiliary transformer causes an orthogonal ΔV variation and thus a voltage phase variation.

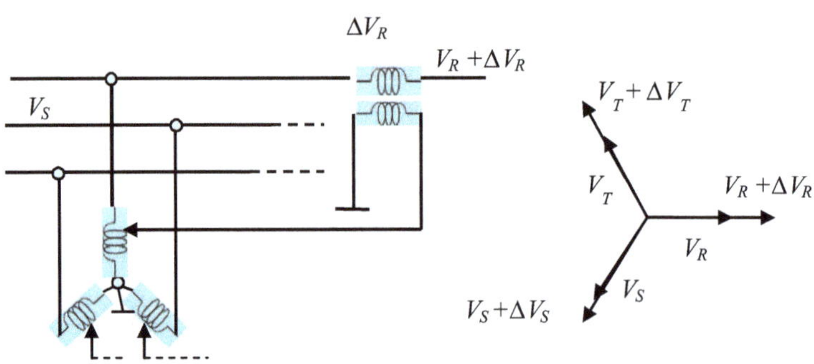

Fig. 2.54 Phase R: basic booster scheme and voltage phasor diagram

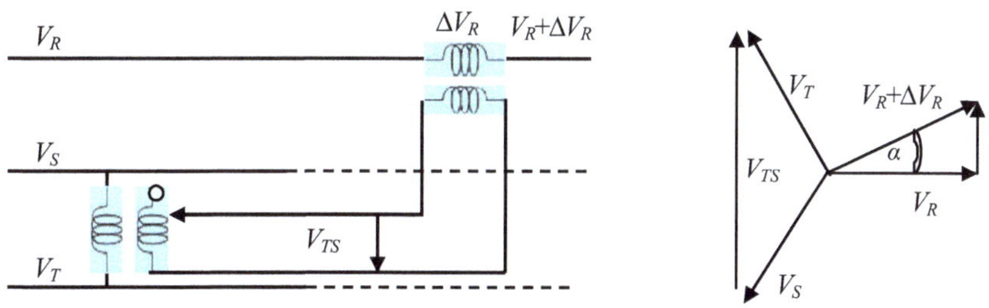

Fig. 2.55 Basic scheme (phase R) of a phase shifter and related voltage phasor diagram

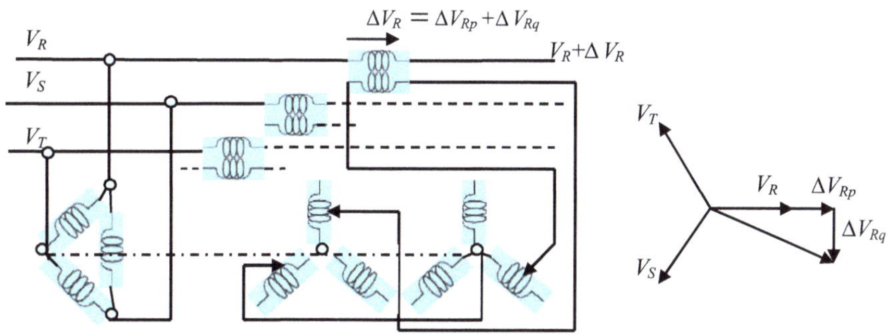

Fig. 2.56 Basic scheme of regulating transformer for controlling voltage amplitude and phase

This variation induces a phase shifting α (positive or negative) and a variation of the voltage amplitude (the smaller the variation, the smaller the phase shifting) between the voltages upstream and downstream of the transformer. Thus, these are complex-ratio transformers, as are ordinary three-phase transformers with different connection modes of the primary and secondary windings and where $\alpha = \pm 30°$. From the phasor diagram in Fig. 2.55 we also deduce that the shifting α is approximately proportional to variation ΔV of the voltage introduced by the transformer.

ii) In-phase and in-quadrature regulating transformers (QBT)

This regulating transformer controls the voltage amplitude and phase that embodies the previous two and is exemplified in Fig. 2.56.

From the latter type, with appropriate measures a pure phase angle regulator (PAR) may be derived if voltage amplitude is kept constant. The regulating transformers shown (mechanically switched) are series regulating transformers, which introduce an additional voltage between two nearby nodes at the same nominal voltage, so creating amplitude and phase voltage variation.

In Table 2.2 the phase diagrams and characteristics for three types of regulating transformers are shown. Notation is as follows: *i*, *o*: input and output terminals of regulating transformer; $\overline{v}_i, \overline{v}_o$: input and output voltage, in p.u.; $\overline{\Delta v}$: additional voltage, in p.u.; α: phase displacement of $\overline{v}_o$ vs. $\overline{v}_i$; β: phase displacement of $\overline{\Delta v}$ vs. $\overline{v}_i$.

Table 2.2 Regulating transformer phase diagrams and characteristics

Summary of results			
IPRT	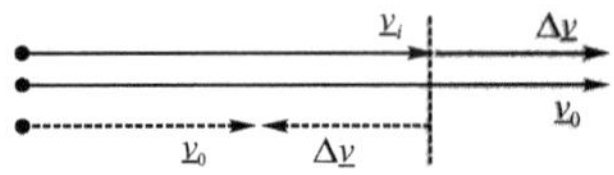	$V_o \neq V_i$ $\alpha = \beta = 0$	Voltage and reactive power flow control in bus where installed
QBT	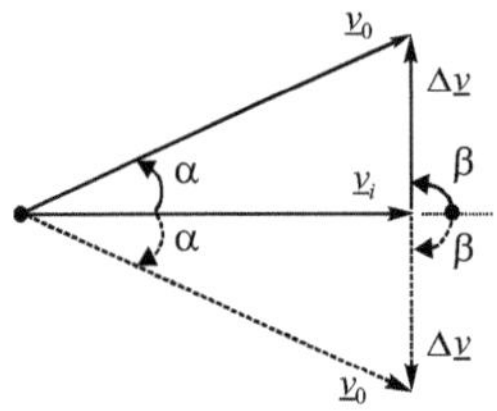	$V_o \neq V_i$ $\beta = \pm\pi/2$ variable α	Control of active (due to $\alpha \neq 0$) and reactive power flows
PAR	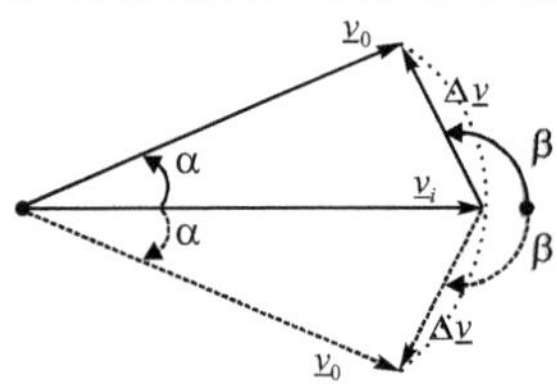	$V_0 = V_i$ variable β and α	

Therefore, a regulating transformer may be represented by the classic single-phase equivalent circuit in p.u. with a complex p.u. turns ratio of

$$\bar{m} = \overline{\left(\frac{N_1}{N_2}\right)} = me^{j\alpha}.$$

QBT and PAR regulating transformers are distinguished from classic transformers in that they can also operate as reactive power sources by controlling the real power through angle α and by controlling the reactive power through voltage amplitude, thereby imposing active and reactive power flow in the line they are operating. Obviously, these are stepping (discrete) controls with very slow dynamics with respect to rotating and static reactive power generators.

2.5 Conclusion

An effort has been made throughout the chapter to introduce the major existing power system resources that can contribute to grid voltage control. They are: rotating (synchronous generator-compensator) or static equipment (SVC, STATCOM, UPFC), which allow continuous fast control of their local voltages and reactive powers, both delivered and absorbed; compensating equipment controlled with continuity such as the TCR, FC-TCR, TSC-TCR; others based on ON/OFF switching controls (MSR, MSC and TSC). All are reactive power sources used to control reactive power delivery/absorption.

The control schemes of these reactive power generators were introduced and important details provided as mainly concerned local voltage control. Evidence was given to the general high speed characteristic of continuous control solutions of local bus voltages and related limitations.

Transformers and their stepping controls were also analysed and evidence given to the fact that these are not reactive power sources except when they are classified as phase-shifting transformers (PSTs). PSTs such as QBTs and PARs are able to control active and reactive power flows through mechanically switched settings. All other transformers control only low voltage values.

All control resources considered here—mainly synchronous generators and compensators together with static equipment (SVC, STATCOM, UPFC), but also compensating equipment (TCR, FC-TCR, TSC-TCR) and lastly shifting transformers (PSTs) such as QBT and PAR—are the primary controllable reactive power resources for the grid voltage control systems introduced in the chapters ahead.

References

1. Arcidiacono V, Ferrari E, Marconato R, Dos Ghali J, Grandez D (March/April 1980) Evaluation and improvements of electromechanical oscillation damping by means of eigenvalue-eigenvector analysis. Practical results in the central Perú power system. IEEE Trans Power App Syst PAS-99(2):769–778
2. Bhavani SP, Sai BCh (2008) Transient stability enhancement of power system using STATCOM. Int J Elec Power Eng 2(4):271–276
3. Eremia M et al (eds) (2006) Electric power systems. Volume 1: electric networks. Publishing House of Romanian Academy, Bucharest
4. Erinmez IA (ed) (1986) Static VAR compensators. CIGRE Working group 38-01, Task force No. 2 on SVC
5. Gyugyi L (1979) Reactive power generation and control by thyristor circuits. IEEE Trans Ind Appl 1A-15(5):532–531
6. Gyugyi L (1992) Unified power flow control concept for flexible AC transmission systems. IEE Proc C 139(4):323–331
7. Hingorani NG, Lazlo G (1999) Understanding facts: concepts and technology of flexible ac transmission systems. Wiley-IEEE Press
8. IEC (2003) IEC Standard 60214-1: tap-changers, 1st edition, Geneva, 2003
9. Kimbark EW (1977) How to improve system stability without risking subsynchronous resonance. IEEE Trans PAS-96(5):1608–1619
10. Marconato R (January 2002) Electric power systems. Volume 1: Background and basic components, 2nd edn. CEI–Italian Electrotechnical Committee
11. Masood T, Abdel-Aty E, Aggarwal RK (November 2006) Static synchronous compensator (STATCOM) modelling and analysis techniques by Matlab & Sat/Fat acceptance tests in the light of commissioning and installation scenarios. 21st International Power System Conference (PSC), Tehran
12. Mathur RM (ed) (1984) Static compensators for reactive power control. Committee on static compensation, canadian electrical association (CEA). Cantext Publications, Winnipeg

13. Metha H et al (May 1992) Unified power flow controller for flexible AC transmission systems. EPRI Flexible AC Transmission System (FACTS) Conference, Boston
14. Miller TJE (1982) Reactive power control in electric systems. Wiley, New York
15. Mori S et al (1992) Development of large static var generator using self-commutated inverters for improving power system stability. IEEE/PES Winter Power Meetings, Paper N 92WM165-1
16. Quazza G, Ferrari E (1972) Role of power station control in overall system operation. In: Handschin E (ed) Real-time control of electric power systems. Elsevier, Amsterdam, pp 215–257
17. Saccomanno F (1992–2003) Electric power systems: analysis and control. Wiley, New York (English version)
18. Therond PG (August 2000) (convenor): Unified power flow controller (UPFC). CIGRE technical brochure. Task force 14.27
19. Vithayathil J et al (February 2004) Thyristor controlled voltage regulators. CIGRE technical brochure, working group B4.35

Chapter 3
Grid Voltage and Reactive Power Control

3.1 General Considerations

A widespread inadequate control of grid voltage and reactive power has become more critical in recent years due to the general trend by system operators and electrical utilities to operate transmission networks as close as possible to their maximum capacity. The need for suitable control solutions capable of dealing with increased power loads and losses, possible grid contingencies and voltage collapse risks has therefore grown in ever tighter and enmeshed networks. Yet, a lack of real-time, closed-loop, automatic coordination of reactive power resources for network voltage control seems as persistent as it is unjustified.

Basically, a system operator controls grid voltage by one of three methods: manual, automatic or a combination of the two. Moreover, the transmission system operator controls voltages on EHV substations, while the distribution system operator controls medium to low voltages. These operators carry out different and complementary control tasks on voltages and reactive powers.

In principle, a transmission system operator's objective is to impose the optimal voltage profile in an EHV system with the goal of high transfer capability, minimum losses and high voltage stability. This can be achieved by controlling the reactive power resources operating at that voltage level by:

- Injection/absorption of reactive power by operating generators;
- Switching compensating equipment on/off;
- Setting voltage set-points of SVCs, STATCOMs and OLTCs;
- Blocking OLTCs when risk increases;
- Paralleling hot reserves, opening/re-energising lines, shedding loads up to the very unconventional employment of other system components, such as the UPFC (unified power flow controller) used to support voltages.

The principal objective of a distribution system operator is to guarantee adequate voltage levels at the load buses, again by controlling the reactive power resources available on the MV-LV side: switching MV-LV compensating equipment on/off and setting MV-LV OLTC set-points. Rarely is on/off switching of HV and MV

S. Corsi, *Voltage Control and Protection in Electrical Power Systems*,
Advances in Industrial Control, DOI 10.1007/978-1-4471-6636-8_3

lines used for this purpose. The use of reactive power reserves by shunt static compensation and dynamic FACTS to improve the power factor of factory loads is widely used to reduce reactive power flows from remote areas. This is an expensive solution, which cannot be considered for overall grid nodes except when strictly necessary. The cheapest solution has to be considered first: minimisation of reactive power injection distances and coordination as much as possible of local reactive power resources. And so, compared to the case of longer-distance reactive power transfer, the effect of voltage drop due to transmitting power at non-unity power factors is minimised, losses are reduced and transmissible active power is increased.

The ideal case, one which results in minimum losses, is obviously one that entails the artificial injection of reactive power at the load buses themselves, thereby compensating the local reactive power demand at 100 %. Even in this unrealistic situation, a great amount of compensating equipment has to be continuously and widely controlled via complex and therefore critical coordination.

Returning to a realistic situation, several types of compensating equipment exist in the transmission grid: those applied at the distribution level are scheduled according to long-term load forecasting and switched on/off mostly by hand. When seen by the EHV grid, this compensating equipment is deemed part of the load and is not a control variable. Apart from this, compensating equipment fulfils the function of reactive power injection/absorption at the voltage levels to which it is applied. Articulating the possible ways these levels can be controlled is crucial to any grid voltage control proposal. Future distributed "smart grids" should act to increase the automatic controllability of compensating equipment at the distribution level.

Load shedding as a response to voltage problems is a widely used practice—one that is in fact often overused. It is more properly linked to the real needs of system protection, to be resorted to only in the case of a high security risk in a system. Even then, load shedding should be the ultimate control action, to be utilised only after the aforementioned control resources are used to their fullest.

Looking again at transmission and distribution control centres: their organisation, available control systems, coordination and controllability of processes—all these are issues that differ from one power system to another. Nevertheless, a common understanding of control centre operation should be, in principle, in line with the following tenets:

- The transmission system should assume the primary role in overall system voltage control because it can strongly affect the voltages of the distribution subgrid it feeds; each low voltage grid, on the other hand, only weakly affects voltages in the transmission network. Therefore, the transmission grid should guarantee high quality and stable voltage control on EHV and HV buses through adequate and timely coordination of available reactive power resources in the requisite amount at the proper location.
- Robust and solid voltages on the transmission grid make it relatively simple to guarantee high quality voltage control at the load sides by minimising control effort (switching) on MV-LV reactive resources and the OLTC. Clearly, each distribution dispatcher should guarantee a fixed compensation, agreeing on the amount at the planning stage with the transmission dispatcher.

Accordingly, the most advanced coordinated and fast control of system voltages should be achieved through operations at the transmission level. It is easy to see why transmission network voltage control is an outstanding distribution network control solution. Without it, a classic distribution network has serious problems linked to:

- Frequent voltage changes;
- Frequent OLTC stepping or, in some cases, an insufficient control margin for regulating voltages on the low voltage sides;
- Pointless reactive power recirculation among the OLTC transformers operating in a given distribution area;
- Too-slow OLTC dynamics with respect to voltage decay speed;
- Negative effect of the OLTC on load voltages when a voltage instability condition is approached.

From here we focus on transmission system voltage control due to its crucial importance; voltage control at the distribution grid is treated in a dedicated chapter, Chap. 10.

Operators have long desired but never achieved centralised real-time voltage control via optimal power flow (OPF). OPF traditionally considers the overall power system and related load flow (LF) computation at a given operating point, one which is based on the most recent state estimation (SE) report. Any optimisation-upgrading step would require a new SE output, whose reliability through an iterative process takes time.

In the modern supervisory control and data acquisition (SCADA) system that is now applied in dispatching control centres, the SE updating time interval is reduced to an order of 5 min. Yet, even a processing time of this duration is too long for transacting a reliable SE update, including implementation of topology checks, when compared to power system voltage dynamics. It does not allow for a truly effective real-time, stabilising voltage control system. On the other hand, faster state estimation is achievable but not advisable, because it is not widely available and its reliability is poor, mainly during contingencies.

For this reason primarily, optimal power flow has been and still is considered a proper forecasting computation, useful for the dispatch of voltage-reactive power but far from a real-time control that every few seconds would request a reliable, complete SE update. Engineers continue to pursue the objective of faster SE. Their quest is for a voltage control system based on OPF that can control all available generators and other on-field resources at the same time via a highly complex control system operating through a widespread, high performance telecommunication network capable of obtaining optimised voltages and losses, quantities deduced from the power system model in use.

It is evident that without fast and reliable SE updating, many complexities stand in the way of practical and reliable control, and there is too high a risk of incorrect computation or convergence problems! The challenge is the design of a simpler, faster control, one able to satisfy the requisites of real-time, closed-loop, automatic and—as much as possible—optimal voltage control. Without it, centralised real-time voltage control via OPF remains unfeasible.

Before we can achieve reasonable and feasible transmission network automatic voltage regulation, we must first reflect on the following fundamentals:

- In a power system, any existing on-field equipment provided with a voltage controller allows voltage support at the bus where it is installed. That is, each piece of equipment is able only to control its local voltage or to inject/absorb reactive power at the bus to which it is connected.
- Regarding transmission network voltage control, a feasible objective is to sustain the voltage not of a single bus but rather all voltages in a given network area, to increase area operation security and efficiency as well as voltage quality. Accordingly, grid voltage control necessarily consists of possible area voltage control by coordination of available equipment in the area, with the aim of sustaining and optimising main area load voltages ahead of load variation and possible contingencies.
- Grid control in general and area voltage regulation in particular is a task for the grid dispatcher, who defines the control strategy according to his organisation, the available on-field resources and system monitoring and extant control solutions. Until today, and in spite of the fact that many operation problems are linked to voltage changes, most system operators control power system voltages by hand: such manual control entails written instructions, telephone communication and remote tele-control.

From minimal control to several fully automatic voltage control systems, a wide variety of mixed manual and automatic solutions are used around the world in power system operation. This section presents some of these area automatic voltage control systems.

A final relevant aspect to highlight in the discussion of automatic voltage control is the difference between the following two types of control:

- Continuous voltage and reactive power control, also called *voltage regulation*;
- Discontinuous, extreme shunt voltage and reactive power control, also called *voltage protection*.

Continuous voltage control operates at any instant to sustain system voltages by controlling reactive power resources in the system with continuity, at amounts necessary to maintain voltages at the desired values. By its nature, this control is automatic. Available resources that allow for continuous control are, synchronous generators, SVCs and FACTS. Other components such as compensating equipment or OLTCs allow discrete control because of their switching actuation, which is limited by the number of switching manoeuvres possible. This discrete control can be combined with the continuous control of the generators, SVCs, STATCOMs and so on to realise new advanced wide-area voltage regulating systems (V-WAR) [1–9].

Discontinuous step control of voltage and reactive power at a given section of a network is often a drastic control measure, resulting in a loss of the continuous feeding of a given load, a generator tripping or lines opening to counteract a given sequence of events. Usually, it is a kind of protecting control tailored to a particular event in a given part of the network, requiring fast voltage/reactive power measurements for recognising on time any dangerous incoming decay. This kind of protecting automatic control is always ready but never works unless the given thresholds are overcome. New, advanced protecting controls are also called *special protection schemes* (SPS), *remedial control schemes* (RCS), or *wide area protection* (WAP) when they relate to a given system area [10, 11].

Regulating voltage control and protecting voltage control schemes are not alternative solutions; rather, they complement one another when they operate in the same grid area as follows:

- Regulating control works continuously to maintain a system far from voltage instability, with a voltage plan optimised to the working conditions at hand;
- Protecting control operates only when the regulating system has performed to its maximum capacity by using all available reactive power resources up to its saturation. It determines the loss of part of a system/load to achieve the security of the process to be maintained.

Area-protecting voltage control is presented in Chap. 11.

3.2 Voltage-Reactive Power Manual Control

So-called "manual" practice for grid voltage control, until now in widespread use by system operators worldwide, typically consists of transmission system operator or independent system operator (TSO/ISO) control centres that dispatch the forecasted reactive power of generating units, scheduling power plant high side voltages, switching shunt capacitors or reactor banks and setting the voltage set-points of OLTC and FACTS controllers (usually by written rules). This solution to the conventional network voltage control problem entails real-time operator decisions based on system monitoring and threshold alarms, operated through commands sent by telephone or tele-command in a kind of "manual coordination" of the available reactive power resources and bus voltage controllers (as mentioned in Chap. 2).

This kind of grid control is nowadays considered quite unsatisfactory because:

- Unit reactive power dispatching and plant high-side voltage scheduling are based on the study of off-line forecasting: actual network operating conditions are often different from forecasted values and therefore unpredictable;
- Voltage set-point coordination is often operated through written operating rules or requested by the system operator only when it is urgently needed. Therefore, untimely or inadequate control can occur in the event of most system dynamic phenomena;
- The skill and ability of the operator to recognise and correctly face grid problems in a timely fashion must be very high and continually put to a severe test.

Often, operators face familiar repeated phenomena, responding with tested countermeasures; but when confronted with novel situations they often fail. Moreover, operators generally are more preoccupied with system security than with system efficiency and optimisation, aspects that are usually overlooked.

3.2.1 Manual Voltage Control by Reactive Power Flow

The ways that a system operator can control reactive power flowing in a given line are few and of approximate effect. Basically, manual control must modify the voltage difference at the edges of the considered line; or, equivalently, it must modify the reactive power delivery/absorption near one or both edges of the line. This can be achieved through employing the following operations at the line edges:

- Changing OLTC transformation ratio;
- Switching compensating equipment on/off;
- Changing voltage set-point of local generators/synchronous compensators or of their reactive power delivery/absorption;
- Changing voltage set-point of the local FACTS.

Due to variable operating conditions and the interactions of a line with its surrounding grid, results vary for any given control, and an iterative process is often required to approximate a desired result. Manual tracking is usually too slow to respond to a system's needs.

Generally speaking, control of reactive power flow is needed in more than one line, and the multi-flow control problem is very complex and has a low probability of successful resolution when addressed manually.

3.2.2 Manual Voltage Control by Network Topology Modification

After all other local reactive power resources are used, a possible though extreme way to control voltage, generally not used, entails network topology modification based on an increase or reduction of the capacity effect provided by electrical lines. In this case, the control strategy basically consists of switching off some low-charged lines in an area of high voltage and, conversely, switching on lines in cases of low voltage after all other local reactive power resources are used. As before, results vary for any given act of control switching, according to the specific grid's operating condition; therefore, this type of control goes in the anticipated direction, but its extent of its effect is not easily predictable.

3.3 Voltage-Reactive Power Automatic Control

Classic automatic voltage control in a power system is provided by a generator's automatic voltage regulators (AVRs), which maintain stator edge voltages at set-point values by fast closed-loop control of generator excitation. Therefore, changing either the load seen by the generator or the AVR voltage set-point value brings about a change in the reactive power delivered or absorbed by the generator. Variations fall inside the generator limiting field defined by the over-excitation limit (OEL) and under-excitation limit (UEL). This classic control, in general application around the world, contributes mainly to the safe and stable operation of the generator, but it does not contribute efficiently to EHV grid voltage support, even if the generator's available reactive power resources could allow more.

Utility and system operators take many approaches to improve voltage control in transmission grids, and many projects have been developed around the world. These approaches generally do not consider voltage-reactive power automatic control or continuous operator control of generator AVRs because practical difficulties arise. In most cases, the approach adopted is limited to a power factor correction based on off-line planning studies; this correction is made through installation of extra shunt capacitors or reactor banks, necessitating a significant investment. When allowed, an automatic control system could require the switching of these components, but it would be under many constraints on the amount of manoeuvres necessary to preserve the life of the components.

The availability in some cases of unit step-up transformers that are OLTC-equipped provides an additional opportunity for network voltage control, provided the regulation system supports a plant's EHV side instead of the generator stator edges. Nevertheless, step-up transformer control is usually manual as well as stepped, and it is slow when automated.

Another common solution is automatic support of power plant high-side voltage through AVR line drop compensation. This practice increases grid voltage support but introduces destabilising interactions between primary voltage regulators.

In recent years the use of FACTS controllers for network voltage automatic support [3.19], mainly SVC and STATCOM, has been seriously considered, even when the related costs do not justify the choice; if extensively applied, FACTS controllers require a coordinated control system similar to that described in § 4.1.4 of chapter 4. Recently, through the impetus of ongoing market liberalisation, some AVR manufacturer solutions include unit reactive power control or power factor control and, in some cases, plant high-side voltage control (HSVC).

Next we present generator AVR control and its evolution up to high-side voltage control. Together with FACTS, generator AVR control best represents true, effective, fast and continuous *bus* voltage-reactive power automatic control. We include a word on OLTC control, how active and reactive power flows impact its performance. Section 4.1.4 of Chap. 4 moves from single-bus to *area* voltage-reactive power control, developing the new subject of grid automatic voltage regulation.

3.3.1 Automatic Voltage Control by OLTC Transformer

The major objective of under-load AVC at the transformer's secondary winding through the motorised tap changer is to sustain voltage at the load side. This control is obtained by an automatic voltage regulator (see Fig. 2.48), based on local low-voltage side measure. Its aim is to automatically sustain the voltage at set-point value, with slow dynamics characterised by delays between subsequent steps and a dominant time constant on the order of tens of seconds.

In general, an OLTC allows tap variation in the field range

$$0.9\ \text{p.u.} \leq m \leq 1.1\text{p.u.}$$

with a large number of steps.

We consider the case of a 32-step OLTC for a 20 % voltage control field. A step covers 5/8 %, that is, 0.00625 p.u. of the nominal voltage. Therefore, the integral control in Fig. 2.48 guarantees a steady state error within the nonlinear insensitivity used in step control. Obviously, such insensitivity is a little larger than 5/8 p.u.

Because of the large number of control steps involved, each characterised by a small voltage drop, a transformer OLTC model is often simplified by a continuous model approximation. From the equivalent scheme in Fig. 2.45 and denoting with "$(..)_0$" the operating point around which linearisation is applied, the model is:

$$V_1 = NV_2 + \frac{N^2(R_S R_L + X_S X_L)}{R_L^2 + X_L^2},\ P_2 + jQ_2 = \frac{V_2^2(R_L + jX_L)}{R_L^2 + X_L^2},$$

$$V_1 = V_2 N + \frac{N^2(R_S P_2 + X_S Q_2)}{V_2^2},$$

$$\Delta V_2 = \left[\frac{V_2}{3NV_2 - 2V_1}\right]_0 \Delta V_1 - \left[\frac{(V_2)^3 + 2N(R_S P_2 + X_S Q_2)}{V_2(3NV_2 - 2V_1)}\right]_0 \Delta N$$
$$- \left(\frac{N^2 R_S}{V_2(3NV_2 - 2V_1)}\right)_0 \Delta P_2 - \left(\frac{N^2 X_S}{V_2(3NV_2 - 2V_1)}\right)_0 \Delta Q_2.$$

Now, because $(3NV - V_1)_0 > 0$ (since $0.9 < (N)_0 < 1.1$), it follows that ΔV_2 increases when V_1 increases and N, P_2 and Q_2 decrease.

Moreover, referring to the OLTC linearised Eq. (2.7), the link between N, V_2 and V_1 is given by

$$\Delta V_2 = -\left[\frac{V_2}{N}\frac{(R_L^2 + X_L^2) - (R_S R_L + X_S X_L)}{(R_L^2 + X_L^2) + (R_S R_L + X_S X_L)}\right]_0 \Delta N + \left[\frac{R_L^2 + X_L^2}{N\left((R_L^2 + X_L^2) + R_S R_L + X_S X_L\right)}\right]_0 \Delta V_1;$$

or, with the obvious meaning of the H,

$$\Delta V_2 = -(H_1)_0 \Delta N + (H_2)_0 \Delta V_1.$$

This equation, depending on R_L and X_L, implicitly takes into account the active and reactive power flows, Q_2, P_2.

Lastly, at the load side the equations linking V_2 and Q_2 and the corresponding linearised model are

$$Q_2 = \frac{V_2^{\,2} X_L}{R_L^2 + X_L^2}, \quad \Delta Q_2 = \left(\frac{2V_2 X_L}{R_L^2 + X_L^2} \right)_0 \Delta V_2 = (H_3)_0 \Delta V_2.$$

So, referencing the scheme of Fig. 2.48, the block diagram of the OLTC continuous linearised model is represented in Fig. 3.1 together with the nonlinear control.

A full, linearised continuous model of the OLTC control loop is shown in Fig. 3.2. Any stepping effect on the controlled variables is neglected by this scheme.

Figure 3.2 shows the added dependence of reactive power transfer Q_2 from V_2 and therefore from N. The complexity of the relationship of V_1 and V_2 to N basically depends on the reactive and active power flows, mainly from Q_2 due to the R_S neglecting value, otherwise as obvious at no-load, this relationship is

$$V_1 / V_2 = N.$$

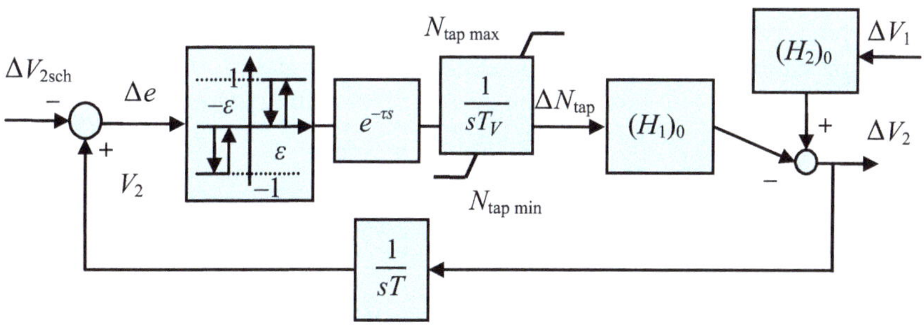

Fig. 3.1 Nonlinear block diagram of OLTC voltage control loop

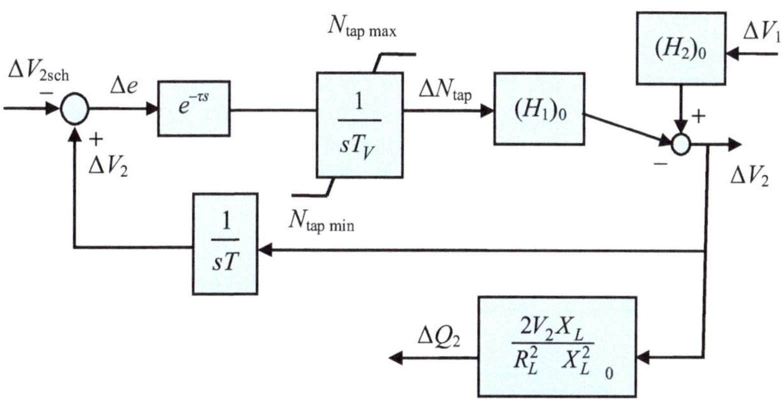

Fig. 3.2 Linear block diagram of OLTC voltage control loop

3.3.2 Automatic Voltage Control (AVR) of Generator Stator Edges

The major objective of synchronous generator stator edge voltage regulation is alternator proper and secure operation. This type of control is obtained via an AVR based on local voltage fast measure feedback, to automatically sustain the voltage at the set-point value, with loop dynamic performance characterised by a dominant time constant value of the order of a few 100 ms up to 1 s, depending on the exciter's characteristics.

AVR control therefore impacts transmission network voltage mainly at a generator's connecting bus. It does so by sustaining local medium voltage there during normal and perturbed operating conditions and by recovering transients following local short circuits. It realises the "primary voltage control" of a generator.

The AVR regulates voltage at the generator's stator by controlling field excitation voltage V_f (Figs. 3.3 and 3.4; [5, 12–16]). The AVR control system also generally includes OEL and UEL limits, power stabilising feedback (PSS) and line drop compensation feedback (compounding) (see § 3.3.3 for more detail).

The main variables seen in Fig. 3.4 are defined as shown:

V_{ref} = primary control loop's voltage reference value;
V_m = measured value of the voltage at the generator's stator;
V_{lim} = control signal provided by the over-excitation limiting loop;
V_{st} = additional signal provided by the PSS;
I_f = field excitation current.

Under normal operating conditions, the OEL, UEL and PSS are open-loop feedbacks (they do not reclose the corresponding V_{lim} and V_{st} signal feedbacks); therefore, the operating model of the AVR is described by $F_v(s)$ transfer function alone. Under perturbed conditions determining the extent of the excitation limits (OEL or UEL), the feedback V_{lim} recloses by overlapping $F_v(s)$. Analogously, under significant and persistent variations of the active power and rotor speed, the PSS feedback recloses by overlapping $F_v(s)$. In critical operating conditions one of the limiting loops and all the other control loops represented in Fig. 3.4 work and interact simultaneously.

To simplify the analysis under normal operating conditions, we consider the alternator excitation system to be energised from an independent supply. In this scenario, AVR dynamics are simply determined by the $F_v(s)$ control law with high

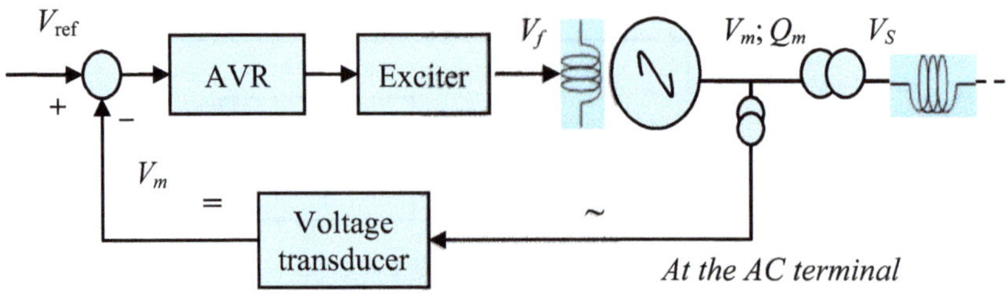

Fig. 3.3 Schematic diagram of the generator voltage control loop

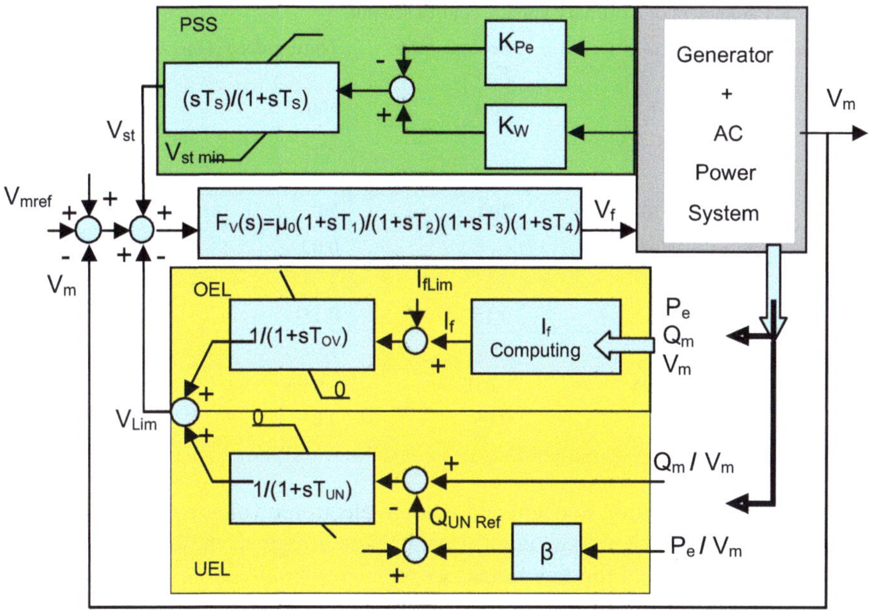

Fig. 3.4 AVR block diagram with control block $F_v(s)$, over- and under-excitation limits (*OEL* and *UEL*) and power system stabilising feedbacks (*PSS*)

frequency zeroes ($1/T_Z$, $z=1$) and poles ($1/T_P$; $p=2, 3, 4$), which allows the controlling range to be very wide, while high values of the amplification factor μ_0 at low frequencies allow a quasi-null steady state error.

The order of transfer function $F_v(s)$ is as follows:

- *Third order*:

$$F_v(s) = \frac{\mu_0(1+sT_{Z1})}{(1+sT_{P2})(1+sT_{P3})(1+sT_{P4})}$$

for systems with exciting dynamo or alternator and rotating diodes. Rough values for the primary voltage control parameters are given in Table 3.1 [5]. A good approximation for a not too low frequency is

$$F_v(s) = \frac{\mu_0 T_{Z1}}{T_{P2}(1+sT_{P3})(1+sT_{P4})}.$$

- *Second order*:

$$F_v(s) = \frac{\mu_0(1+sT_{Z1})}{(1+sT_{P2})(1+sT_{P3})}$$

Table 3.1 Parameters of a primary voltage control scheme

Parameter	First order $F_v(s)$	Second order $F_v(s)$	Third order $F_v(s)$	Unit of measure
μ_0	400	3000	500	p.u./p.u.
T_{Z1}	0.8–1.5	2	2	s
T_{P2}	–	0.05	0.05	s
T_{P3}	5–20	200	20	s
T_{P4}	–	–	0.02	s
T_S	3	3	3	s
K_{pe}	0.15	0.15	0.15	p.u./p.u.
K_ω	15	15	15	p.u./p.u.
$V_{st\ min}$	−0.05	−0.05	−0.05	p.u.
$V_{st\ max}$	0.05	0.05	0.05	p.u.

for systems with exciting dynamo and modern electronic voltage regulator. Rough values for the primary voltage control parameters are given in Table 3.1. A good approximation for a not too low frequency is

$$F_v(s) = \frac{\mu_0 T_{Z1}}{T_{P2}(1 + sT_{P3})}.$$

- *First order*:

$$F_v(s) = \frac{\mu_0(1 + sT_{Z1})}{1 + sT_{P3}}$$

for static excitation systems in absence of excitation current feedback. Rough values for the primary voltage control parameters are given in Table 3.1. A good approximation for a not too low frequency is

$$F_v(s) = \frac{\mu_0 T_{Z1}}{T_{P3}}.$$

Linear Analysis of Generator Voltage Control Loop

$F_v(s)$ control parameter values are tuned to achieve a given steady state accuracy in the voltage control loop as well as a proper dynamic behaviour in terms of stability and response speed. Good accuracy requires a high static gain, whereas loop cut-off frequency is representative of control speed.

Standard analysis of the generator voltage control loop looks at the case of a generator feeding an infinite bus (infinite short-circuit power, representing a very large, interconnected power system) or a load Z_C through a step-up transformer of reactance X_T and a line of reactance X_L. The equivalent scheme in Fig. 3.5 represents the system feeding an infinite bus.

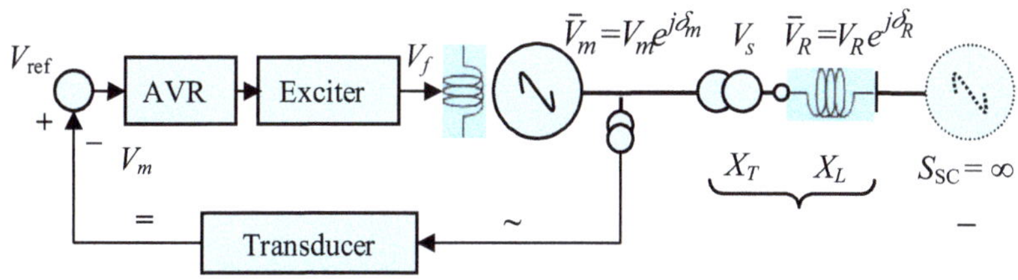

Fig. 3.5 Generator connected to an infinite power bus

According to the Park transform [16, 17], the generator model represented by a d (direct)-q (quadrature) coordinate system can be summarised with the following p.u. equations:

$$V_q = a(s)V_f - x_d(s)I_d, \quad V_d = x_q(s)I_q, \quad I_f = b(s)V_f + \frac{(A_{NN}P_f^*)s}{\Omega_N a(s)I_d}, \tag{3.1}$$

where A_N=alternator nominal apparent power; Ω_N=nominal rotor speed; P_f^* =excitation power at no-load and stator nominal voltage. Moreover,

$$s = d/dt, \quad \overline{V_m} = V_d + jV_q, \quad \overline{I_m} = I_d + jI_q,$$

$$\bar{A} = \overline{V_m I_m}^* = (V_d + jV_q)(I_d - jI_q) = P_m + jQ_m.$$

Accordingly, a general result at around nominal speed and overlooking subtransient time constants due to the machine damping windings is

$$V_q \approx \frac{V_f}{1+sT'_{d0}} - x_d \frac{1+sT'_d}{1+sT'_{d0}} I_d, \quad V_d \approx x_q \frac{1+sT'_q}{1+sT'_{q0}} I_q, \tag{3.2}$$

where T'_d =direct-axis transient short-circuit time constant; T'_{d0} =direct-axis transient open-circuit time constant; T'_q =quadrature-axis transient short-circuit time constant; x_d=direct axis synchronous reactance; x_q=quadrature axis synchronous reactance [11, 16, 17]. Typical parameters values are shown in Table 3.2.

Table 3.2 Typical parameters of a synchronous machine model

Parameter	Turbo-alternator	Hydraulic generator
x_d	1.9 p.u.	1.1 p.u.
T'_d	1.1 s	1.9 s
T'_{d0}	7–10.0 s	6.0 s
x_q	1.7 p.u.	0.7 p.u.
T'_q	0.04 s	0.07 s

Case of Generator Delivering Reactive Power Only

In this case,

$$I_q = V_d = 0, \quad \overline{A} = \overline{V_m}\overline{I}_m^* = jV_q I_d = jQ_m.$$

Consequently, operational reactance $x_q(s)$ from

$$V_d = x_q(s) I_q$$

is noninfluential. Conversely, as regards $x_d(s)$, we assume only field windings acting along the circuit axes.

The relationship between stator current, voltage and excitation voltage is thus defined by (see also [13]):

$$V_q = \frac{V_f}{1 + sT'_{d0}} - x_d \frac{1 + sT'_d}{1 + sT'_{d0}} I_d.$$

From the above equation, the following linearised model is obtained:

$$\Delta V_m = \Delta V_q = \frac{\Delta V_f}{1 + sT'_{d0}} - x_d \frac{1 + T'_d}{1 + sT'_{d0}} \Delta I_d.$$

From

$$\overline{I_m} = \frac{\overline{V_m} - \overline{V_R}}{jx_e} \tag{3.3}$$

we get

$$I_d = \frac{V_q - V_R \cos(\delta_R - \delta_i)}{x_e}.$$

For constant amplitude V_R (imposed by the infinite bus) and $\delta_R - \delta_i \approx 0$ because of a nil transferred active power, we obtain:

$$\Delta I_d = \frac{\Delta V_q}{x_e} = \frac{\Delta V_m}{x_e}.$$

Furthermore, defining

$$x'_d \triangleq x_d \frac{T'_d}{T'_{d0}},$$

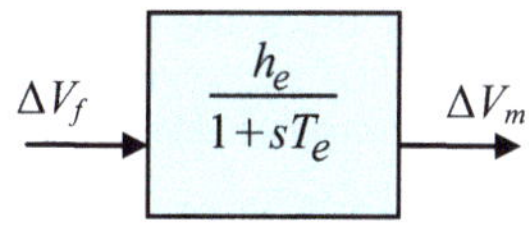

Fig. 3.6 Linearised block diagram linking generator and excitation field voltages

Eq. (3.3) becomes

$$\Delta V_m = \frac{x_e}{x_e + x_d} \cdot \frac{1}{1 + \left(\frac{x_e + x'_d}{x_e + x_d}\right) sT'_{d0}} \Delta V_f \tag{3.4}$$

Or

$$\Delta V_m = \frac{h_e}{1 + sT_e} \Delta V_f \tag{3.5}$$

with

$$h_e \triangleq \frac{x_e}{x_e + x_d}, \quad T_e \triangleq T'_{d0} \cdot \frac{x_e + x'_d}{x_e + x_d}.$$

A block diagram (Fig. 3.6) represents Eq. (3.5).

We note that an expression similar to (3.5) can be obtained as well when the generator feeds a load, as is shown next.

Case of Generator Feeding a Load

In the case of a generator feeding a load Z_C in Fig. 3.7, and with the objective of demonstrating the link between the generator and its reactive power, we suppose load Z_C to be linear, purely reactive and with impedance $Z_C \approx jX_C$, meaning the machine operates as a compensator (i.e., the delivered real power is zero).

So, the stator voltage lies entirely on the quadrature axis; that is,

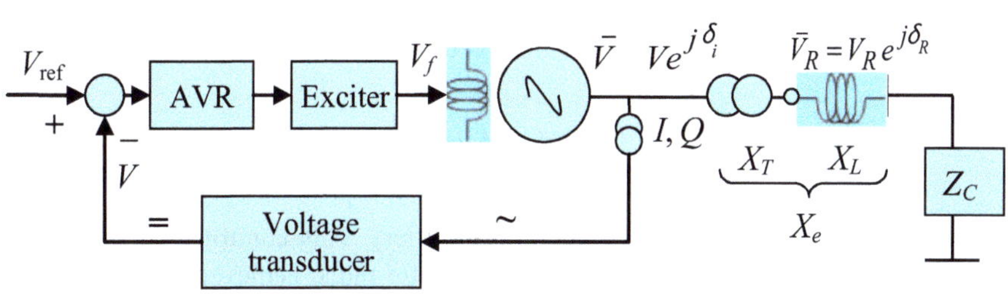

Fig. 3.7 After Recovering VR at the extreme top-right

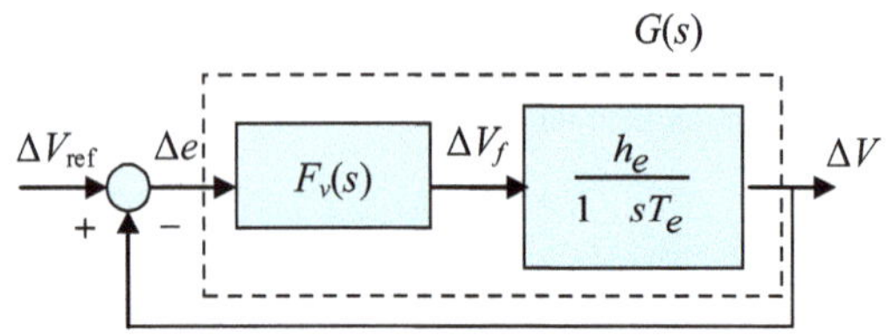

Fig. 3.8 Block diagram of primary voltage control loop of an on-load generating unit

$$V_d = 0, \quad V_q = V, \quad I_d = I, \quad I_q = 0.$$

In this case,

$$\bar{I} = \frac{\bar{V}}{j(x_e + x_C)} = \frac{jV_q}{j(x_e + x_C)} = \frac{V_q}{x_e + x_C} = I_d.$$

Moreover, $\delta_i = \delta_{Ri} = \pi/2$; therefore,

$$\Delta I_d = \frac{\Delta V_q}{x_e + x_C}.$$

Therefore, Eq. (3.5) is confirmed with the parameters:

$$h_e = \frac{x_e + x_c}{x_e + x_c + x_d}, \quad T_e = T'_{d0} \cdot \frac{x_e + x_c + x'_d}{x_e + x_c + x_d}.$$

A similar result can be achieved when the load is resistive (and thus the real power delivered is different from zero) and when the generator is connected to an infinite bus system also delivering active power [13].

Taking into account Eq. (3.5), using for simplicity V, Q instead of V_m, Q_m, the block diagram of the voltage control loop of a generating unit is as displayed in Fig. 3.8, whose forward transfer function is

$$G(s) = F_v(s) \cdot \frac{h_e}{1 + sT_e}. \tag{3.6}$$

Under steady state conditions ($t \to \infty$, $s \to 0$): $F_v(0) = \mu_0$.

For a static exciter, a good approximation of $F_v(s)$ at operating frequency is

$$F_v(s) = \mu_T = \mu_0 \frac{T_{Z1}}{T_{P3}}$$

(the static gain of the voltage regulator amplifier). This control loop is usually designed to have a cut-off frequency of about 4/5 rad/s, achieved with a high μ_T value, guaranteeing a very low steady state error.

Alternator Steady State Operating Condition

Under steady state operating condition ($s=0$), where values of the various quantities are indicated by the superscript °, Eq. (3.4) becomes

$$\Delta V^{\circ} = \frac{x_e}{x_e + x_d} \Delta V_f^{\circ}$$

whereas
$$\Delta V_f^{\circ} = F_v^{\circ}(s)\left(\Delta V_{\text{ref}}^{\circ} - \Delta V^{\circ}\right) = \mu_0\left(\Delta V_{\text{ref}}^{\circ} - \Delta V^{\circ}\right)$$

Therefore, error e° in Fig. 3.8 assumes the value

$$e^{\circ} = \Delta V_{\text{ref}}^{\circ} - \Delta V^{\circ} = \Delta V^{\circ} \frac{x_e + x_d}{x_e \mu_0},$$

$$e^{\circ}(\text{p.u.}) = \frac{\Delta V_{\text{ref}}^{\circ} - \Delta V^{\circ}}{\Delta V^{\circ}} = \frac{x_e + x_d}{x_e \mu_0}.$$

To obtain a high steady state accuracy on the voltage control loop, e°(p.u.) must be lower than a very small ε:

$$\frac{x_e + x_d}{x_e \mu_0} < \varepsilon.$$

Wishing to obtain a steady state voltage error lower than 0.5 % when passing from no-load to full-load operation ($\varepsilon < 0.005$), and assuming $x_e \approx 1.0$ p.u., the static gain μ_0 assumes values of about

$$\mu_0 > 200 \text{ p.u./p.u. for hydro units;}$$
$$\mu_0 > 400 \text{ p.u./p.u. for thermal units.}$$

It is interesting to note that high static gain in voltage regulators provides, at low frequencies, an effect similar to what is provided by an integral control law. Therefore, in the absence of additional effects provided by the PSS, line drop compensation, OEL or UEL, the generator terminal voltage is practically equal to its set-point value.

Alternator Dynamic Behavior

Stability and speed of response of the voltage control loop can be easily achieved using the classic Bode diagrams of amplitude and phase of the open-loop transfer function. For example, the hydraulic unit with

$$x_e = 1.0 \text{ p.u.}, \quad x_d = 1.0 \text{ p.u.}, \quad x_d' = 0.3 \text{ p.u.}, \quad T_{d0}' = 7.0 \text{ s}$$

given
$$h_e = 0.5 \text{ p.u. and } T_e = 4.55 \text{ s.}$$

In the case of rotating exciters with modern voltage regulators, the second order $F_v(s)$, given by

$$F_v(s) = \mu_0 \frac{1+sT_{Z1}}{(1+sT_{P2})(1+sT_{P3})},$$

with static gain $\mu_{0=}$1000 p.u./p.u. and time constant values as given in Table 3.1, determines a control margin of about 90° and a cut-off frequency near 2 rad/s.

In the case of static exciters of modern turbo-alternators, no single stability problem is recognised, as we show in the next example, the turbo-alternator unit with

$$x_e = 1.0 \text{ p.u.}, \quad x_d = 2.0 \text{ p.u.}, \quad x_d' = 0.3 \text{ p.u.}, \quad T_{d0}' = 7.5 \text{ s.}$$

Therefore,

$$h_e = 0.33 \text{ p.u.}, \quad \text{while} \quad T_e = 3.3 \text{ s.}$$

Such quantities are within the ranges

$$0.33 < h_e < 1.0, \qquad 3.2 < T_e < 7.5,$$

where the upper bounds reflect the no-load condition while the lower bounds refer to full-load operation. As a consequence, the time constant T_e is always greater than the time constant T_{Zl} and the lower than the time constant T_{P3} of the first order $F_v(s)$:

$$F_v(s) = \mu_0 \cdot \frac{1+sT_{Z1}}{1+sT_{P3}}.$$

As the cut-off frequency is certainly greater than $1/T_{Zl}$ and thus much greater than $1/T_e$, we can make use of the following high frequency approximation of $G(s)$:

$$G(s) = \mu_0 h_e \cdot \frac{T_{Z1}/T_{P3}}{sT_e}.$$

This means the loop control margin is on the order of 90°.

The loop cut-off frequency ω_v is derived directly from

$$|G(s)| = 1.0 \approx \frac{\mu_T h_e}{\omega_v T_e},$$

which gives

$$\omega_v \approx \frac{\mu_T h_e}{T_e}.$$

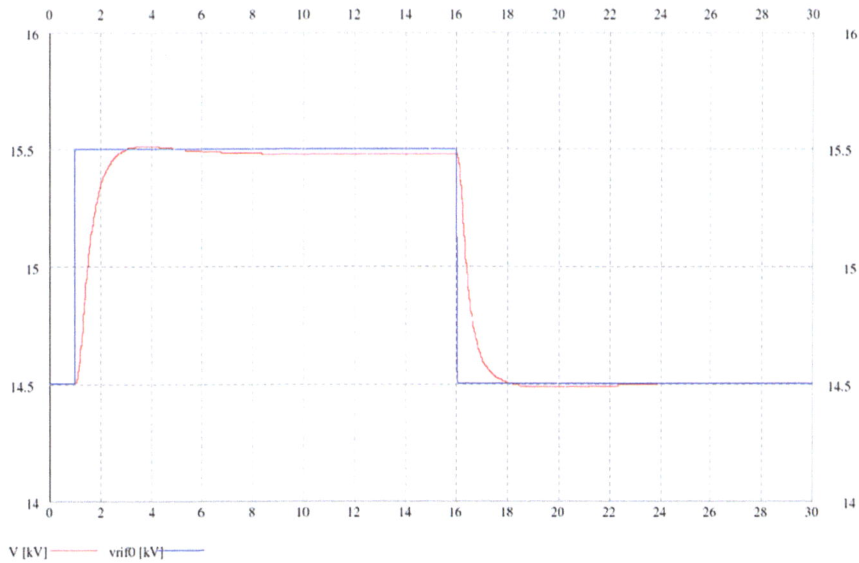

Fig. 3.9 Generator voltage control loop transients following AVR set-point steps

The transient gain μ_T, also called the "dynamic gain", thus accounts for the response speed of the AVR control loop: the higher the μ_T the faster the control loop.

Taking into account that the usual value of T'_{d0} is 7.0 s, the transient gain μ_T should be close to 50.0 p.u./p.u. Accordingly, the parameters of the static exciter should have the following values:

$$\mu_0 \approx 400.0 \text{ p.u./p.u.}, \quad T_{Z1} \approx 1.5\,\text{s}, \quad T_{P3} \approx \frac{\mu_0}{4} \cdot T'_{d0}\text{s}.$$

Based on these values, in response to a step variation of the voltage set-point, the generator voltage reaches the new imposed value, with near-zero steady state error, in an aperiodic way after about 1–2 s, as shown in Fig. 3.9.

3.3.3 Automatic Voltage Control by Generator Line Drop Compensation (Compounding)

Objective of Compounding

Figure 3.10 shows a closed-loop automatic control at the generator level consisting of a reactive power feedback (α_c) on the input of the generator's AVR. Depending on the sign of feedback α_c, the objective of compounding could be

- To support an increase of the local HV bus bar voltage (V_S);
- To produce an increase in generator voltage control stability.

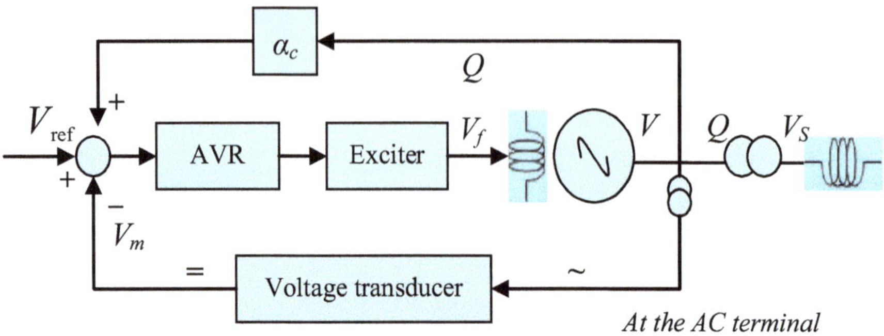

Fig. 3.10 Schematic control diagram of generator line drop compensation

Therefore, α_c provides an additional feedback on the automatic voltage control loop that modifies the basic AVR control objective, allowing deviations from a rigorous generator stator edge voltage control at the set-point value. Analysis of this additional feedback, overlapping the classic generator voltage control loop, is shown next with the use of the generator linear model under AVR control, as was previously described.

Link Between Voltage and Reactive Power

Consider the equivalent scheme in Fig. 3.5 of a generator connected to an infinite bus, using for simplicity from now on V and Q instead of V_m and Q_m. The reactive power provided by the generator is given, with the obvious meaning of the symbols, by

$$Q = \frac{V^2 - VV_R\cos(\delta_R - \delta_i)}{X_e}.$$

Now, assuming $P=0$ (the machine operates as a compensator of reactive power), to which corresponds $\delta_R = \delta_i$:

$$Q = \frac{V^2 - VV_R}{X_e},$$

and with $V \approx V_R \approx 1$ p.u. at the considered operating point:

$$\Delta Q \approx \frac{\Delta V}{X_e}. \tag{3.7}$$

Equation (3.7) is also a good linear approximation of the system seen in Fig. 3.5 when $P \neq 0$ and $V \approx V_R \approx 1$ (Fig. 3.11).

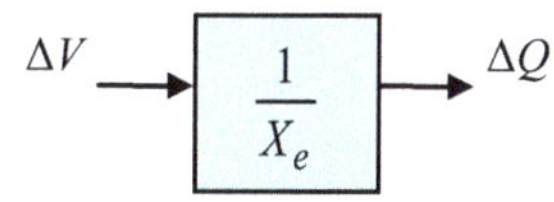

Fig. 3.11 Linearised block model linking generator variations on voltage and reactive power

Line Drop Compensation (Compounding)

As pictured in Fig. 3.10, AVR compounding is a reactive power feedback on the generator voltage control loop represented by the block diagrams of Figs. 3.6 and 3.8. This simplified model and the corresponding hypothesis can also be the basis for the analysis of the line drop compensation control loop.

The block diagram of Fig. 3.12 illustrates the dynamic link between the generator voltage set-point and the reactive power delivered by the generator described by Eq. (3.7):

The signal $\alpha_c \Delta Q$ is called the "compound signal"; coefficient α_c represents the compound factor, with $\alpha_c > 0$ indicating "positive" compounding; and $\alpha_c < 0$ indicating "negative" compounding.

With high enough μ, at steady state the error $\Delta e \approx 0$; therefore,

$$\alpha_c \Delta Q + \Delta V_{\text{ref}} - \Delta V = 0.$$

With a constant voltage set-point value, we have

$$\alpha_c \Delta Q - \Delta V = 0.$$

Then, coefficient α_c represents the slope:

$$\alpha_c = \Delta V / \Delta Q.$$

Clearly, α_c has the dimension of reactance. So, under steady state conditions, the generator stator edge voltage is given by

$$V = V_{\text{ref}} + \alpha_c Q. \tag{3.8}$$

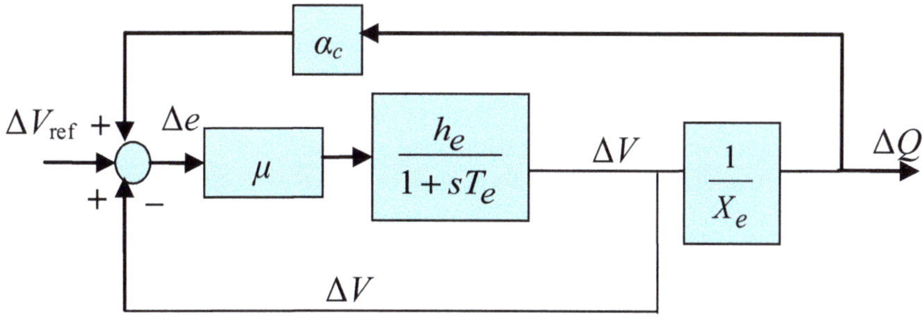

Fig. 3.12 Block diagram of line drop compensation control loop

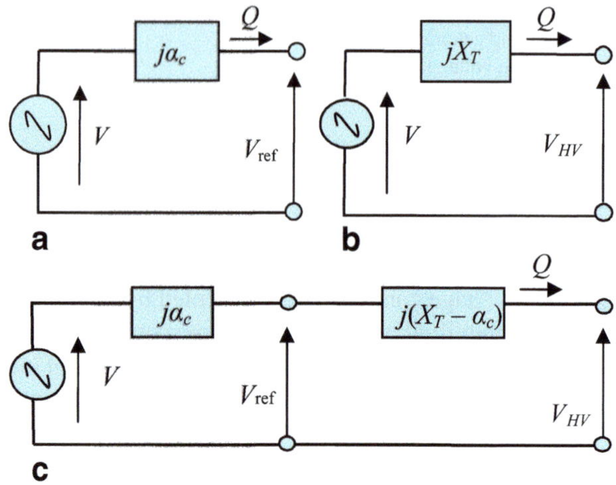

Fig. 3.13 Single circuit equivalent for (**a**) compound action, (**b**) step-up transformer and (**c**) the two combined

The equivalent circuit in Fig. 3.13a represents Eq. (3.8), while Fig. 3.13b represents the generator connected through the step-up transformer to the HV bus bar. Hence, Fig. 3.13c is obtained from Fig. 3.13a, b by assuming α_c to be positive and less than the X_T reactance. Under steady state conditions, the point where the voltage amplitude is kept constant at the V_{ref} value is thus inside the step-up transformer.

Under this real hypothesis and upon variations of reactive power Q, the voltage drop in a portion of the step-up transformer is completely offset, precisely as if its reactance had been diminished from X_T to $X_T - \alpha_c$. Of course, if $\alpha_c = X_T$, the entire drop in the transformer will be offset. In this situation, controlled voltage will be V_{HV} at the transformer's HV terminals, rather than what the voltage is at the generator's terminals, as is true in the case without compound action. From this point we note that the V_{HV} compounding control cannot be achieved in practice because, as will be seen later, it generally corresponds to an unstable operating point.

Equation (3.8) provides the linear voltage-reactive power steady state characteristic for positive and negative α_c (solid lines in Fig. 3.14), which gives a clear picture of the line drop compensation effect.

With line drop compensation, the reference voltage of the generator AVR is therefore the voltage at the stator terminals under no-load conditions (zero reactive load) only.

Analogous to the turbine steady state characteristics (governor speed, real power), the compound effect is equivalent to the introduction of a drop, or "statism", in the voltage regulator, since the presence of negative compounding (i.e., positive statism) would mean stator voltage decreases while load increases, that is, assuming a negative α_c value. Conversely, a positive value of α_c provides positive compounding (i.e., negative statism).

Regarding the value of α_c, since $V_{max} = 1.05 \div 1.1$ p.u. and assuming $V_{ref} = 1.0$:

- At full real power $(Q_{max} \approx 0.5 \text{ p.u.}) : \alpha_c \approx 0.1 \div 0.2\text{p.u.}$;
- At the technical minimum $(Q_{max} \approx 0.7 \text{ p.u.}) : \alpha_c \approx 0.07 \div 0.14\text{p.u.}$

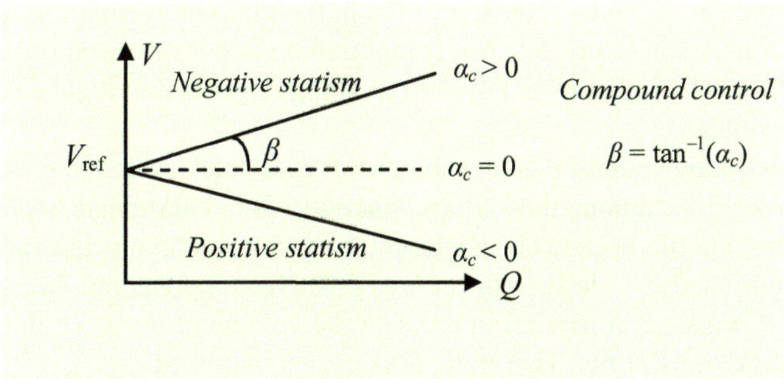

Fig. 3.14 Voltage-reactive power steady state characteristics with line drop compensation

Hence, even a reactance X_T greater than about 0.13 p.u. might be offset.

Adopting the most conservative condition ($\alpha_c \approx 0.07 \div 0.14$ p.u.), offsetting will be somewhere between $X_T/2$ and X_T, and as we will see next, a large offsetting will compromise the generator voltage control loop stability. On the contrary, a consequence of X_T offsetting is an increase in the short-circuit power of power plant HV buses where compound action is introduced. This increase makes the generator node stronger.

Line Drop Compensation and Stability

Complete offsetting of the reactance X_T of a step-up transformer ($\alpha = X_T$) would be equivalent to high side voltage control, where the generator would regulate the power plant voltage at the local HV bus.

Unfortunately, line drop compensation does not allow for HSVC. To better understand this point we need to refer to the possible power station configurations given in Fig. 3.15.

First Configuration Case

In power plants with low capacity units, all units may be parallel-connected to a single MV bus (Fig. 3.15a). In this instance, since that the voltage to be controlled

Fig. 3.15 Generators parallel-connected to (**a**) MV bus; (**b**) HV bus

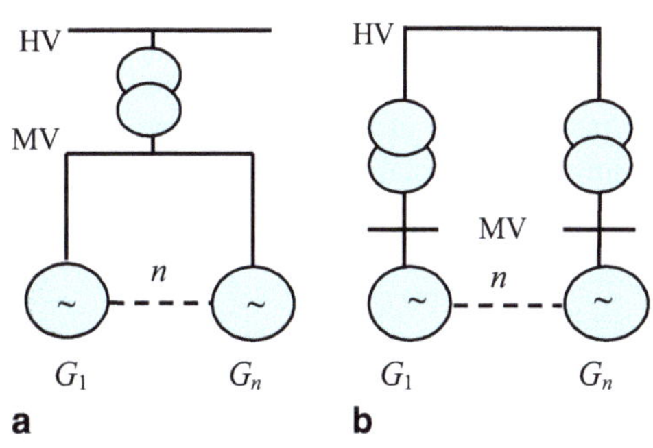

is the same for all units, referring to the horizontal voltage-reactive power steady state characteristic (zero statism), compounding is out of operation and there will be n AVR (n control loops in Fig. 3.8) working in parallel to regulate the same MV bus bar voltage.

This working condition is unstable for a "real pole". Moreover, the distribution of reactive power among the various generators is indeterminate under steady state conditions, but this second consideration is not relevant in practice due to the instability mentioned. Similarly in the case of positive compounding, because there will be n AVR working in parallel to regulate the voltage of the same internal point of the MV/HV transformer. Therefore, HSVC is not allowed.

In conclusion, stability in Fig. 3.15 case (a) compulsorily *requires* the use of a negative compound at each AVR in a way that allows each generator to regulate the voltage at an internal point of its stator winding reactance. This determines, with appropriate offset value of that reactance, enough electrical distance between any two points inside the generators, while voltages are automatically regulated by plant AVRs at set-point values.

Generators with stator edges in parallel require $\alpha_c < 0$.

Second Configuration Case

In power plants with high-capacity units, generators are usually paralleled to a single HV bus (Fig. 3.15b). In this scheme, the previous drawback to stability is overcome since each voltage regulator is sensitive to the corresponding voltage of its MV bus bar. Furthermore, the step-up transformer reactance provides an adequate electrical distance between MV buses.

In this instance a positive compound is now possible, but the step-up transformer reactance should not be fully offset (i.e., high side voltage control is not allowed); if it were, V_{HV} ideally would be kept constant at the value of V_{ref} by n AVRs with horizontal steady state characteristics (V_{HV}, Q). However, such a scheme would determine voltage instability due to the existence of parallel voltage control loops. Obviously, with partial offset the higher the positive compounding the lower the stability margin. This is also the reason compound factors should be within the range mentioned before: $\alpha_c \approx 0.07 \div 0.14\,\text{p.u.}$

For assessing the compound effect on the stability of a generator's primary voltage control loop (when it operates alone at a power station), the case of a unit feeding a reactive load is considered (see Fig. 3.7). In Fig. 3.12 the corresponding dynamic model of the line drop compensation control loop is shown. In this case, equivalent external reactance seen by the stator edge of the generator is considered: $x_e = x_T + x_L + \alpha_c$. The block diagram in Fig. 3.12 becomes as that shown in Fig. 3.16a or its equivalent, represented in Fig. 3.16b

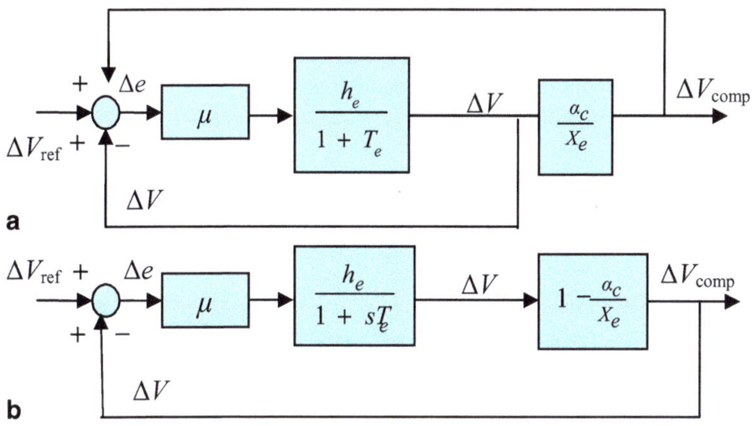

Fig. 3.16 Block diagram of the generator voltage control loop of a unit with (**a**) compound action and (**b**) the equivalent

This means that in the case $1-\alpha_c/x_e<0$, feedback becomes positive with the consequent instability (due to the real pole) of the voltage control loop. On the contrary, with the feedback gain positive but lower than 1.0, the speed of the voltage control loop is reduced due to the loop's lower static gain.

In practice, with $\alpha_c>0$, compensation is usually less than 50% of transformer reactance. Therefore, compounding does not allow real voltage control of the bus's HV side while still contributing to this objective.

Figure 3.15 case (b) shows as easily that, in practice, a lot of negative compounding can be found, thus improving AVR stability margins in spite of supporting local HV side bus voltages; but in so doing, compounding fails its original objective and is therefore less effective than a classic generator voltage control loop for sustaining V_{HV}.

Line Drop Compensation Simplified Feedback

Equation (3.8) shows that compounding feedback comes from the reactive power. In practice, feedback often comes from $I\sin\phi$, with $\tan\phi = Q/P$; Eq. (3.8) therefore becomes

$$V = V_{ref} + \alpha_c \frac{Q}{V}$$

or

$$V^2 - V_{ref}V - \alpha_c Q = 0.$$

Then

$$V = \frac{V_{ref}}{2} + \sqrt{\left(\frac{V_{ref}}{2}\right)^2 + \alpha_c Q}.$$

In agreement with Eq. (3.8), this shows that if reference voltage V_{ref} and compound factor $\alpha_c > 0$ are constant for all operating conditions, then voltage at the alternator terminals increases with the delivered reactive power.

If the compound signal is

$$V_c = \alpha_c \cdot \frac{Q}{V} \quad \text{and} \quad Q = \frac{V^2}{x_e},$$

then

$$V_c = \frac{\alpha_c}{x_e} V.$$

Linearising gives

$$\Delta V_c = \frac{\alpha_c}{x_e} \cdot \Delta V.$$

This link is the same as in Fig. 3.16a. The dynamic analysis is thus made and the stability conclusions which are reached when feedback comes from Q are still valid when feedback comes from $I \sin \phi$.

To summarise:

- Line drop compensation with positive gain contributes to the support of the HV voltage by regulating voltage at an internal point of the unit step-up transformer. The compensation is usually less than 50 % of the transformer reactance.
- Line drop compensation with negative gain contributes to the stability of the AVR control loop by regulating voltage at an internal point of the generator stator winding.

3.3.4 Generalities on Automatic High Side Voltage Control at a Substation

HV-EHV bus bar voltage closed-loop continuous control at a local substation of a power plant can be properly achieved by local generators. Conversely, in the case of load/grid substations, automatic continuous voltage control can be obtained by local FACTS devices like the SVC, STATCOM and UPFC, which operate on reactive powers delivered or absorbed by the substation.

Shunt capacitor and reactor switching in practice are seldom operated; they are generally for recovering control margins at local generators and/or FACTS and so are not used for fine and continuous high side voltage control.

Analogously, OLTC transformers, when finalised to their low voltage side control due to slow stepping, do not significantly contribute to high side voltage control. In principle, they could contribute, but it would require a change in their task and therefore their control logic to sustain their high side voltages. Even in this unusual case they would be the last to operate, after generators and FACTS, due to their slow speed dynamics and delays.

For these reasons, we will continue to limit our analysis of high side voltage control to generators, SVC, STATCOM and UPFC.

Voltage Control at a Substation

The receiving-end voltage of a transmission line is, at any instant, a function of the values assumed by either the equivalent generator and equivalent line impedance Z_L seen by the line bus considered or the equivalent generator transformer impedance Z_T or the local load and its power factor.

The magnitude of receiving-end voltage V significantly and speedily changes depending on these parameters, as shown by Fig. 3.17. Achievable constant voltage regulation by a local variable VAR source at the load bus is also represented by the horizontal line $V = V_{ref}$.

Lag-lead effects due, respectively, to load inductive and capacitive characteristics combined with operating compensating equipment and FACTS at the load bus determine the various shapes of the V-P curves.

Large voltage variation is not normally tolerable in a power system. Under-voltages cause degradation in load performance, particularly in induction motors, and in protection interventions, whereas over-voltages cause transformer and motor saturation with consequent harmonic generation and possible insulation problems.

Load changes and switching of power system components can cause significant voltage variation in system buses, especially in the case of a weak power system.

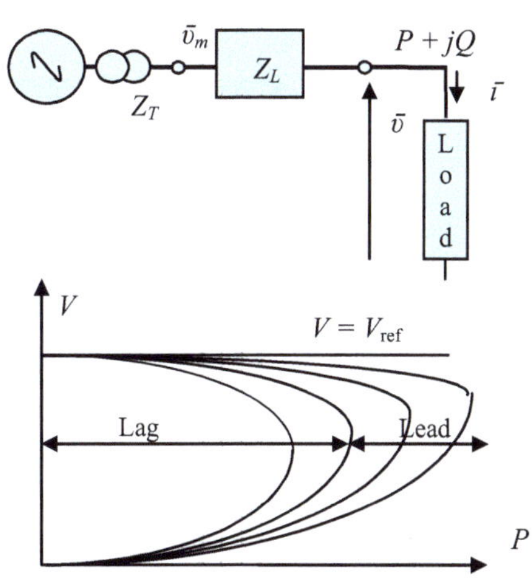

Fig. 3.17 Thevenin equivalent seen by the load bus and related V-P curves

In the extreme case, when power demand at the receiving end exceeds maximum transmittable power, local voltage may continue to decrease, eventually reaching collapse, as shown in Fig. 3.17.

3.3.5 Automatic High Side Voltage Control at a Power Plant

Principal Scheme

An innovative power plant automatic voltage control (that is, HSVC) of the local HV side bus bar is shown in Fig. 3.18. It represents a nonconventional power station control able to coordinate the reactive powers of operating generators in the plant through a high side voltage regulator (HSVR) aimed to control local substation HV bus bar voltage in a closed loop.

This kind of automatic regulation can maintain voltage V_S at the imposed value V_{Sref} by a continuous coordinated control of the reactive powers of the operating generators in the plant. An analysis of the control scheme requires an appropriate model of the process, as introduced next.

Model of the Power Plant

We consider a power station plant with n generators in parallel on the same EHV bus bar, which is connected to a prevailing power system through an equivalent reactance X_e, as represented in Fig. 3.19.

In Fig. 3.19, the notations are:

S_{ni} = MVA apparent nominal power of ith unit,
S_n = MVA apparent nominal power of power station,
V_n = kV nominal voltage of power station,

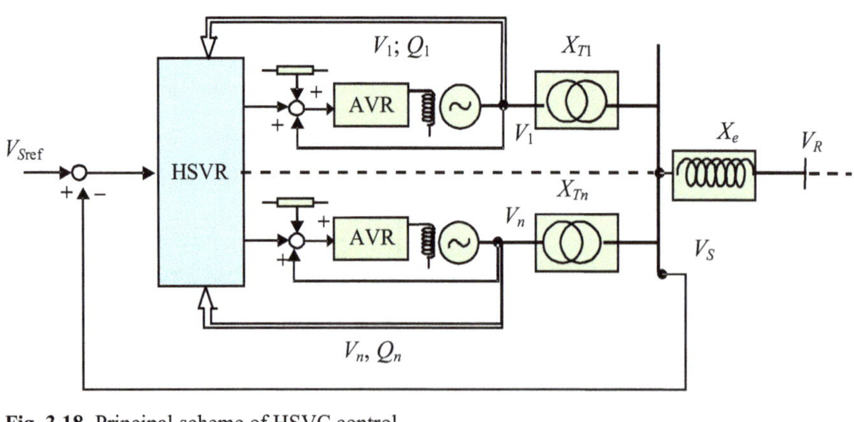

Fig. 3.18 Principal scheme of HSVC control

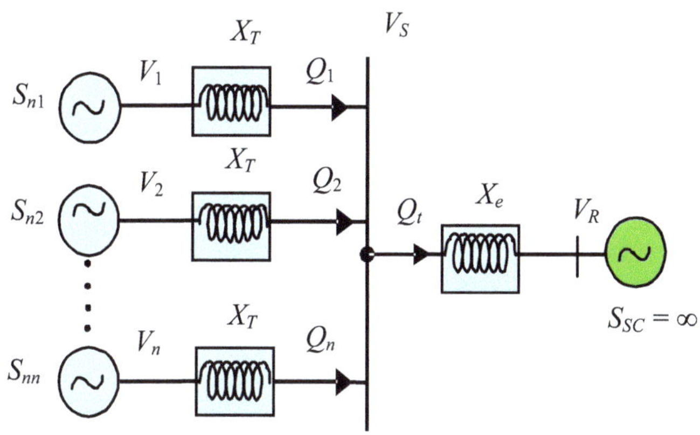

Fig. 3.19 Equivalent representation of a power station connected to a large grid

V_i = voltage of ith generator, in p.u. of V_{ni},
Q_i = reactive power of ith unit, in p.u. of S_{ni},
Q_t = total reactive power of power station, in p.u. of S_n,
V_S = voltage of power station EHV bus bar, in p.u. of V_n,
V_R = voltage of grid bus bar, in p.u. of V_n,
X_T = reactance of unit transformer, in p.u. of $(V_{ni})^2/S_{ni}$,
X_e = equivalent reactance of external grid, in p.u. of $(V_n)^2/S_n$,
and

$$\bar{V}_i = V_i e^{j\theta i}, \quad i = 1,\ 2, \ldots, n,$$
$$\bar{V}_S = V_S e^{j\theta S}, \quad (V_i \text{ and } V_S \text{ in p.u.}),$$
$$\delta_i = \vartheta_i - \vartheta_S, \quad i = 1, 2, \ldots, n.$$

Now, referring to the reactive power of each unit:

$$Q_i[\text{p.u.}] = \frac{V_i^2 - V_i V_S \cos\delta_i}{X_T}, \quad i = 1, 2, \ldots$$

By linear approximation around the working point, under the usual assumptions

$$V_{i(0)} = V_{S(0)} \quad \text{and} \quad \delta_{i(0)} = 0,$$

it turns out that:

$$\Delta Q_i = \frac{V_{S(0)}}{X_T}(\Delta V_i - \Delta V_S), \quad i = 1,\ 2, \ldots, n \tag{3.9}$$

$$\Delta Q_t = \frac{V_{S(0)}}{X_e}(\Delta V_S - \Delta V_R).$$

Moreover, under the assumption of negligible reactive power losses on X_T,

$$\Delta Q_t = \frac{1}{S_n}\sum_{k=1}^{n}\Delta Q_k S_{nk} = \frac{V_{S(0)}}{X_e}(\Delta V_S - \Delta V_R). \tag{3.10}$$

From Eqs. (3.9) and (3.10) we obtain, respectively,

$$\Delta V_S = \Delta V_i - \frac{X_T}{V_{S(0)}}\Delta Q_i, \quad i = 1,2,\ldots,n,$$

$$\Delta V_S = \frac{X_e}{V_{S(0)}}\frac{1}{S_n}\sum_{k=1}^{n}\Delta Q_k S_{nk} + \Delta V_R.$$

From the last two equations the following can be deduced:

$$\Delta V_i - \Delta V_R = \frac{X_T}{V_{S(0)}}\Delta Q_i + \frac{X_e}{V_{S(0)}}\frac{1}{S_n}\sum_{k=1}^{n}\Delta Q_k S_{nk}, \quad i = 1,2,\ldots,n,$$

$$S_n\left(\Delta V_i - \Delta V_R\right) = \frac{X_T S_n + X_e S_{ni}}{V_{S(0)}}\Delta Q_i + \frac{X_e}{V_{S(0)}}\sum_{k=1,k\neq i}^{n}\Delta Q_k S_{nk}, \quad i = 1,2,\ldots,n.$$

In matrix form,

$$\begin{bmatrix}\Delta V_1 - \Delta V_R\\ \vdots \\ \Delta V_n - \Delta V_R\end{bmatrix} = \underbrace{\begin{bmatrix}a_{11} & \cdots & a_{1n}\\ \vdots & \ddots & \vdots\\ a_{n1} & \ldots & a_{nn}\end{bmatrix}}_{\mathbf{A}}\begin{bmatrix}\Delta Q_1\\ \vdots\\ \Delta Q_n\end{bmatrix},$$

where

$$\mathbf{A} = [a_{ij}], \text{where } a_{ii} = \frac{X_T S_n + X_e S_{ni}}{V_{S(0)} S_n}, \qquad a_{ij} = \frac{X_e S_{nj}}{V_{S(0)} S_n}, \quad i \neq j, \quad i,j = 1,\ 2,\ldots,n. \tag{3.11}$$

In compact form, the result can be also written as

$$\Delta V - \Delta V_R = \mathbf{A}\cdot\Delta Q, \quad \Delta Q = \mathbf{A}^{-1}(\Delta V - \mathbf{I}_{\mathbf{col}}\Delta V_R),$$

where

$$\Delta V = \begin{bmatrix}\Delta V_1\\ \Delta V_2\\ \vdots\\ \Delta V_n\end{bmatrix}, \quad \text{while} \quad \mathbf{I}_{\mathbf{col}} = \begin{bmatrix}1\\1\\ \vdots\\ 1\end{bmatrix} = \left(\mathbf{I}_{\mathbf{row}}\right)^{\mathrm{T}}.$$

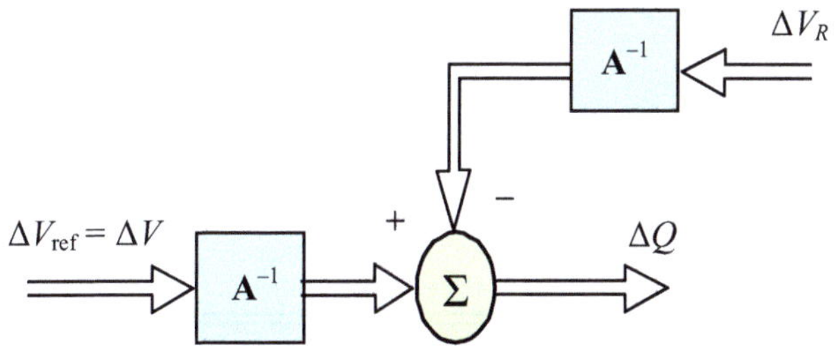

Fig. 3.20 Linear model of a power plant with *n* generators connected to an equivalent grid

Therefore, the model result is a matrix with coefficient values dependent on operating point and on parameters X_e and X_T. Both **A** and $\mathbf{A}^{-1}$ are full matrices, therefore the reactive power of each unit depends on the voltages of all the other units inside the power station. Figure 3.20 shows the block diagram of the considered linear model.

In this scheme, it is also assumed $\Delta V_{ref} = \Delta V$ under the assumption that the primary voltage control loops have negligible dynamics with respect to HSVC control. This is confirmed by the design of the HSVC control dynamics.

ΔQ in Fig. 3.20 represents the overall reactive power delivered/absorbed by the power station units as a whole; any change determines a bus bar voltage variation according to

$$\Delta V_S = \frac{X_e}{V_{S(0)}} \cdot \frac{1}{S_n} \sum_{k=1}^{n} \Delta Q_k S_{nk} + \Delta V_R. \tag{3.12}$$

Eq. (3.12) provides the additional step in the scheme of Fig. 3.21, showing the link between the generator voltage at the power station and the local EHV bus bar voltage V_S.

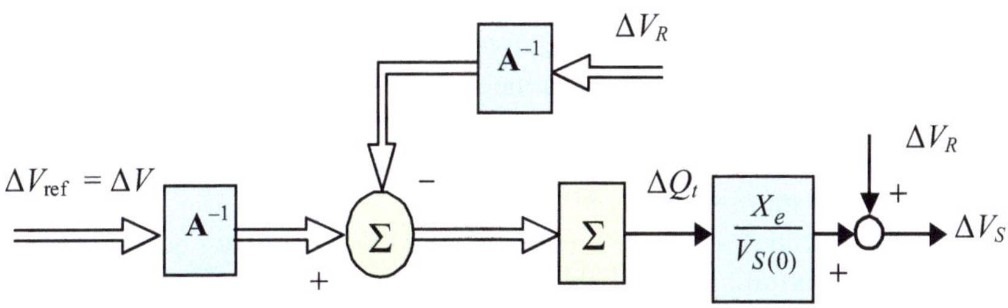

Fig. 3.21 Linear equivalent model of power plant with n generators connected to equivalent grid: link between generators and EHV bus bar voltage

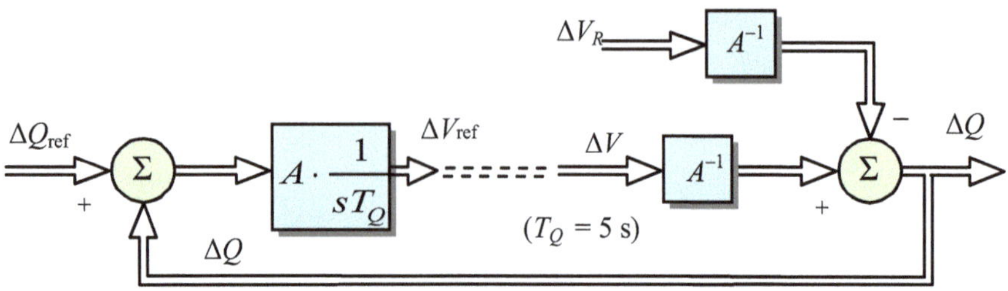

Fig. 3.22 Block diagram of a power station's reactive power control

High Side Voltage Regulator

Starting from the simplified system model shown in Fig. 3.21, power station reactive power and voltage control loops are now introduced with the assumption that they are slower with respect to the AVR voltage control loop (the dominant time constant of the AVR control loop is assumed to be on the order of 0.5 s; see § 3.3.2.1).

The generator reactive power control loop, while not common in practice, should in principle regulate the reactive power of its generator without interfering with the reactive powers of the remaining generators operating at the same power station; that is, without negatively interacting with them. Even when all the generators in a power station are provided with autonomous reactive power control loops, their consistent dynamic interaction is still often the cause of poor or variable performance throughout the power station. Nevertheless, engineers seek to control all generator reactive power in a way that pre-empts dynamic interaction between them. This is a priority and a pressing need in plants that have a large number of generators.

A type of "power station centralised control" such as the one modelled by the block diagram in Fig. 3.22 is a possible solution. This is a centralised, noninteracting control scheme of integral type that allows dynamic decoupling among unit reactive power control loops as well as reactive power absorption/delivery of each generator in accordance with set-point values. ΔV_{ref}, ΔQ_{ref} and ΔQ are the vectors of variables; control matrix A is given by Eq. (3.11).

The resulting diagonal control, when X_e is properly identified, allows a first order dynamic at each reactive power control loop, each characterised by a time constant T_Q on the order of 5 s (see Fig. 3.23 of the integral reactive power control loop following a set-point Q_{ref} step variation). The corresponding variations of generator AVR voltage set-point and stator voltage show the steady state error due to the AVR control law of proportional (P) type.

This type of centralised reactive power control moves the plant generators' reactive power by controlling the variables of the set-point vector ΔQ_{ref} in unison. This useful control, which brings the operating points of the plant generators together, from the over- (Q^+_{lim}) up to the under- (Q^-_{lim}) excitation limits, can be effectively used for EHV bus bar voltage regulation.

Recognising that automatic EHV voltage regulation is the primary objective of any power system operation, a possible control scheme for power station high side voltage regulation is proposed in Fig. 3.24.

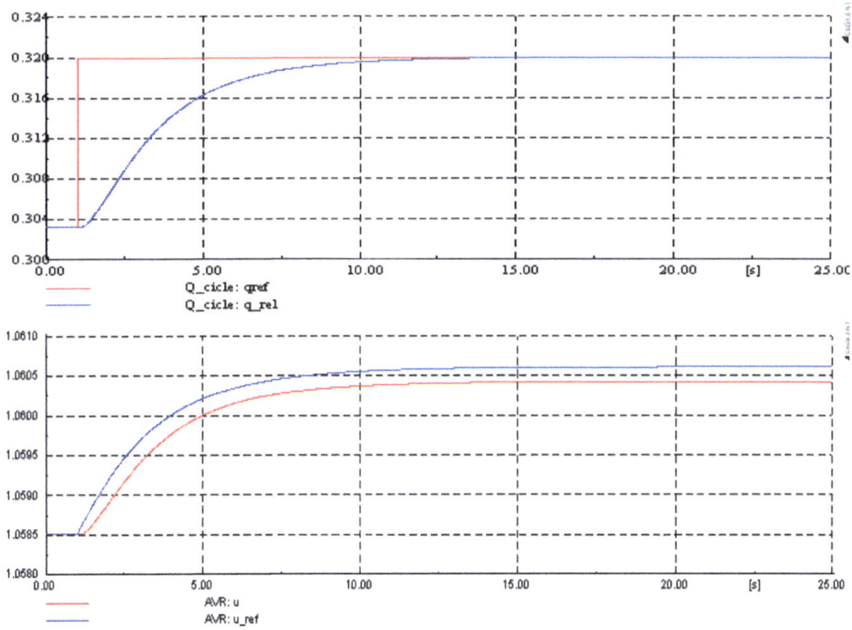

Fig. 3.23 Set-point step response of a generator reactive power control loop

In this figure, an external voltage control loop overlaps the generator reactive power loops previously considered with slower dynamics. A proportional-integral (PI) control law characterises the proposed control scheme. "Proportional" refers to a requirement that large step perturbations in the grid be suddenly covered. The output q of the control block is called the "reactive power level", which ranges from $+1$ to -1 in p.u. of the generator over- and under-excitation limits, respectively, given by the column matrix linking Δq to ΔQ_{ref} in the block diagram (Fig. 3.24). Equation (3.12) describes the link between ΔV_S and ΔQ.

The external voltage control loop, in order to be dynamically decoupled with respect to the inner reactive power control loops, must have a dominant time constant of about $T_s = 50$ s. Toward this end, we require a proper design of the coefficients K_P and K_I of the PI control law.

Coefficient K_P can be computed by an assuming instantaneous compensation between the ΔQ variation determined by a grid perturbation ΔV_R at $t = 0^-$ and the ΔQ_{ref} provided at $t = 0^+$ by the ΔV_S variation corresponding to ΔV_R. From this assumption comes the following value of K_P:

$$K_P = \frac{V_{S(0)}}{X_T} \frac{1}{\frac{1}{S_n} \sum_{k=1}^{n} Q_{\lim k} S_{nk}},$$

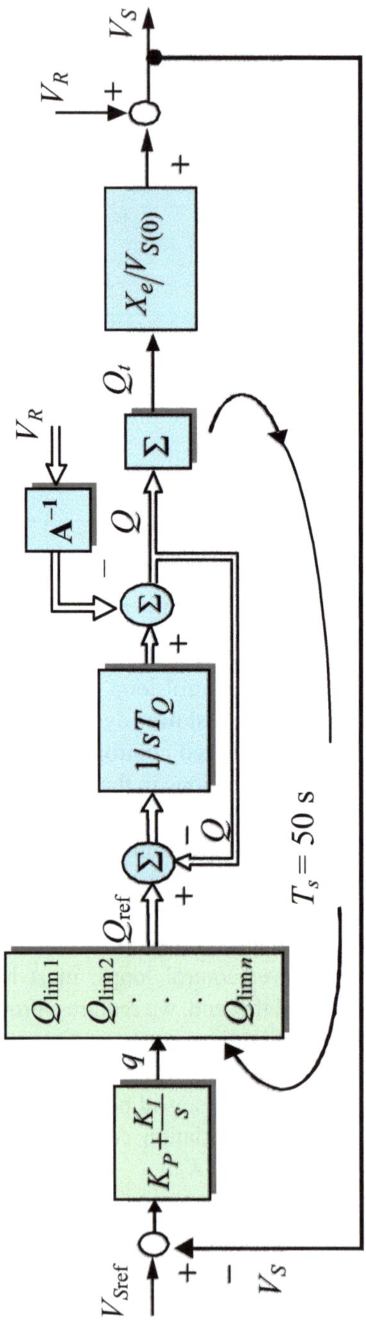

Fig. 3.24 Block diagram of a power station EHV bus bar voltage regulation

with X_t in the p.u. of the power station (i.e., the equivalent transformer).

If we consider all units to be the same size (S_{ni}), then

$$K_P = \frac{V_{S(0)}}{X_T} \cdot \frac{1}{\frac{1}{n}\sum_{k=1}^{n} Q_{\lim k}}, \tag{3.13}$$

We now compute K_P. To simplify the analysis, we consider the equivalent generator of the power station shown in Fig. 3.19. This equivalent generator, of size S_n, reactive power Q_t and stator voltage V, is connected through an equivalent reactance X_T to local HV bus bar (with voltage V_S), which sees the remaining network by the equivalent reactance X_e connected to an infinite bus (V_R).

Equations (3.9) and (3.10) describe reactive power variation in this simplified equivalent scheme. From these equations come the following:

$$\Delta V_S = \Delta V - \frac{X_T}{V_{S(0)}} \Delta Q_t,$$

$$\Delta V_S = \frac{X_e}{V_{S(0)}} \Delta Q_t + \Delta V_R.$$

From these, the system model that links V, V_R and V_S is achieved:

$$\Delta V - \Delta V_R = \frac{X_T + X_e}{V_{S(0)}} \Delta Q_t,$$

$$\Delta V_S = \frac{X_e}{V_{S(0)}} \Delta Q_t + \Delta V_R.$$

Next, we show the simplified equivalent representation of the control system, Fig. 3.25, coherent with the hypotheses of very fast inner control loops with respect to the outer loop on V_S ($Q_{\lim}$ is in p.u. of S_n).

We assume an instantaneous compensation between the ΔQ variation determined by a grid perturbation ΔV_R at $t = 0^-$ and the ΔQ_{ref} provided at $t = 0^+$ by the ΔV_S variation corresponding to ΔV_R; that is:

$$\Delta Q_{i(0)} = \Delta Q_{\text{ref}(0)}, \quad \Delta Q_{i(0)} = -\frac{V_{S(0)}}{X_T + X_e} \Delta V_R,$$

$$\Delta Q_{\text{ref}i(0)} = -Q_{\lim} K_P \left[\Delta V_R + \frac{X_e}{V_{S(0)}} \left(-\frac{V_{S(0)}}{X_T + X_e} \Delta V_R \right) \right].$$

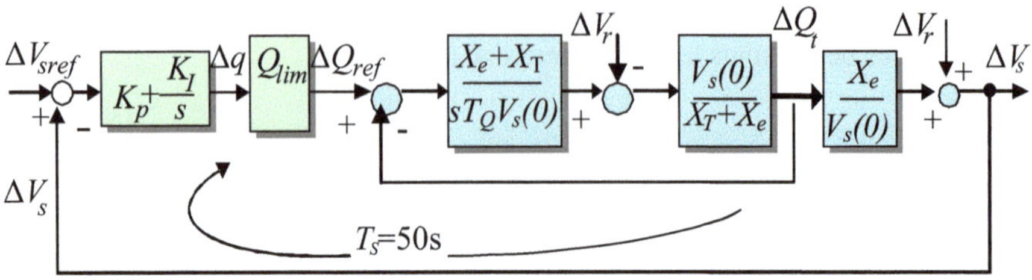

Fig. 3.25 Simplified block diagram of EHV bus bar voltage control loop, with a power station represented by an equivalent generator

The value of proportional coefficient K_P is then obtained:

$$K_P = \frac{V_{S(0)}}{X_T} \cdot \frac{1}{Q_{\lim}}. \tag{3.14}$$

Considering n generators as in (3.13):

$$K_P = \frac{V_{S(0)}}{X_T} \cdot \frac{1}{\frac{1}{S_n}\sum_{k=1}^{n} Q_{\lim k} S_{nk}}.$$

We now come to coefficient K_I, which must be able to impose a first order dynamic to the outer control loop of Fig. 3.25, with a dominant time constant of $T_S = 50$ s, when K_P assumes the above imposed value (3.14).

Therefore,

$$K_I = \frac{K_P + V_{S(0)} S_n / X_e \sum_{k=1}^{n} Q_{\lim k} S_{nk}}{T_S} = \frac{V_{S(0)}}{T_S} \cdot \frac{S_n}{\sum_{k=1}^{n} Q_{\lim k} S_{nk}} \cdot \frac{X_e + X_T}{X_e X_T}. \tag{3.15}$$

In fact, referring to Fig. 3.25, the system "pole" at the closed loop is given by

$$s = -\frac{Q_{\lim} K_I \cdot X_e / V_{S(0)}}{1 + Q_{\lim} K_P \cdot X_e / V_{S(0)}} = -\frac{1}{T_S}.$$

Then

$$K_I = \frac{K_P + V_{S(0)} / X_e Q_{\lim}}{T_S} = \frac{V_{S(0)}}{T_S Q_{\lim}} \cdot \frac{X_e + X_T}{X_e X_T}.$$

Equation (3.15) has therefore been achieved by considering n generators instead of the equivalent one.

In conclusion, and different from line drop compensation control, with a high side voltage regulator it is possible to regulate at the set-point value the voltage of the local HV bus bar with a PI control law that characterises the HV transient by a high speed response dynamic in the first part and slower response in the second part. This is shown in the traces of Figs. 3.26, 3.27 and 3.28.

Lastly, though beyond the scope of this book, we mention that with a high side voltage regulator it is also possible to increase voltage stability during a black start-up manoeuvre, mainly at the initial voltage launch phase [18]. This is accomplished

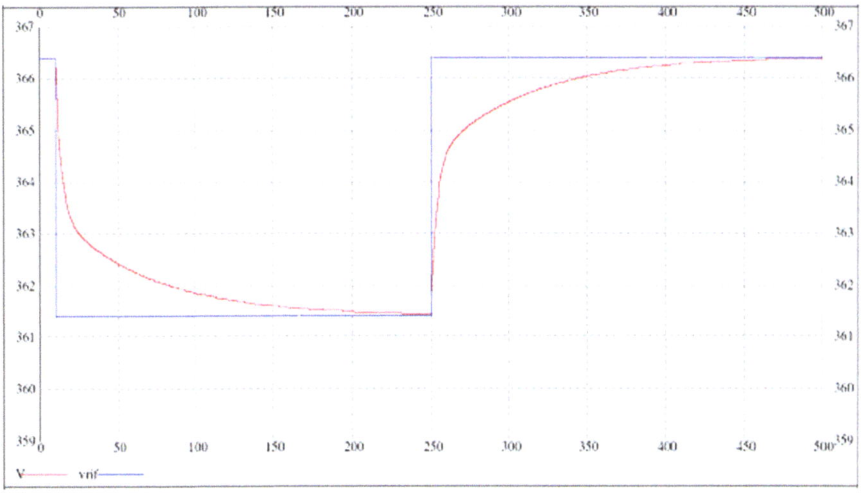

Fig. 3.26 Transients of power plant high side bus bur voltage V_S following step variations of HSVC voltage set-point V_{Sref}

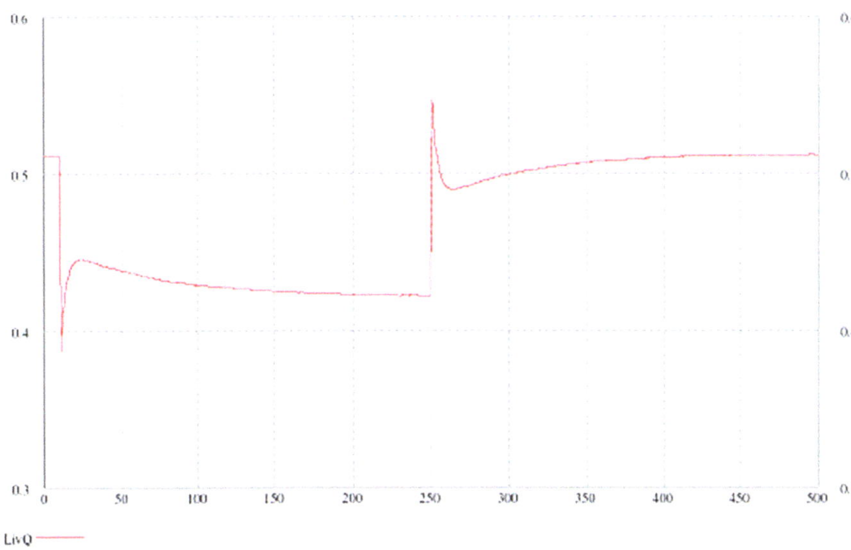

Fig. 3.27 Transients of HSVC output (q level) following step variations of HSVC voltage set-point V_{Sref}

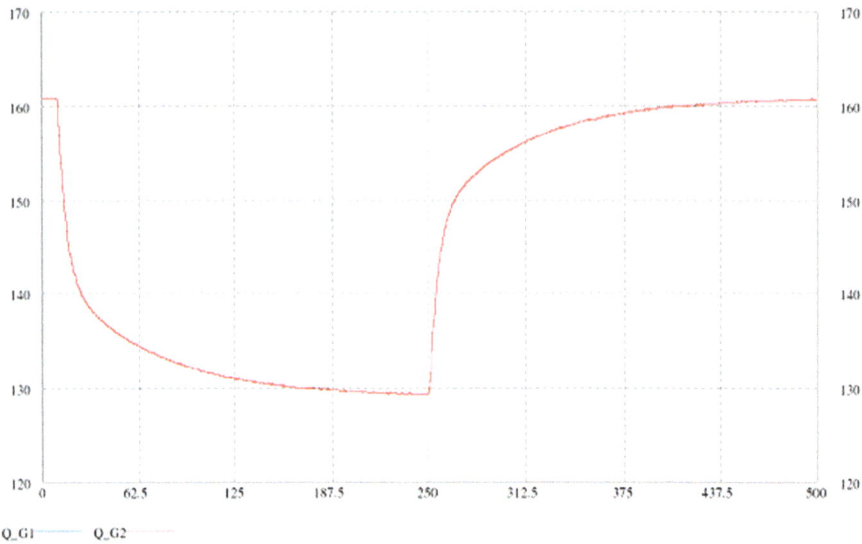

Fig. 3.28 Power plant with two generators. Overlapped transients of generator reactive powers following step variations of HSVC voltage set-point $V_S$ref

by an operation with a proper control at the power station used to energise the no-load line.

3.3.6 Automatic Voltage-Reactive Power Control by SVC

The main static VAR compensator (SVC) task, already introduced in Chap. 2, § 2.3.4, is that of reactive power delivery or absorption aimed at increasing power transmission capability by maintaining or controlling specific parameters of the electrical power system; mainly, the voltage at the local bus or reactive power flow in the lines linked to the bus.

SVC Voltage Regulation

As already seen in Fig. 2.25, basic SVC control operates the SVG as a perfect terminal voltage regulator:

- Amplitude V_T of terminal voltage v_T is measured and compared with voltage reference V_{ref}.
- The e error (ΔV_T) is processed and amplified by a PI controller to provide current reference I_{ref} for the SVG. Therefore, I_{comp} is closed-loop controlled via input I_{ref} so that V_T is maintained with continuity and precisely at the level of the reference voltage V_{ref} in the face of power system and load changes.

In practice, the static VAR compensator is not used as a perfect terminal voltage regulator; rather, the terminal voltage is allowed to vary in proportion to the compensating current, as seen in Fig. 2.26, the SVC static characteristic.

The desired terminal voltage versus output current characteristic of the SVC can be established by an additional control loop that uses one of the SVC control auxiliary inputs, as shown schematically in Fig. 3.29.

A signal proportional to the amplitude of the compensating current $\gamma\ I_{comp}$ is derived and summed to the fixed reference V_{reff} with an ordered polarity such that the capacitive current provides a negative effect on the actual reference V_{ref}, which increases conversely with the inductive current. The actual reference V_{ref} controlling the terminal voltage thus becomes:

$$V_{ref} = V_{reff} - \gamma I_{comp}, \tag{3.16}$$

with $\gamma > 0$ and $I_{comp} > 0$ when delivered by the SVG.

The regulation slope γ is defined as

$$\gamma = \left| \frac{\Delta V_{C\max}}{I_{C\max}} \right| = \left| \frac{\Delta V_{L\max}}{I_{L\max}} \right|,$$

where (see also Fig. 3.30):

ΔV_{Cmax} = deviation (decrease) of terminal voltage from its nominal value at maximum capacitive SVCcurrent,

ΔV_{Lmax} = deviation (increase) of terminal voltage from its nominal value at maximum inductive SVCcurrent,

I_{Cmax} = maximum capacitive compensating current,

I_{Lmax} = maximum inductive compensating current.

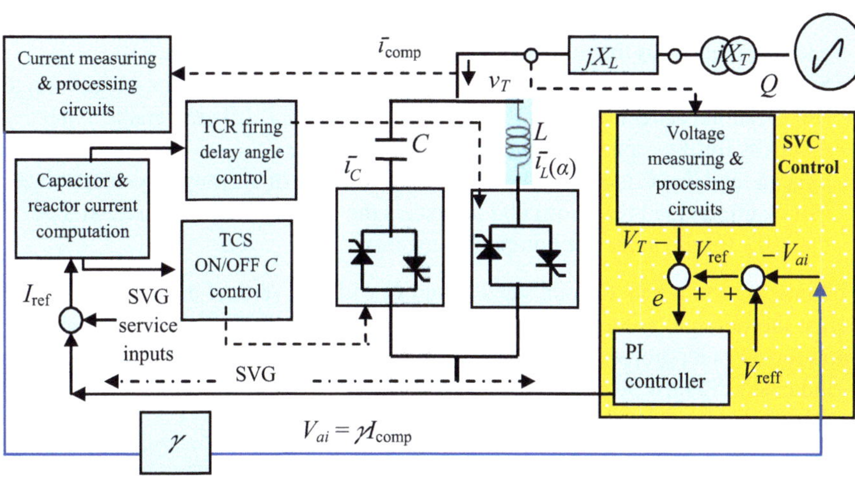

Fig. 3.29 Implementation of classic *V-I* slope of SVC static characteristic by overlapping a control loop that changes reference voltage in proportion to output current

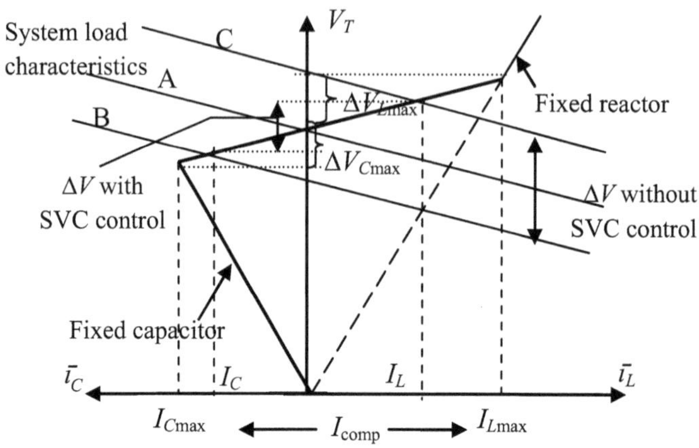

Fig. 3.30 V-I steady state characteristic of an SVC

The V_{ref} Eq. (3.16) indicates that the amplitude of the terminal voltage V_T due to the additional feedback decreases from the fixed, no-load value with increasing capacitive current, in accordance with the γ slope, whereas it grows with increasing inductive compensating current until the maximum capacitive or inductive compensating current, respectively, is reached. For further terminal voltage changes, the output current of the compensator becomes similar to that obtainable with a fixed capacitor or reactor.

A typical SVC characteristic—terminal voltage versus output current, with a given slope—is represented in Fig. 3.30. The same figure also shows particular load characteristics of the AC system (voltage versus reactive current linear characteristics, assuming for each a constant voltage at the equivalent generator and Z as a constant reactance $X=X_L+X_T$):

1. System load line A intersects the SVC *V-I* characteristic at the nominal (reference) voltage; thus the output current of the compensator is zero.
2. System load line B is below line A due to a decrease in the power system voltage (for example, generator outage); it intersects the SVC *V-I* characteristic at a point that calls for the capacitive compensating current I_C.
3. System load line C is above line A due to an increase in the power system voltage (for example, load rejection) and intersects the SVC *V-I* characteristic at a point that asks for the inductive compensating current I_L.

The intersection points of load lines B and C with the vertical axis of the voltage define the terminal voltage variations without compensation. The terminal voltage variation ΔV with drop compensation is entirely determined by the regulation slope as indicated in Fig. 3.30.

SVC Voltage Control Drop

Assuming the SVC voltage control loop operates by a PI control law, the regulation slope at steady state is determined by the condition:

$$e(\infty) = 0,$$

where

$$e(\infty) = -\gamma I_{\text{comp}} + V_{\text{reff}} - V_T = 0, \quad \gamma > 0.$$

At small variations with $\Delta V_{\text{reff}}=0$,

$$\gamma \Delta I_{comp} + \Delta V_T = 0$$

$\gamma = -\Delta V_T / \Delta I_{comp}$ is the slope of the linear dependence between V_T and I_{comp}. Therefore:

- When current variation is due to a capacitive effect, $\Delta I_{\text{comp}}>0$ and $\Delta V_T<0$;
- When current variation is due to an inductive effect, $\Delta I_{\text{comp}}<0$ and $\Delta V_T>0$.

In terms of reactive power:

$$V_T = --\gamma I_{\text{comp}} + V_{\text{reff}}, \quad Q_{\text{comp}} = V_T I_{\text{comp}}, \tag{3.17}$$

and multiplying through by V_T:

$$(V_T)^2 = -\gamma V_T I_{\text{comp}} + V_{\text{reff}} V_T = -\gamma Q_{\text{comp}} + V_{\text{reff}} V_T.$$

Linearising:

$$2V_T \Delta V_T = -\gamma \Delta Q_{\text{comp}} + V_{\text{reff}} \Delta V_T + V_T \Delta V_{\text{reff}},$$

$$\Delta V_T = \frac{-\gamma \Delta Q_{\text{comp}} + V_T \Delta V_{\text{reff}}}{2V_T - V_{\text{reff}}} = -\gamma_q \Delta Q_{\text{comp}} + \frac{V_T \Delta V_{\text{reff}}}{2V_T - V_{\text{reff}}}.$$

In the first approximation:

$$V_T = -\gamma_q Q_{\text{comp}} + V_{\text{reff}}.$$

Therefore:

- When reactive power is due to a capacitive effect,

  $Q_{\text{comp}} > 0$ and $V_T = V_{\text{reff}} - \gamma_q\, Q_{\text{comp}} < V_{\text{reff}}$;

- When reactive power is due to an inductive effect,

  $Q_{\text{comp}} < 0$ and $V_T = V_{\text{reff}} - \gamma_q \, Q_{\text{comp}} > V_{\text{reff}}$.

The compounding effect can also be understood through Thevenin's equivalent electric circuit representing the power system seen by the SVC.

SVC in the Case of Capacitive Current
Equation (3.17) can be represented by Fig. 3.31, the effect of SVC voltage control in the case of capacitive current.

Organising the buses by placing bus V_T at the far right of the equivalent circuit seen by the SVC, the model corresponding to Eq. (3.17) can be changed to that shown in Fig. 3.32, due to the fact that

$$V + ZI_{\text{comp}} = V_T = V_{\text{reff}} - \gamma I_{\text{comp}} = V_{\text{reff}} - \gamma_q Q_{\text{comp}}$$

Therefore,

$$V + (Z + \gamma) I_{\text{comp}} = V_T + \gamma I_{\text{comp}} = V_{\text{reff}}, V_T < V_{\text{reff}}.$$

This corresponds to fixing voltage V_{reff} inside the SVC at a point having as a distance from the V_T bus a γ value reactance (see Fig. 3.33). This also corresponds to a $-\gamma$ increase in SVC equivalent capacitive reactance X_C (that is, a reduction of the SVC capacitor). Obviously, the greater the increase in γ the less the V_T bus is controlled at the V_{reff} value.

With $\gamma = X_C$, the SVC drop has a slope equal to, but opposite in sign from, its fixed capacitor *V-I* link. With this positive slope (see Fig. 3.30) there is a risk that the SVC operating point will be reached on the capacitive side only, thereby showing its inadequacy with respect to SVC dimensioning and local grid characteristics.

In general, the following condition must be verified:

$$|\gamma| \ll |X_C|.$$

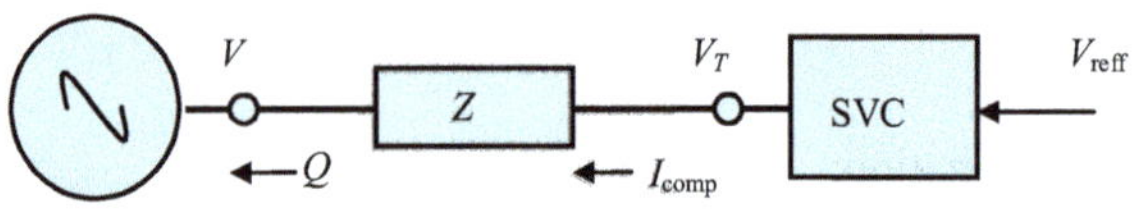

Fig. 3.31 Equivalent electric circuit of SVC with capacitive current and regulating voltage drop

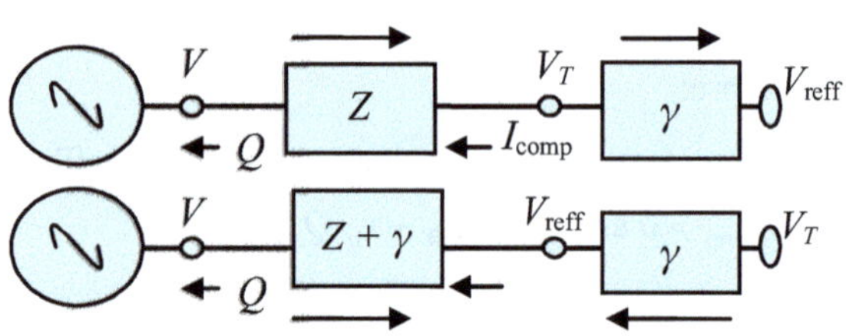

Fig. 3.32 Equivalent circuit of SVC with regulating voltage slope and VT bus at the extreme end with reactive power injected by SVC

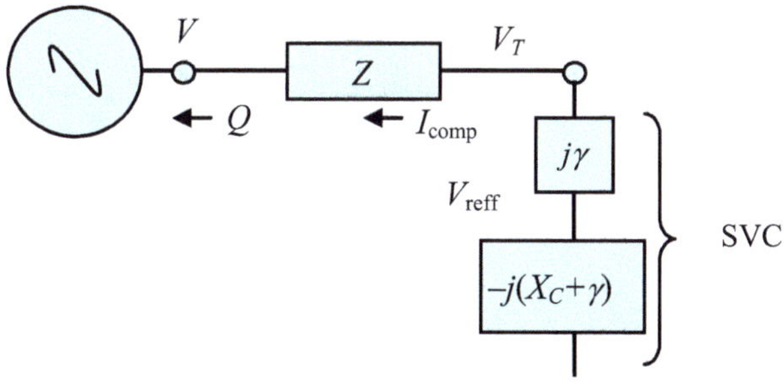

Fig. 3.33 SVC equivalent circuit with capacitive current: regulating voltage drop determines the internal point of the SVC regulated at the V_{reff} value

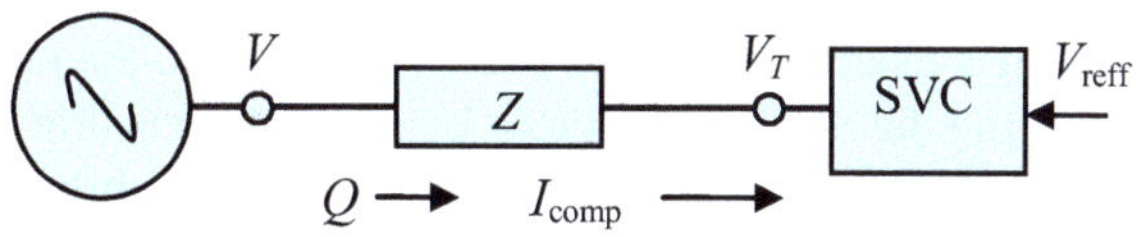

Fig. 3.34 Equivalent electric circuit of SVC with inductive current and regulating voltage drop

SVC in the Case of Inductive Current

Equation (3.17) can be also represented as in Fig. 3.34: the case of SVC inductive current.

Organising the buses by placing bus V_T at the extreme right (that is, the bus where the SVC is connected to the grid), the equivalent circuit can be modified to the one shown in Fig. 3.35, due to the fact that

$$V - ZI_{\text{comp}} = V_T = V_{\text{reff}} + \gamma I_{\text{comp}} = V_{\text{reff}} + \gamma_q Q_{\text{comp}}$$

Therefore,

$$V - (Z + \gamma) I_{\text{comp}} = V_{\text{reff}},\ V_T > V_{\text{reff}}.$$

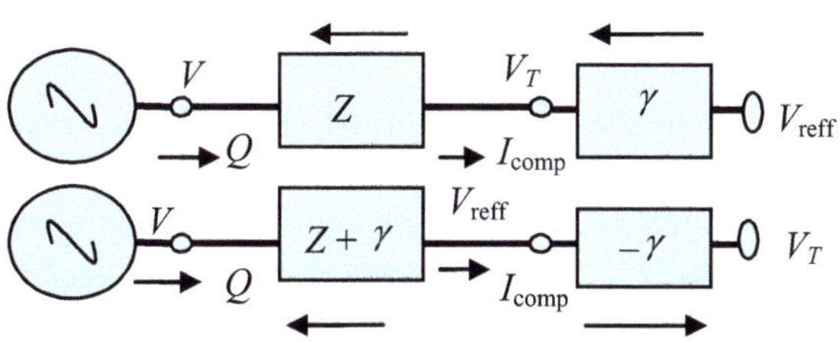

Fig. 3.35 Equivalent circuit of SVC with regulating voltage drop and V_T bus at extreme end with reactive power absorbed by SVC

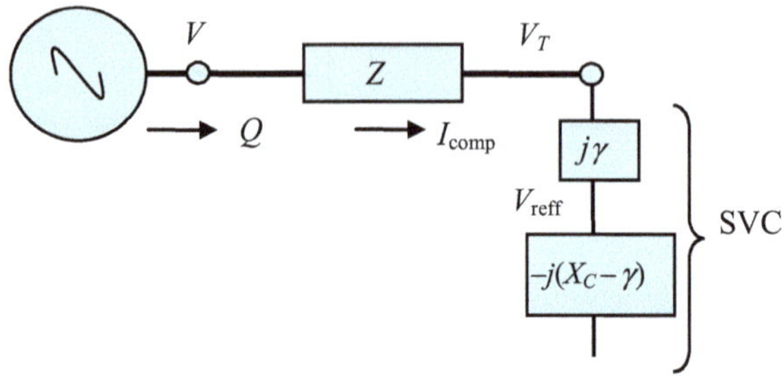

Fig. 3.36 Equivalent circuit of SVC with regulating voltage drop determining internal point of SVC regulated at V_{reff} value, with inductive current

This corresponds (see Fig. 3.36) to fixing the voltage value V_{reff} at an internal point of the equivalent X_L reactance of the SVC; that is, fixing it inside the SVC at a point having a reactance value of γ at a distance from the V_T bus. This also corresponds to reducing the SVC reactance X_L by γ (i.e., a reduction of the SVC reactor).

Obviously, the greater the increase in γ, the higher the V_T bus voltage with respect to V_{reff} value.

The extreme edge, with γ compensating the whole X_L value ($\gamma=X_L$), corresponds to fixing the lower reactor edge at the V_{reff} value. Obviously, this objective is not feasible when the reactor is ground-connected unless $V_{ref}=0.0$.

With this extreme drop (see Fig. 3.30) the SVC operating point simply moves along the fixed reactor characteristic with a consequent fast saturation of the voltage control, thereby showing its inadequacy with respect to SVC dimensioning, local grid characteristics and practical values that are used for V_{reff}.

Accordingly, the following condition must always be verified:

$$|\gamma| \ll |X_L|.$$

The limiting conditions on γ can be written as

$$\gamma \ll \min\{X_C; X_L\}.$$

Dynamic Behaviour of SVC

Dynamic behaviour of an SVC in the normal compensating range can be characterised by the basic block diagram with transfer functions linked to the control scheme of Fig. 3.29. Figure 3.37 shows the equivalent system (seen by the SVC) delivering reactive power (I_{comp} is negative in the case where the SVC absorbs reactive current).The following equation represents the simplified instantaneous model, through which local SVC control dynamic is analysed:

$$V + ZI_{comp} = V_T.$$

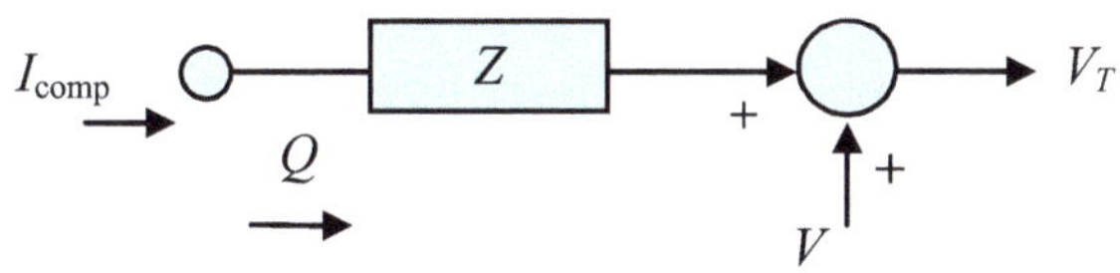

Fig. 3.37 Block diagram of the equivalent grid seen by SVC

On the SVC side, the here analysed dynamics in the linear operating range must be modelled as being derived directly from the basic SVC control scheme of Fig. 3.29. The previously introduced FC-TCR dynamic model (Chap. 2, § 2.3.3.3 FC-TCR) can be used for TCS-TCR- and SVC-type VAR generators, as well, when operated around a fixed TSC capacitor.

The transfer function between the amplitude of the fundamental reactor current $I_{LF}(\alpha)$ and its reference is given by

$$G(s) = ke^{-T_d s},$$

where:

s	=	Laplace transform operator,
k	=	constant gain,
T_d	=	transportation lag, corresponding to firing delay angle α.

The transfer function $G(s)$ can be written in a simplified form by a first order approximation for the exponential term:

$$G(s) \approx k\frac{1}{1+sT_d}.$$

Continuous control of $I_{LF}(\alpha)$ is obtained by computing total capacitive current I_C linked to V_T voltage value and TSC operating susceptance B_C.

In Fig. 3.38, error ε_i represents the TCR's required inductive current; then the slope of the nonlinear characteristic with I_{Lmax} saturation is tan $\vartheta = 1/k$. The SVC total current I_{comp} is the algebraic sum of the TCR output $I_{LF}(\alpha)$ with I_C.

The dynamic link between the SVC current reference I_{ref} and the SVC output voltage V_T is obtained by joining the last two block diagrams (Figs. 3.37 and 3.38). Now the missing external voltage control loop can be overlapped with a PI regulator compounded by additional feedback γ, which imposes the slope to the SVC voltage control drop characteristic.

The complete dynamic scheme, shown in Fig. 3.39, also includes two measuring filters F1 and F2, TCS current limit I_{Cmax} and a switching logic that changes the value of the B_C parameter according to the ε_i values and an due delay after the given thresholds are overcome. The PI regulator parameters largely determine the speed and stability of the voltage control loop.

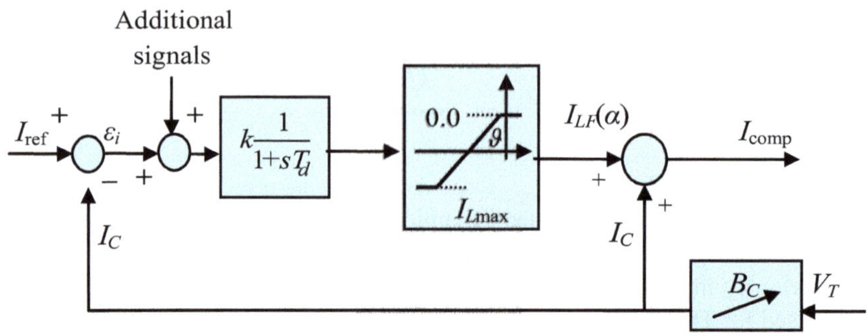

Fig. 3.38 Block diagram of SVC total current including TCR current computation and control

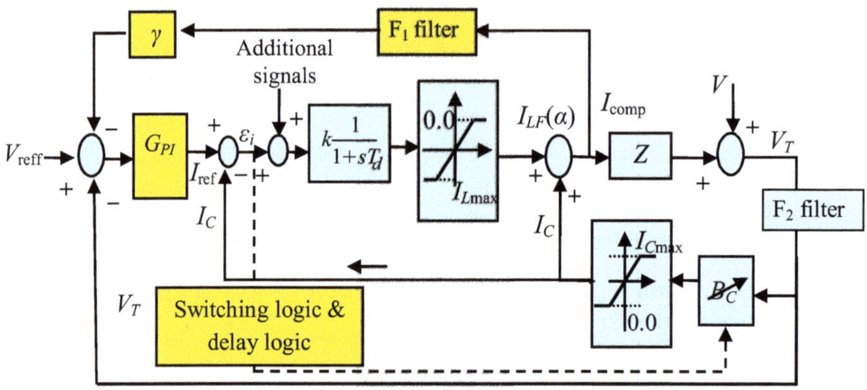

Fig. 3.39 Block diagram of SVC voltage control loop with additional compounding feedback γ realising the voltage drop characteristic

In the figure,

- F_1 and F_2 can be represented by first order transfer functions:

$$F_{1,2}(s) = \frac{1}{1+sT},$$

in which time constant T is on the order of 10–15 ms;

- T_d is the SVG transportation lag on the order of 3–6 ms;
- γ is the regulation slope, typically 1–3 %.

A further elaboration of the Fig. 3.39 scheme gives the block diagram of Fig. 3.40, where the linear block imposing current saturations I_{Lmax} and I_{Cmax} also represents, by a simplified continuous modelling (instead of a discrete nonlinear model), the B_C switching logic and delay.

This is an acceptable simplification, also taking into account the linear operation of the TCR around each operating point fixed by TCS switching.

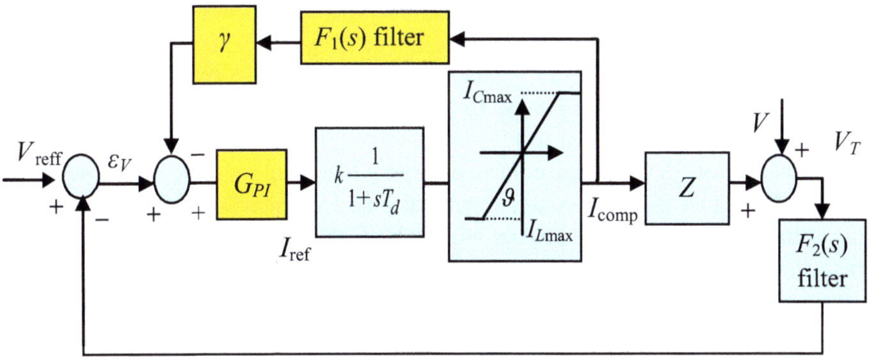

Fig. 3.40 Block diagram of SVC voltage control loop with additional compounding feedback γ realising the voltage drop characteristic

The PI voltage regulator is dimensioned in proportional (K_P) and integral (K_I) parameters in such a way that the transfer function between I_{comp} and the voltage error $\varepsilon_V = V_{reff} - F_2(s)V_T$ is

$$\frac{I_{comp}}{\varepsilon_V} = G_1(s) \approx \frac{1}{\gamma} \cdot \frac{1+sT_1}{1+sT_2},$$

with $T_1 = K_P/K_I \ll T_2 \approx 20$–$50$ ms (T_2 depending on parameters T_d, γ, K_I and K_P).

We should note that the SVC dynamic behaviour and stability is a function of the power system equivalent impedance Z being an integral part of the feedback loop. For this reason, control is usually optimised for maximum expected system impedance (minimum short circuit capacity). This means the voltage control response time will be somewhat longer if system impedance is decreased. With practical SVC, the worst case response time is typically in the range 30–80 ms.

From the closed loop transfer functions:

$$\frac{V_T}{V_{reff}} = \frac{G_1(s)Z}{1+G_1(s)ZF_2(s)}, \quad \frac{V_T}{V} = \frac{1}{1+G_1(s)ZF_2(s)}, \quad \frac{V_T}{I_{comp}} = \frac{Z}{1+G_1(s)ZF_2(s)}.$$

System eigenvalues (poles) given by the characteristic polynomial clearly depend on the values of the γ and Z parameters; but, they also depend on the static and dynamic characteristics of $G_1(s)$, $F_2(s)$.

Under steady state conditions ($s \to 0.0$) the above equations become

$$\frac{V_T}{V_{reff}}(\infty) = \frac{Z}{\gamma F_1 + F_2 Z} = \frac{Z}{\gamma + Z},$$

$$\frac{V_T}{V}(\infty) = \frac{\gamma F_1}{\gamma F_1 + F_2 Z} = \frac{\gamma}{\gamma + Z},$$

$$\frac{V_T}{I_{comp}}(\infty) = \frac{Z\gamma F_1}{\gamma F_1 + F_2 Z} = \frac{Z\gamma}{\gamma + Z}.$$

These results confirm that as slope becomes smaller ($\gamma \rightarrow 0.0$) terminal voltage remains constant with respect to variations in V (V_T depends in an integral way on V_{reff} only). Therefore, any change in I_{comp} due to V recovers V_T to the value imposed by $V_{reff.}$

Similarly, with increasing slope ($\gamma \gg Z$) terminal voltage becomes unregulated and assumes values very near to V.

Transients of SVC voltage control against a load step increase are shown next (Fig. 3.41). At $t=50$ s, the system's general load experiences a step increase, and the voltage magnitude at the bus bar with an SVC drops. In very little time the SVC recovers the controlled voltage at a value a bit smaller than set-point level (green). To accomplish this, the SVC operates through its voltage control loop by decreasing the reactive power absorbed by its inductor and thereby increasing the voltage output, but doing so according to the static characteristic drop.

Dynamic Behaviour of SVC Reactive Power Control

In order to handle dynamic VAR disturbances, an SVC may be equipped with switchable VAR reserves or automatic operating point setting control. The objective is to impose SVC reactive power output according to a reference value that can be manually or automatically changed.

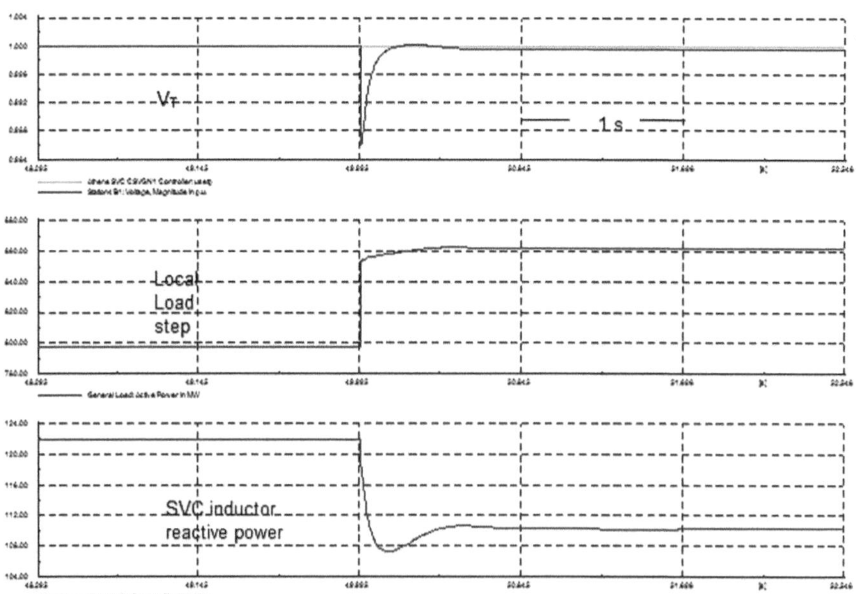

Fig. 3.41 SVC voltage control loop traces following a local load step variation in the presence of additional compounding feedback γ realising the voltage drop characteristic

Control Scheme Overlapping the Voltage Regulation

When a possible disturbance results in a new SVC operating point with a steady VAR output, the SVC VAR control, at a fixed VAR set-point value, effectively changes the voltage reference of the control loop of Fig. 3.29 in order to slowly bring the VAR output back to the set reference value. The response time of this reactive power control loop is slow, so as not to interfere with the faster voltage regulation or any fast stabilising or auxiliary functions that might be included in the overall VAR output control.

A possible scheme for implementing a basic VAR reserve control is shown in Fig. 3.42. The magnitude of SVC output reactive power $V_T I_{\text{comp}}$ is measured and compared against the reference Q_{ref} (assumed positive to inject reactive power into the grid). The error signal is sent to an integral regulator with a large time constant and added to the fixed voltage reference V_{reff} summing junction. This forces the voltage regulator input signal to change until the difference between the actual SVR output VARs and the Q_{ref} disappears.

The above described operation of the SVC reserve control can be illustrated by a trajectory on the static characteristic in Fig. 3.43 as follows:

- Suppose operating point A is modified by a sudden change in system load characteristic with a consequent drop in amplitude of terminal voltage V_T, to which an instantaneous reduction on the capacitive current also corresponds.
- Voltage drop ΔV_T from A to B forces the output current to increase from I_{CB} to I_{CC} via the fast voltage regulator up to working point C on the *V-I* curve along the new system load characteristic.
- Furthermore, since $I_{CC} \cdot V_{TC} > Q_{\text{ref}}$, an error signal ΔQ is generated, determining the slower Q recovery by reducing voltage set-point up to the final equilibrium point in D.

In terms of a linear dynamic model, the SVC block diagram, including the reactive power control, can be defined by overlapping the voltage control loop described in Fig. 3.40, with additional reactive power control of integral type (Fig. 3.44). This scheme assumes the delivered reactive power has positive sign, which also depends on the voltage V_T and the operating equivalent reactance X provided by the SVC: $X > 0.0$ capacitive effect; $X < 0.0$ inductive effect:

$$Q = \frac{V_T^2}{X}, \quad \Delta Q = \frac{2(V_T)_0}{X} \Delta V_T.$$

According to these assumptions, the SVC dynamics, including the reactive power control, can be analysed by the linear control model of Fig. 3.45.

With practical SVC, the reactive power control loop time response is slower and decoupled with respect to the voltage control loop, being characterised by a dominant time constant in the range 1–5 s. The parameters γ and Z mostly contribute to the voltage control loop performance, and μ and X contribute to fixing the reactive power control loop speed.

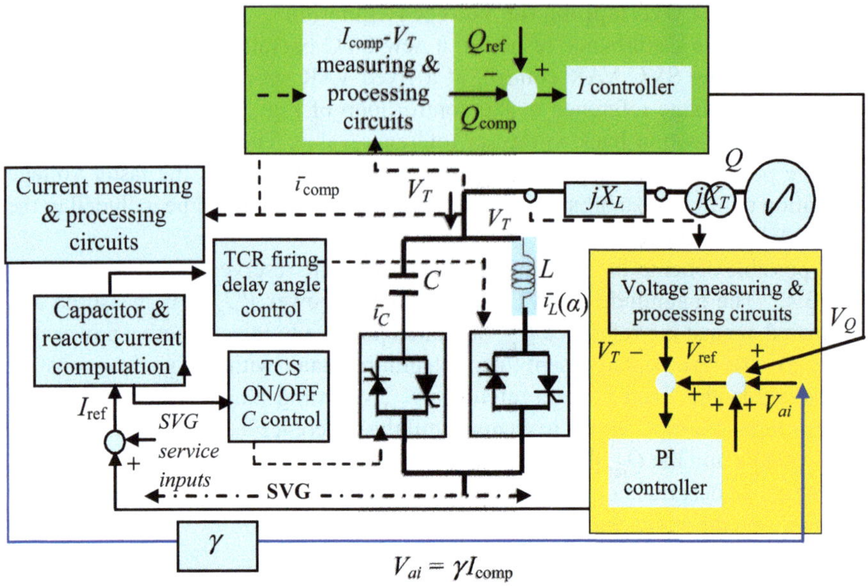

Fig. 3.42 Implementation of SVC reactive power control by overlapping voltage with drop control with a slower and integral control loop that imposes reactive power output by changing reference voltage in proportion to integral of reactive power error

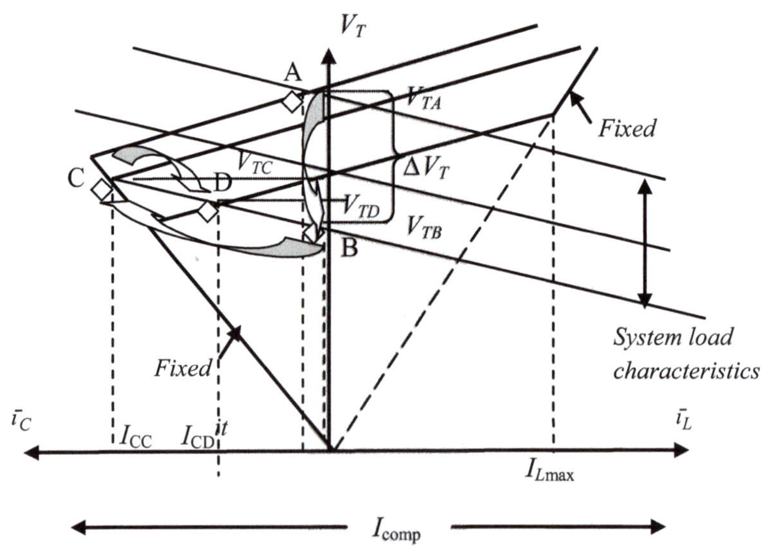

Fig. 3.43 Trajectory of SVC operating point following sudden change (from A to B, instantaneously) after a disturbance, recovered by voltage control loop first (from B to C–fast control), followed by reactive power control loop (from C to D—slow control)

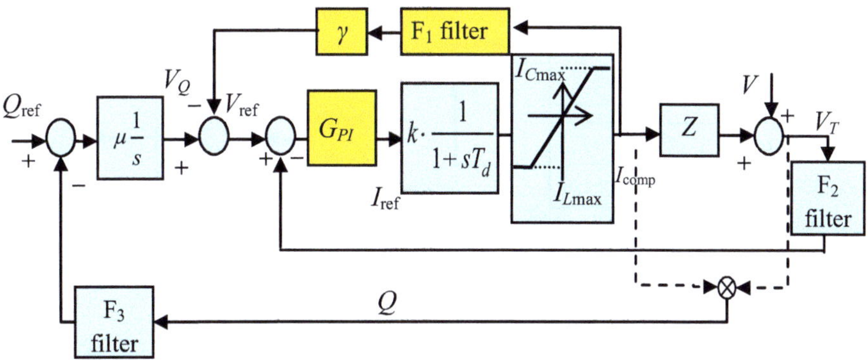

Fig. 3.44 Block diagram of SVC reactive power control loop that overlaps the faster voltage control with additional compounding feedback γ realising the voltage drop characteristic

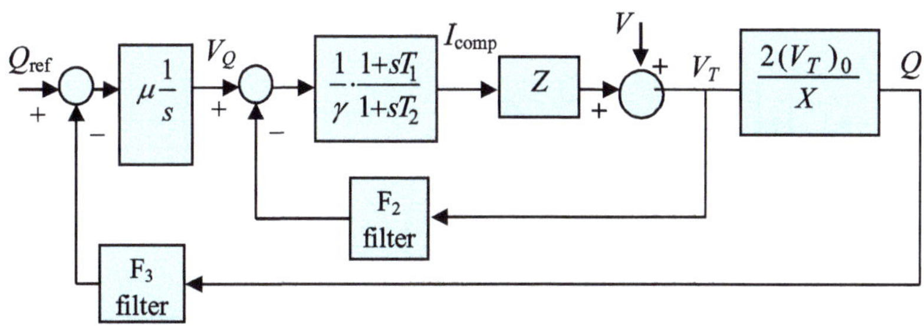

Fig. 3.45 Linear model of the SVC reactive power control loop that overlaps the faster voltage control with additional compounding feedback γ realising the voltage drop characteristic

The slower dynamic typically works when the SVC contributes to transmission network-coordinated voltage control or secondary voltage regulation (SVR), which is described in Part II of this book.

In short, the simplified linear model of the SVC can be represented with first order transfer functions for both the voltage and reactive power control (Fig. 3.46). The latter, which is usually switched off when the SVC is operating the local voltage control, can be put into operation in the case where the SVC contributes to SVR.

Control Scheme with Reactive Power Loop in Place of Voltage Regulation
In this alternative control scheme (see Fig. 3.47) a manual or automatic switch selects the operating reactive power or the voltage control loop. This solution simplifies model analysis but requires output tracking between the two integral regulators in a way that achieves bumpless switching between the two control loops.

The linear model through which the reactive power control loop dynamics are analysed is shown in Fig. 3.48.

This is a simple integral control loop with a dominant time constant $T_Q \gg T_d$. Therefore, the control gain μ_Q is dimensioned in a way to achieve T_Q in the range of

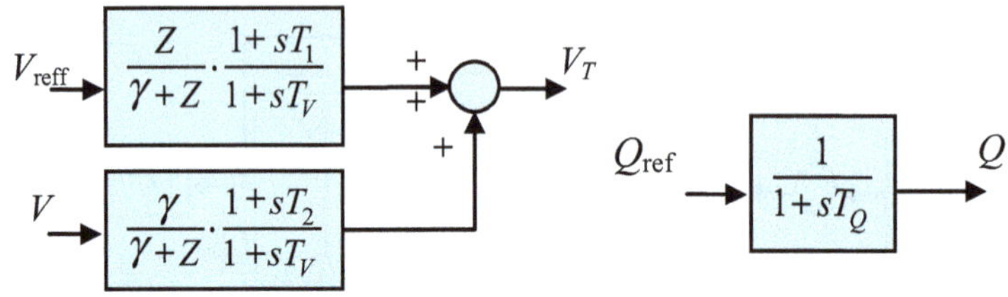

with $T_V = 30\text{–}80$ ms, $\quad T_V = \dfrac{\gamma T_2 + ZT_1}{\gamma + Z}, \quad T_Q = 1\text{–}5$ s

Fig. 3.46 Simplified linear models of SVC voltage and reactive power control loops in presence of additional compounding feedback γ realising the voltage drop characteristic

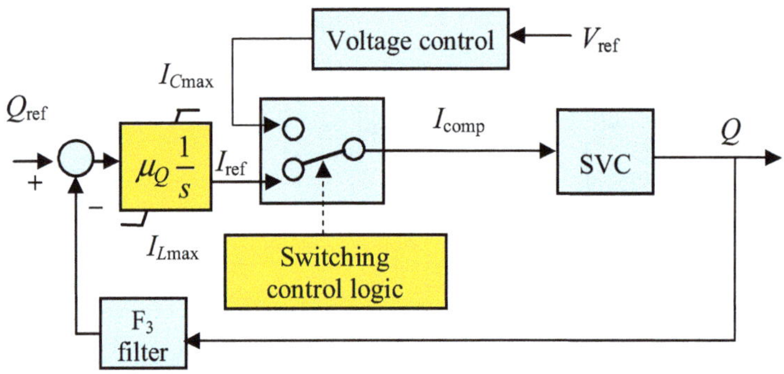

Fig. 3.47 SVC reactive power control to be activated by switching from the voltage control

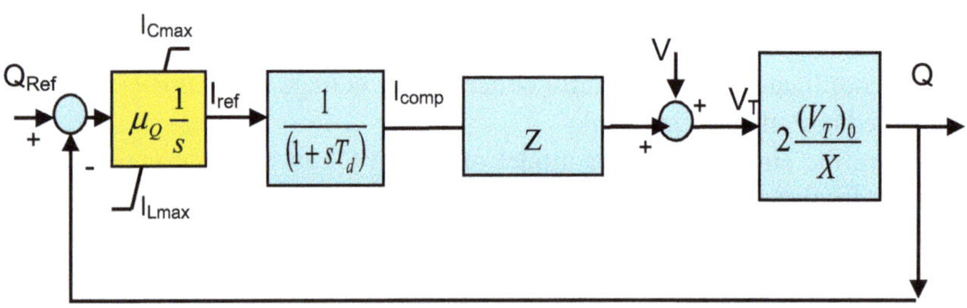

Fig. 3.48 SVC reactive power linear control to be activated by switching from the voltage control

1–5 s around the normal operating point of both the equivalent impedance Z seen by the SVC and the SVC internal reactance X:

$$T_Q \approx \frac{X}{2\mu_Q Z (V_T)_0}$$

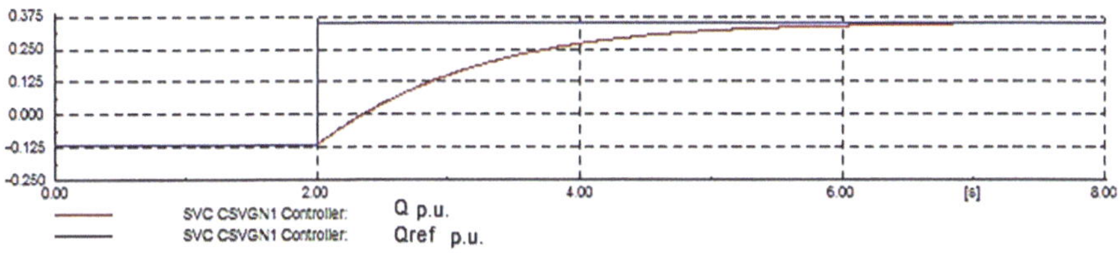

Fig. 3.49 SVC reactive power control loops traces following a set-point step variation

As in the case of the previously considered scheme, the simplified linear model of SVC reactive power control can be represented by a first order transfer function:

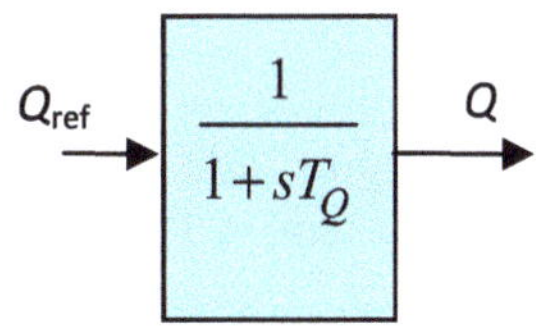

Figure 3.49 shows SVC reactive power loop traces following a step variation of the Q_{ref} set-point.

At $t=2$ s, Q_{ref} has a step increase asking for more reactive power to be delivered. According to the designed speed, the SVC achieves without error the new required reactive power. In order to do this, the SVC operates through its reactive power control loop by decreasing the reactive power absorbed by its inductor according to an integral control law that fixes the dominant time constant to about 5 s.

3.3.7 Automatic Voltage-Reactive Power Control by STATCOM

The main task of a static compensator (STATCOM), as introduced in Chap. 2, § 2.3.5, is reactive power delivery or absorption so as to increase the power transmission capability by maintaining or controlling specific parameters of the electrical power system, mainly voltage at the local HV bus or reactive power flow in lines linked to the bus.

Analogous to the SVC, STATCOM control requirements can be derived from a power system's functional compensation needs, namely:

1. Direct voltage support to improve grid operation and prevent voltage instability;
2. Power system transient and dynamic stability improvements to increase the first swing stability margin and provide damping for power oscillations.

In the next section, we discuss how STATCOM functional control requirements fully satisfy the aspect of voltage control given in the first item above. In Chap. 8 we analyse control functionality linked to stability, as mentioned in the second.

STATCOM Grid Voltage Regulation

We refer to the STATCOM model with PWM control in Fig. 2.28 and to the related grid voltage control system introduced by the block diagram of Fig. 2.29. More specifically, as in the case of the SVC control system:

- Amplitude V_T of terminal voltage v_T is measured and compared with reference voltage V_{ref}
- Error ΔV_T is processed and amplified by a PI controller to provide the current reference I_{ref} (or Q_{ref}) for the voltage source inverter (VSI). Therefore, I_{comp} (Q_{comp}) is closed-loop controlled via input I_{ref} so that V_T is maintained with continuity and precisely at the level of the V_{ref} by coping with lower system and load changes.

The current error determines the amplitude variation of the VSI voltage output and therefore the magnitude and polarity of the reactive current to be drawn by the STATCOM inverter from the AC system. A PI controller is used to speed up response and reduce steady state error.

V_{DC} control is generally slower than VSI AC output control. Accordingly, the ϑ input to the PWM modulator can be considered constant during AC current and voltage transients, mainly when their control loops are very fast (i.e., with a dominant time constant on the order of tens or hundreds of milliseconds).

In practice, the STATCOM is not used as a perfect terminal voltage regulator. As for the SVC, a possible reactive current feedback on the voltage reference value can be used (see Fig. 3.50) to determine a droop characteristic in the voltage regulation, as was introduced by static characteristics pictured in Figs. 2.30 and 2.31.

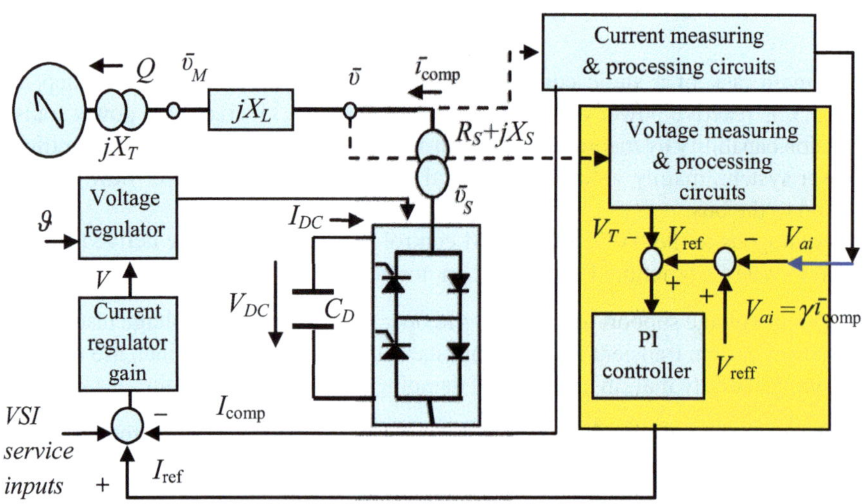

Fig. 3.50 Classic *V-I* slope implementation of STATCOM static characteristic by overlapping a control loop, changing the reference voltage in proportion to output current

A signal proportional to the amplitude of the compensating current γI_{comp} is derived and added to the fixed reference V_{reff} with an ordered polarity such that the capacitive current provides a negative effect on the actual reference V_{ref}, which conversely increases with inductive current. The actual reference V_{ref} controlling the terminal voltage thus becomes

$$V_{ref} = V_{reff} - \gamma I_{comp},$$

with $\gamma > 0$ and $I_{comp} > 0$ when delivered by the VSI.

The regulation slope γ is defined by

$$\gamma = \left| \frac{\Delta V_{C\max}}{I_{C\max}} \right| = \left| \frac{\Delta V_{L\max}}{I_{L\max}} \right|,$$

where (see also Fig. 3.51):

ΔV_{Cmax} = deviation (decrease) of terminal voltage with capacitive VSI current from zero to the maximum,

ΔV_{Lmax} = deviation (increase) of terminal voltage with inductive VSI current from zero to the maximum,

I_{Cmax} = maximum capacitive compensating current, transiently reachable,

I_{Lmax} = maximum inductive compensating current, transiently reachable.

The equation for V_{ref} indicates that the amplitude of the terminal voltage V_T, due to additional feedback, decreases from the fixed no-load value with increasing capacitive current, according to slope γ, whereas it grows with increasing inductive

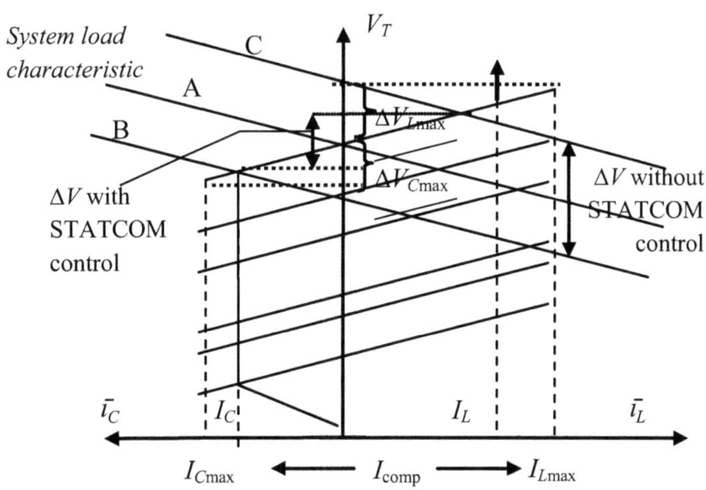

Fig. 3.51 *V-I* steady state characteristic of STATCOM

compensating current until the maximum capacitive or, respectively, inductive compensating current is reached.

A typical STATCOM characteristic, that of terminal voltage versus output current, with a given slope, is shown in Fig. 3.51, together with the particular load characteristics of the AC system (voltage versus reactive current linear characteristics, assuming for each of them a constant voltage at the equivalent generator and Z as a constant reactance $X=X_L+X_T$). We observe:

1. System load line A intersects STATCOM V-I characteristic at nominal (reference) voltage, thus, the output current of the compensator is zero.
2. System load line B is below line A due to a decrease in power system voltage (for example, generator outage) and intersects the STATCOM V-I characteristic at a point that calls for the capacitive compensating current I_C.
3. System load line C is above line A due to an increase in power system voltage (for example, load rejection) and intersects the STATCOM V-I characteristic at a point that calls for the inductive compensating current I_L.

The points of intersection of load lines B and C with the vertical axis (voltage) define the terminal voltage variations in the case of no compensation. The terminal voltage variation ΔV with drop compensation is entirely determined by the regulation slope as indicated in Fig. 3.51.

STATCOM Voltage Control Drop

Assuming the STATCOM voltage control loop operates by a PI control law, the regulation slope at steady state is determined by the condition $e(\infty)=0$, where

$$e(\infty) = -\gamma I_{\text{comp}} + V_{\text{reff}} - V_T = 0, \quad \gamma > 0.$$

Considering small variations $\Delta V_{\text{reff}}=0$,

$$\gamma \Delta I_{\text{comp}} + \Delta V_T = 0.$$

$\gamma = -\Delta V_T / \Delta I_{\text{comp}}$ is the slope of the linear dependence between V_T and I_{comp}. Therefore:

- When current variation is due to a capacitive effect, $\Delta I_{\text{comp}}>0$ and $\Delta V_T<0$;
- When current variation is due to an inductive effect, $\Delta I_{\text{comp}}<0$ and $\Delta V_T>0$.

In terms of reactive power:

$$V_T = -\gamma I_{\text{comp}} + V_{\text{reff}}, \tag{3.17}$$

$$Q_{\text{comp}} = V_T I_{\text{comp}},$$

and multiplying through by V_T:

$$(V_T)^2 = -\gamma V_T I_{\text{comp}} + V_{\text{reff}} V_T = -\gamma Q_{\text{comp}} + V_{\text{reff}} V_T.$$

Linearising:

$$2V_T \Delta V_T = -\gamma \Delta Q_{\text{comp}} + V_{\text{reff}} \Delta V_T + V_T \Delta V_{\text{reff}},$$

so

$$\Delta V_T = \frac{-\gamma \Delta Q_{\text{comp}} + V_T \Delta V_{\text{reff}}}{2V_T - V_{\text{reff}}} = -\gamma_q \Delta Q_{\text{comp}} + \frac{V_T \Delta V_{\text{reff}}}{2V_T - V_{\text{reff}}}.$$

In the first approximation:

$$V_T = -\gamma_q Q_{\text{comp}} + V_{\text{reff}}.$$

Therefore:

- When reactive power is due to a capacitive effect, $Q_{\text{comp}} > 0$ and $V_T = V_{\text{reff}} - \gamma_q Q_{\text{comp}} < V_{\text{reff}}$;
- When reactive power is due to an inductive effect, $Q_{\text{comp}} < 0$ and $V_T = V_{\text{reff}} - \gamma_q Q_{\text{comp}} > V_{\text{reff}}$.

STATCOM with Capacity Current

We illustrate the case of the STATCOM with a capacity current through the Thevenin equivalent electric circuit of the system seen by the STATCOM (Fig. 3.52), and interpret the effect of the voltage control.

Organising the buses by placing bus V_T at the extreme right of the equivalent circuit seen by the STATCOM, the model corresponding to Eq. (3.17) is changed to what is shown in Fig. 3.53, assuming

$$Z = R + jX \approx jX, \quad Z_S = R_S + jX_S \approx jX_S,$$

and due to the fact that

$$V + XI_{\text{comp}} = V_T = V_{\text{reff}} - \gamma I_{\text{comp}} = V_{\text{reff}} - \gamma_q Q_{\text{comp}} = V_S - X_S I_{\text{comp}}.$$

Therefore,

$$V_S - (X_S - \gamma) I_{\text{comp}} = V_{\text{reff}}, \quad V_T = V_{\text{reff}} - \gamma I_{\text{comp}} = V_{\text{reff}} - \gamma_q Q_{\text{comp}},$$
$$V_T < V_{reff}.$$

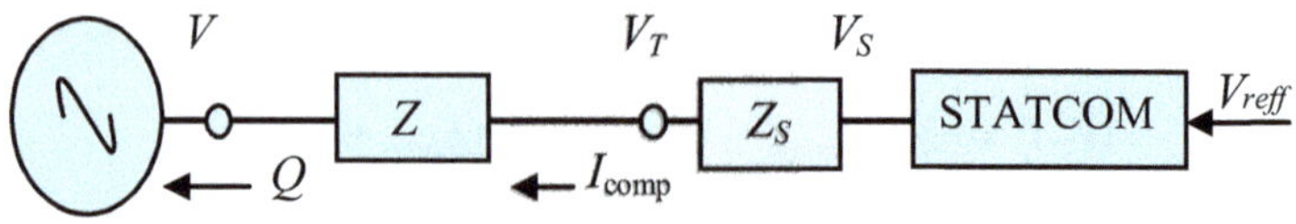

Fig. 3.52 Equivalent electric circuit of STATCOM with capacitive current and regulating voltage drop

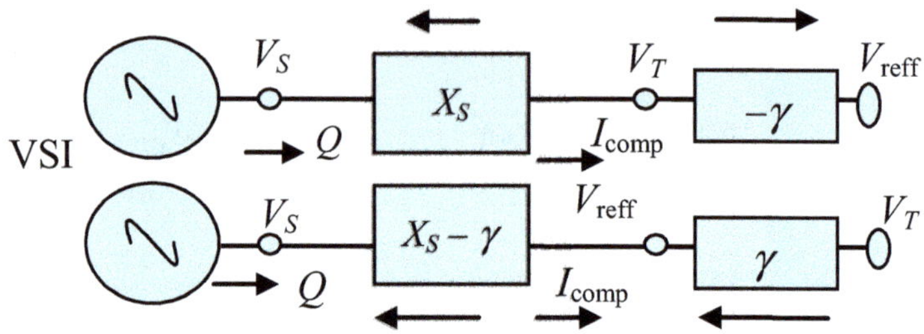

Fig. 3.53 Equivalent circuit of STATCOM with regulating voltage slope and V_T bus at extreme end with reactive power injected by VSI

This corresponds (see Fig. 3.54) to fixing the voltage value V_{reff} inside X_S at a point having as a distance from the V_T bus a γ value reactance. Obviously, the more γ increases, the less the V_T bus is controlled at the V_{reff} value.

With $\gamma = X_S$, the STATCOM drop reaches its maximum slope. With this high positive slope (see Fig. 3.51) the STATCOM operating point risks being reached on the capacitive side only, therefore showing its inadequacy with respect to STATCOM system capability and local grid characteristics.

In general, the following condition has to be verified:

$$\gamma \ll X_S$$

STATCOM with Inductive Current

Equation (3.17) can also be represented as in Figs. 3.55 and 3.56, the case of STATCOM inductive current.

Organising the buses by placing the bus V_T at the extreme right, that is, the bus at which the STATCOM is connected to the grid, the equivalent circuit can be modified as follows in Fig. 3.56, due to the fact that

$$V - X\,I_{comp} = V_T = V_{reff} + \gamma I_{comp} = V_{reff} + \gamma_q Q_{comp} = V_S + X_S I_{comp}$$

Therefore,

$$V_S + (X_S - \gamma)I_{comp} = V_{reff,} \quad V_T = V_{reff} + \gamma I_{comp} = V_{reff} + \gamma_q Q_{comp},$$
$$V_T > V_{ref}.$$

This corresponds (see Fig. 3.57) to fixing the voltage value V_{ref} inside Z_S at a point having as a distance from the V_T bus a γ value reactance. Obviously, the more γ increases the higher the V_T bus voltage is with respect to the V_{ref} value.

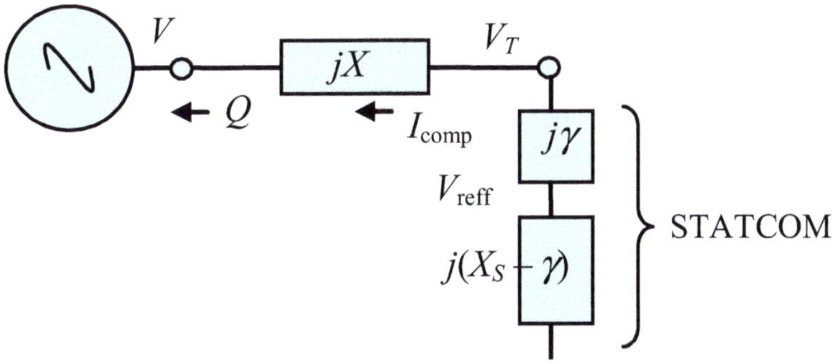

Fig. 3.54 Equivalent circuit of STATCOM with regulating voltage drop determining internal point of XS regulated at V_{reff} value, with capacitive current

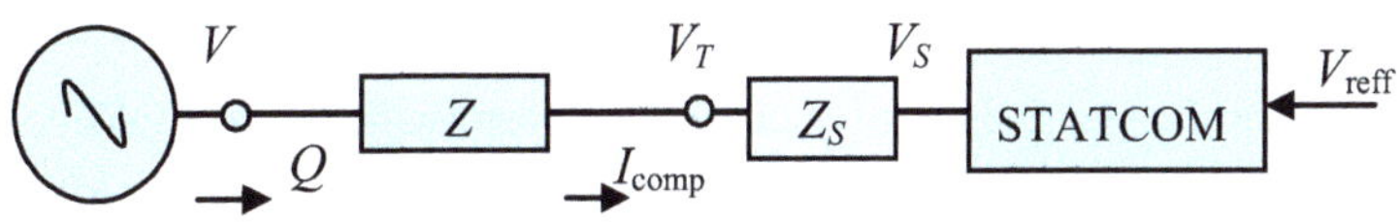

Fig. 3.55 Equivalent electric circuit of SVC with inductive current and regulating voltage drop

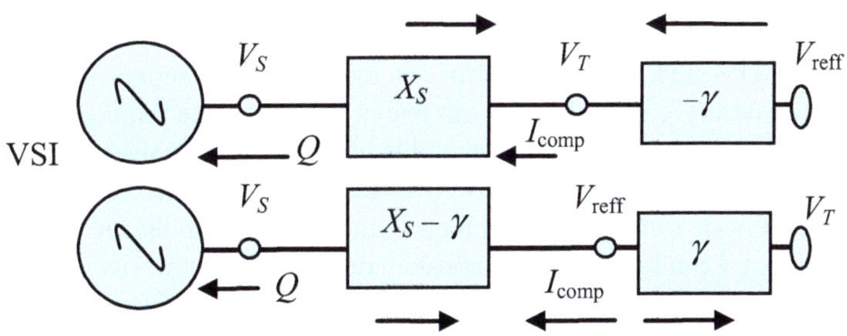

Fig. 3.56 Equivalent circuit of a STATCOM with regulating voltage drop and V_T bus at extreme end, with reactive power absorbed by the STATCOM

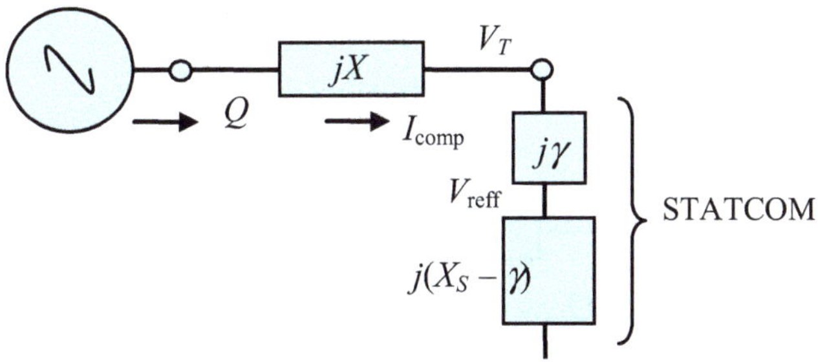

Fig. 3.57 Equivalent circuit of STATCOM with regulating voltage drop determining internal point of STATCOM regulated at V_{reff} value with inductive current

With $\gamma = X_S$, the STATCOM drop reaches its maximum slope. With this highly positive slope (see Fig. 3.51) the STATCOM operating point risks being reached on the inductive side only, thereby showing an inadequacy with respect to STATCOM system capability and local grid characteristics.

Accordingly, the STATCOM must always verify, either with capacitive or inductive currents, that

$$\gamma \ll X_S.$$

Dynamic Behaviour of the STATCOM

STATCOM dynamic behaviour in the normal compensating range can be characterised by the basic transfer function block diagram linked to the Fig. 3.50 control scheme.

We consider the Fig. 3.52 equivalent system seen by the STATCOM delivering reactive power (I_{comp} is negative in the case where the STATCOM absorbs reactive current); the following equation represents its simplified instantaneous model (Fig. 3.58), through which we analyse the local STATCOM dynamic.

$$V + X\, I_{\text{comp}} = V_T$$

On the STATCOM side, the basic control scheme in Fig. 2.29 shows the combination of two contemporary control loops, with one at the V_{DC} value required for a proper and continuous VSI operation. The second is on the reactive power output (Q_{comp})/current (I_{comp}), allowing for grid support in terms of voltage/reactive power delivery. Usually, the AC side's current control loop is much faster than the DC side's voltage control, so input ϑ can be considered constant during current transient analysis.

Moreover, the VSI transfer function between the $Q_{\text{ref}}/I_{\text{ref}}$ input and the $V_S\angle\vartheta$ output is very fast with respect to the grid current and voltage control and can be simply represented as instantaneous by a gain or by a first order model, but one that has a very small time constant T_d, always assuming constant input ϑ.

Therefore, the transfer function between the amplitude of the fundamental output voltage V_S and the fundamental reactor current I_{LF} is given by

$$G(s) \approx k\frac{1}{1+sT_d}.$$

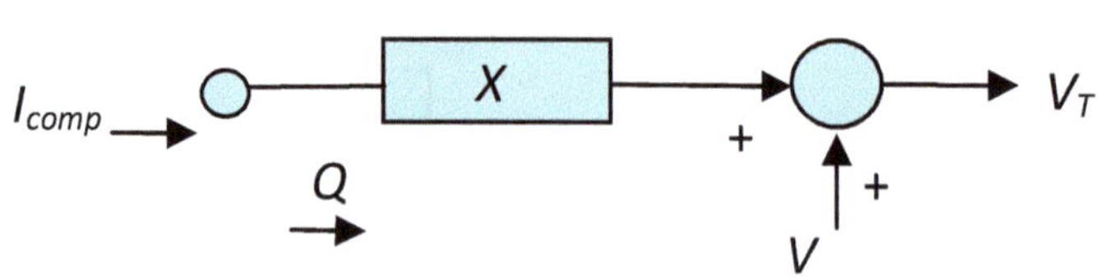

Fig. 3.58 Block diagram of the equivalent grid seen by STATCOM

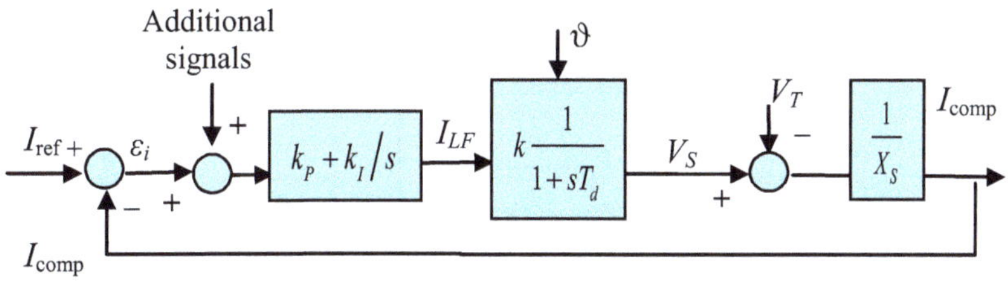

Fig. 3.59 Block diagram of the STATCOM current control loop

Continuous control of I_{comp} is obtained by imposing the amplitude of the fundamental voltage output V_S, defined by the PWM, which takes into account the phase angle ϑ required by the V_{DC} control loop. In Fig. 3.59, the error current ε_i is the input to a PI regulator that defines for a given ϑ value fundamental current reference I_{LF} and, therefore, fundamental voltage output V_S.

The dynamic link between VSI current reference I_{ref} and output voltage V_T is obtained by joining together the last two block diagrams (Figs. 3.58 and 3.59). Now the missing external voltage control loop can be overlapped with another PI regulator compounded by the additional feedback γ that imposes the slope to the STATCOM voltage control drop characteristic. The complete dynamic scheme, shown in Fig. 3.60, should include measuring filters on the feedback measurements. PI regulator parameters largely determine current and voltage control loop speed and stability. Obviously, the inner control loop is faster (tens to hundreds of millisecond) than the outer control (a few seconds).

In the figure:

- T_d is the VSI transportation lag, on the order of 1–2 ms;
- γ is the regulation slope, typically $\ll X_S$;
- I_{Cmax}, I_{Lmax} are maximum values of the VSI delivered and absorbed currents, respectively.

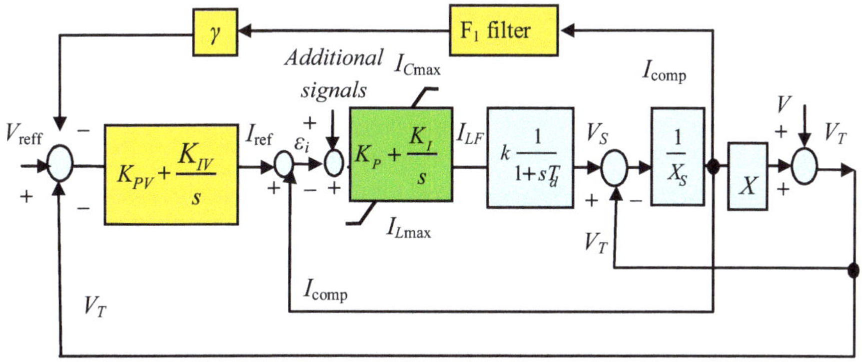

Fig. 3.60 Block diagram of STATCOM voltage control loop with additional compounding feedback γ realising the voltage drop characteristic

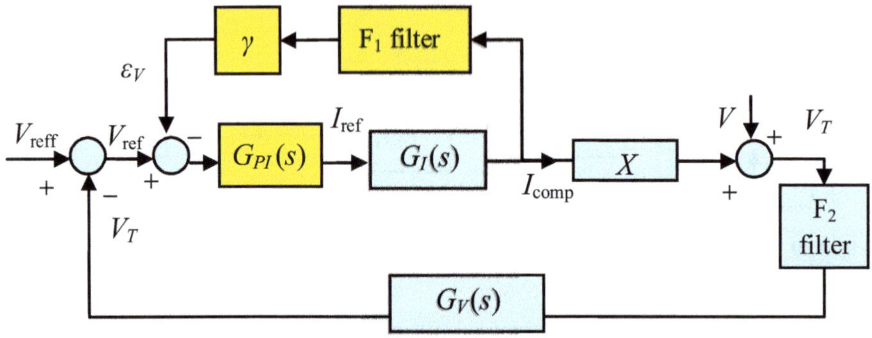

Fig. 3.61 Block diagram of the STATCOM voltage control loop with additional compounding feedback γ realising the voltage drop characteristic

A further elaboration of the Fig. 3.60 scheme gives the block diagram where the current control loop is represented by a first order $G_I(s)$ transfer function (with unitary static gain), while the V_T cumulative feedback on the input summing junction is provided by a resultant second order transfer function $G_V(s)$ (with unitary static gain) (Fig. 3.61).

The PI inner current control loop is faster than the PI outer voltage control loop; their dominant time constants are usually in a ratio of 10:1. In the case of high speed control, current and voltage loop dynamics are on the order of tens and hundreds of milliseconds, respectively. Conversely, in the case of transmission network voltage support, the two dynamics can be slowed to seconds and tens of seconds, respectively.

Proportional (K_{PV}) and integral (K_{IV}) parameters of the voltage regulator also contribute to the parameter values of the transfer function between I_{comp} and voltage error $\varepsilon_V = V_{reff} - G_V(s)V_T$, characterised by a static gain $1/\gamma$:

$$\frac{I_{comp}}{\varepsilon_V} = G_C(s) \approx \frac{1}{\gamma} \cdot \frac{(1+sT_1)(1+sT_3)}{1+sb+s^2a},$$

with $T_1 = K_P / K_I$, $T_3 = K_{PV} / K_{IV}$, a and b depending on T_1, T_3, γ and $G_I(s)$ time constants.

We should note that STATCOM dynamic behaviour and stability is a function of power system equivalent impedance, X being an integral part of the feedback loop. For this reason, control is usually optimised for maximum expected system impedance (minimum short circuit capacity). This means voltage control response time will be somewhat longer if system impedance is below the maximum value. With a practical STATCOM, the worst case response time is typically in the range of tens of milliseconds.

From the closed loop transfer functions:

$$\frac{V_T}{V_{\text{reff}}} = \frac{G_C(s)X}{1+G_C(s)XF_2(s)G_V(s)}, \quad \frac{V_T}{V} = \frac{1}{1+G_C(s)XF_2(s)G_V(s)},$$

$$\frac{V_T}{I_{\text{comp}}} = \frac{X}{1+G_C(s)XF_2(s)G_V(s)}.$$

System eigenvalues (poles) given by the characteristic polynomial clearly depend on γ and X parameter values, but they also depend on the static and dynamic characteristics of $G_C(s)$, $F_2(s)$ and $G_V(s)$.

Under steady state conditions ($s \to 0.0$), the above equations become:

$$\frac{V_T}{V_{\text{reff}}}(\infty) = \frac{X}{\gamma_1 + F_2 X} = \frac{X}{\gamma + X}, \quad \frac{V_T}{V}(\infty) = \frac{\gamma}{\gamma + F_2 X} = \frac{\gamma}{\gamma + X},$$

$$\frac{V_T}{I_{\text{comp}}}(\infty) = \frac{X\gamma}{\gamma + F_2 X} = \frac{X\gamma}{\gamma + X}.$$

These results confirm that as slope becomes smaller ($\gamma \to 0.0$), terminal voltage remains constant with respect to variations in V (V_T depends in an integral way on V_{ref} only). Therefore, any change in I_{comp} due to V recovers V_T up to the value imposed by V_{ref}.

Similarly, as slope increases ($\gamma \gg X$), terminal voltage becomes unregulated and assumes values very close to V.

Dynamic Behaviour of the STATCOM Reactive Power Control

In order to handle dynamic VAR disturbances, a STATCOM may be equipped with an automatic operating point setting control. The objective of this control is to impose STATCOM reactive power output according to a reference value that can be manually or automatically changed.

Control Scheme Based on Proper Tuning of Existing Current Control Loop

This case is linked to the control scheme already shown in Fig. 3.50, in which the inner control loop refers to the VSI output current. As mentioned before, this is also equivalent to a reactive power control loop (substituting I_{ref} with Q_{ref}) and can be tuned in a way that satisfies the required Q_{comp} slower dynamics with respect to the usually faster I_{comp} performance.

Obviously, fixing the reactive power according to the set-point Q_{ref} value (manually or automatically) requires the outer voltage control loop be open.

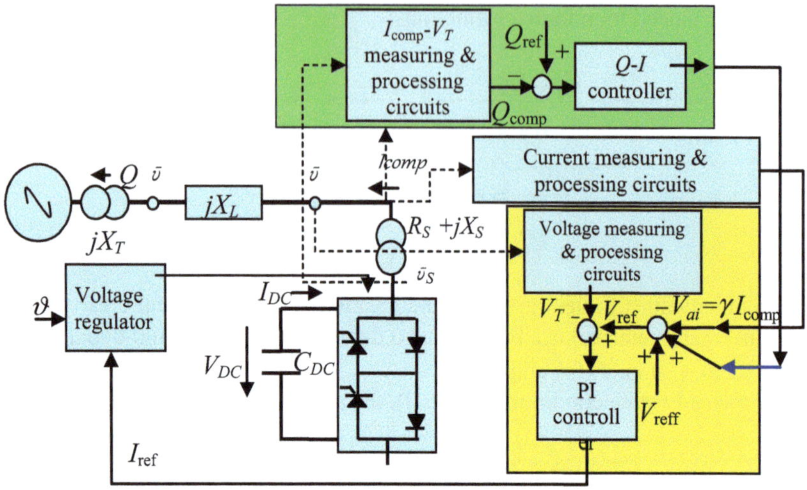

Fig. 3.62 Implementation of STATCOM reactive power control by the overlapping of the voltage (with drop) control with a slower, integral control loop that imposes reactive power output by changing HV bus reference voltage in proportion to the integral of reactive power error

Control Scheme Based on Overlapping Existing Voltage Regulation

In the case when the STATCOM inner control loop is on V_S while the outer loop is on V_T, the reactive power control requires an additional external reactive power control loop, despite the above simple scheme. This is shown in Fig. 3.62.

At a fixed VAR set-point value, when a possible disturbance results in a new STATCOM operating point with a steady VAR output, the STATCOM VAR control effectively changes the voltage reference V_{ref} in order to bring the VAR output slowly back to the set reference value. The response time of this reactive power control loop is usually slow (a few seconds) so as not to interfere with the faster inner voltage regulation or any fast stabilising or auxiliary functions that might be included in the overall VSI output control.

The magnitude of the STATCOM output reactive power $V_S I_{comp}$ is measured and compared against reference Q_{ref} (positive with reactive power injected into the grid). The error signal is sent to an integral regulator with a large time constant and added to the fixed voltage reference V_{ref}. This control forces the V_T voltage regulator input signal to change until the VSI output VARs, Q_{comp} and Q_{ref} become equal.

The above described operation of the STATCOM reserve control can be illustrated by a trajectory on the static characteristic in Fig. 3.63:

Suppose operating point A is modified by a sudden change in the system load characteristic with a consequent drop in amplitude of terminal voltage V_T, to which an instantaneous reduction on inductive current also corresponds.

- Voltage drop ΔV_T from A to B forces the input current to further reduction via the fast voltage V_T regulator, up to working point C on the *V-I* curve, where the reactive power input is near nil.

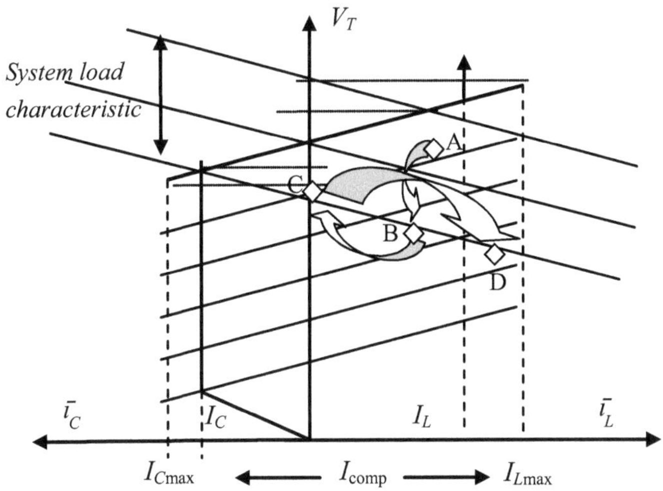

Fig. 3.63 Trajectory of STATCOM operating point following a sudden change (from *A* to *B* instantaneously) after a disturbance, recovered by the voltage control loop first (from *B* to *C*, fast control), followed by the reactive power control loop (from *C* to *D*, slow control)

- Furthermore, keeping Q_{ref} unchanged, an error signal ΔQ is generated, thus determining the slower Q input recovery by reducing the voltage set-point up to the final equilibrium point at D.

This trajectory also shows the reactive power control loop, by itself, does not appear to be the best solution to HV grid control because it could compromise the bus bar voltage profile.

In terms of the linear dynamic model, the STATCOM block diagram including the reactive power control can be achieved by overlapping the voltage control loop described in Fig. 3.61 with additional reactive power control, of integral type, introduced in Fig. 3.64. This scheme assumes positive sign on the delivered reactive power, which also depends on voltages V_T and V, as well as on operating equivalent reactance X:

$$Q_{\text{comp}} = \frac{V_T^2 - V_T V}{X}, \quad \Delta Q_{\text{comp}} = \frac{2(V_T)_0 - V_0}{X} \Delta V_T = \frac{(V_T)_0}{X} \Delta V_T.$$

Accordingly, the STATCOM dynamics, including the reactive power control, can be analysed by the linear control model in Fig. 3.65.

With a practical STATCOM, the time response of the reactive power control loop is slower and decoupled with respect to the voltage control loop, characterised by a dominant time constant in a range of 1–5 s. Parameters γ and X contribute mostly to voltage control loop performances, while μ and X contribute to fixing the reactive power control loop speed. The slower dynamic is typically required when the STATCOM contributes to the transmission network-coordinated voltage control or secondary voltage regulation (SVR), as described in Part II of this book.

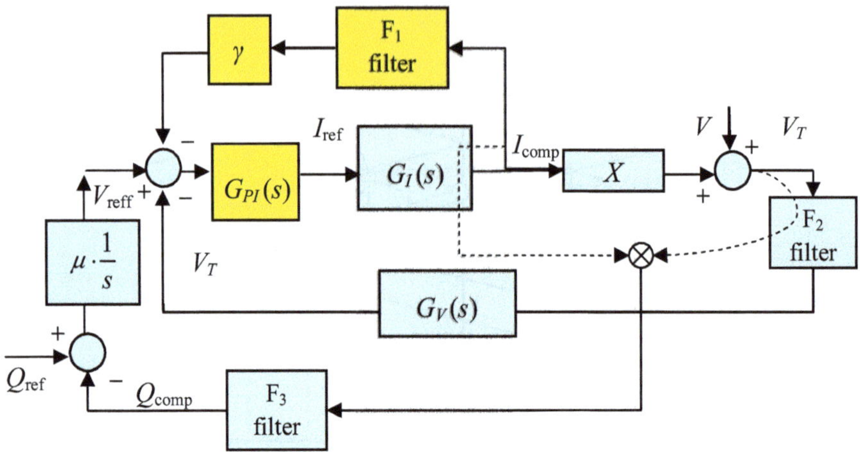

Fig. 3.64 Block diagram of STATCOM reactive power control loop overlapping the faster voltage control with additional compounding feedback γ realising the voltage drop characteristic

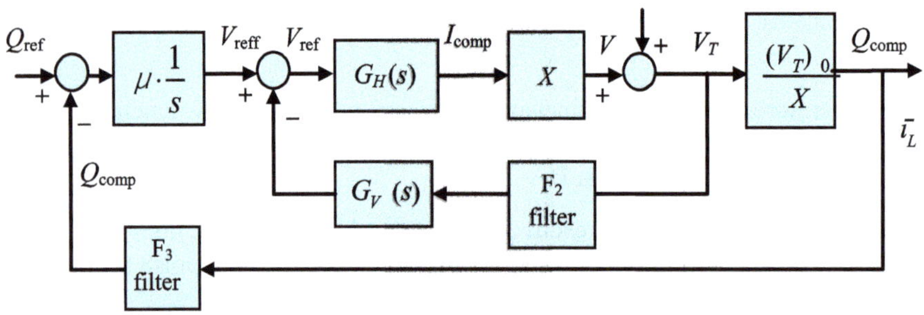

Fig. 3.65 Linear model of the STATCOM reactive power control loop overlapping the faster voltage control. The additional compounding feedback γ realising the voltage drop characteristic can be easily added per Fig. 3.64

To sum up, the simplified linear model of the STATCOM (Fig. 3.66) can be represented with high order transfer functions for both voltage and reactive power controls. The second control, usually switched off when the STATCOM is operating on the local voltage only, can be put into operation in case the STATCOM contributes to SVR.

Control Scheme with Reactive Power Loop in the Place of the Voltage Regulation

In this alternative control scheme (see Fig. 3.67), a manual or automatic switch selects one of the reactive power or voltage control loops. Analogously to SVC, this possible solution simplifies the model analysis but requires the output tracking between the two integral regulators in a way that achieves bumpless switching between the two control loops.

The linear model through which we analyse the reactive power control loop dynamics in Fig. 3.67 is shown in Fig. 3.68.

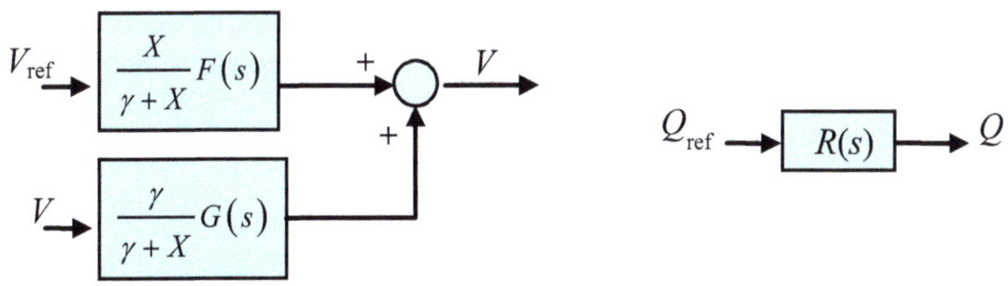

Characterised by dominant time constants:
$G(s)$, $F(s)$: $T_V = 30$–100 ms; $R(s)$: $T_Q = 1$–5 s.

Fig. 3.66 Simplified linear models of the STATCOM voltage and reactive power control loops in presence of additional compounding feedback γ realising the voltage drop characteristic

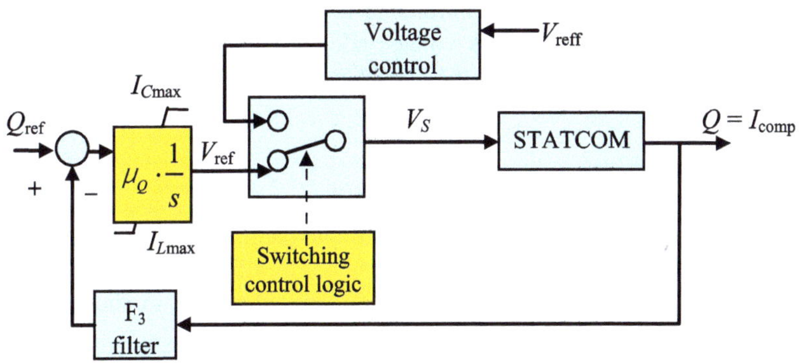

Fig. 3.67 STATCOM reactive power control to be activated by switching from voltage control

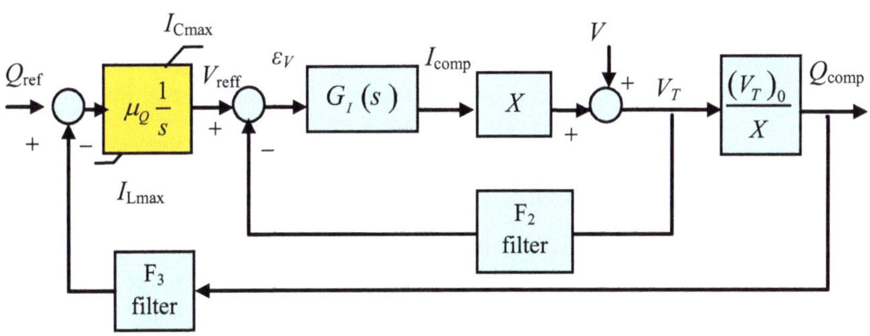

Fig. 3.68 SVC reactive power linear control to be activated by switching from voltage control

This is a simple integral control loop with a dominant time constant T_Q determined by the integrator. Therefore, control gain μ_Q is dimensioned in a way that achieves $T_Q = 1$–5 s around the normal operating point of both the equivalent impedance Z seen by the STATCOM and the VSI transformer reactance X_S.

As in the case of the previous scheme, the simplified linear model of STATCOM reactive power control can be represented by a first order transfer function, as in the figure:

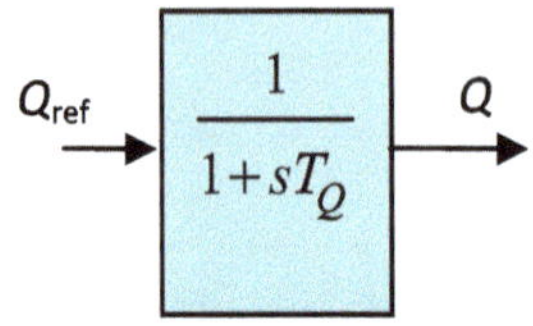

3.3.8 Automatic Voltage-Reactive Power Control by UPFC

As already seen in Chap. 2, § 2.3.6, the unified power flow controller provides an output voltage that is fully controllable in amplitude and phase, which can be "vectorially" added to network voltage allowing independent control of both reactive and active power flows in the line where the UPFC is operating.

UPFC Control Schemes

UPFC control can be seen as subdivided into two parts: converter system control and power system control. Converter system control includes current balance between converters, DC side voltage control, AC output control of shunt converter voltage and series converter voltage control.

In keeping with the scope of the book, we are more interested in UPFC power system control, involving execution of regulating functions that support local grid performance. As mentioned previously, these functions essentially consider: active and reactive power flow control on the transmission line operated by the UPFC and voltage regulation at the UPFC HV bus.

Starting from these baseline controls, more complex functionalities, like transient stability and voltage stability, can be pursued by UPFC, as already introduced in § 2.3.6.

Two control parameters are required for simultaneous active and reactive power control by each UPFC converter:

- In the case of the shunt UPFC branch, active power exchange with the grid primarily depends on the phase shift of the converter output voltage with respect to the local HV bus voltage. The reactive power flow of the shunt converter is instead controlled by varying the amplitude of the converter output voltage.
- In the case of the series UPFC branch, the active and reactive power flows in the transmission line are influenced by the phase angle and amplitude of the series injected voltage. Therefore, the active power controller can significantly affect the level of reactive power flow and vice versa.

In principle, with three control variables (i.e., Q_S of the shunt converter and the module and phase of phasor V_{pq}) it should be possible to regulate the V_T amplitude, the Q_{pq} delivered by the H bus to the receiving-end bus and the active power P transferred to the receiving-end bus.

Obviously, there will be strong interaction among these control loops, requiring a proper control system design. Consideration of the strong dependence between reactive power flow control in the line (Q) and reactive power control at the sending end (Q_t) (the local bus under regulation by the shunt inverter), a simplification must be made by selecting one of the two reactive power controls to be maintained in operation (see Fig. 2.41).

Furthermore, a high decoupling of active and reactive power current control is only possible through sophisticated control laws of predictive type.

UPFC Shunt Converter Control

Figure 3.69 shows the block diagram of the dynamic model seen in Fig. 2.41, where it is shown that the shunt converter has essentially two control loops: the AC bus voltage and the DC internal voltage controls. The components of converter output along the direct and quadrature axes are controlled on the basis of shunt current measure (I_S), and computation of its direct and quadrature component conversion. Through a PI control law the I_S direct axis component contributes to the DC voltage control and therefore to the active power control.

Another independent PI control branch operates on the error on the I_S quadrature axis component contributing to reactive power control, that is, to AC bus voltage control. In Fig. 3.69, measurements V_T and I_S come from the AC grid model, whereas the V_{DC} measurement comes from the converter DC side.

UPFC Series Converter Control

Again referring to Fig. 2.41, the series converter has essentially two control loops, modelled through the block diagram in Fig. 3.70 and related to the active and reactive power flow in the transmission line. The components of converter output (V_{pq}) along the direct and quadrature axes are controlled on the basis of the shunt current measure (I) and computation of its direct and quadrature components.

Through PI control laws, the I direct axis component I_d contributes to active power P regulation and therefore to transmission angle control. Another independent control branch through a PI control law on the error on the I quadrature axis component I_q contributes to line reactive power flow regulation and therefore to the local AC bus voltage control V_H. Alternative to Q control is V_H direct control, achievable by substituting Q and Q_{ref} with, respectively, V_H and $V_{H\text{ref}}$.

In conclusion, as far as grid voltage support is concerned, the UPFC can make a useful contribution either by operating the local high side voltage control or by controlling the amount of reactive power transferred along the line to the receiving-end bus.

The production/absorption of UPFC reactive power within its own circular capability makes power transfer control possible. From this perspective the UPFC

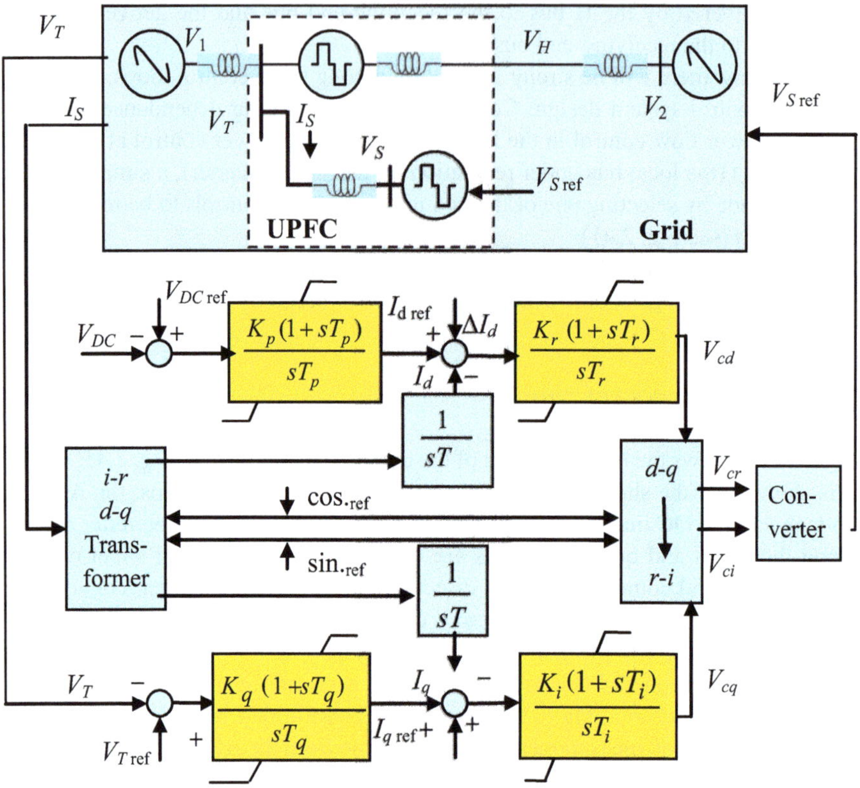

Fig. 3.69 Control scheme of UPFC shunt converter

can be considered a generator in terms of delivery or absorption of reactive power. Furthermore, the UPFC, like a generator, can contribute to secondary voltage regulation. For this to happen, the additional information needed (with respect to the equivalent generators seen by the UPFC) is the line direction to which the required reactive power is delivered or from where it is absorbed.

The UPFC general control scheme in Fig. 2.41 gives a practical realisation of the previously mentioned grid voltage support functionalities. It shows the two converters' distinct controls, illustrating how dynamic interaction between them is unavoidable. Shunt converter control requires local voltage and reactive power real-time measurement to operate V_S amplitude regulation through I_S-Q_S control.

At its side, a series converter needs more information to control the module and phase of the injected V_{pq} voltage, namely, the angular phase δ between V_H and V_T and their modules (or equivalent information on the electrical variable phasor angle); and fast measurements of active P and reactive Q power transfers along the line, with positive or negative sign corresponding to the flow direction.

P regulation, as the main UPFC functionality, is always working and substantially imposes σ angle real-time variations to the injected V_{pq} series voltage. V_T and

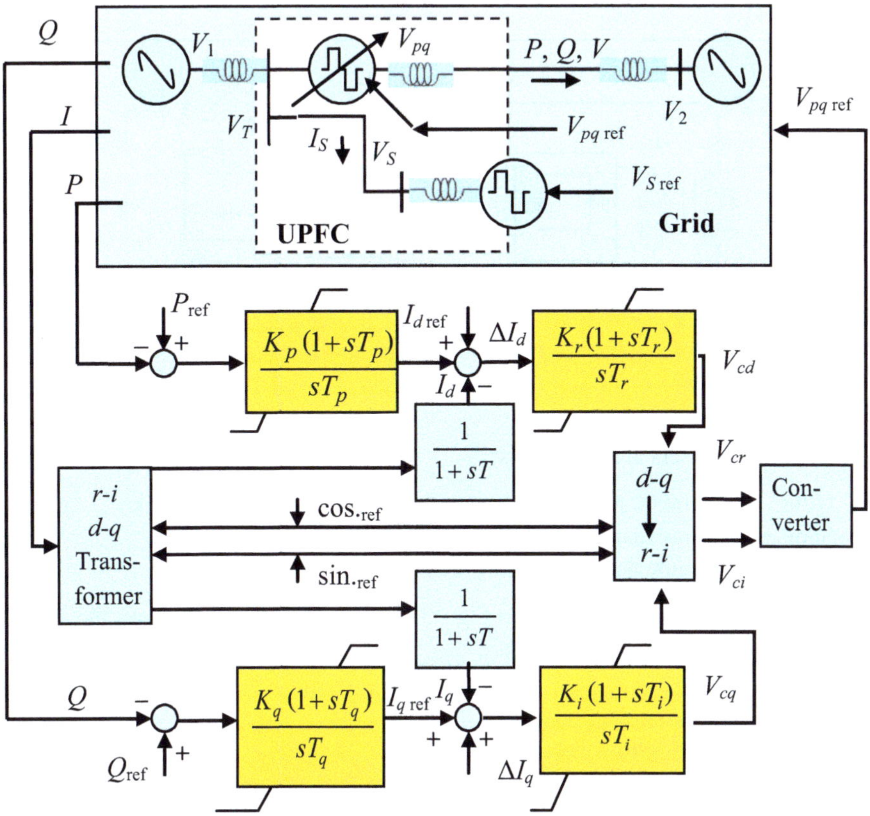

Fig. 3.70 Control scheme of UPFC series converter

Q transfer controls alternate with each other because the injected V_{pq} series voltage amplitude can be controlled to achieve a given Q or a given V_H, but both cannot happen at the same time. Moreover, the two simultaneous control loops, V_T and V_H, of the two converters are not allowed due to their strong electrical coupling: one must be excluded. Lastly, the Q control could have a continuous dynamic interaction with the Q_S control (if it is operating), with possible oscillating transients between the two very fast control loops, unless adequate control parameter tuning exists.

Hence, the V_T and Q transfer regulations can simultaneously be achieved by controlling V_S with the shunt converter and V_{pq} with the series converter, as shown in the following tests.

UPFC Dynamic Behaviour

We consider the case of double transmission lines with a UPFC, in Fig. 3.71, and present some dynamic tests showing the transient response of UPFC voltage and reactive power control loops.

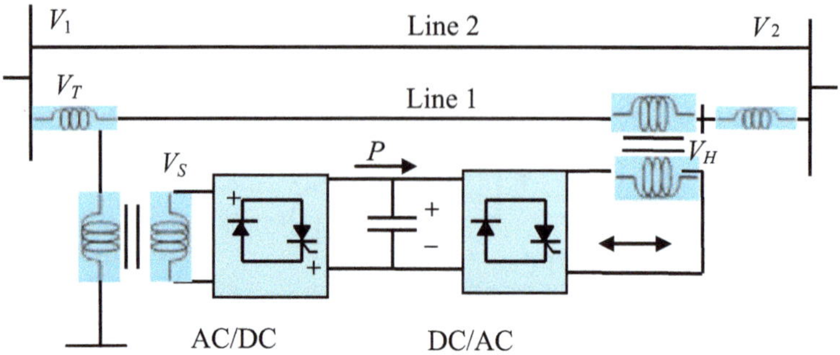

Fig. 3.71 Considered grid for tests presented on UPFC voltage and reactive power flow control loops

UPFC parameters are as follows:

- UPFC size = 100.00 MVA; C_{DC} = 1250.00 μF; V_{DC} nominal = 40.00 kV;
- Short-circuit impedance series line: 15 %;
- Short-circuit impedance shunt line: 25 %.

Referring to the control schemes in Figs. 3.69and 3.70; Table 3.3 shows the UPFC control parameters used in the tests.

Test on P Control Loop

P_{ref} in the series converter control system is increased in a 3-MW step variation, thus determining the fast transients in all UPFC control loops. Figure 3.72 shows all three transients of P, Q and V_S as being in the closed loop, with P forced to increase while Q and V_S are maintained at the constant values which are required by unchanged set-points. All the transients are very fast (a few milliseconds), with Q and V_S maintained (I control) at 0.0 and 0.99 p.u., respectively, before the P step, while P increases by the required amount of 3 MW.

Table 3.3 UPFC control parameter values

Parameters of UPFC control	Series converter	Shunt converter
K_r	0.1	0.1
T_r	0.17	0.001
K_i	0.1	0.1
T_i	0.17	1.0
T	0.01	0.01
K_P	5	100
T_P	0.1	0.6
K_q	5	20
T_q	0.1	0.001

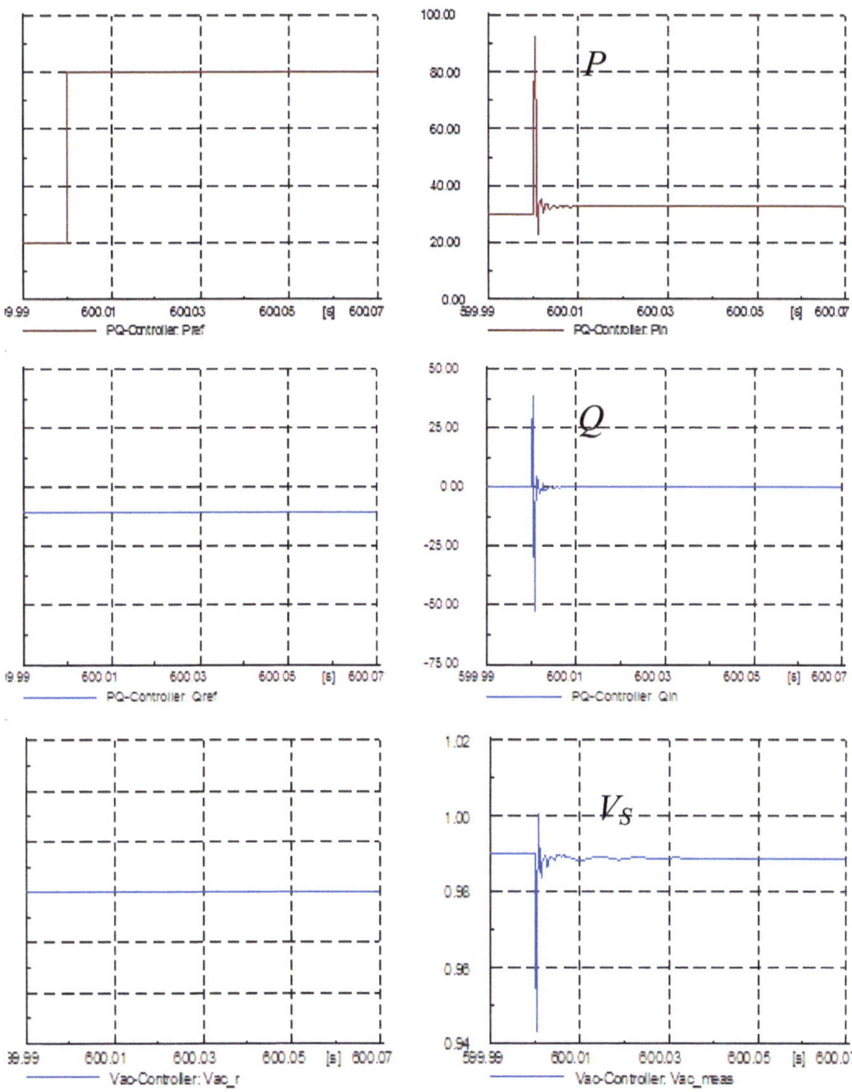

Fig. 3.72 UPFC transients following P_{ref} step increase: traces of P_{ref} and P together with Q_{ref}, Q and with V_{Sref}, V_S (V_{ac})

Test on Q Control Loop

Q_{ref} in the series converter control system increases by a 3-MVAR step variation, thus determining the fast transients in all UPFC control loops. Figure 3.73 shows all three transients of P, Q and V_S in the closed loop, with Q forced to increase, while P and V_S are maintained at the constant values required by unchanged set-points, except for small differences due to the compounding effect. All the transients are

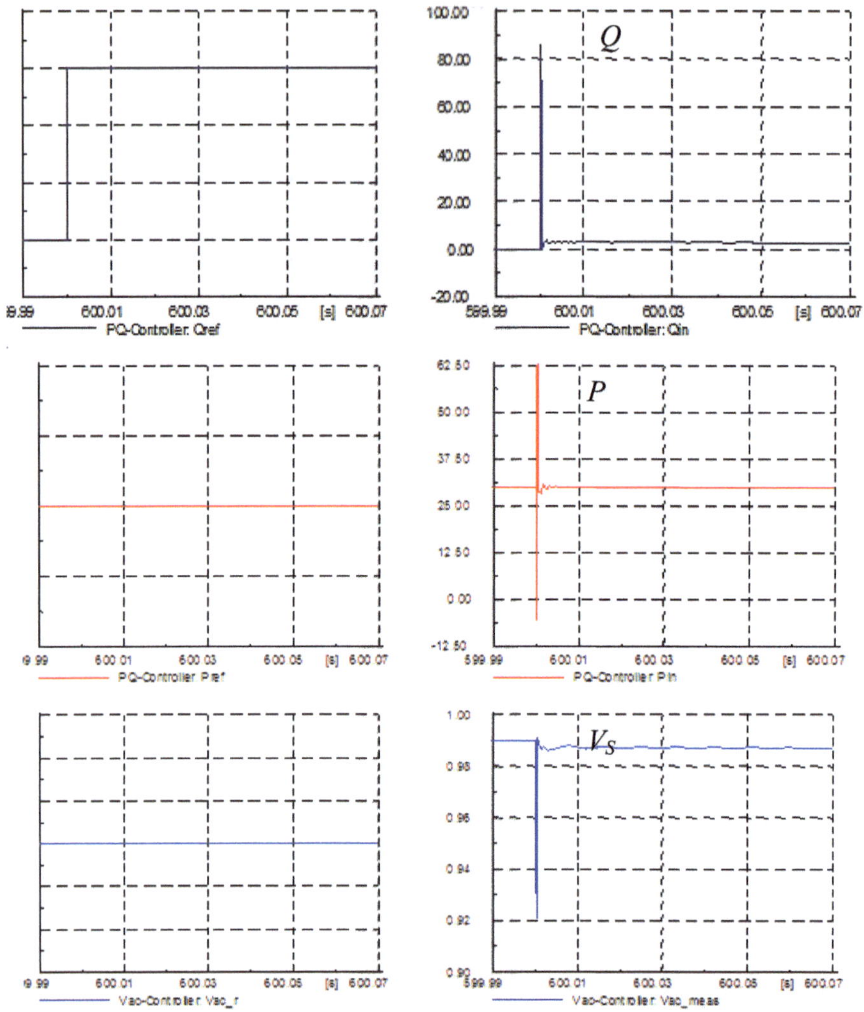

Fig. 3.73 UPFC transients following Q_{ref} step increase: traces of P_{ref} and P together with Q_{ref}, Q and with $V_{S\text{ref}}$, V_S (V_{ac})

very fast (a few milliseconds), with P and V_S maintained at the values before the Q step, while Q increases by the required amount.

Test on V_S Control Loop at the Shunt Converter

$V_{S\text{ref}}$ in the shunt converter control system is increased by a 0.5 % step variation, therefore determining fast transients in all the UPFC control loops. Figure 3.74

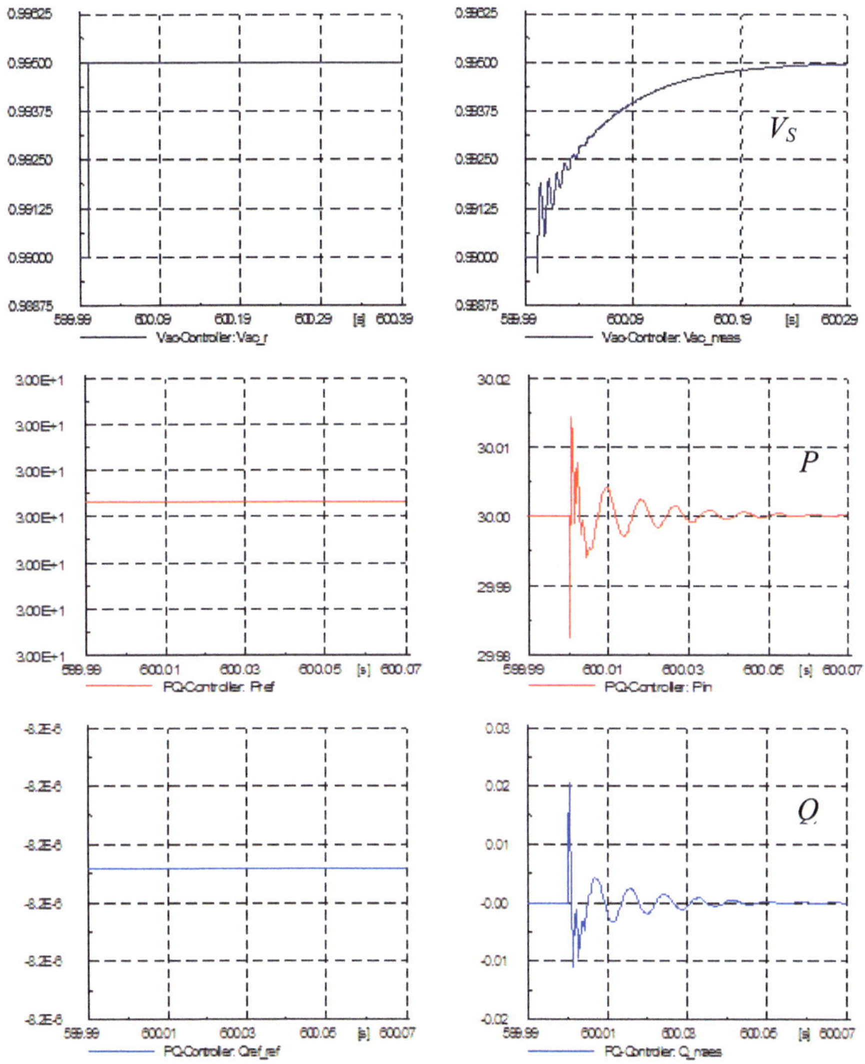

Fig. 3.74 UPFC transients following $V_S$ref step increase: traces of P_{ref} and P together with Q_{ref}, Q and with V_{Sref}, V_S (V_{ac})

shows all three transients of P, Q and V_S in a closed loop, with V_S forced to increase, while P and Q stay at the constant values required by unchanged set-points.

All these transients are less fast than before, due to the slower dominant time constant (about 100 ms) of the V_S control loop. Also, in this case the values of P and Q remain unchanged after damped oscillation, while V_S increases by the required 0.005-p.u. amount.

3.4 Conclusion

The main concerns of Chap. 3 are the control characteristics and dynamic performance of high speed voltage and reactive power control loops of those principal power system components operating on the grid buses to which they are connected. The results presented are based mostly on simplified but essential modelling and dynamic analysis of continuous closed-loop automatic control.

We mentioned, in order, the different speeds that continuous voltage control rotating synchronous generators/compensators have (dominant time constant of 400–800 ms) as compared to the faster static equipment (SVC, STATCOM, UPFC), allowing speedier dynamics by a factor of 10. Moreover, it was shown that voltage closed-loop controls are usually combined with a line drop compensation based on additional feedback from the current or reactive power flow.

In the case of synchronous generators/compensators, this compounding is not a mandatory functionality; it can be finalised to sustain the HV bus bar (positive compounding) or to stabilise generator dynamics (negative compounding). Conversely, in the case of SVC, STATCOM and UPFC, drop compensation must be in operation and based only on negative feedback. This determines the expected slope of the operating static characteristic inside the allowed current field and therefore avoids a too-fast saturation of the voltage control loop.

Line drop compensation was also presented, linking it to a physical interpretation. The following cases were considered:

- Voltage regulation at an intermediate point inside the generator's transformer elevator in the case of positive compounding;
- Voltage regulation at an internal point of the generator stator winding in the case of negative compounding;
- Voltage regulation at an internal point of the SVC, far from the HV bus bar by a reactance equal in value to the compounding feedback gain, in the case of negative compounding;
- Voltage regulation at an intermediate point inside the STATCOM transformer elevator in the case of negative compounding;
- Voltage regulation at an intermediate point inside the transformer elevator of the UPFC shunt converter in the case of negative compounding.

These physical considerations are very useful and will be revisited in the analysis of transmission network voltage control problems.

In addition to voltage control, the system components considered can provide fast closed-loop reactive power controls which deliver to or absorb from the grid. Their dynamics have often been defined by design criteria based on time-decoupling of overlapped control loops. That is, the outer reactive power control loop is designed to be slower than the inner voltage control loop and is generally characterised by a dominant time constant of about 5 s. Such a performance complies with the dynamic design of a transmission grid voltage control that is based on available and controllable reactive power resources in the field.

We should note that the proposed dynamic models are often simplified through linear representation as characterised by a dominant real mode, therefore allowing extreme simplified models of the first order. Doing so minimises the dynamic performance order of the voltage/reactive power control loops we considered, something that is also easily achievable in a real power system if correct tuning of control loop parameters is pursued.

Differences on field from the first order performances shown are possible but are mostly due to inadequate maintenance of the control loops mentioned, combined with fulfilment aspects that introduce disturbing and useless nonlinearity into the control solution.

Having covered all that is needed concerning single bus bar voltage-reactive power through local reserves, we move on to Part II and look closely at grid area voltage and reactive power control.

References

1. Corsi S, Pozzi M, Sabelli C, Serrani A (2004) The coordinated automatic voltage control of the Italian transmission grid, part I: peasons of the choice and overview of the consolidated hierarchical system. IEEE T Power Syst 19(4):1723–1732
2. Corsi S, Pozzi M, Sforna M, Dell'Olio G (2004) The coordinated automatic voltage control of the Italian transmission grid, part II: control apparatus and field performance of the consolidated hierarchical system. IEEE T Power Syst 19(4):1733–1741
3. Corsi S, Chinnici R, Lena R, Bazzi U et al (1998) General application to the main Enel's power plants of an advanced voltage and reactive power regulator for EHV network support. CIGRE conference
4. Martins N, Corsi S (eds) (2005) Coordinated voltage control in transmission systems. CIGRE Technical Brochure, Task Force 38.02.23, June 2005
5. Paul JP, Leost JY, Tesseron JM (1987) Survey of secondary voltage control in France: present realization and investigations. IEEE T Power Syst 2:505–511
6. Lefebvre H, Fragnier D, Boussion JY, Mallet P, Bulot M (2000) Secondary coordinated voltage control system: feedback of EdF. Proceedings IEEE/PES summer meeting, Seattle, July 2000
7. Sancha JL, Fernandez JL, Cortes A, Abarca JT (1996) Secondary voltage control: analysis, solutions, simulation results for the Spanish transmission system. IEEE T Power Syst 11(2):630–638
8. Corsi S, Arcidiacono V, Cambi M, Salvaderi L (1998) Impact of the restructuring process at Enel on the network voltage control service. Bulk Power System Dynamics & Control, IREP-IV, Santorini, August 1998
9. Corsi S, Cappai G, Valadè I (2006) Wide area voltage protection. CIGRE Paper B5–208, Paris
10. Corsi S (2009) Wide area voltage regulation and protection. IEEE PowerTech Conference, Bucharest, June 2009
11. Marconato R (2002) Electric power systems (Background and basic components), vol 1, 2nd edn. CEI–Italian Electrotechnical Committee, Milan
12. Marconato R (2004) Electric power systems (Steady state behaviour, controls, short-circuits and protection systems), vol 2, 2nd edn. CEI–Italian Electrotechnical Committee, Milan
13. Eremia M (ed) (2006) Electric power systems (Electric networks), vol 1. Romanian Academy Publishing House, Bucharest

14. Ferrari E, Floris R, Saccomanno F (1969) Stability limits of turbo-alternators and static exciter with different control structures. Part 1: stability analysis at small variations. LXX AEI annual meeting, Rimini
15. Arcidiacono V, Ferrari E, Marconato R, Saccomanno F (1976) Analysis of factors affecting the damping of low frequency oscillations in multivariable systems. CIGRE paper 32–19
16. Kimbark EW (1956) Power system stability, vol 3. Wiley, New York
17. Saccomanno F (1992–2003) Electric power systems: analysis and control. Wiley, New York (English version)
18. Corsi S, Pozzi M (2003) Multivariable new control solution for increased long lines voltage restoration stability during black startup. IEEE T Power Syst 18(3):1133–1141
19. Vithayathil J et al (2004) Thyristor controlled voltage regulators. CIGRE technical brochure, Working Group B4.35, February 2004

Part II
Wide Area Voltage Control

Introduction to Part II

In Part II, our main subject is enlarged, with grid wide area voltage control taking the place of the single bus voltage control discussed in Part I. It is useful to remember the important contribution of rotating generators to local voltage support: they usually constitute a large reactive power reserve in a power system, a feature not always fully utilised in the improvement of system operation security and efficiency.

A rational, effective and minimal control effort clearly requires the alignment of rotating generators inside a power station or among power stations in a given grid area, in order to sustain local voltages by avoiding dynamic/oscillating interaction among them.

It is also essential we remember the characteristics, powerfulness and limits of the high side voltage controller. The HSVC allows operation of a power station as an equivalent generator that controls the local HV bus bar voltage. Nevertheless, when more than one power station is electrically coupled, the control efforts of the corresponding HSVCs must be coordinated externally to avoid conflicting controls.

Analogously, very fast SVC, STATCOM and UPFC voltage and reactive power controls confirm them as strongly effective in local bus control, but again, they require an external coordination with any other voltage controls in the same area of influence to avoid dynamic conflicts. Moreover, the negative compounding of this equipment does not allow the local HV bus voltage control at their set-point values due to the negative or positive differences they introduce, depending on whether inductive or capacitive current flows through them.

Having acknowledged the strength, speed and limits of the local controls of the main power system component, we can more easily describe the prime control objective of the grid voltage: to complement the main local controls with a few slower voltage and reactive power controls dedicated to grid areas. Obviously, an additional task is to maintain unchanged the dynamic characteristics of the local controls they overlap.

More precisely, to increase power system stability and efficiency as well as to simplify and improve power system operation and protection, it is undoubtedly necessary that we at least find and analyse control solutions for the entire power system or large subsystems of it.

Chapter 4
Grid Hierarchical Voltage Regulation

4.1 Structure of the Hierarchy

4.1.1 Generalities

Hierarchical systems based on HV grid subdivision into areas and on automatic co-ordination of each area's reactive power resources aiming to control local voltages have been investigated in Europe (mainly in Italy and France) since 1980. These systems are collectively termed either *coordinated voltage regulation* (CVR), to highlight the required coordination among area control resources, or they are also called *secondary and tertiary voltage regulations* (SVR and TVR), to emphasise the different layers of the control hierarchy. Reference studies and applications come from Italy [1–9] and France [10–12], followed by Belgium [13, 14], Spain [15, 16] and more recently by United States, Brazil, Taiwan, South Korea, Romania and South Africa [17–22]. An international CIGRE task force investigated the subject and, in 2005, published an extensive report [23].

After 1990, based on experimental applications in *voltage wide-area regulating systems* (V-WAR), certain European countries (primarily Italy and France) employed general applications of V-WAR as their national real transmission system. These projects lasted many years, for a great many reasons: their novelty; SCADA-EMS linked updates at dispatcher control centres; unbundling of transmission and generation companies; and the growing emphasis at the beginning of 2000 on energy market rules in spite of improvements in power system control.

With changing utility organisation, and under the impetus of the energy market liberalisation, hierarchical voltage control systems are becoming stronger and more appreciated [23–44], mainly where they are already operating, but elsewhere, too, as knowledge about them grows. System operators, in fact, recognise that SVR and TVR simultaneously permit both the simplification of automatic control of transmission network voltages overall by increasing system efficiency and stability and the distinction of the contributions of different participants to voltage ancillary service in correct, simplified ways.

S. Corsi, *Voltage Control and Protection in Electrical Power Systems*,
Advances in Industrial Control, DOI 10.1007/978-1-4471-6636-8_4

Meanwhile, North American interest in V-WAR or more simplified local power plant solutions, like high side voltage control, has grown since 2000 [17]. At Bonneville Power Administration, wide-area voltage control/protection has been under testing on the site since that time, based on the coordination of generators or load tripping, reactive power switching, TCSC/SVC modulation, power-plant high side voltage schedule and OLTC tap-changing [18]. The US–Canada Blackout in 2003 gave a significant push to V-WAR studies and development [43] under the encouragement of NERC and power system authorities, and a number of projects were thus started (PJM-GM 2011 [44]). EPRI also entered into promotion of V-WAR, proposing to electrical companies in 2011 a project on feasibility studies for advanced voltage control (AVC). Since 1990, Brazil has considered SVR concepts [19]; there, voltage control is proposed for critical distribution areas, too, based on a guiding methodology of OLTC assessment and coordination [20]. More recently, China has shown great interest in SVR applications, as it has declared in CIGRE documents, and consequently has been implementing steps in this direction [36, 37]. In the East–Far East region, emerging countries such as Taiwan, South Korea and Malaysia have already performed feasibility studies aimed at applying SVR to their transmission grids. South Africa has also deeply studied the application of SVR to increase the security and efficiency in its ESKOM power system.

Progress and trends toward improvements in transmission network voltage control require, at the beginning of the new millennium and under the impetus of worldwide *Smart Grids* development, an important evolution coming mostly from still operating "manual control" to innovative "automatic control", through simple, effective closed-loop regulating systems, managed and supervised directly by electrical power system dispatching centres.

Cost/benefit analyses strongly support this innovation [25–27]. Moreover, because voltage control is mainly a local problem, any feasible solutions must consider automatic coordination of local reactive power resources, primarily those of generators and compensators but also of shunt capacitors and reactors, SVCs, STATCOMs, UPFCs and OLTCs. For this reason, the objectives of voltage ancillary service (quality and security improvements in network operation) can be pursued through a decentralised voltage control system which coordinates resources at each power system area/region. Such coordination requires data and signal exchanges between regional dispatcher and local plants/substations. The best policy was understood: "The more exchange there is of real-time electrical data in accordance with power system dynamics, the more improvement there will be in a voltage control system's performance and effectiveness".

The benefit of network voltage control in terms of grid efficiency is more strongly linked to inter-area coordination rather than local control coordination, requiring effective data and signal exchange among regional dispatching centres and the central/national system control centre. In addition, exchange of measurements with neighbouring utilities (e.g., edge-bus voltages and tie-line reactive power flows) and coordination of mutual control actions is also very important for reducing system losses.

Within the framework of energy sector liberalisation and ancillary market competition, on-line and real-time monitoring of the performance of actual EHV control systems also represents a challenging opportunity for a proper and undoubtable recognition of a power plant's contribution to voltage service [26]. Definite improvements that result from coordinated "automatic" real-time voltage regulation can then be summarised as follows:

- Power system operation quality is improved, in terms of reduced variations around the defined voltages profile across the overall grid;
- Power system operation security is enhanced, in terms of larger reactive power reserves kept available by generating units for facing emergency conditions [28];
- Transfer capability of the power system is improved, in terms of increased active power levels transmissible, with reduced risk of voltage instability and collapse [29];
- Power system operation efficiency is enhanced, in terms of active loss minimisation, reduction of reactive flows and better utilisation of reactive resources;
- Controllability and measurability of voltage ancillary service is simplified, in terms of defining functional requirements and performance monitoring criteria [26].

Voltage and reactive power control of an electrical grid requires geographical and temporal coordination of many on-field components and control functions achievable by a hierarchical control structure. A real-time and automatic voltage control system can, in fact, be basically structured in three hierarchical levels: *primary* (component control), *secondary* (area control) and *tertiary* (power system control and optimisation) levels.

Figure 4.1 gives a preliminary spatial view of the three overlapping hierarchical levels of a voltage -reactive control system. It also shows the interaction of the *tertiary* level with the not-real-time and off-line forecasting level based on state estimation and optimal power flow (OPF). This scheme distinguishes real-time levels with automatic closed-loop voltage and reactive power controls from day-before or short term optimal forecasting computation (necessarily delayed with respect to real-time power system operating conditions). In doing so, it offers clarity that helps us recognise relevant differences between real-time and forecasting levels.

It often happens that the *tertiary* voltage (closed-loop) control is confused with the static optimisation problem of voltage-reactive power, which must be considered (due to its long delay with respect to a system's operating conditions) as open-loop control or off-line forecasting related to system operation scheduling.

The most commonly employed OPF objective function is power system loss minimisation, which forecasts, by the use of not-real-time data, the generators' reactive power scheduling in order to maintain appropriate voltage levels within a power system's normal operating limits.

Obviously, higher performances are obtained with an automatic closed-loop voltage control that minimises losses in real time (such as by TVR) in comparison with a system operation based simply on a forecast computation linked to past working points (e.g., by OPF). In fact, a present grid operating condition could be, at times, very different from a forecasted one, mainly during critical operating conditions or in case of a great delay in the computation of a reliable state estimation.

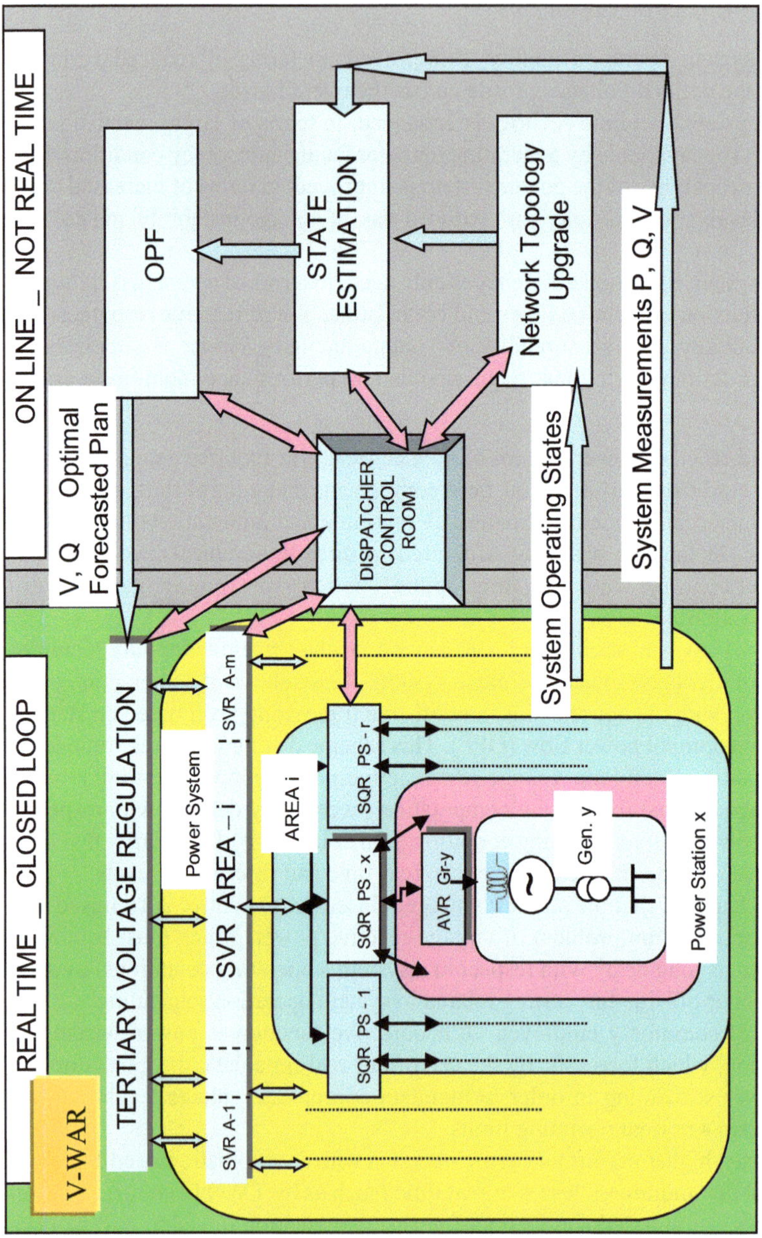

Fig. 4.1 From not-real time to real time and closed loops (Tertiary, Secondary and Primary Levels) voltage and reactive power control levels in power systems; Monitoring + manual control: ⟺; Automatic closed-loop control ⟺.

The "not-real-time" OPF voltage-reactive power issue is, in any event, a useful input reference for TVR computing, as shown in Fig. 4.1 and later fully described. In Fig. 4.1, the SVR level is subdivided into two parts: the decentralised voltage regulation in the system areas (SVR) overlaps the power station layer (secondary reactive power regulation, SQR), which controls the rotating generators, SVC, STATCOM and UPFC reactive powers.

A dispatcher can interfere with the main control levels, mostly with not-real-time levels. He can also switch off the TVR and manually define SVR pilot node voltage set-point values, but in this case he renounces on-line real time system optimisation as well as the stability benefits deriving from TVR. On the contrary, dispatcher manual control inside the SVR level is to be avoided and is very dangerous for system security due to the criticality of the manual reactive power control at the high control speed provided by SVR.

In other words, SVR should be fully automatic, while a manual TVR can be managed by the dispatcher's operator who, so doing, renounces the high reliability and efficiency that automatic TVR provides.

4.1.2 Basic SVR and TVR Concepts

The basic concepts from which SVR grew in Europe are summarised here to aid the understanding of the proposed control system and the reasons for its structure, performance features and advantages:

a. The idea of automatically controlling in real time hundreds of transmission bus voltages is too complex, critical, unreliable and therefore unrealistic and uneconomical;
b. Generating unit reactive power is, obviously, the main resource already available on-field, is low cost and is simple to control for network voltage support;
c. A realistic, simple voltage control system should consider the dominant buses only (a small number among the strongest ones), so allowing a sub-optimal but feasible and reliable control solution;
d. The dominant bus "pilot node" idea becomes solid when joint buses are assumed to be those having high electrical coupling and voltages close to each other within a "regulation area";
e. The control structure, depending on the grid's subdivision into "regulated areas", automatically and as much as possible independently regulates each pilot node voltage;
f. The control resource is essentially based on the reactive powers of the largest units in the area ("control plants"), which mainly influence the local pilot node voltage. The basic principle of TVR comes from the need to increase system operation security and efficiency through a centralised coordination of the SVR decentralised structure;

g. Pilot node voltage set-points must be adequately updated and coordinated with dynamics slower than SVR transients, considering the real condition of the overall grid and avoiding pointless and conflicting interarea control efforts;
h. Pilot node voltage set-points can be computed and updated in real time, considering the global control system structure and its real-time measurements;
i. Pilot node voltage set-points must be optimised in real time for minimising grid losses, always preserving control margins. This can be achieved by updating the optimal forecasted (not real-time) plan according to a real-time system working condition.

It must be pointed out that, notwithstanding the objective of minimising control system complexity, the effort for achieving an effective hierarchical control system is still relevant when a large transmission network is involved, as confirmed by experiences and existing applications already undertaken.

On the one hand, a new power plant apparatus (such as SQR [6]) is needed for controlling the reactive power production of generating units at each power station, but also of synchronous compensators and FACTS, by following the control requests of local the bus bar HSVC or the remote pilot node voltage regulator (SVR/RVR) and taking into account the instantaneous available capability of plant generators or compensators.

On the other hand, a specific regional/central dispatcher regulator (called *secondary voltage regulator*, SVR or *regional voltage regulator*, RVR) [4] is required for automatically maintaining pilot node voltages at their scheduled values, controlling by fast telecommunications new power plant apparatuses (such as SQR), turning reactor banks and shunt capacitor on/off, and ordering OLTCs and FACTS controllers set-points.

Lastly, a new voltage and reactive power optimising regulator (the TVR) is required at the national/utility control level, for coordinating and updating, on-line and in *real time*, all pilot node voltage set-points (see also Fig. 4.2).

All these special, unconventional control apparatuses require a specific design. Moreover, telecommunication speed for data exchange among *primary, secondary* and *tertiary* levels is high, on the order of one-second delays (all included) between any two levels and should require specific/dedicated telecommunication apparatuses and media.

4.1.3 Primary Voltage Regulation

Primary regulation involves controlling the local voltage of synchronous generators and synchronous/static compensators with the major objective of allowing correct and secure operation of the equipment. Obviously, primary voltage control impacts the transmission network, mainly at the MV buses to which generators and compensators are connected, by sustaining the local medium voltages during normal and perturbed operating conditions. Control actions are based on local measurements and aimed to bring out the voltage at the set-point value automatically, with a dynamic performance characterised by a dominant time constant value within

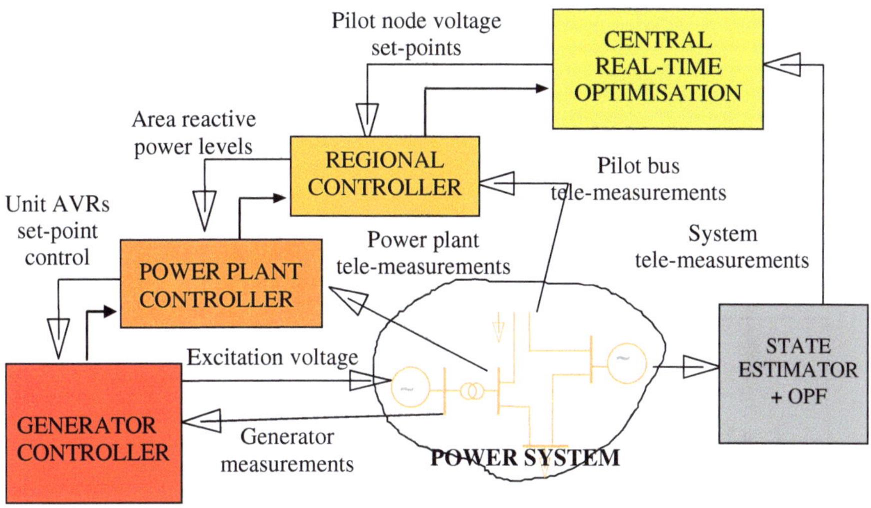

Fig. 4.2 A hierarchical structure for transmission network voltage control

a few hundred milliseconds up to one second: this compensating control must be considered a high-speed local voltage regulation.

An *automatic voltage regulator* (AVR) realises the primary voltage control of a generator (in Fig. 4.3 the unit controller). The AVR regulates voltage at the generator's terminal by controlling field excitation voltage V_f (Fig. 4.3). The AVR's basic functionality was already described at length in § 2.3.1. 2.3.2, 3.3.2, 3.3.3.

Synchronous machine operation at one of the limits, over excitation (OEL) or under excitation (UEL), recloses the correspondent V_{lim} (V_{oex} or V_{uex}) signal feedback defined by an integral regulator:

- OEL input compares actual excitation current I_f with its maximum value I_{flim}. Under steady-state operation, I_f is lower than I_{flim}, and the integral regulator doesn't participate in the primary voltage control. When I_f is greater than I_{flim}, the integral regulator generates a V_{lim} negative value that reduces the excitation voltage of the synchronous generator.

A certain transient level of the excitation windings' overload can be allowed based on the slow dynamics of the thermal phenomena. Accordingly, the stator current limiting value is transiently increased. In this way, excitation current I_f can reach very high values for a short time, being limited by the generator thermal limits (rotor and winding heating, etc.).

- UEL input compares the reactive power Q with a reference Q_{vref}. Under steady-state operation Q is greater than Q_{vref} and the integral regulator doesn't participate in the primary voltage control. When Q is lower than Q_{vref}, the integral regulator generates a V_{lim}- negative value that increases the excitation voltage of the synchronous generator.

Control actions are based on local measures and aimed at preventing operating points from overcoming generator thermal limits, with a dynamic performance

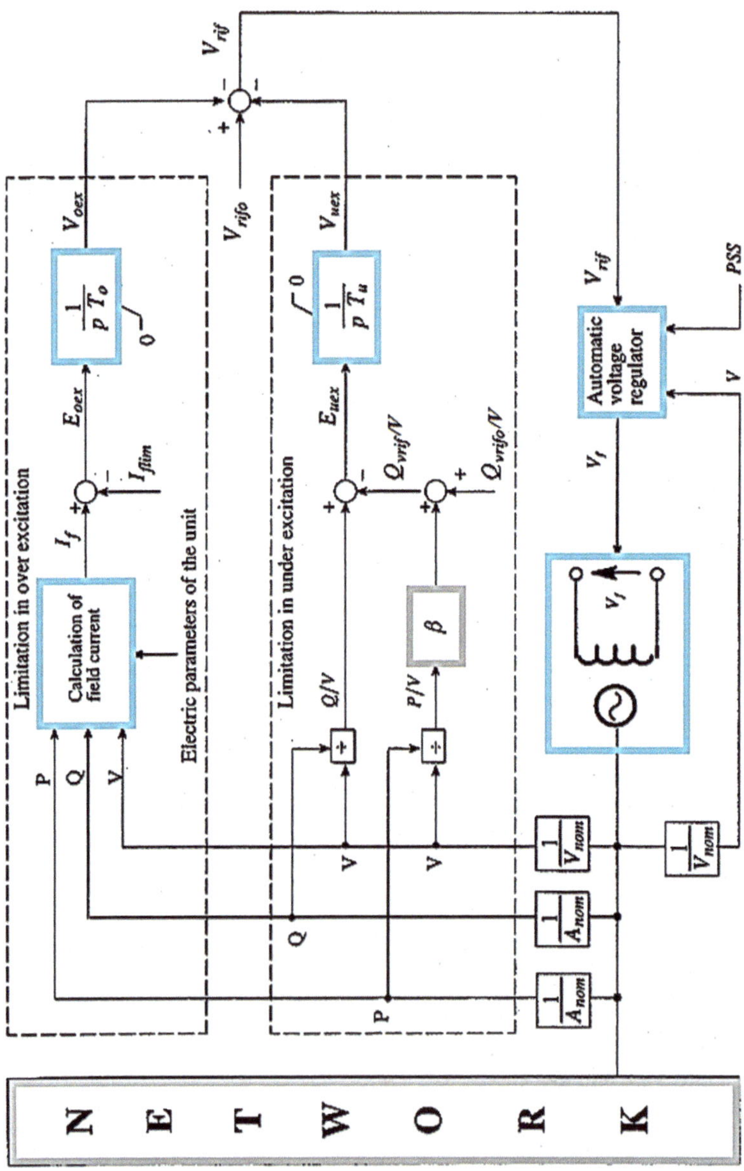

Fig. 4.3 Structure of generator AVR including over- and under-excitation limits (OEL and UEL)

Table 4.1 OEL and UEL parameters

Parameter	Value	Unit of measure
$T_{\lim} = T_o; T_u$	10	s
OEL-V_{limmin}	0	–
UEL-V_{limmax}	0	–
β	Limit drop	–

characterised by a dominant time constant value from within a few seconds to some twenty-plus seconds.

In a hierarchical automatic voltage control system, the role played by OEL and UEL limits is very important, and the shape of their curve and loop dynamics must be carefully reconstructed and taken into account by the power station regulator who controls the generators' reactive power. In fact the generator operating point must be maintained inside the operating limits during normal and perturbed operating conditions, thus avoiding any generator thermal stress and wasted control effort due to possible differences between the real and the not well-reconstructed AVR limits.

Rough values for OEL and UEL parameters are given in Table 4.1.

What follows is an example of the OEL and UEL limiting curves as shown in Fig. 4.4, illustrating the voltage dependence of the OEL curve, which shifts to the right as voltage is reduced.

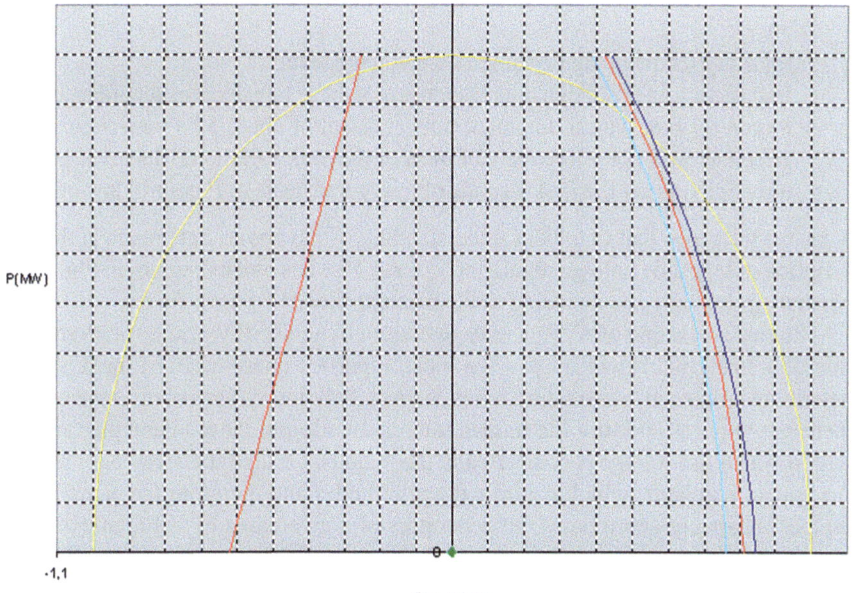

Fig. 4.4 Curves of over- and under-excitation generator limits, respectively, at *right* and *left* sides of the *P, Q-plane*, where the "Gen. A_n" power *circle* (shown as a *yellow* trace) is also shown

4.1.4 Secondary Voltage Regulation: Architecture and Modelling

Principle of Secondary Voltage Regulation

Secondary voltage regulation has, as its first objective, the automatic voltage control at a power system's main transmission buses (i.e., the most important load buses) by controlling the largest available reactive power resources on site. Therefore, primary (see § 3.3 and § 4.1.3) and secondary voltage controls have different and sometimes opposite aims.

Secondary voltage control plays an important role both during normal operating conditions and in front of contingencies:

- In normal grid operation, it ensures:
  - Maintenance of network voltages at a specified value and reduction in their variations;
  - Increase in dispatch control efficiency;
  - Coordination of real-time controls of reactive power resources;
  - Dynamic performance of first-order type to HV voltage transients, with a dominant time constant of about 50 s.
- Under disturbed conditions, secondary voltage regulation:
  - Offers timely controls of generated/absorbed reactive powers in the perturbed area;
  - Speedily recovers the perturbed area voltage level;
  - Imposes a first-order dynamic response to voltage transients in accordance with PI control law, with a dominant time constant of about 50 s (an*I*-control law effect) as well as fast recovery of most of the peak variations (due to large perturbation) during the first seconds of heavy transients (a *P*-control law effect).

A useful reference that helps our understanding of the above statements is the description of high side voltage regulation in § 3.3.5 and its block diagram in Fig. 3.24, when the pilot node is considered to be the controlled HV bus in figure.

The basic principle of SVR is voltage control of a wide HV grid through regulation of a small number of buses—the most important ones—each of them able to determine voltage in surrounding buses, so each defining its area of influence. SVR therefore requires splitting the transmission network into "low-interacting areas", within which the voltage is controlled in the main bus, called the area "pilot node". A regional regulator (which controls the pilot nodes and therefore the areas in the region) separately coordinates the generators of a given area by automatically adjusting their reactive powers to regulate the voltage of the area pilot node.

Analogously to *high side voltage regulation* (HSVR), pilot node voltage regulation consists of closed-loop control of the pilot node voltage through a PI law control, which defines an area reactive power level "q", the reactive powers of all the control power plants in the area. The secondary voltage regulator inputs the instantaneous voltage measure of the area pilot bus and compares it with the pilot node voltage set-point, determining instant by instant the reactive power level to be sent to the control power plants in the area. The reactive power level "q" therefore determines the alignment of each area's generating units, contributing in proportion to their capabilities to total area reactive power.

The automatic voltage and reactive power control of a transmission network considers the hierarchical structure shown in Fig. 4.5, where the control apparatuses are now apparent:

- In this control structure, the first hierarchical level (the primary level) consists of conventional generator voltage regulators (AVRs). These make it possible to take fast-control action in the face of local perturbations (for instance, short circuits near a generator) and thereby collectively determine the "primary" voltage regulation of the network.
- The second hierarchical level consists of power station SQR regulators, which achieve the reactive power required by the RVR or the SVR regulator at a higher hierarchical level (see the next point, on the regional controller), by operating on the primary voltage control set-points.
- The third hierarchical level consists of a slower SVR (or a few RVRs if the grid is subdivided into more than one region: for example, the case of a national dispatcher operating on-field through regional dispatchers), which regulates in an integral way the voltage of the pilot nodes by controlling the reactive power of participating power stations to the second hierarchical level.

The switching of compensating equipment such as capacitor banks and shunt-reactors or the blocking of OLTC tap-changers is part of SVR control action. It operates at each area on the local switching resources only when needed, according to the area control margin value, given by the difference of the real-time value of area reactive power level "q" with respect to its $+1$ or -1 p.u. limits. Proper thresholds of the "q" value habilitate area on/off switching according to pre-defined sequences. As for area OLTC automatic blocking, it also can be linked to SVR if the "voltage instability indicator" used is based on the SVR trend in the area [24, 39–41].

An area is defined by the bus set of the network in which voltages, for normal perturbation, are close to the voltage of a pilot node. The SVR receives its pilot nodes' voltage tele-measurements and sends the area reactive level signals separately to control power plants in each area. The reactive power level signal is the reference for the reactive power regulators (SQRs) of the power stations controlled in the area, and with respect to this reference signal, power stations deliver/absorb reactive power in proportion to their reactive capability limits; in this way all the control generators of one area have the same reactive power margin with respect to the reactive bounds. The combined control actions of the SVR (or RVRs) and SQRs determine the "secondary" regulation of regional network voltage.

As is well known, the success of the above described control scheme mainly depends on the way pilot nodes and control generators are chosen and on the coordination of the RVRs (SVR) set points possibly by a central controller.

The criterion used for locating pilot nodes is based on an intuitive assumption that they must be chosen from among the strongest in the grid. Therefore, these nodes are able to impose the voltage values of the load nodes that are electrically

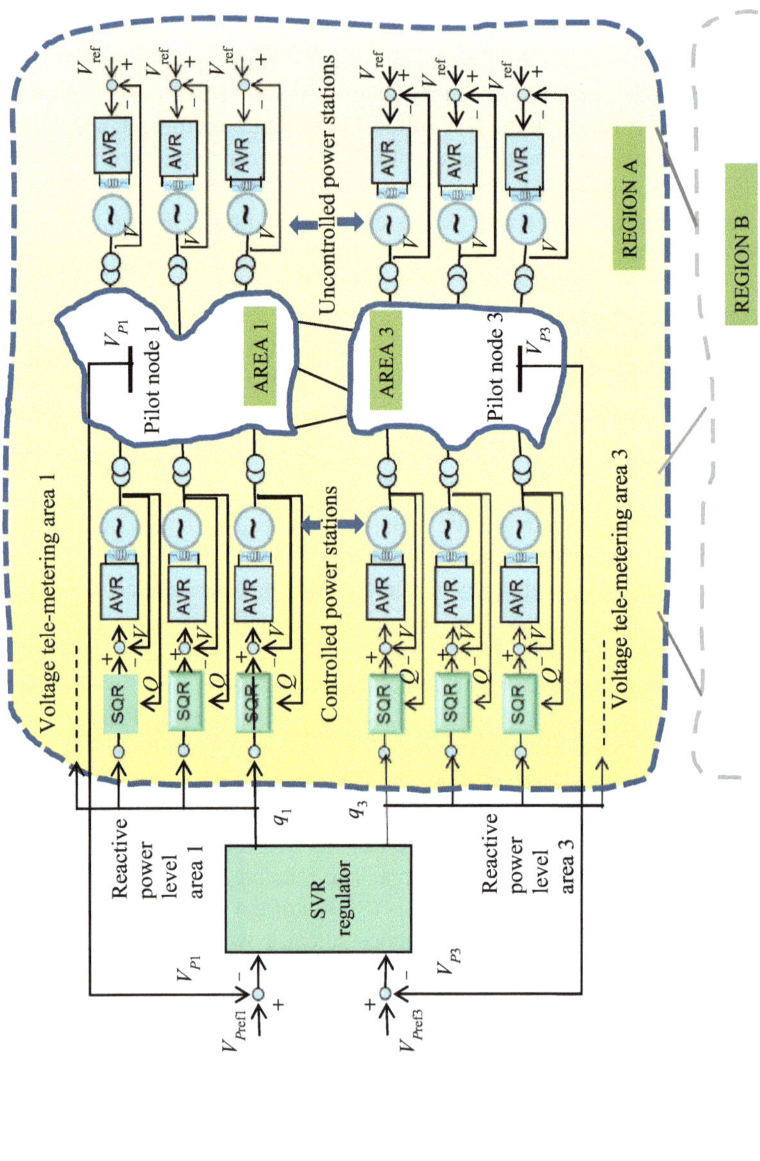

Fig. 4.5 Schematic of hierarchical structure of a secondary voltage control system

close to them. This criterion includes a constraint on electrical coupling between pilot nodes: such a coupling must be lower than a pre-established limit. In this way, in addition to avoiding problems of dynamic interaction between secondary voltage control loops, the criterion can prevent excessive exchange of reactive power among adjacent pilot nodes due to differences, even if small, between the voltage values imposed by the regulating system.

From the above considerations, it is clear that the starting point of the proposed control scheme is to avoid the relatively low electrical coupling between adjacent areas, that is, between two adjacent pilot nodes. This is not a mandatory issue with regard to performance choice, as will be shown later; nevertheless, it is strongly recommended when and where it is achievable. The case of area selection with high electrical coupling among pilot nodes requires a special "decoupling-adaptive" control.

Concerning control generators: these are chosen from among those inside each control area strongly affecting the voltage of the local pilot node. In this way, each area would be sufficiently autonomous in terms of resources to be used for local control needs and because of low electrical coupling with surrounding areas. A good selection of areas and control generators should also make evident those areas which may merit investment in new controllable reactive power resources.

Dynamic Model of Secondary Voltage Control System

A power system having connections with neighbouring grids represented by equivalent grids can be described by a linear model around an operating point written as follows:

$$\begin{bmatrix} \Delta P_1 \\ \Delta Q_1 \\ \vdots \\ \Delta P_g \\ \Delta Q_g \\ \Delta P_{g+1} \\ \Delta Q_{g+1} \\ \vdots \\ \Delta P_n \\ \Delta Q_n \end{bmatrix} = \begin{bmatrix} \frac{\partial P_1}{\partial V_1} & \frac{\partial P_1}{\partial \theta_1} & \cdots & \frac{\partial P_1}{\partial V_g} & \frac{\partial P_1}{\partial \theta_g} & \frac{\partial P_1}{\partial V_{g+1}} & \frac{\partial P_1}{\partial \theta_{g+1}} & \cdots & \frac{\partial P_1}{\partial V_n} & \frac{\partial P_1}{\partial \theta_n} \\ \frac{\partial Q_1}{\partial V_1} & \frac{\partial Q_1}{\partial \theta_1} & \cdots & \frac{\partial Q_1}{\partial V_g} & \frac{\partial Q_1}{\partial \theta_g} & \frac{\partial Q_1}{\partial V_{g+1}} & \frac{\partial Q_1}{\partial \theta_{g+1}} & \cdots & \frac{\partial Q_1}{\partial V_n} & \frac{\partial Q_1}{\partial \theta_n} \\ \vdots & \vdots & \ddots & \vdots & \vdots & \vdots & \vdots & \ddots & \vdots & \vdots \\ \frac{\partial P_g}{\partial V_1} & \frac{\partial P_g}{\partial \theta_1} & \cdots & \frac{\partial P_g}{\partial V_g} & \frac{\partial P_g}{\partial \theta_g} & \frac{\partial P_g}{\partial V_{g+1}} & \frac{\partial P_g}{\partial \theta_{g+1}} & \cdots & \frac{\partial P_g}{\partial V_n} & \frac{\partial P_g}{\partial \theta_n} \\ \frac{\partial Q_g}{\partial V_1} & \frac{\partial Q_g}{\partial \theta_1} & \cdots & \frac{\partial Q_g}{\partial V_g} & \frac{\partial Q_g}{\partial \theta_g} & \frac{\partial Q_g}{\partial V_{g+1}} & \frac{\partial Q_g}{\partial \theta_{g+1}} & \cdots & \frac{\partial Q_g}{\partial V_n} & \frac{\partial Q_g}{\partial \theta_n} \\ \frac{\partial P_{g+1}}{\partial V_1} & \frac{\partial P_{g+1}}{\partial \theta_1} & \cdots & \frac{\partial P_{g+1}}{\partial V_g} & \frac{\partial P_{g+1}}{\partial \theta_g} & \frac{\partial P_{g+1}}{\partial V_{g+1}} & \frac{\partial P_{g+1}}{\partial \theta_{g+1}} & \cdots & \frac{\partial P_{g+1}}{\partial V_n} & \frac{\partial P_{g+1}}{\partial \theta_n} \\ \frac{\partial Q_{g+1}}{\partial V_1} & \frac{\partial Q_{g+1}}{\partial \theta_1} & \cdots & \frac{\partial Q_{g+1}}{\partial V_g} & \frac{\partial Q_{g+1}}{\partial \theta_g} & \frac{\partial Q_{g+1}}{\partial V_{g+1}} & \frac{\partial Q_{g+1}}{\partial \theta_{g+1}} & \cdots & \frac{\partial Q_{g+1}}{\partial V_n} & \frac{\partial Q_{g+1}}{\partial \theta_n} \\ \vdots & \vdots & \ddots & \vdots & \vdots & \vdots & \vdots & \ddots & \vdots & \vdots \\ \frac{\partial P_n}{\partial V_1} & \frac{\partial P_n}{\partial \theta_1} & \cdots & \frac{\partial P_n}{\partial V_g} & \frac{\partial P_n}{\partial \theta_g} & \frac{\partial P_n}{\partial V_{g+1}} & \frac{\partial P_n}{\partial \theta_{g+1}} & \cdots & \frac{\partial P_n}{\partial V_n} & \frac{\partial P_n}{\partial \theta_n} \\ \frac{\partial Q_n}{\partial V_1} & \frac{\partial Q_n}{\partial \theta_1} & \cdots & \frac{\partial Q_n}{\partial V_g} & \frac{\partial Q_n}{\partial \theta_g} & \frac{\partial Q_n}{\partial V_{g+1}} & \frac{\partial Q_n}{\partial \theta_{g+1}} & \cdots & \frac{\partial Q_n}{\partial V_n} & \frac{\partial Q_n}{\partial \theta_n} \end{bmatrix} \begin{bmatrix} \Delta V_1 \\ \Delta \theta_1 \\ \vdots \\ \Delta V_g \\ \Delta \theta_g \\ \Delta V_{g+1} \\ \Delta \theta_{g+1} \\ \vdots \\ \Delta V_n \\ \Delta \theta_n \end{bmatrix} \tag{4.1}$$

where:

$Q_i =$ reactive power at node i,
$P_i =$ active power at node i,
$V_i =$ voltage magnitude at node i,
$n =$ number of grid nodes,
$g =$ number of generator nodes in grid,
$m = n - g =$ number of load nodes in grid.

The linearised model can be also written as:

$$[\Delta V] = \left[S_Q\right][\Delta Q] + \left[S_p\right][\Delta P],$$

where:

$\Delta V =$ vector of voltage modules variations,
$\Delta Q =$ vector of reactive power injection variations,
$\Delta P =$ vector of the active power injection variations,
$n =$ number of HV buses in a considered power system,
$S_Q, S_p =$ sensitivity matrices: $S_Q \in R^{n \times n}$ and $S_p \in R^{n(n-1)}$.

Taking into account decoupling between active and reactive powers and separating the equations describing the dependence in a power system of reactive powers on voltages, system Eq. (4.1) becomes:

$$\begin{bmatrix} \Delta Q_1 \\ \vdots \\ \Delta Q_g \\ \Delta Q_{g+1} \\ \vdots \\ \Delta Q_n \end{bmatrix} = \begin{bmatrix} \frac{\partial Q_1}{\partial V_1} & \cdots & \frac{\partial Q_1}{\partial V_g} & \frac{\partial Q_1}{\partial V_{g+1}} & \cdots & \frac{\partial Q_1}{\partial V_n} \\ \vdots & \ddots & \vdots & \vdots & \ddots & \vdots \\ \frac{\partial Q_g}{\partial V_1} & \cdots & \frac{\partial Q_g}{\partial V_g} & \frac{\partial Q_g}{\partial V_{g+1}} & \cdots & \frac{\partial Q_g}{\partial V_n} \\ \frac{\partial Q_{g+1}}{\partial V_1} & \cdots & \frac{\partial Q_{g+1}}{\partial V_g} & \frac{\partial Q_{g+1}}{\partial V_{g+1}} & \cdots & \frac{\partial Q_{g+1}}{\partial V_n} \\ \vdots & \ddots & \vdots & \vdots & \ddots & \vdots \\ \frac{\partial Q_n}{\partial V_1} & \cdots & \frac{\partial Q_n}{\partial V_g} & \frac{\partial Q_n}{\partial V_{g+1}} & \cdots & \frac{\partial Q_n}{\partial V_n} \end{bmatrix} \cdot \begin{bmatrix} \Delta V_1 \\ \vdots \\ \Delta V_g \\ \Delta V_{g+1} \\ \vdots \\ \Delta V_n \end{bmatrix}, \tag{4.2}$$

Furthermore, distinguishing the EHV nodes between the "*g*" called generation buses and the "*m*" for load buses, we can write:

$$\begin{aligned} [\Delta Q_G] &= -([B_{GG}][\Delta V_G] + [B_{GL}][\Delta V_L]), \\ [\Delta Q_L] &= -([B_{LG}][\Delta V_G] + [B_{LL}][\Delta V_L]), \end{aligned} \tag{4.3}$$

with the obvious meaning of the symbols:

$$[\Delta Q] = -[B][\Delta V], \tag{4.4}$$

where Q and V respectively represent reactive power and voltage vectors of the overall system. This matrix equation allows a simplified but precise enough analysis of the links between the voltage variations on buses and injected reactive powers.

Denoting by n the total number of grid buses:

- $[\Delta Q]_{n\times 1}$ is the vector of the injected reactive powers into the grid buses;
- $[\Delta V]_{n\times 1}$ is the vector of the voltages in the grid buses;
- $[B]_{n\times n}$ is the symmetric matrix of grid node susceptances, in p.u. (including lines and transformers);
- Matrix $-[B]$ represents the sensitivity of the injected reactive powers with respect to the voltages.

From (4.3) it is possible to obtain $[\Delta V_L]$ as

$$[\Delta V_L] = -[B_{LL}]^{-1}[\Delta Q_L] - [B_{LL}]^{-1}[B_{LG}][\Delta V_G].$$

Substituting $[\Delta V_L]$ into the remaining (4.3) equation:

$$[\Delta Q_G] = -\{[B_{GG}] - [B_{GL}][B_{LL}]^{-1}[B_{LG}]\}[\Delta V_G] + [B_{GL}][B_{LL}]^{-1}[\Delta Q_L].$$

Therefore, the system model becomes:

$$\begin{cases}[\Delta V_L] = [H][\Delta V_G] + [X_{CC}][\Delta Q_L] \\ [\Delta Q_G] = -[B_{\text{eq}}][\Delta V_G] + [D][\Delta Q_L]\end{cases}, \tag{4.5}$$

where the introduced matrices are defined as follows:

$$\begin{aligned}
&[X_{CC}] = -[B_{LL}]^{-1} \\
&[H] = -[B_{LL}]^{-1}[B_{LG}] = [X_{CC}][B_{LG}] \\
&[D] = [B_{GL}][B_{LL}]^{-1} = -[B_{GL}][X_{CC}] = -[H]^{\mathrm{T}} \\
&[B_{eq}] = [B_{GG}] - [B_{GL}][B_{LL}]^{-1}[B_{LG}] = [B_{GG}] + [B_{GL}][H] \overset{\text{def}}{=} -[C].
\end{aligned} \tag{4.6}$$

From (4.4) it is possible to deduce:

$$[\Delta V] = [S_Q][\Delta Q].$$

Without loss of generality, the power system buses can be divided simply into generation buses (Q_G, V_G) and load buses (Q_L, V_L):

$$[\Delta V] = [S_{QG} \vdots S_{QL}] \cdot \begin{bmatrix} \Delta Q_G \\ \Delta Q_L \end{bmatrix}.$$

Furthermore, it is necessary to distinguish between the generator buses allocated as control power station buses (Q_{GC}, V_{GC}) and those representing uncontrolled power station buses (Q_{GU}, V_{GU}):

$$[\Delta V] = [S_{QC} \vdots S_{QU} \vdots S_{QL}] \cdot \begin{bmatrix} \Delta Q_{GC} \\ \Delta Q_{GU} \\ \Delta Q_L \end{bmatrix}. \tag{4.7}$$

The uncontrolled power stations operate under primary voltage regulation only, therefore:

$$[\Delta Q_{GU}] = [K_U] \cdot [\Delta V_{GU}] = [K] \cdot [\Delta V] = [0 \quad K_U \quad 0] \cdot \begin{bmatrix} \Delta V_{GC} \\ \Delta V_{GU} \\ \Delta V_L \end{bmatrix}.$$

By substitution, Eq. (4.7) can be rewritten, with the obvious meaning of the symbols, as:

$$[\Delta V] = [S_C \vdots S_L] \cdot \begin{bmatrix} \Delta Q_{GC} \\ \Delta Q_L \end{bmatrix}. \tag{4.8}$$

SVR Control Structure

As already seen in Chap. 3 and recalled by Fig. 4.6, at the power station level, a centralised, noninteracting SVR control scheme of integral type allows dynamic decoupling among power plant unit reactive power control loops, as well as reactive power absorption/delivery of each generator in accordance with set-point values. The following ΔV_{ref}, ΔQ_{ref}, ΔQ are power station vectors of variables, whereas control matrix **A** is given by Eq. (3.11).

Therefore, at the power station level, unit reactive powers can be controlled by a dynamic decoupling control law (**A**/sT_Q) implemented into a centralised (at power station level) control apparatus (SQR) reclosing all the power station's reactive power control loops.

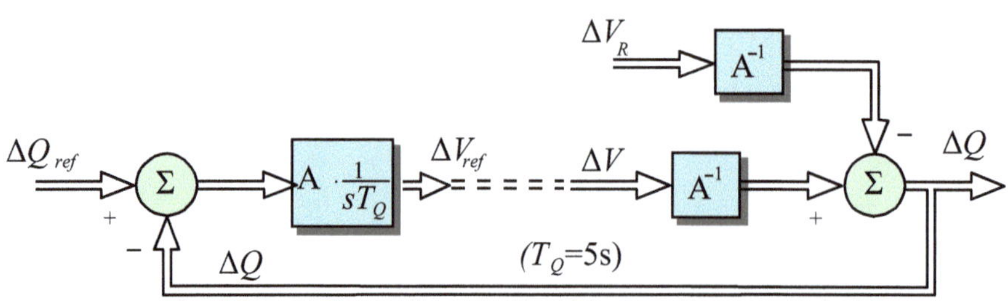

Fig. 4.6 Block diagram of power station's noninteracting reactive power control loops

Analogously, a decoupling, noninteracting control law can be applied to pilot node voltage control loops in the case of pilot node selection with high electrical coupling among them.

Let us designate by z the number of pilot nodes in the region. From (4.8), giving evidence in it of the reactive power contributions for each pilot node, the equation can be rewritten as:

$$\begin{bmatrix} \Delta V_p^1 \\ \Delta V_p^2 \\ \Delta V_p^z \end{bmatrix} = [S_{Cp} \vdots S_{Lp}] \begin{bmatrix} \Delta Q_{GC}^1 \\ \Delta Q_{GC}^2 \\ \Delta Q_{GC}^z \\ --- \\ \Delta Q_L \end{bmatrix}, \tag{4.9}$$

where vector ΔQ_{GC}^k represents the reactive power variation of control power plants in area k. More precisely, because each controlled power station must operate as an equivalent unit:

$$\Delta Q_{GC}^k = \text{Diag}\{\alpha_i\}\left|\Delta V_{\text{ref}\,C}^k - \Delta V_{GC}^k\right|, \tag{4.10}$$

where $\alpha_i \triangleq \frac{(V_{GCi}^k)_0}{x_i}$ is a constant that depends on operating point and unit transformer reactance. From Eq. (4.3):

$$\Delta V_{GC}^k = S_C^k \Delta Q_{GC}^k + S_L^k \Delta Q_L. \tag{4.11}$$

By substituting Eq. (4.11) into Eq. (4.10) it is possible to obtain, for the kth area, the dependence of the control power plant reactive powers on the set-points of the corresponding AVRs and the load variations:

$$\Delta Q_{GC}^k = \text{Diag}\{\alpha_i\}\cdot\left|\Delta V_{\text{ref}C}^k - S_C^k \Delta Q_{GC} - S_L^k \Delta Q_L\right|,$$
$$\left[\Delta Q_{GC}^k\right] = \left[H^k\right]\left[\Delta V_{\text{ref}\,C}^k\right] + \left[D^k\right]\left[\Delta Q_L\right].$$

For the overall grid, the equation becomes:

$$\left[\Delta Q_{GC}\right] = \left[H\right]\cdot\left[\Delta V_{\text{ref}\,C}\right] + \left[D\right]\left[\Delta Q_L\right].$$

Substituting this result into (4.9), the model of the schematic representation of the system under control in Fig. 4.7a is revealed. Matrices $\boldsymbol{S}_{Cp}$ and $\boldsymbol{H}$ are, in general, block diagonal dominant.

Starting from the Fig. 4.6 model, we note that the SVR control scheme of secondary voltage regulation has to include:

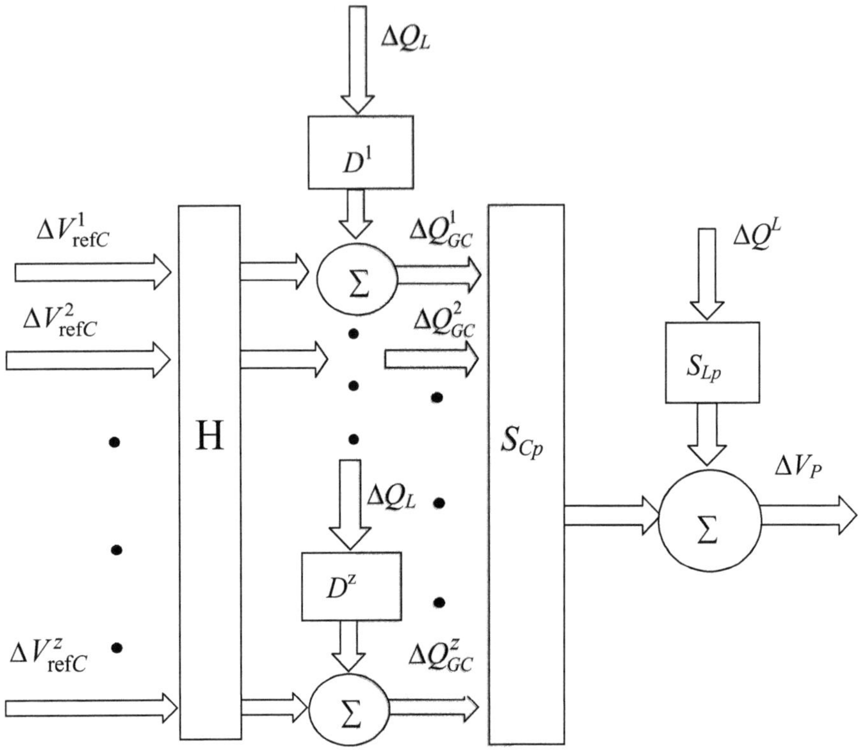

Fig. 4.7a Schematic representation of the pilot node voltages linearized dependence from load variation and power stations control

1. A power plant reactive power regulator (SQR), which provides the inner control loop to achieve the desired reactive power at the controlled power station;
2. A secondary voltage regulator (SVR) which provides the outer control loop with the objective of achieving a desired voltage profile across the grid through regulating the voltage of the pilot nodes.

A schematic representation of the secondary voltage regulation control structure is given in Fig. 4.7b, under the acceptable simplifying assumption of neglecting primary voltage control loop dynamics.

From here, g_k means number of controlled generators associated with the kth pilot node, and $g = \Sigma g_k$ is the total number of controlled generators in the grid. In the generic kth area, the reactive power regulator of the ith control generator is assumed to be of purely integral type. Its reference signal is proportional to the reactive power level q_k of the area considered. The proportionality coefficient is given by the reactive capability limit $Q^k_{\text{lim}i}$ of the ith generator.

The reactive power level q_k supplied by the pilot node voltage regulator is defined by a proportional-integral control law applied to a linear combination of the differences between the secondary voltage references and the corresponding pilot node voltages.

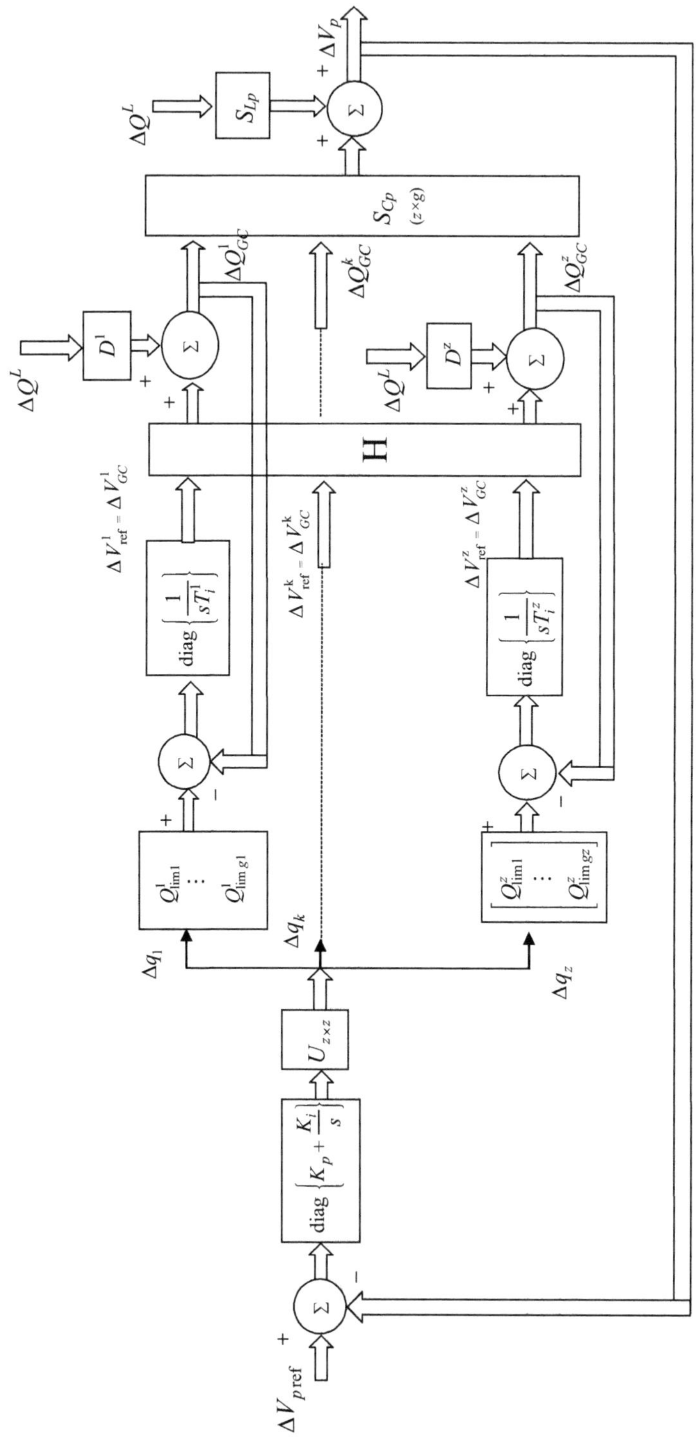

Fig. 4.7b Block diagram of the Secondary Voltage Regulation

The dynamic design of the control system consists of the computing of the integrators' time constants T_i^k in the power station's reactive power control loops (considered before), as well as of the SVR's PI control parameters and coefficients of the control matrix $\boldsymbol{U}$.

From a practical point of view, the dynamic behaviour of the controlled system is simplified by making the superimposed control loops dynamically time-decoupled. This means the response time constant of a control loop has to be dominant with respect to the time constants of its internal loops (time-decomposition). In this connection, the response time constant $T_Q = 1/\omega_c$ (ω_c being the cut-off frequency of the reactive power control loop) of the power plant reactive power control loop must be chosen sufficiently higher than the response time constant of the primary voltage control loops and sufficiently lower than the desired dynamic response (the time constant) of the pilot node voltage regulation.

Referring to Fig. 4.7b, T_i^k can be selected with good approximation as follows[1]:

$$T_i^k = \frac{(H_{kk})^{ii}}{\omega_c}.$$

Bearing in mind the chosen time decoupling among pilot node voltage control loops, analysis of the slower dynamic modes associated with the main loops of a secondary voltage control system may be made on the basis of the block diagram in Fig. 4.8. This block scheme derives from Fig. 4.7b by neglecting the low response time constants of the power plant reactive power control loops and denoting by $Q_{\text{lim}}(g \times z)$ the block diagonal matrix of the control generators' reactive capability limits. Therefore, the control matrix **U** synthesis must be made with the aim of ensuring a reduced dynamic interaction between the voltage regulations of the individual pilot nodes: the dynamics of each pilot node should be characterised by one dominant time constant only:

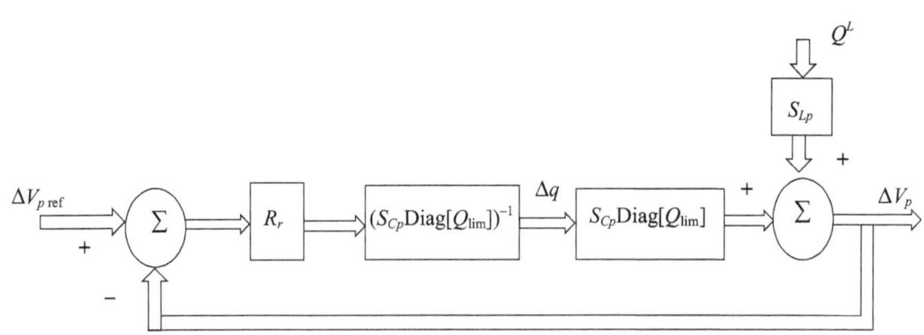

Fig. 4.8 Pilot node voltage control loops in the region

[1] The more H is a dominant diagonal matrix, the higher the dynamic decoupling is among power station reactive power control loops and their cut-off frequency nearer the desired ωc.

$$\mathbf{U} = (S_{Cp}\ \text{Diag}\ [Q_{\text{lim}}])^{-1}.$$

It is important we realise that the assumption: Δq(desired) $=\Delta q$(obtained), not considering additional dynamics in the model, is based on following hypotheses: *The reactive power loops at SQR level are faster than the pilot node voltage loops, which are also dynamically decoupled thanks to the U control law:* essentially, $\mathbf{R}_{\mathrm{r}}$ is a diagonal matrix:

$$\mathbf{R}_{\mathrm{r}} = \text{Diag}\left[K_{pi} + \frac{K_{ii}}{s}\right]$$

Generally speaking, the considered complete noninteraction among pilot nodes voltage control loops requires a full control matrix **U** and therefore a centralised control system structure. In practice, due to the criterion used for selecting pilot nodes, one which is based on reduced electrical coupling among them, the system linear model is characterised by a $\mathbf{U}^{\mathbf{P}}$ matrix with a dominant diagonal or block diagonal coefficients:

$$\mathbf{U}^{\mathbf{P}} = (S_{Cp}\ \text{Diag}\ [Q_{\text{lim}}]).$$

This implies that a strictly diagonal control matrix **U** (that is, a fully decentralised control) or when necessary, a block diagonal control matrix **U** (each block representing the pilot node subset controlled by a regional voltage regulator (RVR) at a regional dispatching centre) is enough to produce a satisfactory SVR dynamic performance.

For computing the values of $\mathbf{R}_{\mathrm{r}}$ control parameters K_{pi} and K_{ii}, we refer to the already seen HSVC dynamic design in § 3.3.5, Eqs. (3.13) and (3.14), under the assumption of a sufficient decoupling among loops of pilot node voltage controls and a correct attribution of meaning to the symbols used.

It must be said, however, calling T_s the desired time constant of each pilot node voltage control loop ($T_s\ \omega_c >> 1$), the more the matrix $\mathbf{W} = \mathbf{U}\ \mathbf{U}^{\mathbf{P}}$ approximates the unit matrix **I**, the better the V_{pi} voltage transient will be characterised by a T_s dominant time constant.

An alternative criterion for computing the control matrix $\mathbf{U}_{(\mathbf{ZXZ})}$ considers minimisation of the following functional based on least squares differences between the **W** matrix coefficients and the pure diagonal matrix **I** $(1/T_{s,\,i})$:

$$\text{Min}\left\{J(U_{r,s})\right\} = \text{Min}\left\{\sum_{k=1}^{z}\left[\left(W_{k,k} - \frac{1}{T_{s,k}}\right)^2 + \sum_{\substack{j=1\\ j\neq k}}^{z}\left(W_{k,j}\right)^2\right]\right\}.$$

Reference Transients

From here, the transients of two separate tests on secondary voltage regulation are shown:

- The test on the SQR reactive power control loops following a q level step variation;
- The test on the SVR pilot node voltage control loop after a voltage V_{Pref} set-point step variation.

These overlapped control loops are clearly shown in Fig. 4.5.

SQR and SVR transients jointly represent the full dynamic characteristics of the secondary voltage regulation. HSVC shows the same characteristics being designed using identical criteria as that of SQR-SVR.

Test on the SQR reactive power control loops

Considering a power station with four generators under SQR control, the test results in Fig. 4.9 show transients following step variations on the reactive power level "q" under the hypothesis of an "open" pilot node voltage control loop.

In fact, in this operating condition all the generators in the power station track the "q" step control with the dynamic characteristics imposed by SQR only: a first-order trend with a 5s dominant time constant. It can be seen that the four generators' reactive powers and voltages move concordantly aligned, tracking the reference control signal step variations, in the top figure. Moreover, under SQR, no dynamic interaction among the reactive power control loops under testing is evident, as expected. The last transient in the bottom figure shows the pilot node voltage variation consequent to an increase of the considered power station's reactive power delivery (in the first part) followed by voltage reduction under the "q" step-down.

Test on SVR pilot node voltage control loops

Considering a power system with 12 pilot nodes under SVR control, the test results in Fig. 4.10 show the transients in six areas (1, 2, 4, 5, 6, 11) following step variations on the Area 2 pilot node voltage set-point. In this case, control loops of both the SQR and SVR operate accordingly to recover the Area 2 pilot node voltage at the new set-point value and to maintain unchanged the pilot node voltages of the remaining Areas 1, 4, 5, 6, 11, as well.

The test demonstrates the proper selection of the grid pilot nodes with light dynamic interaction among their voltage control loops. In fact, step variations at the Area 2's voltage set-point determine significant changes in the corresponding pilot node voltage, as well as small transient effects on the other set-points (Fig. 4.10 top).

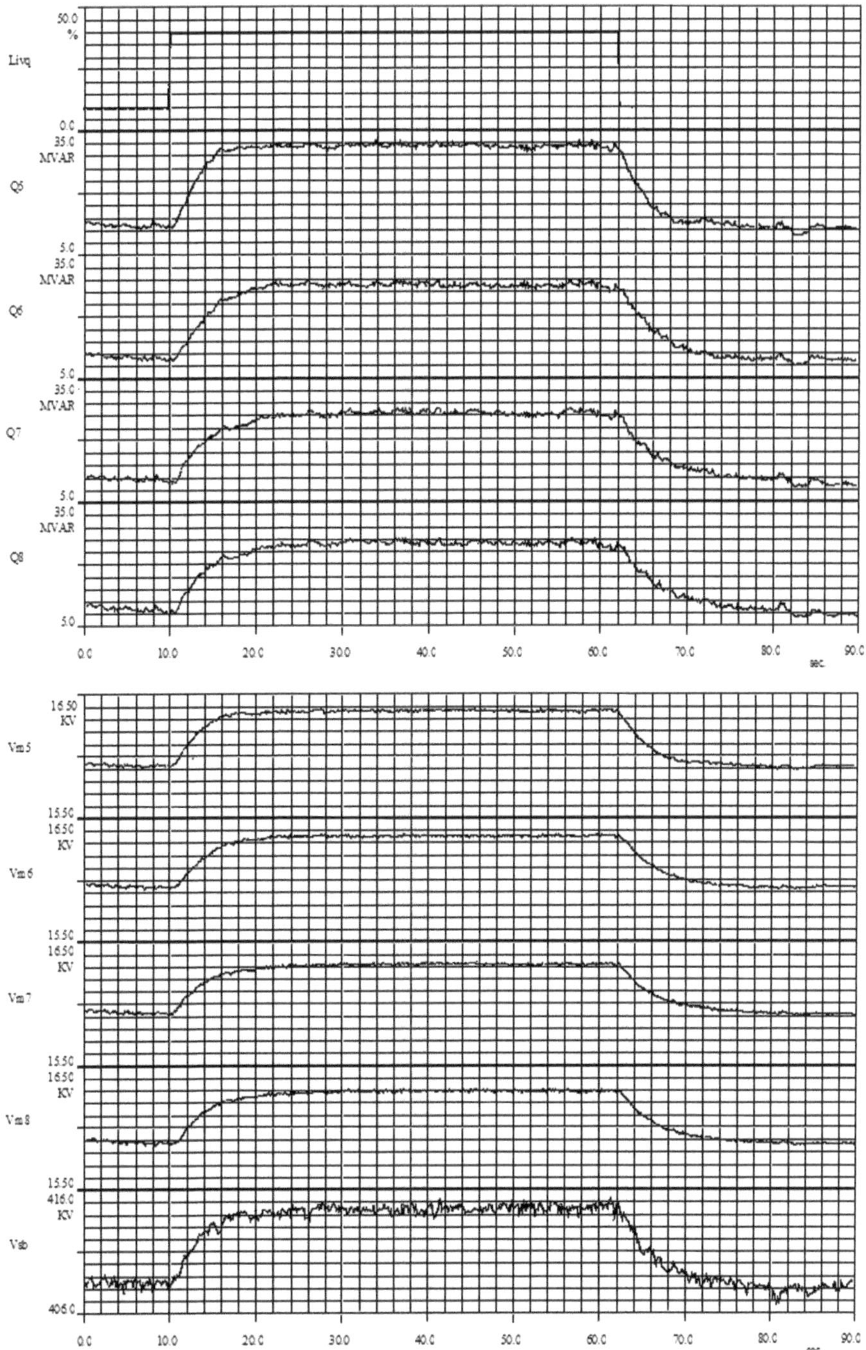

Fig. 4.9 Step response of reactive power control loops in a real power station, showing a dominant time constant of about 5 s. V_{Sb} represents the local HV pilot bus bar voltage transient under SQR control only

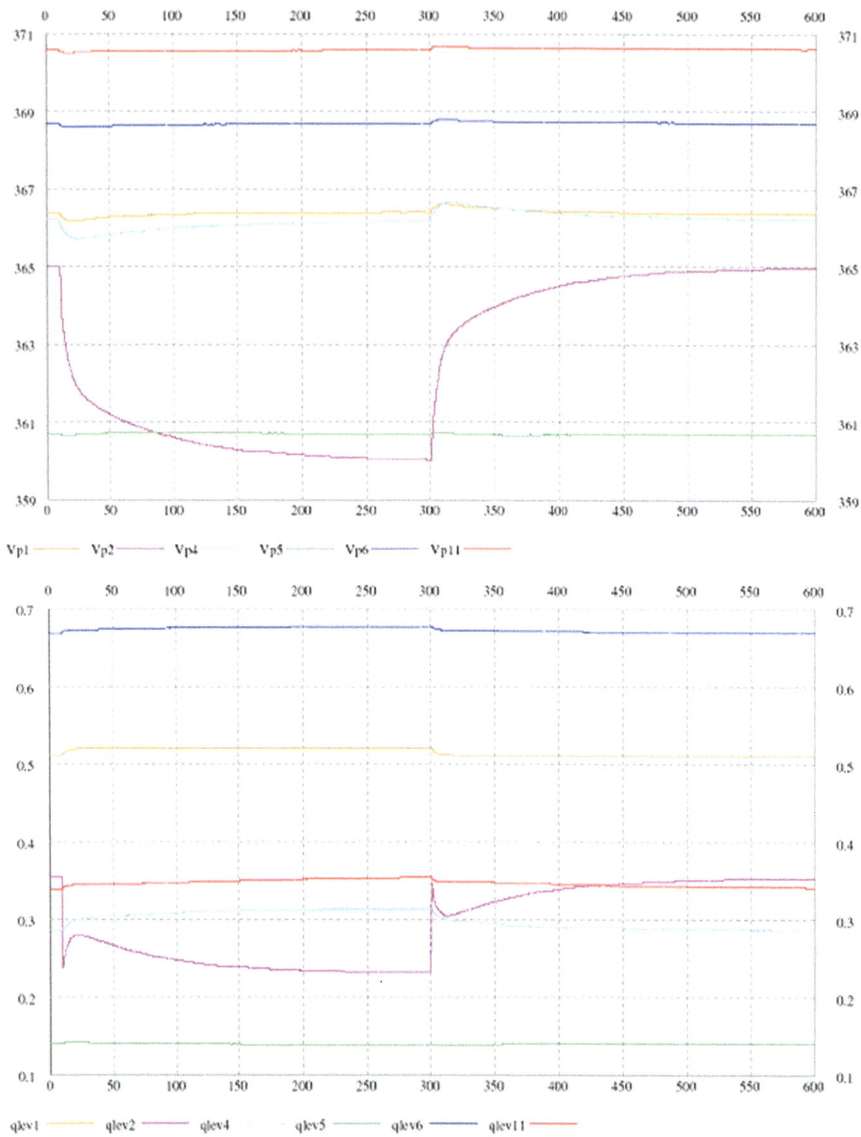

Fig. 4.10 At the *top*, the dynamic response of the pilot node voltages V_{pi} following the set-point V_{ref} step variations at Area 2 only. At the *bottom*, the corresponding area reactive power control levels with the violet q_2 showing the area with the largest control effort

The Area 2 reactive power control level confirms this result, significantly changing with respect to the other area control levels, which conversely remain almost constant.

Again, the SQRs impose on all power system control generators the tracking of the corresponding control levels q_i with a dynamic performance of a 5s dominant

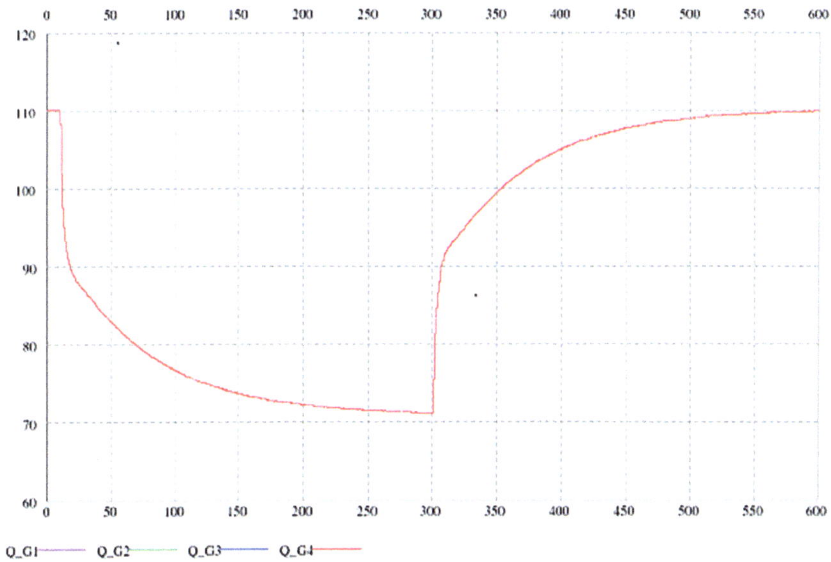

Fig. 4.11 At a power station under control, the overlapped transients of four generators allowing Area 2 pilot node voltage set-point tracking the $V_{\text{ref}\,2}$ step variations

time constant. In its turn, the SVR imposes on all the SQRs the tracking of the corresponding pilot node voltage set-point with a dynamic performance of a 50s dominant time constant (in the second part of the transient; the first part being strongly influenced by the proportional control parameter).

The combined effect of the two layers of regulators gives as a result in Area 2 the four generators' reactive powers, in Fig. 4.11, which move concordantly aligned, tracking the reference control signal q_2 variations shown at the bottom of Fig. 4.10. Lastly, under SVR no dynamic interaction among the pilot node voltage control loops being tested are seen (Fig. 4.10 top).

The test results shown make evident that the appropriate operation of secondary voltage control within each area depends mostly on the way grid pilot nodes and control areas are selected, as was introduced at the start of the chapter.

A control area is properly defined if the following conditions are met:

- With pilot node voltage maintained unchanged at the set-point constant value, the voltages at the other nodes in the area have small variations, even when the local load significantly changes;
- Voltage control within a control area does not significantly influence voltages in the other areas;
- Control resources in each control area should be, as much as possible, able to maintain unchanged the pilot node voltage under normal and disturbed grid operating conditions.

In real power systems the above conditions can be satisfactorily achieved at a high percentage. But this does not exclude particular cases where for some reason the above conditions are not strictly fulfilled by the considered power system. These cases require specific solutions:

- Significant dynamic interaction between some areas can still be reduced by designing a noninteracting control law;
- An area having few control resources gives evidence that new compensating equipment needs to be installed.

Identification of control areas and pilot nodes can be achieved through electrical distance based methods. The proposed algorithm (described later) is based on the following steps:

i. Choose network pilot nodes by selecting those with largest short circuit power;
ii. For each selected pilot node, determine the corresponding area according to an electrical distance method;
iii. Verify, by using the reactive power balance, whether area reactive power sources can supply area demand;
iv. Verify if the voltage variation at the pilot node is representative of voltage variation at the other area buses;
v. Verify that:

- Distances between the pilot node and remaining system buses confirm, in the presence of SVR, the identified area subdivision;
- Electrical distances between the pilot node of each area and neighbouring area pilot nodes are considerably large.

4.1.5 Tertiary Voltage Regulation

The basic idea of TVR derives from the need for a system's operating security and efficiency to increase through centralised real-time coordination of the decentralised SVR structure:

- Pilot node voltage set-points must be adequately updated and coordinated online and in real time, with dynamics slower than SVR, by consideration of the real operating condition of the overall grid and by avoiding pointless and conflicting SVR inter-area control efforts;
- To this end, pilot node voltage set-points can be computed and updated in real time simply by use of the SVR control system operating conditions that give reliable, synthetic, timely information on what is going on at the overall system: "SVR controls that are active on the physical process and the pilot node measurement feedback provide, at any instant, an undoubtable figure of the most important essential happenings in the real process";

- Therefore, pilot node voltage set-points can be optimised in real time to effectively minimise grid losses while still preserving the control margin by simply referring to the "grid equivalent" real-time system model, based on few but very reliable and significant data on control variables the SVR is able to provide to TVR.

The TVR control level is therefore aimed at optimising nationwide voltages by a "suboptimal real-time control". This involves determining moment by moment the pilot node voltage set-point values by minimising the differences of the measured pilot node voltages with respect to their historical references or off-line forecasted values, always maintaining the control margin in each area. Having a proper selection of SVR areas, this simplified TVR optimisation is able to achieve a safe and efficient closed-loop system control by a slower than SVR dynamic performance.

Therefore, the tertiary loop represents the continuous computing of a wide-area, real-time, updated, optimal voltage plan, applied to the grid through the global coordination of automatic control actions achieved by an SVR. The main TVR objectives are these: (i) the management, at a low speed, of the reactive power flow between the power system areas, accomplished by minimising power system losses; (ii) the increasing of the power system's controllability and stability.

Tertiary control has until now been performed mostly manually by dispatchers, with poor results because neither real-time nor real optimal control can be achieved this way. When automated, thus becoming on-line and real-time (as in Figs. 4.1 and 4.2), TVR would have a control scheme like the one in Fig. 4.12, characterised by an outer control loop (overlapping the SVR) with a dominant time constant of around 5–10 min and a sampling rate of under 1 s.

The pilot node voltage set-point $|V_{\text{Pref}}(t)|$ vector can be provided automatically by TVR output, operating in real time and closed loop; otherwise, with the TVR out of service, pilot node voltage daily trends can be computed by not-real-time and off-line OPF (for remote automatic or manual provisional dispatching of the (day-before-computed) SVR set-point plan); or, it can be selected by the regional dispatcher according to his experience (i.e., by manual setting of SVR set-points).

TVR real-time optimisation defines the most appropriate pilot node voltage set-point $|V_{\text{Pref}}(t)|$ vector for secure/efficient operation on the basis of an integral law (4.12) of the real-time optimised vector $|V_{\text{Pref}}(t)|$ representing the best increment to be actuated on $|V_{\text{Pref}}(t)|$ according to the minimisation of the TVR objective function (4.13) [7]:

$$\left|V_{P\text{ref}}(t)\right| = \frac{1}{T_T}\left[\left|\int_0^t \Delta V_{P\text{ref}\,i}(\tau)d\tau\right|\right]_{V_{P\min}}^{V_{P\max}} + \left|V_{Pi}(0)\right|, \tag{4.12}$$

where T_T is the gain of the integral regulator fixing the TVR closed-loop dominant time constant at 5–10 min. Called $[S]$, the sensitivity matrix is between area reactive levels $\Delta_{q\text{LEV}}$ and pilot node voltages ΔV_{Pref}:

$$[\Delta V_{P\text{ref}}] = [S][\Delta q_{\text{LEV}}].$$

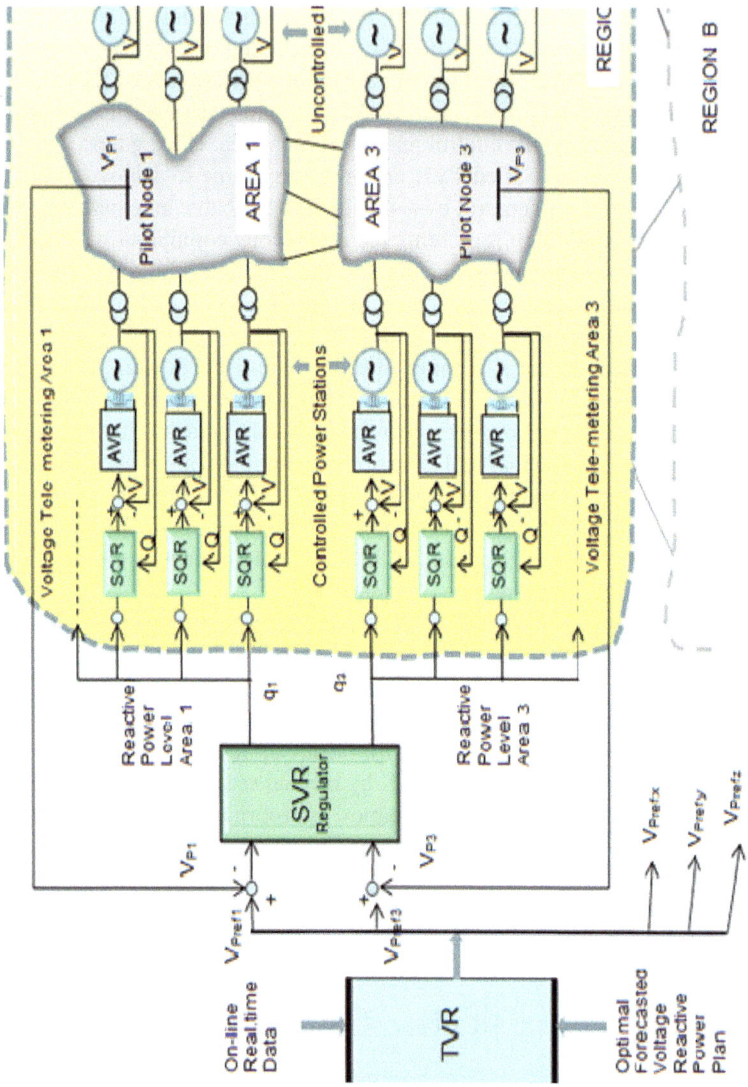

Fig. 4.12 Schematic diagram of the hierarchical structure of a secondary and tertiary voltage control system

Equation (4.12) imposes the dynamics of the TVR control loop by integrating the result $|V_{P\text{ref}}(t)|$ of the "TVR objective function minimisation": Min(OF), where OF is based on the actual network state estimation and the forecasted optimal voltage and reactive power plan:

$$OF = \left[V_P + \Delta V_{P\text{ref}} - V_P{}^0\right]^{\mathrm{T}} \mathbf{Q}^2 \left[V_P + \Delta V_{P\text{ref}} - V_P^0\right] + \left[q_{\text{LEV}} + S^{-1}\Delta V_{P\text{ref}} - q_{\text{LEV}}^0\right]^{\mathrm{T}} \mathbf{R}^2 \left[q_{\text{LEV}} + S^{-1}\Delta V_{P\text{ref}} - q_{\text{LEV}}^0\right], \tag{4.13}$$

where $[V_P]$, $[q_{\text{LEV}}]$ are the vectors of real-time measurements of the pilot node voltages and area reactive power levels; $[V_P^0]$, $[q_{\text{LEV}}^0]$ are the vectors of optimal forecasted pilot node voltages and area reactive power levels (coming from the "state estimation" and the OPF block); $\mathbf{Q}^2$, $\mathbf{R}^2$ are weight matrices whose selection allows us to attribute importance to pilot node voltage differences (**Q**) rather than to the effort of control area reactive power levels (**R**).

In conclusion, Eqs. (4.12) and (4.13) together represent the TVR control functionality that can be computed in real time because it is significantly dependent on real-time measurements. In principle, $[V_P^0]$, $[q_{\text{LEV}}^0]$ could remain unchanged at given constant values if power system state estimation and/or OPF are not working. In fact, their updated forecasting is not mandatory but simply a help.

The compromise reached by TVR when the available optimal forecasted plan does not fit the real situation well (obviously, there regularly is more or less of a discrepancy in this respect), should properly consist of the achievement of the highest voltage plan consistent with real operating conditions, which minimise network losses as much as is feasible. To obtain this result it is necessary system controllability be preserved, even if close to the limits, so as to avoid the disastrous consequences of open loop operation. Under this condition, in fact, uncontrolled voltages cause undesired heavy reactive power flows, which increase system losses and worsen operation efficiency.

OF optimisation, based on Min(OF), achieves the objective of a real-time compromise between control effort and voltage difference with respect to optimal forecasted values through their weighted and combined minimisation. Therefore, the proposed TVR is the correct and necessary completion of a hierarchical automatic real-time voltage control system.

Moving from the TVR control level to the higher voltage level where OPF is computed, real-time control is necessarily lost because OPF requires state estimation (SE) and its correct updating, which, even if achieved every 5 min, is too delayed for tracking the power system voltage dynamics (by OPF).

In conclusion, at the level above TVR, voltage control can be of a forecasting type only. Moreover, OPF computing would require, in the presence of SVR, the expected upgrade in system model constraints, taking into account SVR structure and area coordination of reactive power resources (see § 4.2.3). The characteristics of the control power station buses should change from a "PV" to a "PQ" type, while pilot node buses become "PV" type. Lastly, OPF computing can be simplified by considering only the pilot node voltages as optimisation output (i.e., a loss minimisation result); this kind of solution is already operating in Italy's national control centre, with output updating every 15 min.

4.2 SVR Control Areas

4.2.1 Procedure to Select Pilot Nodes and Define Control Areas

The criterion proposed for choosing pilot nodes is based on the intuitive understanding that they must be selected from among the strongest buses. In fact, they must also be able to impose voltage on their surrounding buses in front of normal perturbations.

Design criteria, based on short-circuit capacities and sensitivity matrix computations, also require electrical coupling among pilot nodes be sufficiently low, to avoid the possibility of dynamic interaction between secondary control loops. With this constraint, in fact, excessive reactive power exchanges among adjacent regulation areas, determined by even small differences between pilot node voltages imposed by the regulating system, are basically prevented. In case network operational requirements should happen to condition a pilot node selection by determining an excessive electrical coupling among regulation areas, secondary and tertiary control laws will decouple dynamic interactions among inner control loops (see § 4.1.4.3 (SVR Control Structure) for SVR dynamic decoupling and § 3.3.5 for SQR dynamic decoupling).

The analytical procedure for selecting pilot nodes consists of a successive reordering of the sensitivity matrix, expressing the dependence of the whole grid's bus voltages (from now on designated "*n* buses") on reactive power injections while primary voltage regulation is operating. The method assumes the HV bus (of load or generation type) having the strongest short circuit capacity as the "pilot node 1". All buses with the highest coupling coefficient with pilot node 1 are assumed to belong to "regulation area 1" and are excluded from subsequent pilot node choices. This procedure, progressively applied, identifies other pilot nodes that are the strongest of the remaining buses and therefore gradually weaker, until the procedure stops due to insufficient short circuit capacity.

Analytical Procedure for Selecting Pilot Nodes

Evidence is given to the (n-g, n-g) sensitivity matrix $[X_{CC}]$ (see Eqs. (4.5) and (4.6)). From now on, g represents the number of generator MV buses. $[X_{CC}]$ is the sensitivity matrix of the voltages $[\Delta V_L]$ of the EHV buses with respect to reactive powers $[\Delta Q_L]$ injected into the same load buses, when generator voltages (MV) are maintained constant $[\Delta V_G]=0$. This is a diagonal dominant matrix with negative coefficients.

- The diagonal coefficients are of the type:

$$(X_{CC})_{hh} = \left(\frac{\Delta V_{L_h}}{\Delta Q_{L_h}} \right)_{\begin{cases} \Delta Q_{L_s}=0 \\ s \neq h \end{cases}}, h, s = n+1, \ldots, n+N. \tag{4.14}$$

- Those outside the main diagonal are of the type:

$$(X_{CC})_{hk} = \left(\frac{\Delta V_{L_h}}{\Delta Q_{L_k}} \right)_{\begin{cases} \Delta Q_{L_s}=0 \\ s \neq k \end{cases}}, h, k, s = n+1, \ldots, n+N. \tag{4.15}$$

It is important we notice the generic diagonal coefficient $(X_{CC})_{hh}$ is the equivalent reactance seen by the "h" bus:

$$\Delta V_{L_h} = (X_{CC})_{hh} \Delta Q_{L_h}$$

Matrix $[X_{CC}]$ plays a fundamental role in *the selection of pilot nodes and related areas*.

1. Matrix rows and columns are reordered to satisfy the condition:

$$\begin{cases} (X_{CC})_{11}^{(1)} < (X_{CC})_{rr}^{(1)} \\ (X_{CC})_{11}^{(1)} > (X_{CC})_{21}^{(1)} > (X_{CC})_{31}^{(1)} > \cdots > (X_{CC})_{N1}^{(1)} \end{cases}$$

with r=2,…, N.

2. The $(n-1)$ ratios are computed:

$$\beta_{ij} = \frac{(X_{CC})_{ij}^{(1)}}{(X_{CC})_{jj}^{(1)}}, \text{ with } i = 1, 2, ..., N; j = 1, 2, ..., Z,$$

where $0 \leq \beta_{ij} \leq 1$; Z is the number of pilot nodes, to be defined at the end.

3. After the lower limit of the "electrical distance" among the pilot nodes is established, we exclude from subsequent selections those buses related to

the first N_1 rows of the $[X_{CC}]^{(1)}$ reordered matrix having coupling coefficient β_{i1} with bus "1" greater than ε_P, i.e.,

$$\varepsilon_P < \frac{(X_{CC})^{(1)}_{\eta,1}}{(X_{CC})^{(1)}_{11}} \leq 1; \quad \eta = 1, \quad \ldots \quad , \quad N_1.$$

4. The remaining $(N-N_1)=n_1$ columns of the $[X_{CC}]^{(1)}$ matrix are reordered in such a way that the new matrix $[X_{CC}]^{(2)}$ satisfies the following inequalities:

$$(X_{CC})^{(2)}_{11} < (X_{CC})^{(2)}_{rr},$$
$$(X_{CC})^{(2)}_{11} > (X_{CC})^{(2)}_{21} > (X_{CC})^{(2)}_{31} > \cdots > (X_{CC})^{(2)}_{N1}$$

where $\left[X_{CC}\right]^{(2)} \in R^{n_1 \times n_1}, \quad r = 2, \ldots, \quad n_1.$

This corresponds to arranging the $(n-N_1)$ remaining buses in order of electrical vicinity, with bus 1, among them, having the highest power.

5. Analogously to step 3, in the steps that follow we no longer consider the first N_2 of the n_1 nodes (that is, the first N_2 rows of the reordered matrix ($[X_{CC}]^{(2)}$) having coupling coefficient with bus "1":

$$\frac{(X_{CC})^{(2)}_{\eta 1}}{(X_{CC})^{(2)}_{11}} > \varepsilon_P$$

So the next N_2 buses excluded are those for which

$$\varepsilon_P < \frac{(X_{CC})^{(2)}_{\eta 1}}{(X_{CC})^{(2)}_{11}} \leq 1; \quad \eta = 1, \ldots, N_2.$$

6. The reordering procedure of the matrices is repeated, starting from the $(n_1 - N_2)=n_2$ remaining nodes, in accordance with the indicated procedure up to the $(Z+1)$th reordering, which is when—among the $n_{(Z-1)} - N_Z \stackrel{\text{def}}{=} n_Z$ remaining buses—the coefficient $[X_{CC}]^{(Z+1)}$ of the reordered matrix $[X_{CC}]^{(Z+1)}$ is greater than a predefined value $1/\gamma$, which represents the minimum admissible value of the short circuit power for a pilot node:

$$(X_{CC})_{11}^{(Z+1)} > \frac{1}{\gamma}.$$

7. The Z pilot nodes are those corresponding to the first row of the matrices:

$$[X_{CC}]^{(1)}, [X_{CC}]^{(2)}, [X_{CC}]^{(3)}, \ldots, [X_{CC}]^{(Z)}.$$

After having defined the pilot nodes, for each bus of the grid the coupling parameter β_{ij} is computed, defined as follow:

$$\beta_{ij} = \frac{(X_{CC})_{ij}}{(X_{CC})_{jj}}, \quad \text{with } i = 1, 2, \ldots, N; \; j = 1, 2, \ldots, Z,$$

where $0 \leq \beta_{ij} \leq 1$.

These are the coefficients of the (N, Z) sensitivity matrix $[B_{RL}]$, which represents the sharing of the N grid buses among the Z areas in which the grid has been subdivided.

The ith bus is linked to area j if it has the highest coupling coefficient with the jth *pilot* node. That is, the ith bus is associated to area j if:

$$\beta_{ij} > \beta_{ik} \quad \forall \, k \neq j.$$

Other formulations for electrical distance between two buses, based on the $[X_{CC}]$ matrix, are also possible.

4.2.2 Procedure to Select Control Generators

After selecting the pilot nodes and corresponding SVR areas, it is necessary to choose the control generators of each area, that is, the generators participating in that area's pilot node voltage control. These control generators are obviously most able to affect the voltage of the pilot nodes to which they are linked due to a high electrical coupling with them and because of their large capability limits.

Control power plant selection also permits us to preventively recognise those regulation areas having a consistent amount of reactive power resources, as well as those areas where reactive power reserves are critical and the pilot node voltage regulation can more easily reach its saturation.

The analytical procedure for selection of control power stations requires a successive reorganisation of the sensitivity matrix, expressing the dependence of pilot node voltages on reactive power injections by generators. The method assumes all generators belonging to "regulation area *i*" and having highest coefficient placed in the "pilot node *i*" row to be potential "control power plants *i*". All potential power stations with the highest product of pilot node voltage sensitivity by station-rated reactive power capability are definitely assigned to be control plants *i*.

Analytical Procedure for Selecting Control Generators

The proposed analysis can refer to system model (4.5) assuming the additional simplifications of a "reciprocal network" model. A reciprocal network is characterised by symmetrical matrices; therefore, in (4.6), assuming $[B]$, $[B_{LL}]$ and therefore $[X_{CC}]$ and $[C]$ to be symmetrical matrices, we find:

$$[H]^{\mathrm{T}} = [B_{GL}][X_{CC}] = -[D].$$

Then

$$[B_{\mathrm{eq}}] = [B_{GG}] - [D][B_{LG}]. \tag{4.16}$$

The simplified system of equations becomes

$$\begin{cases} [\Delta V_L] = -[D]^{\mathrm{T}}[\Delta V_G] + [X_{CC}][\Delta Q_L \\ [\Delta Q_G] = -[B_{\mathrm{eq}}][\Delta V_G] + [D][\Delta Q_L]] \end{cases} \tag{4.17}$$

Here after, we recall the already defined matrices:

$$\begin{cases} [X_{CC}] = -[B_{LL}]^{-1} \\ [D] = -[B_{GL}][X_{CC}] \\ [B_{\mathrm{eq}}] = [B_{GG}] - [D][B_{LG}] \overset{\mathrm{def}}{=} -[C]. \end{cases} \tag{4.18}$$

From these equations it is also possible to obtain the following links representing voltage variation with respect to injected reactive power:

$$\begin{cases} [\Delta V_L] = -[S_{LG}][\Delta Q_G] + [S_{LL}][\Delta Q_L] \\ [\Delta V_G] = -[S_{GG}][\Delta Q_G] + [S_{LG}]^{\mathrm{T}}[\Delta Q_L]. \end{cases} \tag{4.19}$$

Links are based on the following definitions:

$$\begin{aligned} &[S_{GG}] = -[C]^{-1}, \\ &[S_{LG}] = -[H][S_{GG}], \\ &[S_{LL}] = [X_{CC}] - [S_{LG}][D]. \end{aligned} \tag{4.20}$$

Control generator selection is based on the matrix $[S_{LG}]$ that represents the sensitivity of EHV load bus voltage vector $[\Delta V_L]$ with respect to vector $[\Delta Q_G]$ of reactive powers injected from the generators. The procedure is based on the re-ordering of the submatrix $[S_{LG}]$ by considering the Z rows corresponding to the pilot nodes and the n columns corresponding to the generation buses. This sub-matrix, called $[S_{RG}]$, represents the sensitivity of the pilot node voltages to the injected reactive powers.

The procedure selects for each $[S_{RG}]$ column the highest coefficient and re-orders the n columns in such a way that the first n_1 are all those having the highest coefficient at the first row, that is, those satisfying the following inequalities:

$$(S_{RG})_{1j} \geq (S_{RG})_{kj}; \quad k = 1,3\ldots, \quad z; \quad j = 1,\ldots, \quad n_1.$$

The second n_2 columns are those having the highest coefficient in the second row, that is, those satisfying the following inequalities:

$$(S_{RG})_{2j} \geq (S_{RG})_{kj}; \quad k = 1,3\ldots, \quad z; \quad j = 1,\ldots, \quad n \ .$$

The procedure continues up to the n_z column ($n_1+n_2+\ldots+n_z=n$). The grouping of the $n_1, n_2, \ldots, n_z$ columns selects the generation buses linked with the pilot nodes 1, 2,…, z. Called A_{nj}^k, the generator nominal power at the jth bus of the kth area, the term $(S_{RG})_{kj} A_{nj}^k$ represents real generator capacity to affect pilot node voltage in the area considered. The control generators of the pilot node voltage of the kth area are selected from those inside the area satisfying the inequality:

$$(S_{RG})_{kj} A_{nj}^k > \alpha_c^k,$$

where $\alpha^k_{\ c}$ is the allowed minimum control capability in the considered kth area.

4.2.3 Power Flow and Optimal Power Flow Computation in the Presence of Secondary Voltage Regulation

To obtain a generalised load flow mathematical model in the presence of secondary voltage control, a real-time equal percentage of generator control effort in each area is required. Therefore, it is necessary to include in the system model a number of constraint equations due to secondary voltage regulation structure:

$$\frac{Q_1^j}{Q_{1\max}^j} = \cdots = \frac{Q_{n_c^j}^j}{Q_{n_c^j \max}^j}, \quad j = 1,\ldots,n_a,$$

where:

n_a = number of SVR areas
n_c^j = number of controlling generators of area j

The subscript "max" denotes the capability limit of each generator at its operating point in over-excitation. The subscript becomes "min" in under-excitation.

Further constraints required by OPF are:

- PQ bus instead of PV bus at each SVR control power station;
- PV bus at pilot nodes instead of PQ bus at all the remaining grid nodes.

4.2.4 Examples of Pilot Node and Control Power Station Selection

The proposed simple methods for selecting SVR pilot nodes and control generators are not computationally heavy and give satisfactory results, once refined with some threshold values and also depending on the particular network's characteristics. For instance, accepting a higher electrical coupling increases the number of pilot nodes but also requires more complex control laws to handle closed-loop interactions and dynamic instability risks. Moreover, this choice requires frequent reselection of pilot nodes, even in the event of small network changes.

On the other hand, excessively low electrical coupling reduces the number of pilot nodes and significantly decouples their control loops while at the same time worsening voltage control quality.

Similarly, the acceptance of excessively low products of sensitivity coefficients with generator rated reactive powers increases the number of control power stations and corresponding reserve margins; however, further, unnecessary, control infrastructures could be required to allow small generator participation and coordination with SVR. In practice, the subdivision of the whole system into regulation areas must be robust and conservative to avoid too-frequent control system reconfigurations in front of network small changes.

The following examples of area selection made in power systems around the world help us to easily recognise the high degree of their robustness in terms of years the same selection is confirmed valid. Moreover, the different shapes of selected SVR areas from one country to another are pointed out to indicate the territory occupied by the grid but also to show specific grid operation criteria such as the use of separated and electrically decoupled buses at the same substation, an operation practice found in the Taiwan HV grid.

In some cases, results coming from automatic procedures must be corrected by taking into account other specific grid characteristics, like those in South Africa, where there exists a very powerful bus without any loads in its potential area. Relevant grid structural changes compromise the mentioned area selection robustness and should require new analysis for checking their impact on pilot nodes, regulation area edges and control power station selection, as well as for adequate retuning of SVR-TVR regulation parameters.

A pre-established number of pilot nodes is not the correct starting point in the mentioned analysis because pilot node number depends on network topology, load and generator location and size, power system operating conditions (such as peak or off-peak load), imposed minimum electrical coupling among pilot nodes and robustness of choice. Therefore, an analysis for pilot node selection has to be pondered deeply and iteratively adjusted by use of an automatic computing procedure. In this way the proper number and location of the pilot nodes is found in accordance with the grid's seasonal changes, with the power system's growth over time, and after very heavy contingencies followed by large topology changes in the grid.

Pilot Nodes and Control Power Stations in Italy

Consolidated studies on pilot node and control power plant selection provided for a subdivision of the Italian power system into 18 regulation areas (see Fig. 4.13). This plan involved the largest thermal and hydropower plants connected to 400/230-kV networks. Data here presented and already published [1, 2] refer to the year 2000 grid, 60,000-MW peak, for a total reactive power capacity of about 20,000 MVAR.

Pilot node selection in the Italian power system was accomplished by use of an automatic tool based on the algorithm described in § 4.2.1.1 (Analytical Procedure for Selecting Pilot Nodes), combined with practical considerations of transmission grid characteristics. The result of the analysis shows that proper and robust selection for the Italian grid is 18 pilot nodes (the red flags). This result, achieved on the basis of case studies and on-field application results, appears adequate to control voltage in the entire grid. Looking at Fig. 4.13, we find a proper and well distributed covering of the map, meaning the voltage automatic control in the overall Italian transmission grid can be achieved simply by controlling the voltages of the 18 well-selected EHV buses.

Usually, pilot nodes are chosen from among the most powerful bus bars in the network because they are most representative of the voltage level at their local area. For the Italian network, bus bars with nominal voltage equal to or lower than 150 kV are not considered in the pilot node selection process because they are not strong enough to impose voltages on surrounding local and higher voltages buses. Accordingly, the pilot node selection algorithm only considered bus bars with nominal voltage of 400 kV and 220 kV.

Pilot nodes should, in principle, be load buses. Therefore, all generating MV bus bars have been excluded from the selection, being characterised by low sensitivity coefficients due to their proximity to generator bus bars, where AVRs fix constant voltage. Each SVR area has at its disposal local control power stations, so each has a different degree of autonomy in controlling area voltage variation by local MVAR. In the Italian SVR, all the area reactive power levels assume in normal operating conditions values lower than 0.3 p.u. This puts in evidence a large area controllability margin for facing large and small perturbations in over- and under-excitation.

Fig. 4.13 Application plan of the Italian hierarchical voltage control system

Other interesting steady-state study results come from the comparison in the Italian system of HV bus voltage variations with and without SVR when faced with a 100 % uniform increase in system loads.

Figure 4.14 shows a comparison of four cases of control operating conditions in the Italian power system:

a. Case of primary voltage regulation (PVR);

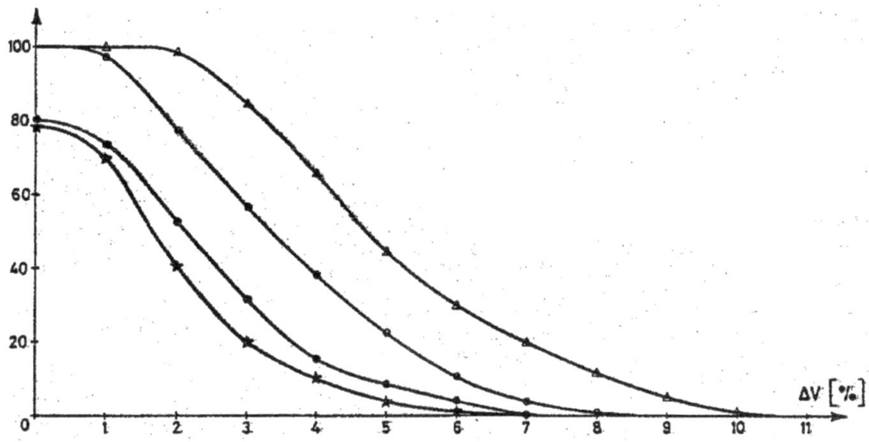

Δ PVR
○ PVR with line drop compensation
● SVR and PVR at the generators not participating in SVR
* SVR and PVR with line drop compensation at generators not participating in SVR

Fig. 4.14 Distribution of Italian bus percentage with voltage variation higher than ΔV after a uniform 100 % increase of total reactive load

b. Case of PVR with line drop compensation;
c. Case of SVR and PVR in generators outside SVR;
d. Case of SVR and PVR with line drop compensation on generators outside SVR.

In summary, the distribution curves in Fig. 4.14 show how the SVR reduces about 40 % of the number of buses having the same voltage variation as occurs with PVR alone (the difference between Δ and ● points). As is obvious, SVR is not able to control voltages at peripheral buses in the area and where controllable reactive power resources are not available or sufficient for the load variation considered.

Pilot Nodes and Control Power Stations in the Taiwan Grid

Pilot nodes, control generators and control area selections for the Taiwan transmission system refer to 2005 and 2009 case studies with the following results:

- 2005 system at peak, base and medium load
  - 10 pilot nodes
  - 32 control generators (power plants)
  - 10 areas.
- 2009 system at peak, base and medium load
  - 10 pilot nodes
  - 34 control generators (power plants)
  - 10 areas.

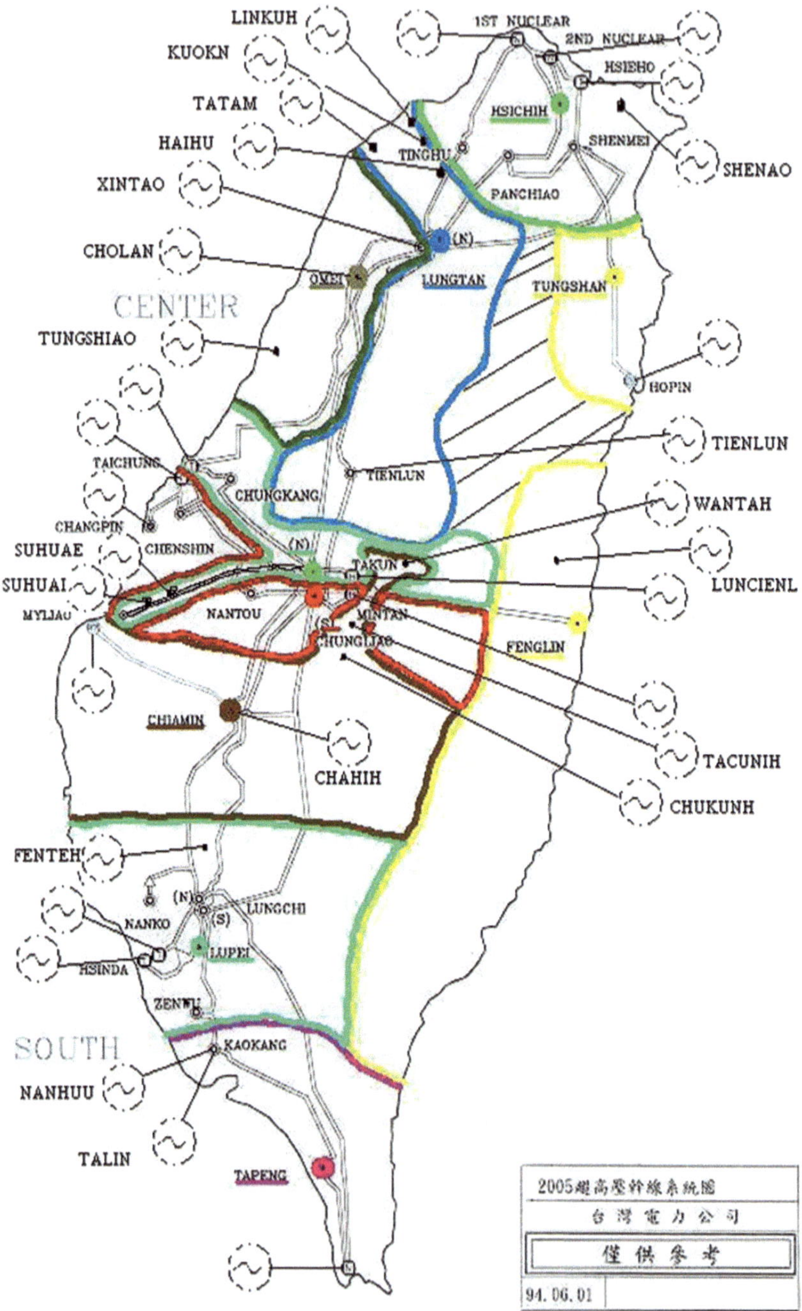

Fig. 4.15 Application plan of the Taiwan 2005 hierarchical voltage control system: pilot nodes and control power stations 1. *HSICHIH*; 2. *LUPEI*; 3. *CHIAMIN*; 4. *CHUNGLIAO-N*; 5. *OMEI*; 6. *CHUNGLIAO-S*; 7. *LUNGTAN N*; 8. *TAPENG*; 9. *TUNGSHAN*; 10. *FENGLIN*

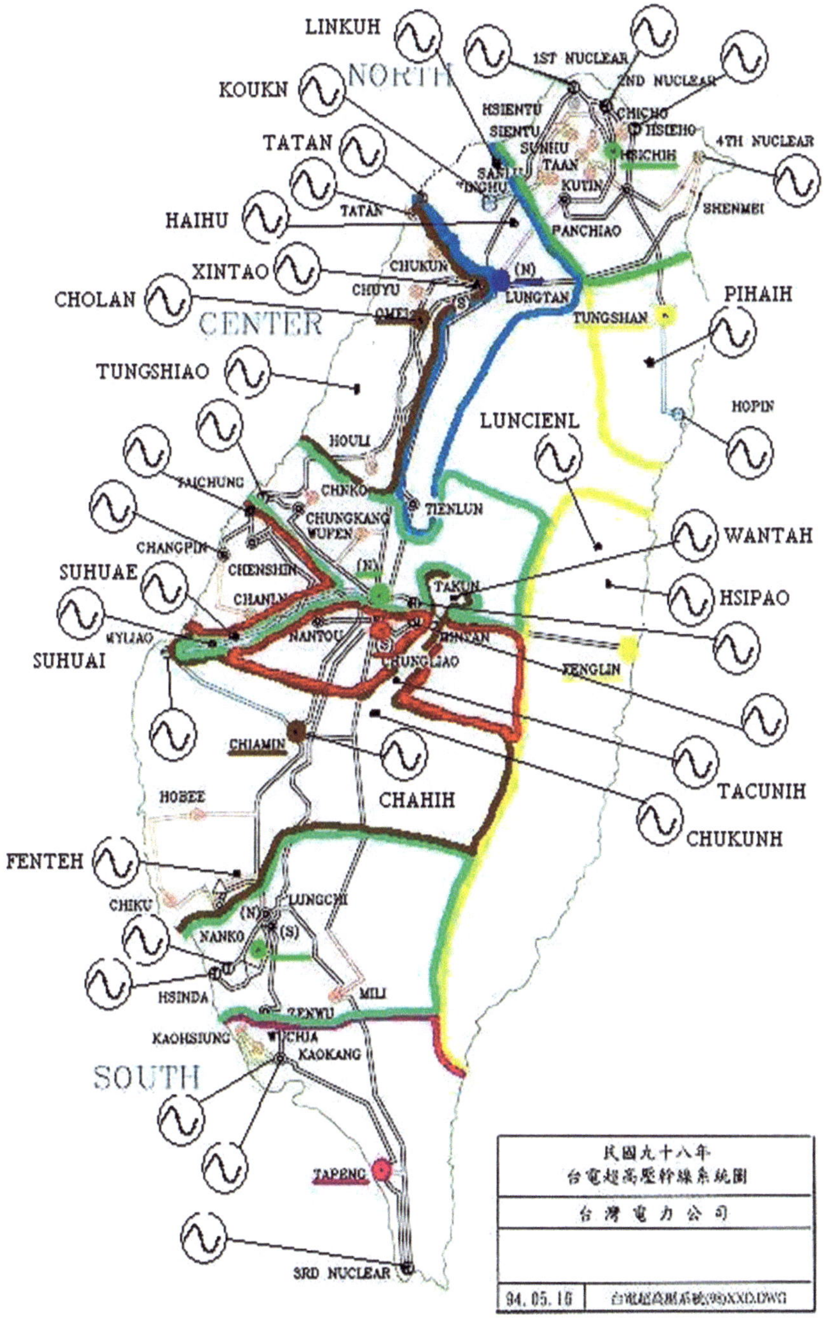

Fig. 4.16 Application plan of the Taiwan 2009 hierarchical voltage control system: pilot nodes and control power stations 1. *HSICHIH*; 2. *LUPEI*; 3. *CHIAMIN*; 4. *CHUNGLIAO-N*; 5. *OMEI*; 6. *CHUNGLIAO-S*; 7. *LUNGTAN-N*; 8. *TAPENG*; 9. *TUNGSHAN*; 10. *FENGLIN*

Pilot node selection in the 2005 case study shows the following result, *which does not change when there is a move from peak to medium and low load* (see Fig. 4.15):

Pilot node selection in the 2009 case study is the same as it was in the previous 4 years. They are (see Fig. 4.16):

Figures 4.15 and 4.16 show a small variation in SVR area edges as well as the singularity of two pilot nodes in the same substation: Chungliao-S and Chungliao-N, due to the continuous operation at separated bus bars feeding independent lines.

Fenglin Area is separated from the others by the North–South mountain chain, and it is the weakest, but it is still selected because it has a very low electrical coupling with the remainder of the power system. Obviously, SVR area selection is also a very powerful result for recognising areas needing additional reactive power resources, like Fenglin and Tungshan.

Pilot Nodes and Control Power Stations in South Korea

This is the case of a meshed grid subdivided into SVR areas. Moving from the 2006 to 2010 South Korea system cases, the operated SVR area selection shows robustness, confirming the validity of the Fig. 4.17 solution for the 2010 case study.

We also note that SVR areas represented by use of an electrical scheme: while it makes evident the lines and substations shared among the SVR areas, it does not allow for a quick glance at the SVR's impact on the territory map. Conversely, the use of a topographical picture as in Fig. 4.17 is certainly more useful and simpler to read.

Pilot Nodes and Control Power Stations in South Africa

The South Africa EHV grid analysis shows an SVR proper and robust structure based on 10 pilot nodes [35]. The result appears adequate to voltage control within the entire grid for the considered case studies related to the years 2007 and 2008.

Bus bars at 220 kV and below are not considered the pilot node selection because they are not strong enough to impose voltage on the surrounding local and higher voltage buses. Accordingly, pilot node selection only considers bus bars with nominal voltages of 765 kV, 400 kV and 275 kV. As already said, pilot nodes should be load buses; therefore, all generating HV bus bars have been excluded from the choice because they are also characterised by low sensitivity coefficients. This is because of their proximity to the generator bus bars having a constant voltage imposed by the AVRs.

The iterative method used to select pilot nodes usually requires, after fulfilling the criteria of the first list of automatic pilot node selection, some adjusting checks on the following:

- Short circuit power of selected buses;
- Proper choice of coupling threshold used among the pilot nodes;

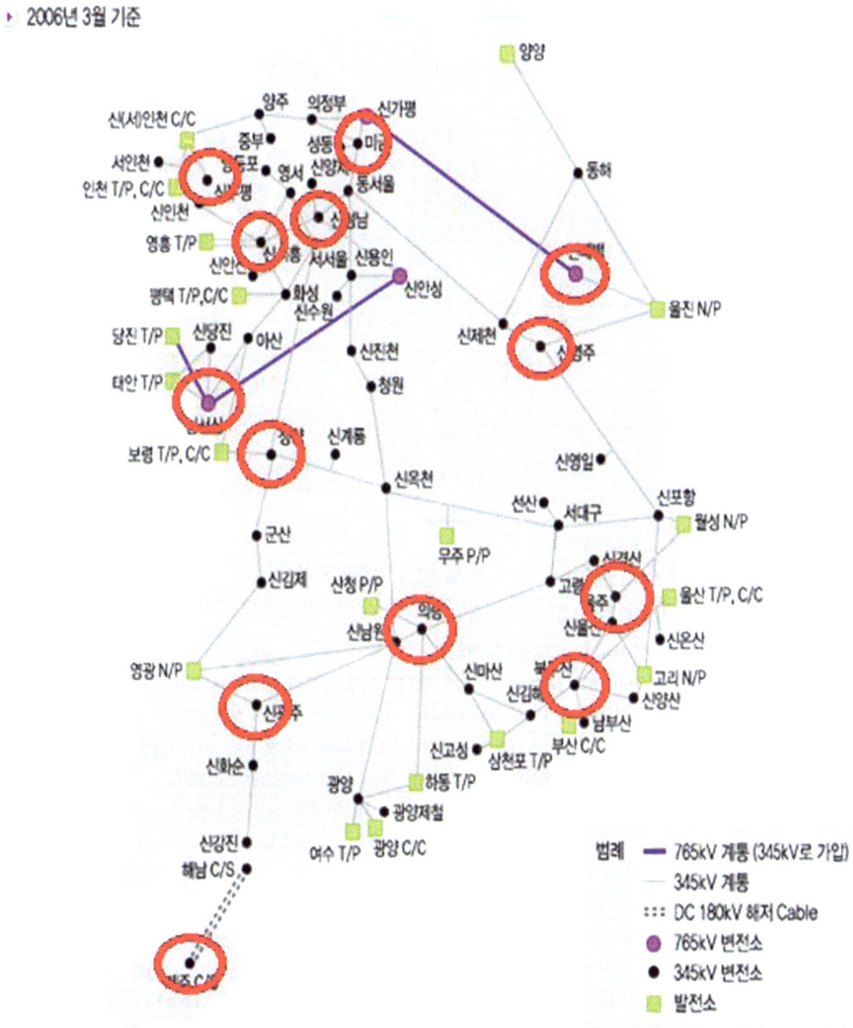

Fig. 4.17 Application plan of the South Korea hierarchical voltage control system: pilot nodes, control power stations and related areas for the 2006 and 2010 EHV grid case studies are shown

- Load location with respect to selected pilot nodes;
- Available control resources for each pilot node.

This detailed analysis, strongly supported by a knowledge of power system characteristics, might require further tests so that the list may eventually be modified/improved.

Also in the South Africa case, the selection of pilot nodes is a combination of the automatic algorithm conditioned by the coupling threshold choice and the selection/inhibition of specific buses. From here a summary view of the steps performed is

Table 4.2 Automatic pilot nodes in 2007 winter peak load scenario; sensitivity 0.08e–4 [p.u./MVAR]

Pilot node	Nominal voltage [kV]	Short circuit power [MVA]	*dV*/*dQ* sensitivity [p.u./MVAR] $\cdot 10^{-4}$
Alpha	400	33,131	0.173340
Zeus	400	27,452	0.290050
Vulcan	400	21,329	0.384950
Apollo	400	19,663	0.446230
Glockner	275	15,599	0.458350
Perseus	400	14,252	0.506710
Muldersvlei	400	9927	0.594050
Pegasus	400	12,812	0.683810
Leseding	400	12,897	0.705860
Spitskop	400	8590	1.094510
Komati	275	6874	1.369940

given to produce a better understanding of analysis peculiarities and the improvement of rough results, automatically provided by an automatic algorithm.

First Pilot Node Automatic Selection

The automatic selection of pilot nodes fixing the maximum sensitivity allowed among pilot nodes at 0.08e–4 [p.u./MVAR] gives the list reported in Table 4.2.

As reported in Table 4.2, the most powerful node in the grid, Alpha, is the first bus bar chosen by the algorithm because it has the lowest sensitivity coefficient. However, this node, as has been recognised, cannot be considered a pilot node because it defines a very small area with unimportant load bus bars associated to it.

The location of nodes in Table 4.2 totally depends on the topology and characteristics of the grid. In the northern section of South Africa (see Fig. 4.18), where the majority of power generation and densest grid topology are present, a consistent number of important pilot nodes are individuated: Alpha, Zeus, Vulcan, Apollo, Glockner, Spitskop and Komati. Even though Komati 275 kV is the HV bus bar of a power station, it is not excluded from the choice because this power station was out of service in the 2007 winter peak load scenario.

In the remaining, very large, part of the country, only four pilot nodes are chosen: (i) Muldersvlei, in the Cape Town area; (ii) and (iii) Perseus, competing with Ruigtevallei, near the important Hydra station); and (iv) Pegasus, in the Natal area. No one pilot node is automatically selected for the Eastern Cape, while Leseding, in the extreme northern province of Limpopo, appears too peripheral and without control power stations (Steel Port power station is decommissioned). In the northwest part of the grid, no pilot node is selected because buses there have low short circuit power and because of the total absence of generators and SVC in the area.

Starting from this list and in accordance with considerations of short circuit power, coupling factor, location of loads and resources available for controlling pilot node voltages, another test is done that excludes the Alpha node from the selection, preferring Perseus bus to Hydra bus and with a small retouching of the sensitivity

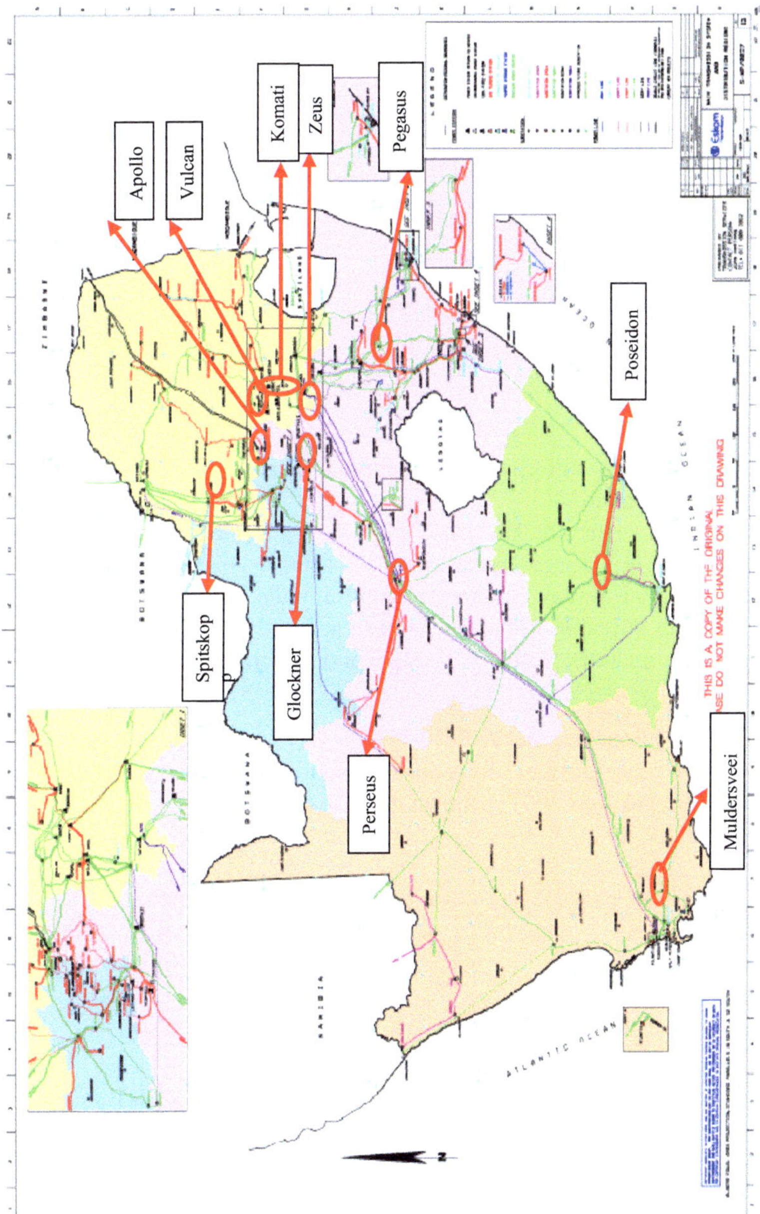

Fig. 4.18 Pilot node selection of the secondary voltage regulation in South Africa

Table 4.3 Automatic pilot nodes in 2007 winter peak load scenario: sensitivity limit among pilot nodes = 0.15e–4 [p.u./MVAR]

Pilot node	Nominal voltage [kV]	Short circuit power [MVA]	dV/dQ sensitivity [p.u./MVAR] · 10^{-4}
Grassridge	220	2026	0.408099
Zeus	400	27,452	0.290050
Vulcan	400	21,329	0.384950
Apollo	400	19,663	0.446230
Glockner	275	15,599	0.458350
Perseus	400	14,252	0.506710
Muldersvlei	400	9927	0.594050
Pegasus	400	12,812	0.683810
Leseding	400	12,897	0.705860
Spitskop	400	8590	1.094510
Komati	275	6874	1.369940

threshold. In Table 4.3 is given the list of pilot nodes automatically selected for the 2007 peak load scenario.

From the list, we see the Eastern Cape now has a pilot node: Grassridge, by choosing Perseus in place of Hydra as the pilot node. This selection appears important for obtaining a more equilibrated control in the southwest, with three pilot nodes (including Poseidon in the selection to support peripheral loads) in a very wide region that includes Northern, Western and Eastern Capes and the Free State south.

Final Pilot Node Selection in the 2007 Winter Peak Load Scenario

From the above considerations, the proposed list of pilot nodes for the South Africa network winter peak load 2007 is shown in Table 4.4.

Table 4.4 Winter peak load 2007 pilot nodes, short circuit power and sensitivities

Pilot node	Nominal voltage [kV]	Short circuit power [MVa]	dV/dQ sensitivity [p.u./MVar] · 10^{-4}
Zeus	400	27,452	0.290050
Vulcan	400	21,329	0.384950
Apollo	400	19,663	0.446230
Glockner	275	15,599	0.458350
Perseus	400	14,252	0.506710
Muldersvlei	400	9927	0.594050
Pegasus	400	12,812	0.683810
Spitskop	400	8590	1.094510
Komati	275	6874	1.369940
Poseidon	400	3855	1.727890

Final considerations:

- There are 10 pilot nodes, 6 of them located in the north, near Johannesburg. This is due to the concentration of generators and loads in this area; 2 of these 6 pilot nodes mainly control the 275-kV grid voltage around Johannesburg;
- Alpha 400 kV, the most powerful node in the network, is not chosen as a pilot node because its associated area is very poor, without loads and generators;
- Pegasus pilot node is chosen because it is located in the middle of an area that includes important load centres such as Invubu, Impala and Durban; this will contribute to maintaining a more homogeneous voltage profile in the area;
- The area of Cape Town with an important amount of generation and load has Muldersvlei 400 kV as the pilot node;
- In the centre-south of the network, a very vast geographic area, there are only 3 pilot nodes due to the restricted number of generator resources and the absence of significant loads (except in the Cape area); these nodes are Perseus, Poseidon and Muldersvlei, all with nominal voltage equal to 400 kV; Perseus is mostly controlled by SVCs inside the area;
- Poseidon has been selected because it is more central than Grassridge with respect to area loads. This choice needs to be confirmed by further static and dynamic tests;
- In the extreme north: Limpopo region, the analysis indicates the possibility of another pilot node at Leseding or Witkop, but the absence of local generation or SVC pre-empts such a proposal. This is another point to be deeply examined through static and dynamic tests.

In what follows, in Fig. 4.18, locations of pilot nodes for the 2007 winter peak load scenario are reported. This selection is also confirmed by the 2008 case study. Generally speaking, pilot node selection and SVR control structure must fit the grid properties best, showing the highest achievable robustness. It should be noted in the following tables that the South Africa summer minimum load is 51 % lower than the winter peak load. The specific differences between summer and winter will be clear from a comparison of Tables 4.5 and 4.6, which summarise 2008 winter peak and 2008 summer minimum load data, respectively.

The 2007 and 2008 winter peak load data are similar. Hence, network operating conditions between summer and winter, and not between consecutive winter peaks, differ the most. Consequently, the robustness of pilot node selection and SVR control structure is mainly tested using the winter 2008 and summer 2008 cases. The

Table 4.5 South Africa power system—2008 peak load main data

	Active value [MW]	Reactive value [MVAR]
Generation	35032.12	6835.19
External infeed	1269.80	163.29
Total load (U)	35496.05	8144.28
Load (Un)	35340.05	7985.34
Load (Un-U)	−155.99	−158.94
Grid losses	805.69	1236.64

Table 4.6 South Africa power system—2008 minimum load main data

	Active value [MW]	Reactive value [MVAR]
Generation	16061.81	703.51
External infeed	1535.10	318.47
Total load (U)	17215.73	3147.96
Load (Un)	17142.10	3033.10
Load (Un-U)	−73.63	−114.86
Grid losses	381.00	−8226.33

performed analyses on pilot nodes and control power plants (conducted on peak and minimum loads for 2008, and on peak load for 2007) resulted in the same selection: they confirm the uncommon high robustness of the SVR control structure of the South Africa HV grid. The selections are:

- The South Africa HV grid is divided into10 SVR areas—see Fig. 4.18;
- The plan includes the largest thermal and nuclear units connected to the 765/400/275-kV grids;
- For the 2007/2008 system at winter peak (system load 35,000 MW):
  - 10 pilot nodes
  - 74 control generators (CG): (19 power stations; max 91 CG)
  - 6 SVCs under control
  - 10 areas.
- For the 2008 summer minimum (system load 17,129 MW)
  - 10 pilot nodes
  - 52 control generators (15 power stations; max 91 CG)
  - 6 SVC.

Tables 4.7 and 4.9 list the pilot nodes (PN) (see Fig. 4.18), control factor (CF) and control generators (CG) of each SVR area. The comparison of the 2007/2008 peak

Table 4.7 Pilot node, control factor and control generators for 2007 and 2008 winter peak

Pilot node (kV)	Control factor	Control generator
Zeus 400	3059	Matla 400 kV; Kriel; Grootvlei
Vulcan 400	1784	Hendrina; Duvha
Apollo 400	1225	Kendal
Glockner 275	2008	Lethabo
Perseus 400	856	Tutuka; Perseus SVC; Hydra SVC
Muldersvlei 400	2810	Koeberg; Palmiet; Ankerlig; Gourikwa
Pegasus 400	1275	Drakensberg; Camden; Majuba; Athene SVC; Impala SVC
Spitskop 400	968	Matimba
Komati 275	1563	Matla 275 kV; Komati
Poseidon 400	211	PortRex; Poseidon SVC

Table 4.8 Reactive power resources available during the 2007 winter peak

Total potential (Po) control generators	48,095 MVA
Total operating (Op) control generators	41,620 MVA
Total over-excitation control reserve	(Po) 24,446 MVAR; (Op) 21,276 MVAR
Total under-excitation control reserve	(Po) 16,848 MVAR; (Op) 14,674 MVAR

Table 4.9 Pilot node, control factor and control generators for 2008 summer minimum

Pilot node (kV)	Control factor	Control generator
Zeus 400	3383	Matla 400 kV; Kriel; Grootvlei
Vulcan 400	2145	Hendrina; Duvha
Apollo 400	1368	Kendal
Glockner 275	2695	Lethabo
Perseus 400	706	Tutuka; Perseus SVC; Hydra SVC
Muldersvlei 400	3200	Koeberg; Palmiet; Ankerlig; Gourikwa
Pegasus 400	1519	Drakensberg; Camden; Majuba; Athene SVC; Impala SVC
Spitskop 400	1207	Matimba
Komati 275	1638	Matla 275 kV; Komati
Poseidon 400	320	PortRex; Poseidon SVC

load results with the 2008 summer minimum load case shows that the pilot node and control power station selections are the same: i.e., robust selection, as expected. The high degree of robustness implies that future expansions should have a limited effect on the SVR scheme. The CF for a given control area (CA) is proportional to the amount of reactive power under control in that area by the electrical distance of the mentioned area's reactive power resources to the local PN. The higher the CF value, the more controllable the CA voltages. Hence, Poseidon (the weakest CA) is characterised by frequent and large reactive power changes. Perseus is the second weakest CA. The value for CF increases with reduction in the power system load. A large amount of reactive power resources (capable of fast reactive power injection into the network) are at the disposal of the South Africa HV grid for network voltage control by SVR (refer to Table 4.8 and 4.10). These reserves (mainly provided by the generators) confirm the SVR potential control of South Africa grid based on their timely use during normal and contingency/emergency conditions to maintain a predefined network voltage profile.

Table 4.10 Reactive power resources available at the 2008 summer minimum

Total potential (Po) control generators	45,491 MVA
Total operating (Op) control generators	28,248 MVA
Total over-excitation control reserve	(Po) 23,171 MVAR; (Op) 14,729 MVAR
Total under-excitation control reserve	(Po) 15,274 MVAR; (Op) 10,184 MVAR

4.2.5 *Examples of Control Apparatuses Required by SVR*

We now give the main characteristics of the new control apparatuses required by a hierarchical voltage control scheme, our main reference being applications in Italy. The pictures shown (already published) refer to control apparatuses that were developed and applied on-field before the year 2000. Moreover, we offer a brief summary of interventions required for the interface of new SVR-TVR control apparatuses with already existing power station equipment (i.e., generator's AVR) and dispatcher control room.

While electronic solutions change with technology's evolution, control functions do not.

The SQR Apparatus: Functional Design and Technological Issues

Since 1985, the field of automatic voltage control has seen the advance of an innovative, microprocessor-based, voltage and reactive power regulator SQR named REPORT [1, 5]. Its rich and sophisticated functionality and friendly operator interface and monitoring capability (see Figs. 4.19 and 4.20) made it ready for on-field application.

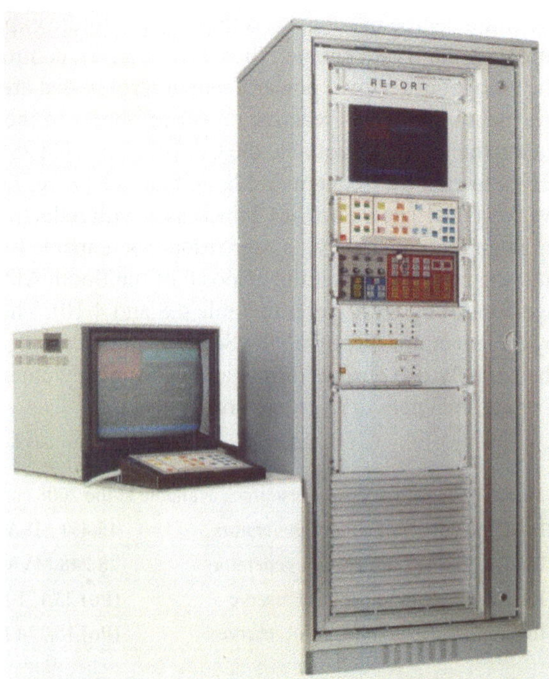

Fig. 4.19 REPORT microprocessor-based apparatus with power station control room monitor and keyboard operator interfaces

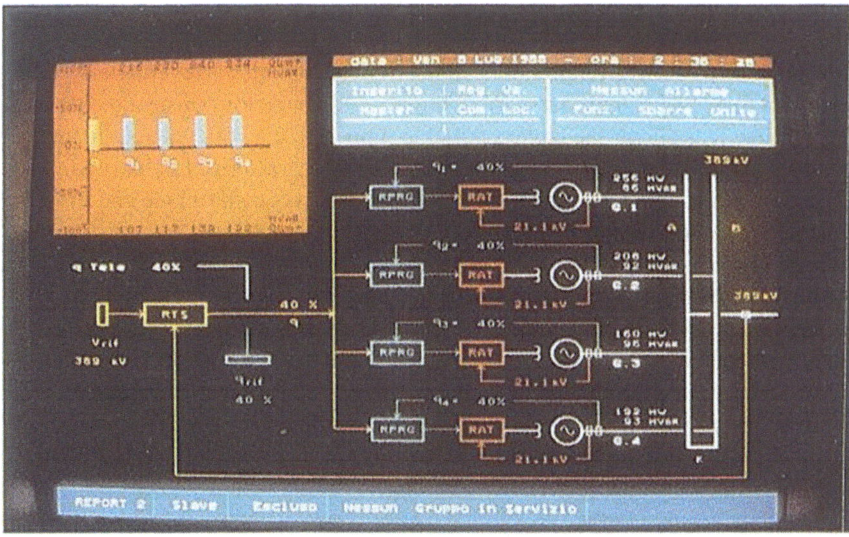

Fig. 4.20 REPORT apparatus—Examples of graphical user interfaces: plant on-line control scheme and voltage profile storing, linked to the calendar, for HSVR mode

REPORT is dedicated to regulating power station reactive power flow or voltage at the local HV bus bar by controlling AVR set-points of local generators and distributing, in a balanced way among all plant generators, the total reactive power generated/absorbed by a power station.

In the first control mode, REPORT regulates the plant generator's reactive powers through the reactive level signal tele-transmitted by the remote RVR. In the second control mode REPORT regulates (operating as an HSVR) power plant EHV bus bar voltage, according to suitable, memorised voltage daily trends (otherwise the set-point is at the disposal of the plant operator).

In both these control modes, REPORT regulates the reactive power of each generator by a closed loop (see Fig. 4.5 and 4.6) overlapping the primary voltage regulation provided by the AVRs. The set-point of each unit reactive power control loop is obtained by multiplying the reactive level signal (see Fig. 4.7b) by the reactive power limit of the considered generator. The unit limits of reactive power generation/absorption are computed, in real time, as a function of unit active power and voltage real values. Such over/under excitation capability limits also take into account the generator cooling system at actual operating conditions.

Main Functional Requirements

A secondary reactive power regulator SQR—a fully automatic control system—has to make use of intelligent algorithms to recognise, in real time and through local information, particular network contingencies (power plant islanding, bus bar isolation, etc.) and correspondingly to choose the most suitable control mode and to adapt regulation parameters. At steady-state operating condition, the reactive level signal is limited between minimum and maximum excitation. Nevertheless, during transients this level can exceed normal limits up to generator transient overloading capabilities. This allows the highest possible support to network voltages in the face of heavy perturbations. SQR dynamic behavior is characterised by two dominant time constants of about 5 and 50 s for unit reactive power control loop and EHV bus voltage control loop, respectively.

It is worthwhile to specify that the above mentioned response time constants concern the ambit of small perturbations while, for large perturbations, suitable gradient limits operate on the speed of reactive power variations according to generator constraints.

An SQR allows, through control parameter setting, the choice of deadbands for unit reactive power and EHV bus voltage control loop errors as well as selection of positive, negative or null static drop of EHV bus voltage regulation, depending on network conditions and on electrical couplings with adjacent SQR regulated nodes. Narrow deadbands, as used, pre-empt the negative effect of an open-loop transient condition, while an acceptable compliance with respect to set-point values is more correctly obtained with adequate static drop in the control law. All the transitions between SQR working states (start-up, shut-down, local bus bar voltage regulation, reactive power level tele-regulation, etc.), either demanded by the operator or ordered by the apparatus internal logic, are carried out through automatic procedures and tracking functions that guarantee bumpless commutations in every situation, so dodging undesirable transients. To avoid a situation where units operate outside their voltage and capability limits, suitable limitations and protections have been implemented in SQR: if one limitation appears on a unit, the action of the corresponding reactive regulator is stopped when the generator tries to go through that limitation.

A powerful performance of SQR comes from the integration, within its software, of a detailed real-time simulation model of the plant-network system. This dynamic model allows closing regulator control loops by a simulated power system, instead of the real power system.

The SQR simulation operating mode could also be at an operator's disposal during normal operation of an actual plant. It could become very useful during apparatus testing, functional checks and control parameter settings, as well as for operator and maintenance-staff training. SQR is also generally provided by rich supervisory and autodiagnostic functions, which continuously control the apparatus' correct running and its field interface effectiveness.

Graphical User Interface Requirements

REPORT is provided with a very friendly operator interface and rich monitoring features (see Fig. 4.20). At the operator's disposal are sophisticated graphic-based screen pages refreshed in real-time (animated pictures, signals and alarms, control parameters, memorised EHV bus voltage daily trends, etc.) and synthetic commands through a dedicated functional keyboard. During normal operation, all control parameters, as well as EHV bus voltage daily trends, can be directly modified via the REPORT user-friendly editor.

Installation Requirements

Placing SQR apparatuses in service is the first and also the heaviest activity, among those related to an SVR multilevel control system practical application. In fact, the large number of installations in power plants requires a certain amount of effort and consistent organisation by utility technicians to manage plant modifications, part of which requires the unit to be out of services. Wiring and interfaces are required towards generator excitation control systems and with the plant control room, where the SQR commands and status signals are available, together with plant subsystem controlled variables.

More precisely, modifications at the AVR for its correct interfacing with the SQR mainly concern the AVR's static, high precision calibrator (at least 12-bit D/A converters) and adoption of static relays with optical insulated up/down commands, to be repeated with a maximum intervention time of 10 ms. Also requested, with SQR in operation, is the automatic exclusion of AVR up/down operator commands, available at the plant control room desk, as well as the opening of the AVR's line drop compensation additional feedback and the signaling of rotor overloadability by AVR. Telecommunication exchange of measurements and commands between the local station and the RVR in the regional dispatcher control room also must be activated.

RVR Apparatus: Functional Design and Technological Issues

RVRs [2] installed at the regional control centres of the Italian transmission system operator (TSO) are lightly integrated with the local SCADA/EMS, basically to achieve communication with the plants. The RVR workstation is connected to the dispatcher's SCADA/EMS control system through a LAN Ethernet with TCP/IP communication protocol, as shown in Fig. 4.21. All control functions of the RVR are implemented inside the workstation. Therefore, the local SCADA performs for the RVR the function related only to data exchange with the controlled plants and communications via LAN Ethernet with the workstation itself.

If requested from the operator it is also possible to start an additional MMI on the EMS main computer through an X-Terminal connection to the RVR session. The RVR software architecture basically consists of a real-time system, cyclically performing activities connected with control functions involving the execution of a predefined sequence of operations within a constant time interval of 500 ms. Inside this time frame, scheduling criteria also allow performance of a portion of less critical activities, such as the updating of the MMI.\RVR software is developed to allow an easy adaptability of control apparatus to the particular configuration of the region it is applied to.

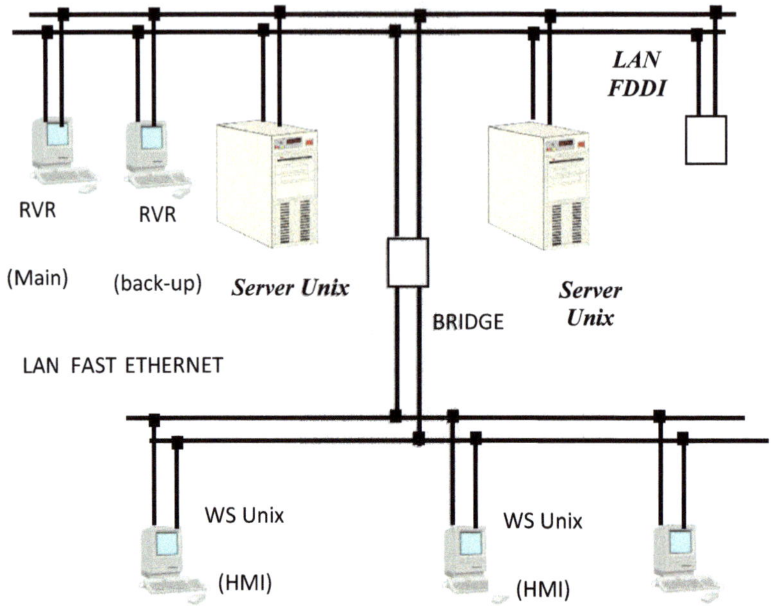

Fig. 4.21 RVR hardware and interconnection architecture

Control and Protection Requirements

A regional voltage regulator (RVR) regulates simultaneously, but with independent and parallel actions, the voltages of its pilot nodes through the remote control of reactive power production of power plants which most affect the voltages of those nodes. To accomplish this task, the RVR employs a separate voltage regulator for each pilot node in the region whose main characteristics are:

- A control law of the proportional-integral type, with a dominant time constant of the voltage control loop kept constant at a 50s value by a control algorithm of the adaptive type. This algorithm considers the number of groups taking part in voltage regulation of the pilot node, of the actual values of their capability curves and of the on-line computed value of the equivalent reactance seen from the pilot node, network side; dominant time constant selection satisfies the need to keep decoupled in time this loop from those it overlaps, therefore determining the dynamic behaviour of the pilot node voltage control loop similar to that of a first-order system.
- A control law proportional coefficient is dimensioned in order to avoid undesired transients of unit reactive powers (see Chap. 3).
- RVR design also achieves full dynamic decoupling and noninteraction among different pilot nodes voltage control loops controlled by the same RVR, avoiding oscillating transients of reactive power between neighbouring areas, mainly when their electrical coupling is not negligible. It is also possible to select a posi-

tive, negative or null static drop of pilot node voltage regulation, depending on local network needs and electrical couplings with adjacent pilot nodes.
- A starting of each pilot node voltage regulator can be controlled by the operator without the need for preliminary manual alignment of control generator voltages and pilot node set-point values.
- Pilot node voltage regulation of one area defines and updates in real time the values of the area reactive power level on the basis of pilot node voltage set-point, which can be defined locally by the manual calibrator (manual local reference); or which comes from the voltage profiles locally stored (automatic local reference); or which is sent by TVR. Tracking functions among pilot node voltage calibrators and corresponding controlled magnitudes enable at any moment bumpless switching between its operation modes.

One or two "vicar" pilot nodes are foreseen for each area to face possible failures of the main pilot node tele-operation equipment. In case of a modification of area network topology, the availability of vicar nodes enables the operator to select that pilot node which performs better than others the area pilot role in the new network topology. The selection of a particular pilot node is made by the RVR operator on the basis of local evaluations of network arrangement and/or by following a request coming from the national voltage regulator (NVR) at the national control room.

The RVR also allows: on-line configuration of the area control system in order to enable the operator to quickly select SVR control power plants and peripheral ones performing local HSVR; selection of substations enabled to SVR automatic control and their involved power components; selection of reactive reserves assigned to SVR; RVR control law parameter adjustment when significant variations occur into the local network.

In a particular network configuration, some control power plants, due to their geographic position, can gravitate to an area close to where they electrically belong. These boundary plants, called "peripheral" in the initial configuration phase, can either participate in tele-control of the pilot node they belong to or of neighbouring ones, as the network configuration varies.

In the RVR automatic local reference operation mode, the set-point value of each pilot node voltage is automatically updated on the basis of a voltage profile associated with the current day and stored in the RVR. For each pilot node in the region, the RVR has three different sets of daily voltage profiles stored on disk and associated with the 365 calendar days after the current day. Each one of these sets differs from others for its origin environment and for profile setting and programming methods. A voltage daily generic profile consists of 96 values corresponding to the set-point values to be implemented every quarter hour. Throughout each 15 min interval, automatic updating of the pilot node voltage reference takes place every minute, according to a ramp trend, connecting the current reference value to that which is foreseen for starting the subsequent interval.

The simulation function of regional voltage regulation can be an RVR tool. Its purpose is to enable the RVR apparatus to operate in a way similar to an actual case without the need of activating its interface to the plants. The simulation function represents, therefore, a particularly useful tool both during the functional test phase and the setting up of the RVR in the laboratory or for operator training. The network of the simulated region must be described with a sufficiently detailed model to consider all significant aspects useful for the reconstruction of phenomena connected with the regulation of network voltages.

To initialise simulation, specific commands are available allowing the setting of parameter values and of some significant electrical variables for calculation of the initial state of the simulated equivalent network. Specific commands allow the simulation of load variation, network perturbation, failure signals from plants or from tele-operation equipment, incorrect values of tele-measures and so on. Interface with the NVR is also simulated.

RVR has two classes of automatic acquisition and storage function of transient phenomena concerned with network voltage regulation:

- The first class relates to transient trends in ordinary operation and consists of recording every 5 min all adequately filtered measures, signals and alarms related to RVR operation. These acquisitions are always active and do not require the presence of any type of trigger since they document the operation of the SVR in ordinary operating conditions and for the previous days.
- The second class concerns the fast transient and allows high density recording of samples (one sample every 500 ms) for some minutes. This storage is oriented to the analysis of perturbations significantly affecting RVR dynamics. Acquisition is enabled by appropriate triggers that recognise heavy disturbed situations in the network and also allow documentation of initial phases of transients before the action of the triggers.

RVR control apparatus includes certain autodiagnostic functions for detecting and signalling possible internal failures or failures located in the interface with the regional EMS. RVR also includes diagnostic functions whose purpose is to acknowledge and signal major failure conditions tied to causes that can be attributed to its interconnected apparatuses (anomalous input signals, lack of execution of sent commands, etc.).

More precisely, RVR functions to:

- Perform consistency tests on I/O magnitudes;
- Detect particular operating conditions in the controlled region (an area in an island-operating condition, area close to voltage instability limit);
- Check the effects of the control actions undertaken, thus allowing diagnosis of possible failures on RVR interface devices.

Graphical User Interface Requirements

All graphic interfaces in the Italian RVR (2000 version) are implemented in the X-Window Motif environment (see Figs. 4.22, 4.23 and 4.24).

MMI pages are shared into different environments, the main one being real-time operation including graphic pages. They are organised according to a level struc-

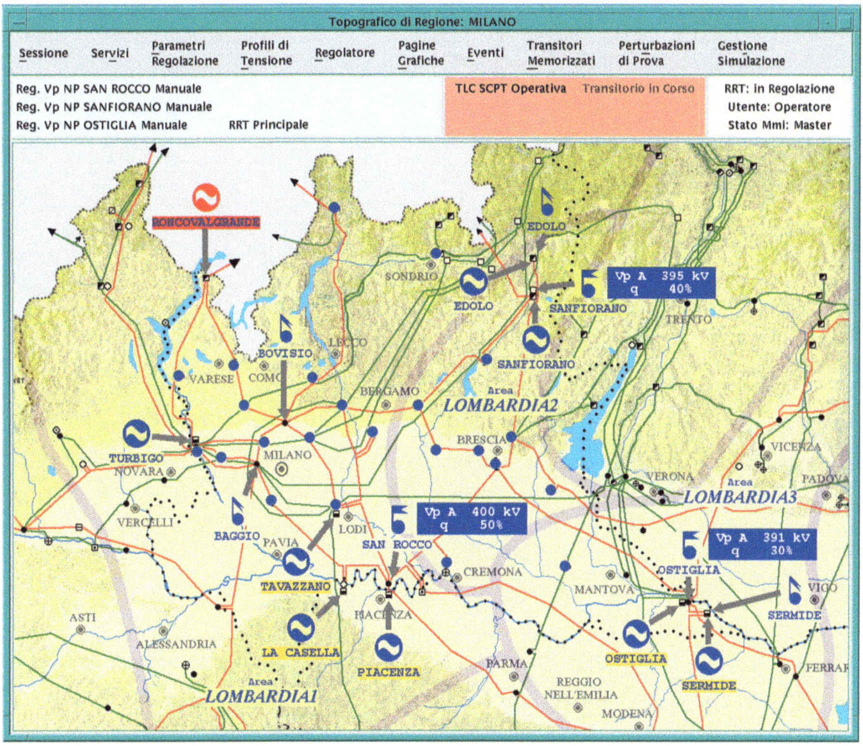

Fig. 4.22 Lombardy region topography with three SVR areas

ture, allowing an easy and organic navigation, starting from a basic page (Fig. 4.22). Pages associated with subsequent levels concern narrower portions of the controlled region and show higher detailed information content through area topographies, area and power plant functional block diagrams and tables. Each page enables RVR commands that are required to control the regulation environment displayed, showing the relevant alarms and signals. All commands can be operated through the mouse.

Additional MMI pages can be shared into the following environments:

- The control system configuration environment, inside which parameters defining control system structure and performance can be displayed, modified and stored;
- The "profiles environment", where the daily profiles of pilot node reference voltages are managed and stored;
  - The display of the "stored transients environment", where it is possible to display transients of recordings already stored on disk;
- The "chronological events environment", showing the sequence of "important" events (alarms, particularly significant signals, etc.) relevant to each area;

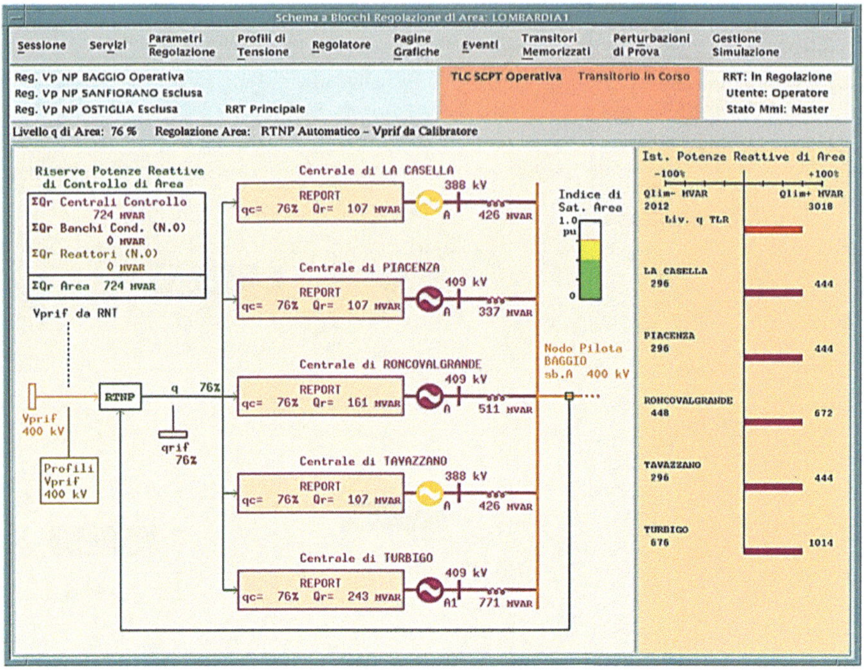

Fig. 4.23 Lombardy region, Baggio area: voltage regulation block diagram

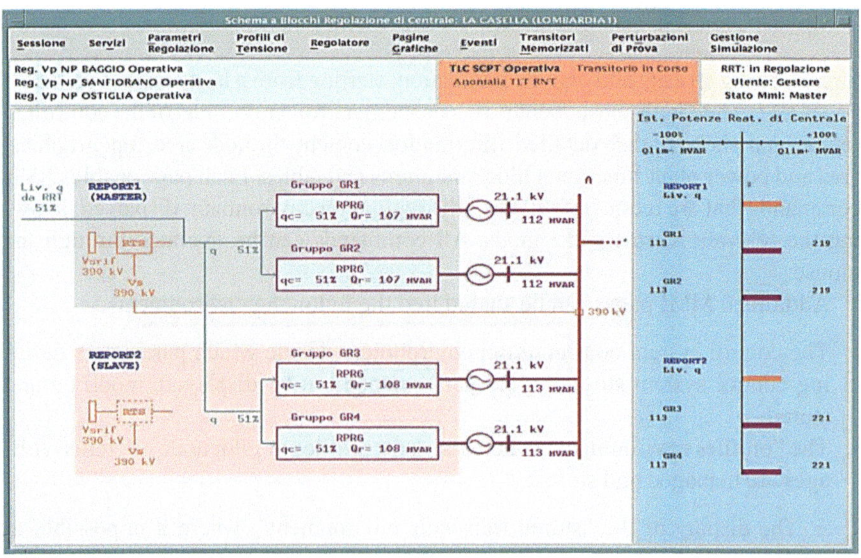

Fig. 4.24 Lombardy region: power plant Q regulation block diagram, with two REPORTs in master/slave configuration

- The “test perturbations environment”, allowing activating canonical perturbations to make tests on system dynamics. For each controlled area one can define: type, amplitude and duration of perturbation; addressed set-point values;
- The “simulation environment”, as duplicate of the operative MMI configured for the simulated region, where specific commands allow one to define the network operating condition and simulated plant (voltage values of bus bars, production of plants, etc.) or to reconstruct different failures of measures and signals received by RVR;
- The “management environment of historical files”, inside which the files residing on disk can be displayed and maintained. These files contain: regulation parameters and configuration, stored voltage profiles, transient recordings, and so on. This environment offers functions required for analysis of these data.

NVR Apparatus: Functional Design and Technological Issues

At the highest level (see Fig. 4.12) a national tertiary voltage regulator (TVR) coordinates in closed loop, for a secure and economic operation, the control action of all RVRs at the national/utility level, establishing the “pilot node voltage” pattern. On the basis of the actual state of the network and the forecasted optimal voltages and reactive powers, the TVR effects slow corrections in order to provide a better balance of reactive power generation levels and control margins among the areas [7]. The increase of the load margin of the transmission network, in the event of critical operating conditions from a voltage collapse point of view, is basically achieved by a proper coordination [8] between the TVR and RVRs finalised to prevent units from being operated at over-excitation limits (a condition linked to the tap-changer reverse action, which can anticipate a voltage-collapse mechanism triggering).

If SVR reactive power control margins are strongly reduced as a consequence of severe perturbation or abnormal load patterns, the TVR attends to the network voltage reduction, progressively renouncing to the optimal short-term planned voltage profile (defined according to economic or security reasons) for a sub-optimal solution. Therefore, the TVR allows power plants controlled by SVR to reach their capability limits only in case transmission network voltages are very low, notwithstanding all network reactive power resources being in operation for voltage support. In this way, a risk reduction of the voltage-collapse triggering, related to over-excitation limit intervention, as well as an increase of the overall loadability of the transmission system are achieved.

In conclusion, coordination between TVR and SVR represents a real opportunity for grid operators to achieve a full exploitation of transmission network power transfer capabilities. This requirement is becoming increasingly important in restructured and liberalised electric energy markets.

Combined with TVR, the NVR’s further main functionality is loss minimisation control (LMC). LMC defines in large power systems the optimal forecasted voltage plan required as input to TVR. This off-line and very slow OPF computing is the main LMC activity based on system state estimation (SE), and it takes into account the constraints determined by the hierarchical structure of the SVR (pilot nodes and control power plants) and its control ties. On the basis of a forecasted state estimation, LMC computes in advance (i.e., the day before) a provisional optimal voltage and reactive power plan, which is stored and used by TVR.

If TVR recognises significant differences between expected and real system operating conditions, it requires the LMC recompute the updated optimal forecasted voltage plan by considering the most recent system state estimation (in the best case, the SE will refer about 5 min before). This delayed OPF will be continuously computed by LMC at every state estimation update and sent to the TVR until the stored and the new optimal forecasted voltage plans resemble each other. In addition, the LMC shows and compares for each area on-progress daily traces of pilot node voltage and required set-point, reactive power levels operating on the plants and optimal forecasted references used by the TVR.

Hardware and Software Architecture Requirements

The design of the NVR suggested the use of a dedicated platform, one that would be properly interfaced with communication subsystems to exchange information with the field data acquisition subsystem. This opened up a vision of NVR integration with the existing SCADA system.

The chosen hardware configuration consists of a workstation with a Digital Unix operating system. Minimally, equipment with memory (2 GB) and hard disk (30 GB) storage must be foreseen, together with the possibility of using one or two (for redundancy) Ethernet/serial communication boards. In addition, conventional backup devices (CDROM, TAPE or DAT) and advanced graphical features (19-in. colour monitor with mouse and keyboard) are required.

The activities necessary for the development of the NVR central unit and its integration with the communication subsystem or the existing SCADA concern, on the one hand, configuration of information exchange with data acquisition subsystems and, on the other, design and development of software code related to the monitoring and control functions, centralised services, a real-time database and man-machine interface.

On an NVR central workstation, a monitoring and control system will need to manage, on-line and in real-time, the following features: data acquisition and communication (DAC); monitoring and control functions (MCF); database administrator (master MDB application ADB); centralised services library (LIB); man-machine interface (MMI).

Database Requirements

The NVR regulation and control system database must fulfil a number of requirements, depending on both the stored data type (static data of the application database and dynamic data for the real-time database), required refreshing times (which determine additional constraints on monitoring and control function performance) and access violation problems (due to the necessity to preserve integrity and coherence of data accessed by different and concurrent processes). These needs are due, for instance, to the fact that a real-time process (RCF) has to exchange data through the database with processes that have lower time requirements (MMI).

4.2.6 SVR Dynamic Performance During Tests in Real Grids

4.2.6.1 Verified Performance and Field Tests

The main results of dynamic performance checking during commissioning field tests are here briefly described in relation to subsystems of Italian coordinated voltage control:

- Plant reactive power regulator (REPORT);
- Pilot node regional voltage regulator (RVR);
- National voltage regulator (NVR).

Basically, commissioning activities are subdivided between the power plant level and the regional control centre level. Some information is also given on the studied performance of tertiary voltage regulation.

REPORT Commissioning

After verification of the coherency between AVR and REPORT limit settings and control parameter adequacy, as for instance the equivalent reactance externally seen by the power station, defined dynamic transients are carried out following step perturbations of the respective set-points.

The results shown in Figs. 4.25 and 4.26 refer to Italy's Edolo hydropower plant, equipped with eight generating/pumping units (160 MVA each). They are related to two different basic test groups:

- The single unit test under REPORT reactive power control: the dynamic response following step variation in the reactive level corresponds to a proper time constant of 5 s (from the top: reactive level, unit reactive power, unit terminal voltage, local EHV bus bar voltage);
- Multiple unit tests under REPORT HSVR control of the local bus bar EHV level: the dynamic response following the step variation in bus bar voltage set-point corresponds to a proper time constant of 50 s and is achieved through a coordinated control of all units (from the top: reactive level, four unit reactive powers, local EHV bus bar voltage).

The adopted design methodology of the REPORT control system includes both "time-decomposition" criteria among overlapped control loops of different hierarchical levels and "noninteractive" control law among control loops at the same hierarchical level. This is a strong guarantee of a high degree of stability in the overall control system: each control loop dynamic performance is comparable with a "first-

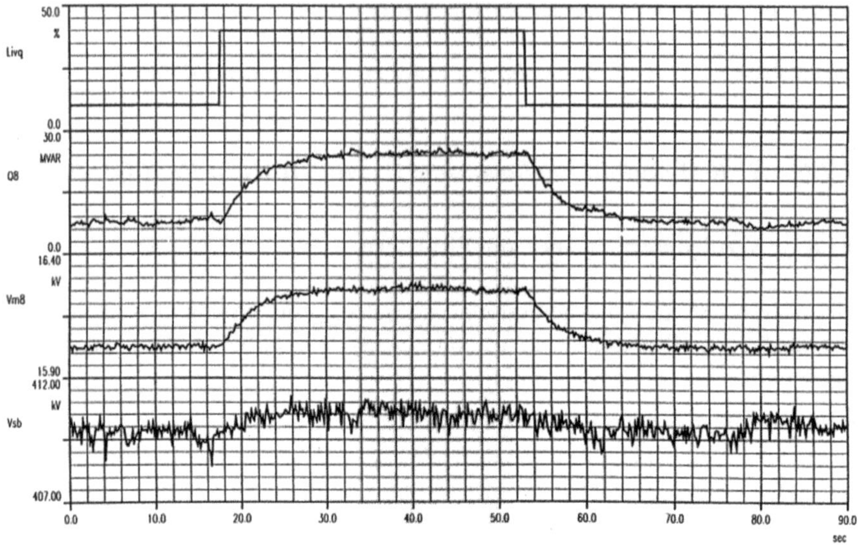

Fig. 4.25 Step response of a generating unit under REPORT reactive power control

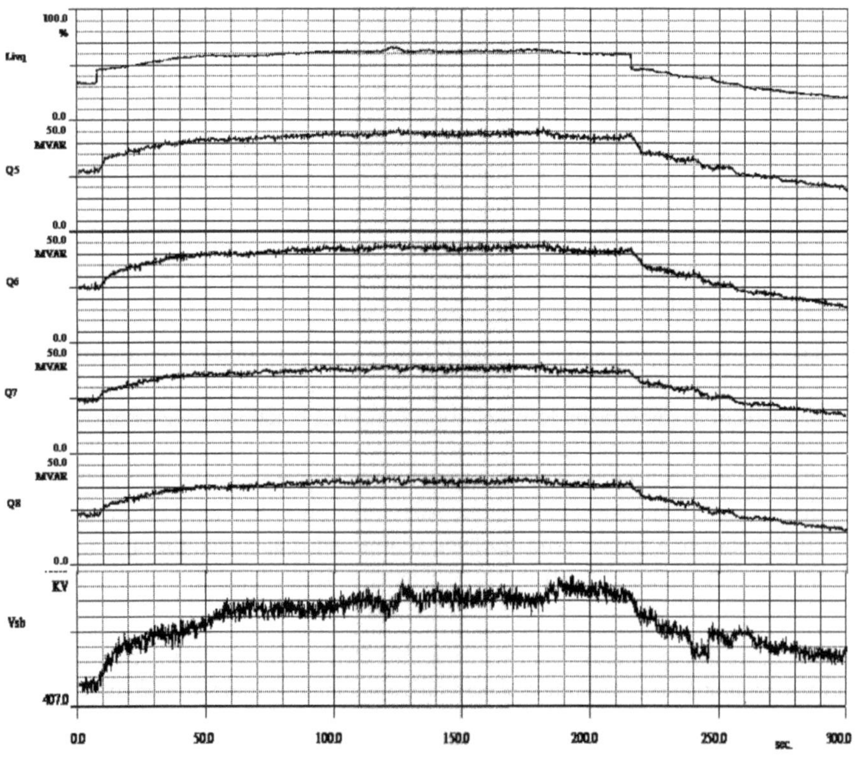

Fig. 4.26 Generating unit step response of REPORT HV bus bar voltage control

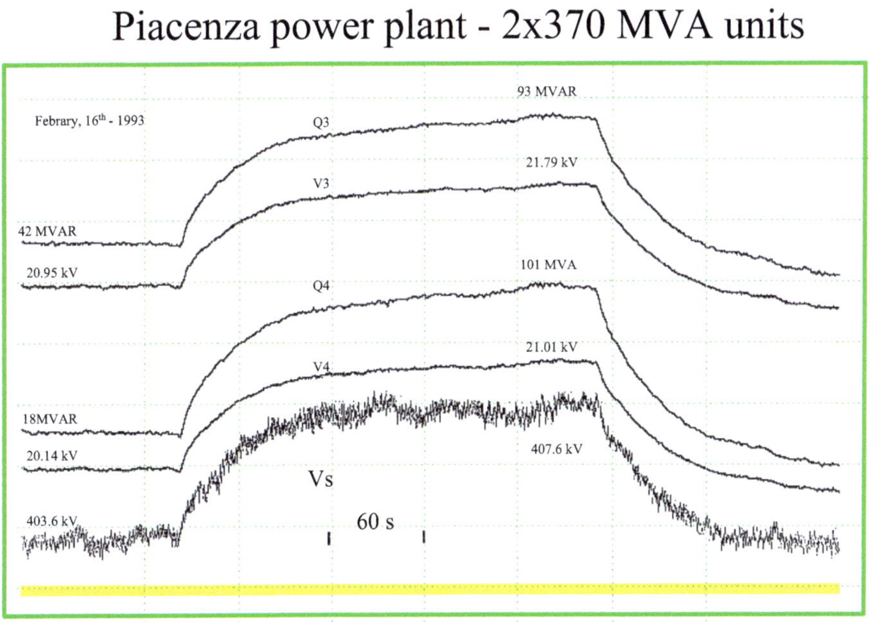

Fig. 4.27 Transient following a 1 % step variation of EHV set-point of power plant voltage and reactive power regulator

order dynamic performance" system characterised by a dominant time constant. HSVR transients in Fig. 4.27 clearly show this concept as achieved by REPORT on-field at the Piacenza power station.

RVR Commissioning

All I/O signals exchanged by the RVR require hardware/software intervention at a regional dispatcher SCADA system for generating the new database variables to be provided or received by RVR, managing either errors in communication or out-of-services system or peripheral device maintenance. Moreover, refreshing time on control loop measurements operated by telecommunication equipment must be reduced to 2 s max.

The following results refer to RVR commissioning activities carried out at the regional dispatcher in Milan during activation of the La Casella thermal power plant, which is equipped with two generating units controlled by a REPORT apparatus.

At activation time, other thermal power plants in the area (Piacenza, Tavazzano and Turbigo) were already in service under RVR control. Dynamic tests have been recorded at both the power plant and control centre levels, using the data acquisition system of the RVR internal data recording feature. The transients in Fig. 4.28 (a step-up followed by a step-down variation on area reactive power level "q", im-

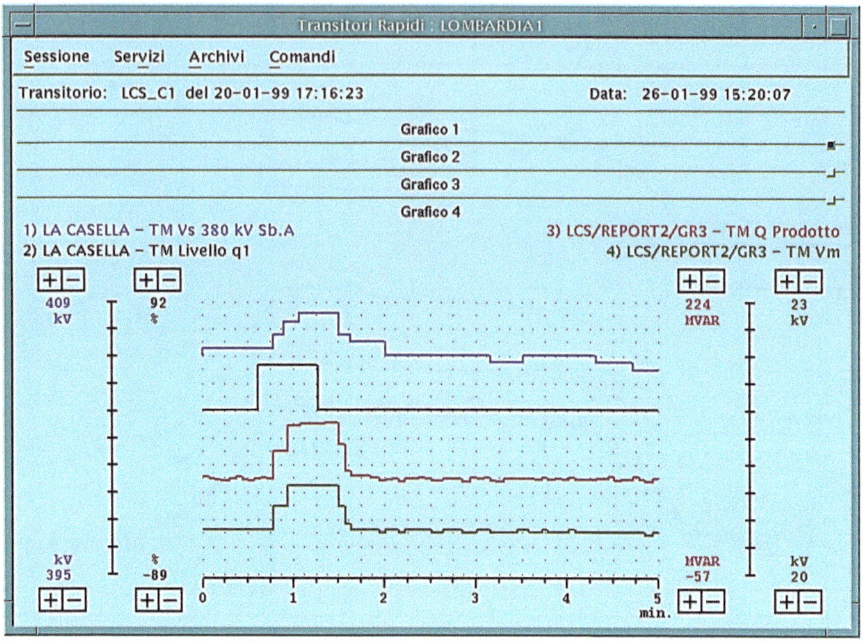

Fig. 4.28 RVR up-down step variation of the area reactive power level

posed by RVR test facilities) and in Fig. 4.29 (an RVR step-up variation of pilot node voltage set-point) makes evident delays between the area reactive level control and power plant unit voltage and reactive power reactions. These are due to telecommunication channel delays in sending/receiving control/measurement signals during tests: the level "q" is sent to the power stations, while the "Q" and "V_m" measurements come from the power stations. Transients show the coordinated and timely reaction of the peripheral power plants following RVR control (from the top: pilot node voltage, reactive power level sent by RVR, reactive power and voltage of one controlled generator at La Casella plant).

Pilot Node Voltage Regulation Test

Figure 4.30 offers evidence of the effectiveness of pilot node voltage regulation in the Lombardy region [10]. Throughout a full day (10 October 1999), the transient of the Baggio pilot node voltage overlaps the constant voltage set-point value (405 kV), maintaining the distance in the range: +1 kV ÷ to −1 kV. Differences are due to operating negative line drop compensation of the pilot node voltage regulation. In the figure, the lower transient represents the area reactive power level. The same transient is represented in Fig. 4.31 together with the reactive powers produced by the Baggio area control power plants (Piacenza, Tavazzano and Turbigo).

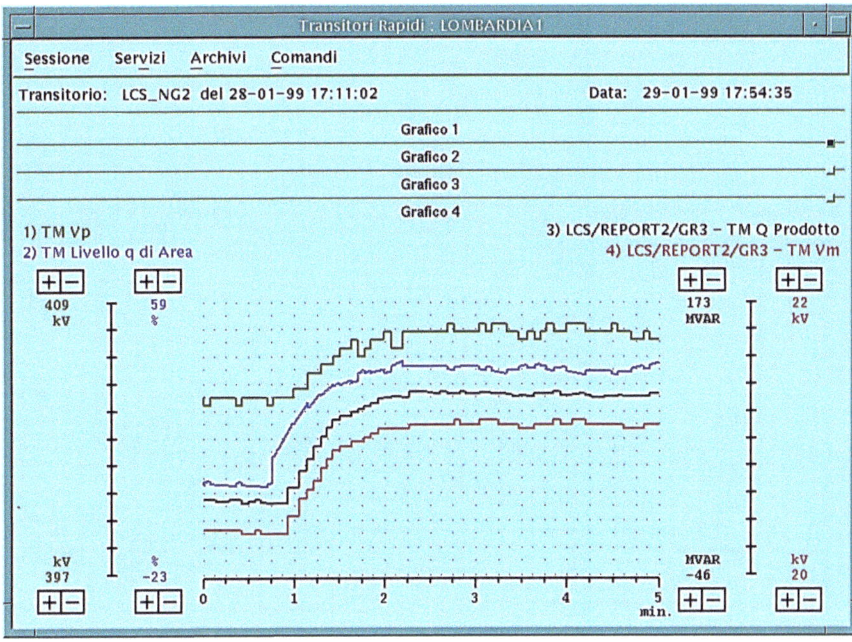

Fig. 4.29 RVR step variation of the pilot node voltage set-point

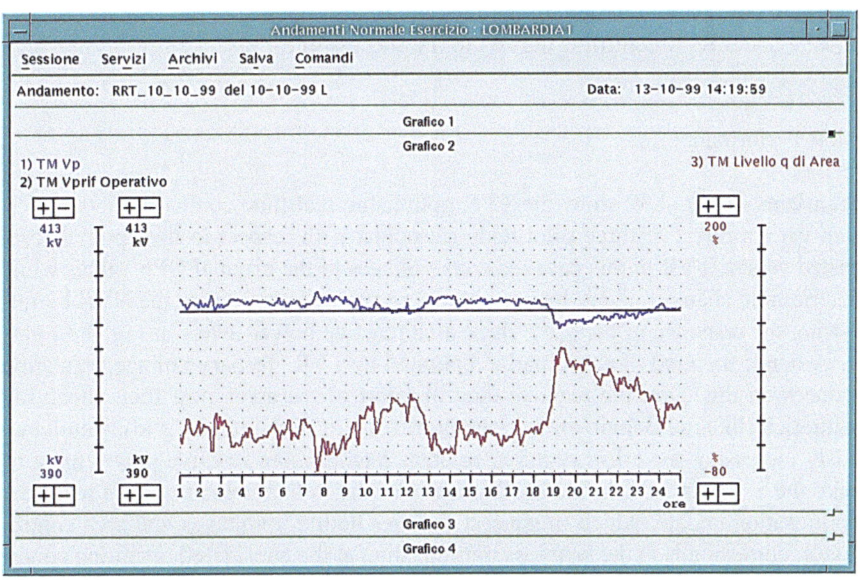

Fig. 4.30 RVR monitoring of a full day, the Baggio area pilot node voltage (*blue*) and set-point (*black*) + reactive power level regulating pilot node voltage, with line drop compensation

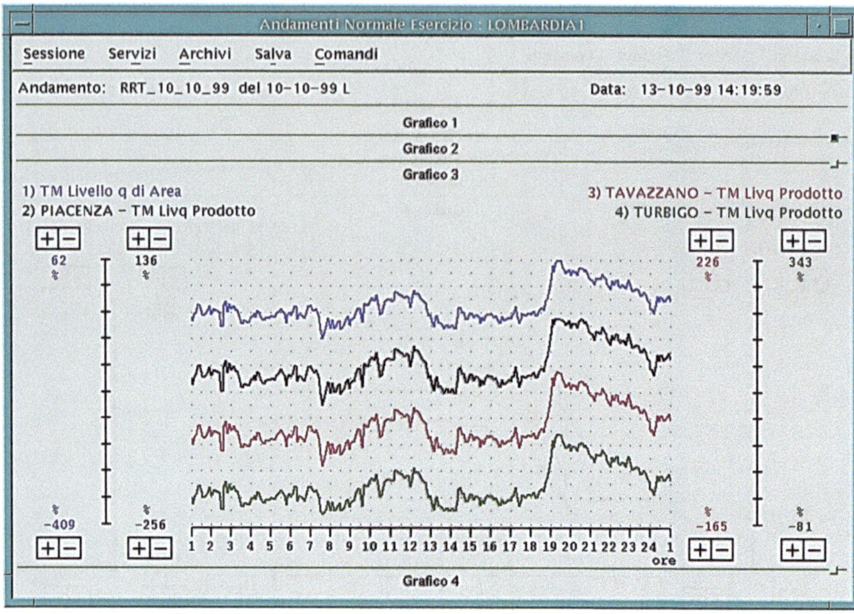

Fig. 4.31 RVR monitoring of a full day, the reactive power of Baggio area power station tracking area reactive power level to regulate Baggio pilot node voltage (Fig. 4.30)

The perfect unison in which all the area control power plants move their reactive powers under RVR control is the key to the success of SVR.

NVR Performance Test

Transients in Fig. 4.32 show the TVR optimising real-time control after the NVR step variation test of three pilot node set-points with respect to the optimal forecasted values. TVR in this case recovers voltages to the original OPF values while determining changes in day-before optimal control efforts. In fact, the NVR by following the response of the same three area reactive power levels in Fig. 4.33 puts in evidence the kind of coordination operated by TVR. Tertiary voltage regulation reduces, in this case, the reactive control effort of the areas near their capability saturation (like the Baggio area, in the northern part of the Italian grid), simultaneously increasing the effort required in areas having more reactive power margins (like the S. Lucia area, in the southern part of the Italian network). This real-time optimisation update, which in general changes both PN voltages and area control levels, corresponds to the best loss minimisation at the considered operating condition that also guarantees a proper control margin.

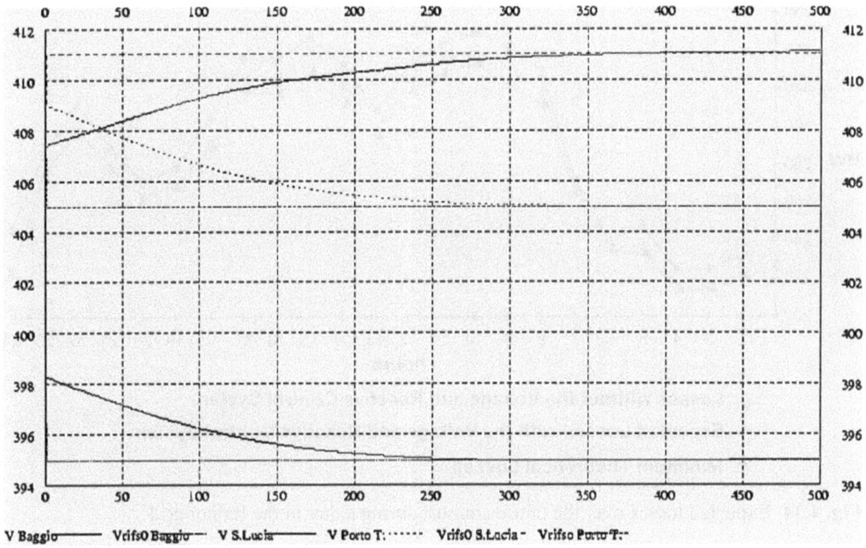

Fig. 4.32 NVR step variation of three pilot node optimal voltage set-points

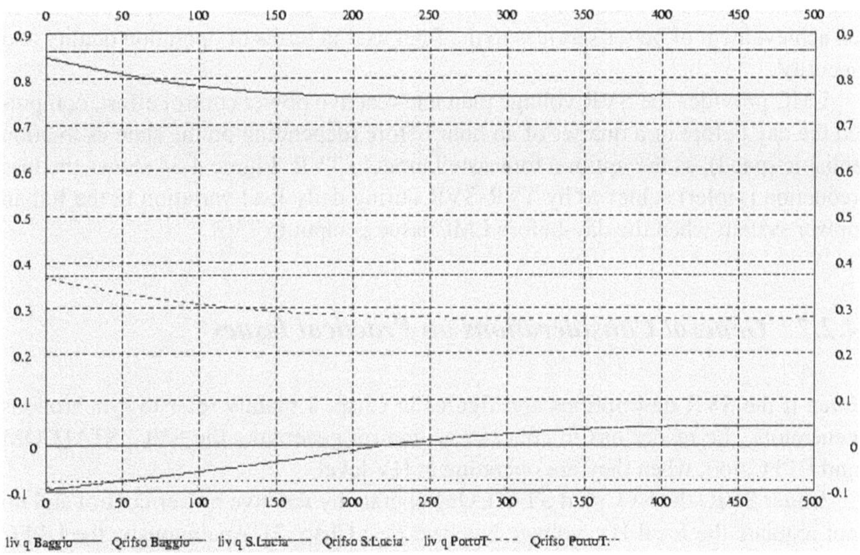

Fig. 4.33 NVR step response of three area reactive power levels

Static Analysis

The main objective of loss minimisation control (LMC) is the achievement of minimum losses in a power system, which is brought about by optimisation of voltage values in the network in such a way as to reduce overall system operating

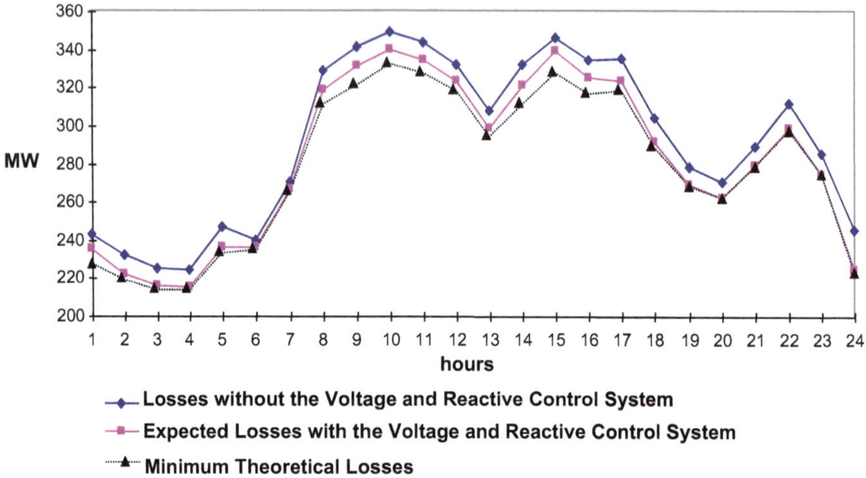

Fig. 4.34 Expected losses over the entire country during a day in the Italian grid

costs. Many static analyses carried out on large power systems have proved that the application of a multilevel control system for grid voltages and reactive power regulation allows a reduction of transmission losses of about 4–6 %, in addition to an achievement of better services to the final user in terms of operation quality and security.

LMC provides the SVR voltage plan and reactive power control effort, computed the day before or a quarter of an hour before (depending on the state estimation reliable speed), as the optimal forecasted input to TVR. Figure 4.34 shows the loss reduction (violet) achieved by TVR-SVR during daily load variation in the Italian power system when the day-before LMC issue is input to TVR.

4.2.7 General Considerations on Practical Issues

Even if the SVR descriptions and figures in Chap. 4 mainly refer to synchronous generators, the reader has to consider as control generators the SVC, STATCOM and UPFC, too, when they are operating at HV level.

Under SVR, the SVC and STATCOM operate by reactive power control and do not regulate the local HV voltage bus bars (see Chap. 3). Analogously, the UPFC works as a line reactive power flow control (Q) by the series converter or as reactive power control (Q_T) at the sending end (a local bus under regulation by shunt inverter): a simplification must be made by selecting which of the two UPFC reactive power controls (see Fig. 2.41) is maintained in the operation. In case the choice is made for line reactive power flow control, then the (Q) flow direction in the series converter is an additional input to be sent by the RVR. Moreover, plant reactive power control loops have to be tuned aligning their dynamics to the 5s dominant time constant that SQR imposes on the rotating generators.

The control apparatuses shown required by a hierarchical voltage control system are obviously reference examples that, in the year 2000, were considered a top solution in terms of technology and functionality. Obviously, one can envision in the changing electronics technology a possible trend from a centralised rack toward a distributed solution with increased peripheral intelligences, linked by optical fibers and local networks.

Considering SQR and SVR functionalities, the very rich description provided here is a good reference, largely covering present and future needs despite the hardware shown. Nevertheless, possible apparent simplification or a liberating cutting out some of supposed minor functionalities could happen, with a consequent risk of:

- Reducing reliability and autodiagnostics;
- Minimising operator monitoring and operation facilities;
- Increasing complexity of software and hardware manual checking and tuning;
- Increasing difficulties and on-field time consumption in pursuit of correct installation and commissioning.

Pilot nodes and control power station automatic computing is a very useful support, even to those having wide knowledge of power system characteristics. The iterative combining of automatically computed solutions with corrective upgrades coming from a deep knowledge of power system electrical characteristics and critical aspects of the operation is the correct way to achieve a very robust and effective SVR structure. An excessive number of pilot nodes should be avoided, as this necessarily increases the number of dynamic interactions among areas and could require too-frequent changes in the SVR control structure.

4.3 Conclusion

The central issue of Chap. 4 is the general description of the hierarchical structure of the automatic closed loop voltage control system of a transmission grid, including design criteria and recommendations for achieving a robust and effective control structure. Characteristics and dynamic performances of SVR and TVR voltage and reactive power control loops were introduced through classic time-response tests.

The results presented are based mostly on simplified but essential linear modelling and dynamic analysis of continuous closed-loop automatic controls. The model simplifications introduced do not determine any significant difference in reconstructed dominant system dynamics due to design characteristics of overlapped control loops. In fact, the first-order dominant dynamic performance of voltage/reactive power control loops considered could be easily achieved in a real power system simply by a correct tuning of control loop parameters.

We mentioned, in order, the slower pilot node continuous voltage control (a 50s dominant time constant) as compared to the faster, inner generator reactive power control (a 5s dominant time constant). The latter, in turn, is slower than the primary voltage control, which allows dynamics that are ten times speedier. Moreover, it was shown how pilot node voltage closed-loop controls are usually combined (as they are for HSVR) with a line drop compensation based on additional negative feedback from the area reactive power level.

Automatic procedures for selection of pilot nodes, related areas and area control generators were also presented, together with application results on some of the largest power systems in the world.

An example of control apparatuses required by SVR and TVR applications through both their hardware and software characteristics was given, as was a detailed description of solutions used in Italy and recommendations for their practical applications. New supports to power system operators provided by these control apparatuses in terms of data recording, real-time monitoring of SVR and TVR operating controls (from control centre to on-field power stations and substations), and simple commands through which operators are able to impose a grid voltage plan and strong autodiagnostics—all were described in detail.

These descriptions are useful for understanding the complexity control apparatuses can possess; they will be revisited in further presentations of transmission network voltage control performance, which must be acknowledged as on-field results of SVR-TVR real applications, and not simply as simulation results.

Possible differences between on-field and first-order performances as shown are due mostly to improper and incompetent maintenance of the control loops mentioned and in combination with fulfilment aspects that introduce disturbing and pointless nonlinearity into control solutions.

References

1. Arcidiacono V (September 1983) Automatic voltage reactive power control in transmission system. CIGRE-IFAC, Survey Paper E, Florence
2. Arcidiacono V, Corsi S, Natale A, Raffaelli C, Menditto V (1990) New developments in the application of Enel transmission system voltage and reactive power automatic control. CIGRE Paper 38/39-06
3. Corsi S, Marannino P, Losignore N, Moreschini G, Piccini G (1995) Coordination between the reactive power scheduling and the hierarchical voltage control of the EHV Enel system. IEEE Trans Power Syst 10(2):686–694
4. Corsi S, Arcidiacono V, Bazzi U, Chinnici R, Mocenigo M, Moreschini G (1996) The regional voltage regulator for Enel's dispatchers. CIGRE Conference Session Group 39
5. Corsi S, Arcidiacono V, Cambi M, Salvaderi L (August 1998) Impact of the restructuring process at Enel on the network voltage control service. Bulk Power System Dynamics & Control, IREP-IV, Santorini, Greece
6. Corsi S, Chinnici R, Lena R, Vannelli G, Bazzi U, Cima E (1998) General application to the main Enel's power plants of an advanced voltage and reactive power regulator for EHV network support. CIGRE Meeting, Paris
7. Corsi S (July 2000) The secondary voltage regulation in Italy, panel session on power plant (high side) voltage control. IEEE/PES Summer Meeting, Seattle
8. Corsi S, Pozzi M, Sabelli C, Serrani A (2004) Coordinated automatic voltage control of the Italian transmission grid, Part I: Reasons of the choice and overview of the consolidated hierarchical system. IEEE Trans Power Syst 19(4):1723–1732

9. Corsi S, Pozzi M, Sforna M, Dell'Olio G (2004) The coordinated automatic voltage control of the Italian transmission grid, Part II: Control apparatuses and field performance of the consolidated hierarchical system. IEEE Trans Power Syst 19(4):1733–1741
10. Paul JP, Leost JY, Tesseron JM (1987) Survey of secondary voltage control in France: present realization and investigations. IEEE Trans Power Syst 2:505–511
11. Lagonotte P, Sabonnadiere JC, Leost JY, Paul JP (1989) Structural analysis of the electrical system: application to the secondary voltage control in France. IEEE Trans Power Syst 4(2):479–486
12. Lefebvre H, Fragnier D, Boussion JY, Mallet P, Bulot M (July 2000) Secondary coordinated voltage control system: Feedback of EdF. IEEE/PES Summer Meeting, Seattle
13. Piret JP, Antoine JP, Stubbe M et al (1992) The study of a centralised voltage control method applicable to the Belgian system. Proc CIGRE, 39–201
14. Van Hecke J Janssens N Deude J Promel F (2000) Coordinated voltage control experience in Belgium. CIGRE Conference, Paris
15. Sancha JL, Fernandez JL, Cortes A, Abarca JT (1996) Secondary voltage control: analysis, solutions, simulation results for the Spanish transmission system. IEEE Trans Power Syst 11(2):630–638
16. Layo L, Martin L, Álvarez M (2000) Final implementation of a multilevel strategy for voltage and reactive control in the Spanish electrical power system. PCI Conference, Glasgow
17. Ilic DM, Liu X, Leung, G, Athans M, Vialas C, Pruvot P (January 1995) Improved secondary-new tertiary voltage control. Proceedings IEEE Winter Meeting, New York
18. Taylor CW, Venkatasubramanian V, Chen Y (May 2000) Wide area stability and voltage control. VII SEPOPE, Curitiba, Brasil
19. Taranto G, Martins N, Martins ACB, Falcao DM, Dos Santos MG (July 2000) Benefits of applying secondary voltage control schemes to the Brazilian system. In: Proceedings IEEE/PES Summer Meeting, Seattle
20. Lemons, FAB, Feijo Jr, WL, Werberich, LC, da Rosa MA (2002) Assessment of a transmission and distribution system under coordinated secondary voltage control. Power Systems Computation Conference (PSCC), Sevilla, Spain
21. Ilic DM, Lang J (July 2011) Complexity of voltage and reactive power dispatch in control centers: From analysis to on-line decision making. Proceedings IEEE/PES Summer Meeting, Detroit
22. Tong J, Souder D, Pilong C, Zhang M, Guo Q, Sun H, Zhang B (July 2011) Voltage control practices and tools used for system voltage control of PJM. Proceedings IEEE/PES SM, Detroit
23. Corsi S, Martins N (eds) (June 2005) Coordinated voltage control in transmission systems. CIGRE Technical Brochure, Task Force 38.02.23
24. Corsi S, Pozzi M, Bazzi U, Mocenigo M, Marannino P (2000) A simple real-time and on-line voltage stability index under test in Italian secondary voltage regulation. CIGRE (Paper), Paris
25. Corsi S, Arcidiacono V, Cambi M, Salvaderi L (August 1998) Impact of the restructuring process at Enel on the network voltage control service. Bulk Power System Dynamics & Control, IREP-IV, Santorini, Greece
26. Corsi S, Pozzi M, Biscaglia V, Dell'Olio G (2002) Fiscal measure of the generators support to the network voltage and frequency control in the ancillary service market environment. CIGRE Meeting, Paris
27. Berizzi A, Sardella S, Tortello F, Marannino P, Pozzi M, Dell'Olio G (August 2001) The hierarchical voltage control to face market uncertainties. Bulk Power System Dynamics & Control, IREP-V Conference, Onomichi, Japan
28. Corsi S, Pozzi M, Marannino P, Zanellini F, Merlo M, Dell'Olio G (2001) Evaluation of load margins with respect to voltage collapse in presence of secondary and tertiary voltage regulation. Bulk Power System Dynamics & Control, IREP-V, Onomichi, Japan

29. Marannino P, Zanellini F, Berizzi A, Medina D, Merlo M, Pozzi M (November 2002) Steady state and dynamic approaches for the evaluation of the loadability margins in the presence of the secondary voltage regulation. MedPower Conference, Athens
30. Corsi S, Pozzi M (2003) Multivariable new control solution for increased long lines voltage restoration stability during black startup. IEEE Trans Power Syst 18(3):1133–1141
31. Eremia M, Petricica D, Simon P, Gheorghiu D (July 2001) Some aspects of hierarchical voltage-reactive power control. IEEE/PES Summer Meeting, Vancouver, Canada
32. Erbasu A, Berizzi A, Eremia M, Bulac C (June 2005) Implementation studies of secondary voltage control on the Romanian power grid. IEEE PowerTech Conference, St. Petersburg, Russia
33. Ilea V, Berizzi A, Eremia M (November 2007) Optimal reactive power flow methodology in power systems with secondary voltage control. 3rd International Conference on Energy and Environment, Bucharest, Romania
34. Ilea V, Bovo C, Merlo M, Berizzi A, Eremia M (June/July 2009) Reactive power flow optimisation in the presence of secondary voltage control. IEEE PowerTech Conference, Bucharest, Romania
35. Corsi S, De Villiers F, Vajeth R (July 2010) Secondary voltage regulation applied to South Africa transmission grid. IEEE/PES General Meeting, Minneapolis
36. Sun H, Zhang B, Guo Q, Yang Z, Wang X, Li H (2003) Design for system-wide optimal voltage control system based on adaptive zone division. Automat Elect Power Syst 27(8):16–20 (in Chinese)
37. Sun H, Guo Q, Zhang B, Wu W, Tong J (July 2009) Development and applications of system-wide automatic voltage control system in China. IEEE PES General Meeting, Calgary, Alberta, Canada, Invited Panel Paper, Number: 09GM0707
38. Brasca C, Chinnici R, Pozzi M, Zaramella M (2000) Practical implementation aspects of a multilevel control solution for voltages and reactive regulation in electrical power systems. PCI Conference, Glasgow
39. Corsi S, Cappai G, Valadè I (2006) Wide area voltage protection. CIGRE Paper B5-208, Paris
40. Corsi S (June 2009) Wide area voltage regulation and protection. IEEE PowerTech Conference, Bucharest, Romania
41. Corsi S (July 2011) Wide area voltage regulation in Italy and related wide area protection. IEEE PES General Meeting, Detroit
42. Corsi S, Pozzi M, Biscaglia V, Dell'Olio G (2002) Fiscal measure of the generators support to the network voltage and frequency control in the ancillary service market environment. CIGRE Meeting, Paris
43. Kundur P, Taylor C (chairs) (2009) Blackout experiences and lessons, best practices for system dynamic performance, and the role of new technologies. IEEE/PES, PSDP Committee Working Group Technical Report
44. Guo Q, Sun H, Tong J, Zhang M, Wang B, Zhang B (July 2010) Study of system-wide automatic voltage control on PJM system. IEEE/PES General Meeting, Minneapolis

Chapter 5
Examples of Hierarchical Voltage Control Systems Throughout the World

Examples of SVR applications throughout the world are briefly presented here according to information available [1]. At present, real, concrete, wide area operating applications exist in Europe, mostly in Italy and France. Several other applications are in progress in Europe, United States and China. Most of the examples presented here refer mainly to results of studies in countries that have shown a clear interest and trust in SVR applications. Obviously, a complete overview of all the real studies and applications of SVR-TVR is not given in this chapter; many are discussed in other chapters of this book.

5.1 French Hierarchical Voltage Control System

5.1.1 General Overview

The coordinated voltage control of the French EHV grid operates at three different levels—primary, secondary and tertiary—which are temporally and spatially independent. Temporal independence means the three overlapped controls do not significantly interact; if they did, the risk of oscillations or instability could increase.

Primary control involves keeping generator stator voltages at their set-point values as according to what was explained in Chap. 3, § 3.3.2, and Chap. 4, § 4.1.3. This control performs partial automatic correction within a few seconds to compensate against rapid random variation in the EHV voltage.

Secondary voltage regulation is basically as presented in Chap. 4, § 4.1.4, with certain differences existing when reference is made to the original dated French SVR solution or when the more recent CSVC control system is considered. Secondary regulation involves splitting up the network into theoretically noninteracting zones, within which the voltage is controlled at the pilot nodes. The original SVR control system automatically adjusts the reactive power of selected generating units to control the voltage at a specific pilot point in the zone, this being considered representative of voltages at all points inside the zone.

S. Corsi, *Voltage Control and Protection in Electrical Power Systems*,
Advances in Industrial Control, DOI 10.1007/978-1-4471-6636-8_5

The more recent *coordinated secondary voltage control* (CSVC) was proposed to overcome certain limitations of the original SVR design and fulfilment. It is used in western France and is operated by automatic and direct adjustment of the AVR voltage set-points of the control generators, thereby optimising pilot node voltages. We will comment on this seeming simplification later.

At the highest level, tertiary regulation is a nationwide voltage-reactive power optimisation function determining the voltage set-points of the pilot points in order to achieve safe and economic system operation. Tertiary regulation in France is still not automated, but if it were it would have a dominant time constant of around 20 min or longer.

Automatic control of HV capacitors becomes necessary when high MVAR amount is required. Automatic control can be carried out at a local level—according to a voltage criterion, for instance, or it can be carried out centrally. The local approach may result in an insufficient and untimely use of all available reactive power sources in the event of an incident, or even due to functional incompatibilities. For this reason, actions have been oriented towards integrating HV capacitor control into the secondary voltage regulation system. As described in § 4.1.4, *integration is governed by the principle that capacitors are switched on a priority basis as soon as the need to increase reactive power generation arises in a given region*. In this way, a large reserve of reactive power can be maintained at the generator level, which is immediately available in the event of an incident. Capacitors are progressively switched on, beginning with those at the lowest voltage level.

In France, wide implementation of secondary voltage regulation began in 1979 [2]. More recently, France's transmission network comprises about 35 control zones, including about 100 thermal generators (conventional fuel and nuclear) and 150 hydraulic generators. Total reactive power capacity available to perform voltage control is estimated at more than 30,000 MVAR.

5.1.2 *Original Secondary Voltage Regulation and Its Limits*

The original French secondary voltage regulation system regulates voltage profile in each selected zone by distributing reactive power from the various regulating generators in the zone. A control system (see also § 4.1.4) comprising two distinct regulation loops is superimposed on the primary loop (AVR) of the regulating generators (Fig. 5.1).

A proportional-integral law is used to calculate a control signal N, also termed the "level" of the zone, from the difference between the set-point value at the pilot node and the voltage effectively measured at a given instant. The level thus indicates the zone's reactive power requirement:

$$N = \alpha \int_0^t \frac{V_C - V_P}{V_n} dt + \beta \frac{V_C - V_P}{V_n}, \tag{5.1}$$

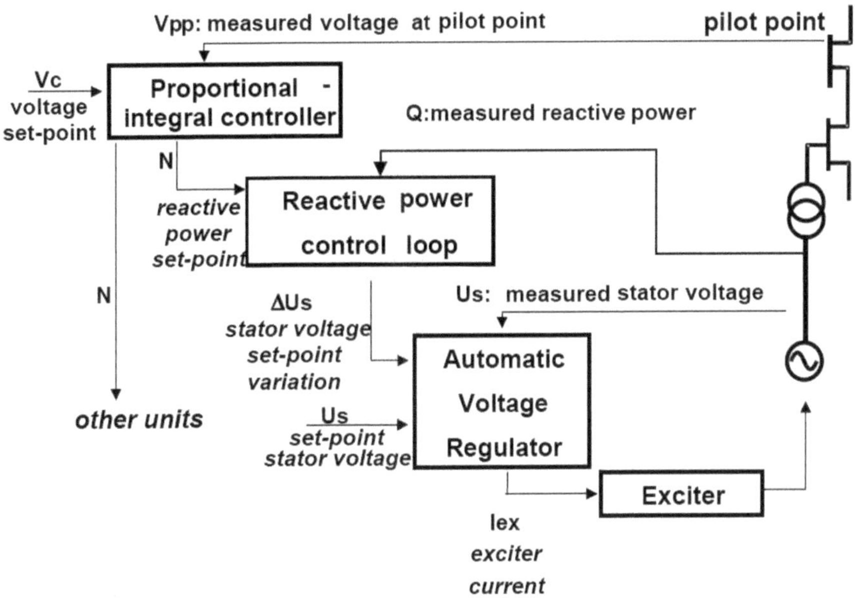

Fig. 5.1 Secondary voltage regulation block diagram [3, 4]

where:

α integral gain
β proportional gain
V_p measured voltage
V_c set-point voltage
V_n nominal voltage

This original SVR control scheme, which has been in operation since the early 1980s, confirms in principle the one shown in Figs. 4.1, 4.2 and 4.5, but its design and performance cannot be compared with those discussed in § 4.1.4 because of the following limitations:

- The control level N of each pilot node is calculated by a dedicated microcomputer located in the zone's regional dispatching centre. Therefore, each pilot node has a dedicated RVR regulator that becomes a zone regulator. This does not allow taking into account possible dynamic interaction among the pilot node controls of the edge zones, unless it is of a complex and burdensome data exchange;
- Very slow dynamic performance affects application: in Fig. 5.2a, the dominant time constant turns out to be greater than 30 min. This is very slow with respect to the declared 3 min and not comparable to the 50s faster and cleaner results shown in § 4.1.4.
- Abnormal transients due to SQR initialisation and a standby period of 5 min to allow reactive power alignment among generator sets;
- Need for an operator to take corrective action on the control level due to transients induced in primary voltage control systems.

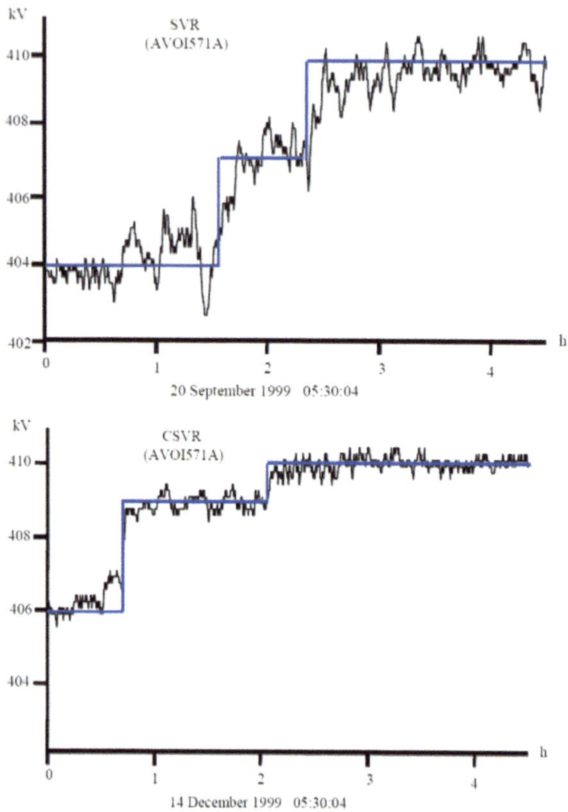

Fig. 5.2 Voltage set-point and voltage response with SVR and CSVC at EdF

Other limitations of France's secondary voltage regulation have been called structural [3, 4]. Still, they appear fully linked with choices made in the design of control apparatuses and in defining the Fig. 5.1 control structure in practice.

From here, some additional critical comments are provided on the reasons for the stated French limitations, by which the decision to move to CSVC is justified:

- If couplings between theoretically independent zones easily change as a result of grid development subsequent to the already implemented SVR, such changes confirm the weakness of the pilot node selection, whose robustness is conversely a mandatory issue. From this point of view, the choice of 35 pilot nodes in France's system appears to be a very high and critical number, easily compromising the robustness of zone selection in front of small system changes;
- SVR, as a suboptimal control, while requiring reactive power alignment of the generating units involved, does not make allowance for excessive demand that might be made on certain units as a result of differences in physical proximity. This limiting feature becomes a true and consistent problem mainly when the selection of SVR control generators is not proper; otherwise, alignment would not determine the relevance of renouncing SVR for the proposed CSVC, as stated;

- If internal reactive power control loops at the generating unit level should become a destabilising factor by amplifying the initial disturbance in the first few instants following certain incidents (generator drop-out, for example), the stability reduction would occur only if the proportional coefficient of the SVR area control law were not properly tuned or if telecommunication between RVR and PQR were too delayed.

Other declared SVR limitations are hardware and software design-related:

- If the control system provides only a partial permission of operating constraints (for example, it does not fully integrate monitoring of permissible voltage limits or generate set operating limits), it would go against the consolidated recommendation on what the SQR has to show to the power station operator. Moreover, the RVR operator should receive by SQR real-time information on generator voltage and reactive power permissible limits together with its operating state under SVR.
- If the use of fixed control loop parameters precludes optimum allowance for operating conditions, the solution of SQR and RVR adaptive control laws surely maintain their closed-loop dynamics at the designed values. Still, adaptive control is not a mandatory issue and might be a requirement in special cases only.

To sum up the situation very briefly, and looking to the § 4.1.4 results, France's SVR would have been capable of very satisfactory results if its control equipment design had been more accurate and had a less critical subdivision of the grid into zones.

5.1.3 Coordinated Secondary Voltage Control (CSVC)

A new voltage control system has been operating in western France since 1998. It is called *coordinated secondary voltage control* (*CSVC*) because regulating signals for neighbouring zones are no longer calculated on an independent basis, as was the case in the country's original SVR system [2], but rather according to the § 4.1.4 description.

CSVC design is based on a layout similar to that used in SVR (§ 4.1.4), with the objective of eliminating the practical limitations commented on above (§ 5.1.2). The basic principle governing a CSVC system continues to be that of regulating pilot node voltages at set-point values. However, the control signal is calculated for a "multiple region", comprising several pilot nodes, and the effect of individual generators on all pilot nodes is correctly taken into account, analogous to the control scheme in Fig. 4.5. The first relevant difference is that CSVC directly computes set-point updates of the primary voltage controls of the generator units by minimising the multivariable quadratic function [3, 4] described next.

This direct control of AVR set-points is a disputable simplification that a study—but not an on-field application—could justify. In fact, on-field direct control is obviously less precise than that provided by the reactive power control loop. This is because of AVR set-point offset in electronic circuits and the small range of voltage controllability (±5 %) with respect to the large reactive power range (100 %)

between the over- and under-excitation limits. Therefore, generator alignment in a power station through direct AVR control is more critical, and furthermore, the complexity of generator coordination and the risk of reactive power recirculation among generators in a power plant are increased. Clearly, through CSVC the design problem of correctly computing the SVR PI control law is skipped, on the other hand the process is open to criticism related to the choice of weights to be used in the quadratic control function shown next.

CSVC obtains generator set-point values by minimising the multivariable quadratic function:

$$\min\{\lambda_v \| \alpha (V_c - V_{pp}) - C_v \Delta U_c \|^2 + \lambda_q \| \alpha (Q_{\text{ref}} - Q) \\ - C_q \Delta U_c \|^2 + \lambda_u \| \alpha (U_{\text{ref}} - U) - \Delta U_c \|^2 \},$$

where:

α	control gain
V_{pp}, V_c	measure and set-point voltage at pilot points
Q, Q_{ref}	measure and set-point reactive power at generating units
U, U_{ref}	measure and set-point generator stator voltage
ΔUc	vector of stator voltage variation
λ_v, λ_q, λ_u	weights for terms in the objective function: pilot point voltage, reactive power, and generator unit stator voltage
C_v	sensitivity matrices linking variations between pilot point voltages and generator stator voltages (network is modelled by sensitivity matrices for coordination between generating sites)
C_q	sensitivity matrices relating generator variations between reactive powers and stator voltages

Network and unit constraints are taken into account at each computation step using the following equations:

$$\begin{aligned} &\|\Delta U_c\| \le \Delta U_{\max}, \\ &a(Q + C_q \Delta U_c) + b\Delta U_c \le c, \\ &V_{pp}^{\min} \le V_{pp} + C_v \Delta U_c \le V_{pp}^{\max}, \\ &V_{ps}^{\min} \le V_{ps} + C_{vs} \Delta U_c \le V_{ps}^{\max}, \\ &V_{THT}^{\min} \le V_{THT} + C_v \Delta U_c \le V_{THT}^{\max}, \end{aligned}$$

where:

a, b, c coefficients of straight lines representing operating diagrams (limits) for generator units (P, Q, U); these diagrams depend on the active power output by the generator unit

V_{pp}, $V_{pp}^{\min}$, $V_{pp}^{\max}$	measure, minimum, maximum voltage at pilot points
V_{ps}, $V_{ps}^{\min}$, $V_{ps}^{\max}$	measure, minimum, maximum voltage at sensitive points
V_{THT}	voltages computed at generator unit EHV output

The control system monitors voltage at a limited number of network nodes, or "sensitive nodes": nodes at which the voltage must be kept between upper and lower limits and not controlled by an integral control law tracking set-point values, as is done for classic pilot points. This is another instance of the better performance given by classic SVR, unless it is operating at a very narrow control band.

Control function minimisation seems an improvement in the area of voltage control: weightings in the objective function may be adjusted to suit different control policies, giving priority to pilot point voltages kept at reference values (high voltage values, for example) or to reactive power generation kept close to the lower limit in order to gain reactive power margins. In practice, this is a minor advantage for many reasons:

i. Substituting integral control of voltage with an optimal control law computed by achieving a compromise between voltage and reactive power through fixed control weightings and matrices does not allow a minimising of system losses at best by giving up a full voltage support either in normal or perturbed working conditions.
ii. In front of system changes, the available control matrices and weightings (defined off-line) can be inadequate for new, unpredicted, operating conditions, with a consequent suboptimal control. To properly change these parameters the multivariable quadratic function has to wait for a state estimation update, which is not compatible with required SVR dynamics.
iii. In the voltage-reactive power control problem, bus voltage is the regulated variable through generator reactive power control. There is no reason for, and some risk exists, for any kind of balance between voltage and reactive power values, mainly when control parameters of the multivariable quadratic function are inadequate to a new operating condition subsequent to contingencies. It is safer with not-updated SE to regulate voltage only.
iv. In practice, EHV voltage weighting choices are higher than those for the other two terms, even at Électricité de France (EdF). This confirms voltages are the main objective (see Fig. 5.2) and have to be controlled by reactive power at a changeable amount depending on the operating state. This also confirms the low utility at the SVR level of simultaneously controlling voltage and reactive power because generator reactive power must be free to move in a timely way between over/under-excitation limits, mainly when control matrices and weightings are not adaptively updated in real time.

In conclusion, CSVC would in practice perform very similarly to the Eq. (5.1) control law and different from the TVR Eq. (4.13). This happens notwithstanding that the two optimising functionalities have very similar structure. This is also true because only very slow dynamics linked with the state estimation update should allow correct computing and updating of the full CSVC control parameters, and such slow dynamics are proper of closed-loop TVR with a dominant time constant of about 10–20 min. This means CSVC is four to six times slower than classic SVR (§ 4.1.4) in order to operate properly and in a reliable way. In practice, CSVC should have only one weight different from zero (reasonably λ_v) if it is to be sufficiently faster and less critically linked to system state estimation. This result is confirmed by the following traces in § 5.1.4.

5.1.4 Performance and Results of Simulations

After years under full time operation in western France, CSVC has gained the confidence of local operators and shown, according to EdF operators, its many advantages comparable with performances referenced in § 4.1.4.

The experimental CSVC system shows a better dynamic response (Fig. 5.2) with respect to the original, very slow SVR in France [3, 4], whose dynamic design does not appear to be optimal.

Voltage Control in Case of Failure and Load Variation

In case of failure (unit or line tripping), primary voltage regulators contribute to voltage regulation enhancement; but sometimes this is neither effective nor sufficient. Consequently, the network remains weakened. After AVR activity, CSVC allows maintenance and restoration of the voltage profile by mobilising and coordinating reactive generations. It is therefore possible to help prevent voltage collapse in the CSVC control area in case CSVC's optimisation is of voltage mostly, as Fig. 5.3 shows.

Figure 5.3 shows the rapid restoration (under 3 min) of pilot node voltage by CSVC after a drop of 5 kV, caused by a tripping of units (λ_v reasonably appears to be the dominating weight in the quadratic control function).

Figure 5.4 compares network voltage performance with primary control alone and with CSVC for a very severe load increase situation (30 % per hour, 60,000 MW of initial load), without modifying the generators' operating schedules. This example shows that the network controlled with CSVC can be loaded with 3000 MW more when compared with a control option that relies only on AVRs.

5.1.5 Final Comments on French Hierarchical Voltage Control Power System

A coordinated secondary voltage regulator system is presented and understood as a solution for overcoming SVR level due to:

- No reactive power control loops, but direct control of primary voltage set-points;

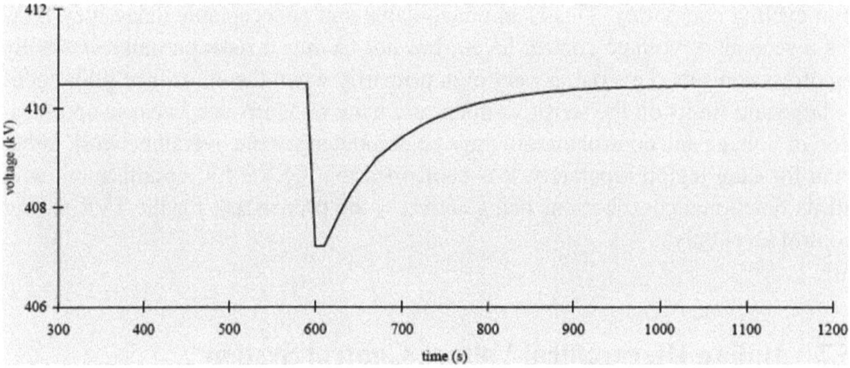

Fig. 5.3 Pilot point voltage during unit tripping

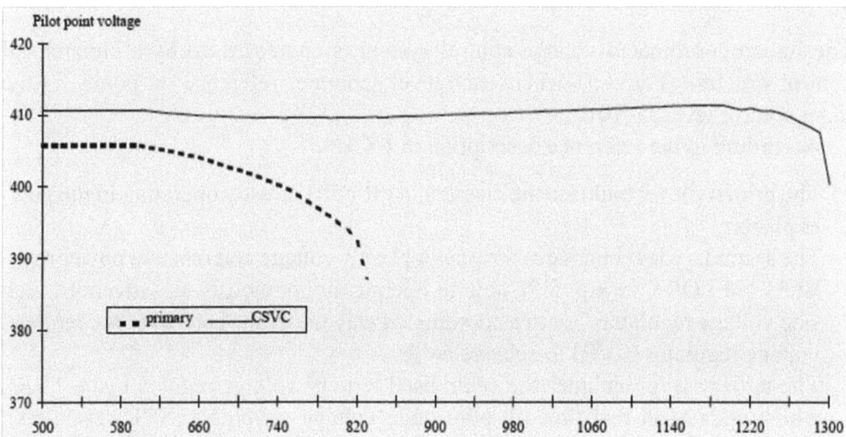

Fig. 5.4 Pilot point voltage

- No pilot node integral control law, but optimisation of pilot node voltages and reactive power control margins, thereby warning of the possibility of merging second and third hierarchical levels by an SVR that includes TVR.

In practice, this is not the correct interpretation. The required control dynamic, one that realises a true secondary voltage control loop, imposes on the CSVC optimisation function the task of correctly referring only to the pilot node voltages (as in the Fig. 5.3 result, with a dominant time constant of about 100 s). Accordingly, CSVC does not differ from classic SVR, which cannot manage control margins and/or losses.

Conversely, the inclusion of the optimal control margin objective asks for continuous updating of control weightings and matrices according to the grid's real operating working conditions. This updating cannot be achieved without the dispatcher's SCADA/EMS control system providing a reliable result of state estimation cycling computing. This is an unavoidable and unacceptable delay for CSVC (as a secondary voltage control level), but not having it risks partially or wholly incorrect controls. This risk is very high primarily when the number of pilot nodes is large and limits on the sensitive nodes are narrow. Moreover, because optimisation of voltage and control margin must be computed for the overall network rather than for each region separately, it is confirmed that CSVC full optimisation, with all its functional contributions being active, is the proper task for the TVR slower control level only.

5.2 Italian Hierarchical Voltage Control System

5.2.1 General Overview

The Italian coordinated voltage control system is characterised by a hierarchical control structure (Fig. 4.5), where each level generates reference set-points for the inner control level [5–10].

According to the reference description in § 4.1.4:

- The primary level includes the classical AVR units already operating in the power plants;
- The secondary level includes the power plant's voltage and reactive power regulators (REPORT, in Fig. 5.5), able to operate autonomously as "advanced high side voltage regulators" or in a coordinated way under the control of the regional voltage regulator (RVR) to achieve SVR;
- The tertiary level includes the centralised tertiary voltage regulator (the TVR), which updates in real time all pilot node voltage set-points (SVR set-points), defined by the solution of an optimisation problem with an objective function that represents a compromise between security and economy (see § 4.1.5).

The three hierarchical levels mentioned are real-time, overlapped, and closed-control loops, requiring a well-defined investigation of dynamics and stability, control apparatus design and control centre and plant adaptations.

Studies for the selection of pilot buses and control power plants resulted in the Italian power system's (60,000-MW peak) subdivision into 18 control areas (see Fig. 4.13). This plan involves the largest thermal and hydro power plants be connected to 400-kV and 230-kV grids, for a total reactive capacity of about 20,000 MVAR.

The Italian hierarchical voltage control system (Fig. 5.5) regulates pilot node voltages in a closed loop through real-time control of the reactive resources that influence those buses most. Regional voltage regulators close the control loops of pilot node voltages, providing each area with a specific reactive power level, one which controls the local power plant's voltage and reactive power regulators (REPORTs). In turn, the REPORT closes the reactive power control loops of the plant units, directly acting on the set-points of the generator AVRs. The RVR also controls capacitor banks, shunt reactors, OLTCs and SVCs to pre-empt saturation of area generators.

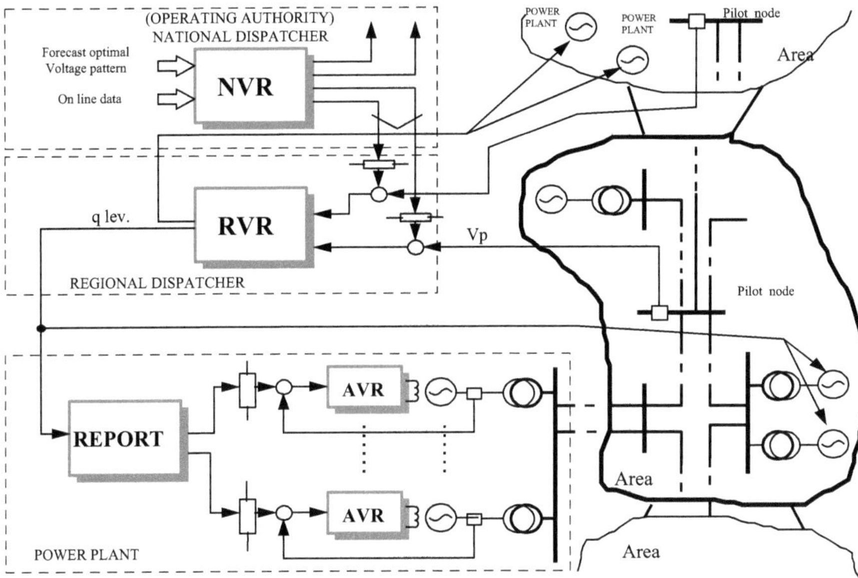

Fig. 5.5 Schematic diagram of the Italian hierarchical voltage control system

AVR rapid control is referred to as *primary voltage regulation* (*PVR*). The combination of the area power plants' REPORT and the regional dispatcher control rooms' RVR implements SVR. At the highest hierarchical control level, a TVR coordinates the RVRs in a real-time closed loop. It establishes on the basis of actual field measurements the current pilot node voltages, which achieve minimum feasible grid losses by slow RVR set-point correction, keeping the system under control at all times. To achieve this further aim, an optimal reactive power flow (ORPF) for loss minimisation control (LMC) computes in short (the day ahead) or very short terms (minutes ahead), the forecasted optimal voltages and reactive levels, starting from the forecasted/current state estimation.

Therefore, TVR minimises the differences between actual field measurements and optimal forecasted references. This computed "compromise" represents the maximum tenable transmission grid voltage plan at any instant. The combination of TVR and LMC makes up the national voltage regulator (NVR), which thus links ORPF forecasting with real-time optimisation of SVR set-points.

The hierarchical voltage control system has different operation modes corresponding to its implementation progress, maintenance interventions and transient or persistent failures:

- Without plant telecommunications, or when the RVR is not operating, REPORT automatically regulates local EHV bus voltage (high-side voltage regulation) according to defined and stored daily trends or the plant operator's voltage set-points, agreed upon by phone with the regional dispatcher;
- Without TSO telecommunications or when the TVR is not operating, the RVR autonomously regulates pilot node voltages of its controlled areas according to stored daily trends or the regional dispatcher's choices;

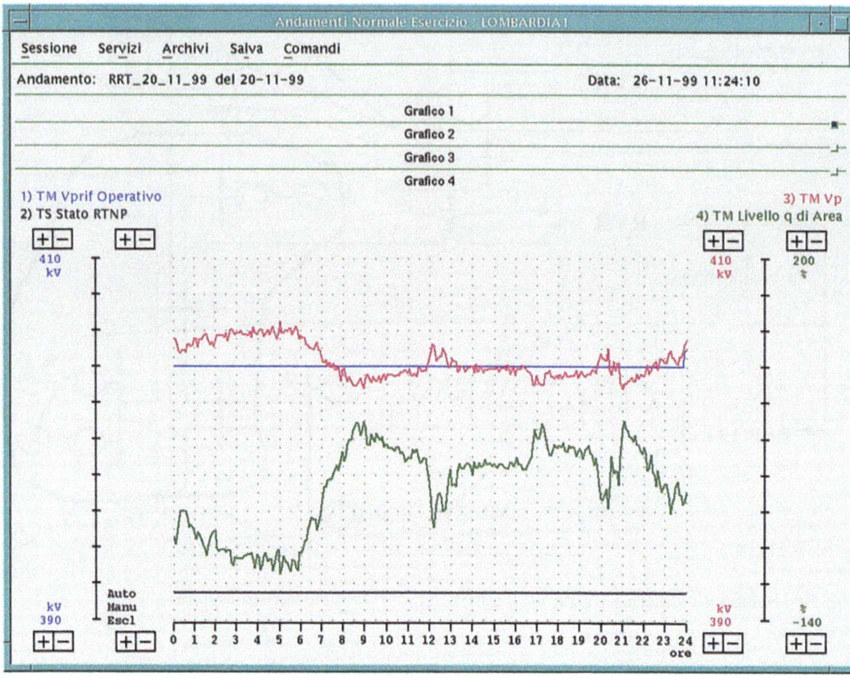

Fig. 5.6 Baggio pilot node, daily voltage profile under the remote control of the Milan RVR by the reactive powers of Piacenza, Turbigo and Tavazzano control power stations

- When the LMC is not operating, the TVR autonomously coordinates the RVRs, assuming, as a reference for optimisation of pilot node voltages and reactive power margins, the available long-term forecasted optimal plan or national control centre operator's manual reference on voltage and reactive power.

In the framework of the voltage control service, the Italian TSO, having completed the application of REPORT apparatuses on all main power plants, as well as of RVR systems in the regional dispatchers' control rooms, can define what the proper voltage service rules—in link with the operating SVR—should be. The LMC has also been activated, providing optimised pilot node voltages at each state estimation upgrade.

5.2.2 Power System Operation Improvement

Voltage Control System Dynamics

Next we present some traces of the SVR operation in the Italian transmission grid, recorded during on-field tests. At the "Baggio" pilot node, the trace of (Fig. 5.6) shows the daily-recorded voltage profile determined by the remote control from the RVR at the Milan dispatcher control room. The on-site test shown here refers to a constant voltage set-point value (blue) for the full day and not to the optimised

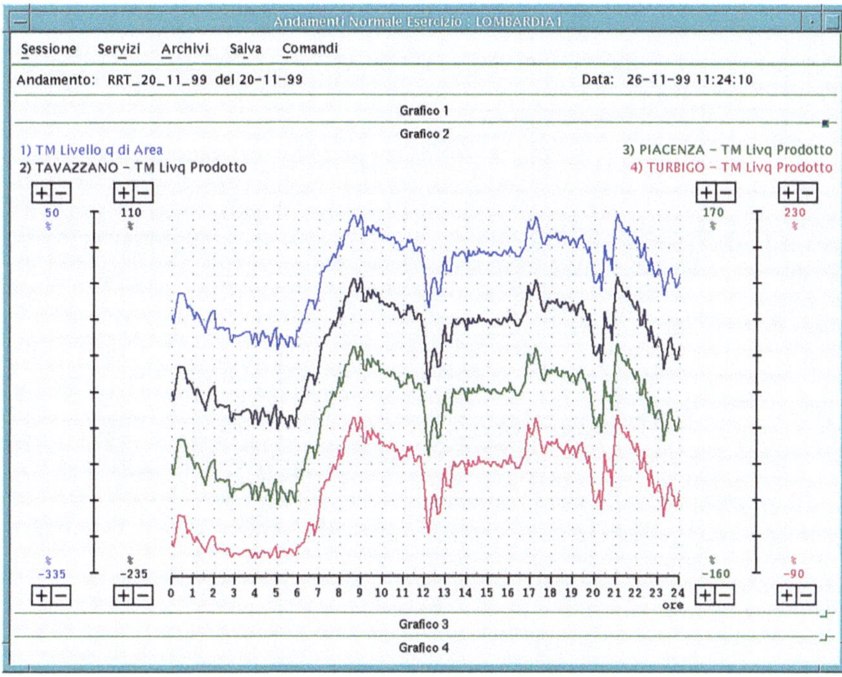

Fig. 5.7 Daily reactive power profiles of Piacenza, Turbigo and Tavazzano control power stations under the remote control of the Milan RVR to regulate the Baggio pilot node as in Fig. 5.6

voltage trend. This set-point choice requires a "compounded" pilot node voltage control loop allowing differences with respect to the set-point value when the control effort is high (violet trace). The RVR control output "q" (green trend) represents the Baggio Area control effort.

Figure 5.7 refers to the correspondent reactive power productions of the three power stations controlling the Baggio voltage. They show a concordant alignment with "q" as expected. The higher control effort happens in under-excitation in the first six hours of the day. The SVR voltage control loop dynamic with a dominant time constant of 50 s is too fast to be recognised in these daily traces.

Voltage Stability Limit Increase

Suitable static and dynamic analyses show that SVR and TVR increase the overall loadability of the transmission system. The study case, the results of which are presented in Fig. 5.8, involves a load ramp increase at several buses in the Rome control area. From this case it is possible to compare simulation results based on a system dynamic model, obtained with SVR alone and with the TVR in service.

The *P-V* trajectory reveals the expected stability-improving effect in terms of both voltage profile and load margins (200-MW and 300-MW margin are increased at "Roma Nord" and "Roma Sud" buses, respectively). Such an increased overall

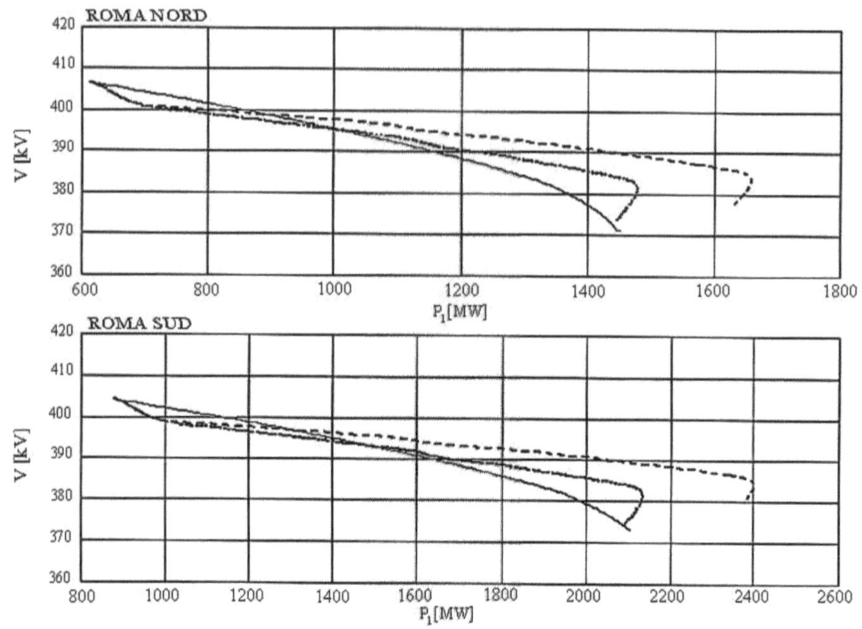

Fig. 5.8 Italian grid load ramp stability margins (dynamic evolution): primary voltage regulation (*continuous line*); secondary voltage regulation (*dotted line*); secondary + tertiary voltage regulation (*dashed line*)

loadability is also demonstrated by computation of suitable off-line and on-line voltage stability indicators [11–13].

The study case whose results are presented in Fig. 5.9 concerns the overall Italian network and involves an increase in load ramps at all buses. In this particular simulation, very short-term reactive power redispatching by LMC has also been simulated: in fact, TVR uses four different optimal pilot bus voltages and area reactive levels computed at every 2000 MW of total load increase. Due to the presence of SVR and TVR, the largest amount of load margin increase achieved is 1500 MW for the overall Italian grid.

Network Loss Reduction

The main objective of LMC and TVR is the achievement of minimum losses in the grid, brought about through short-term optimisation of network voltage values (LMC) in order to update, in quasi-real time, the TVR reference in order to continuously reduce overall system operation costs. Many static analyses conducted on the entire Italian network in recent years (Fig. 5.10) have demonstrated that the

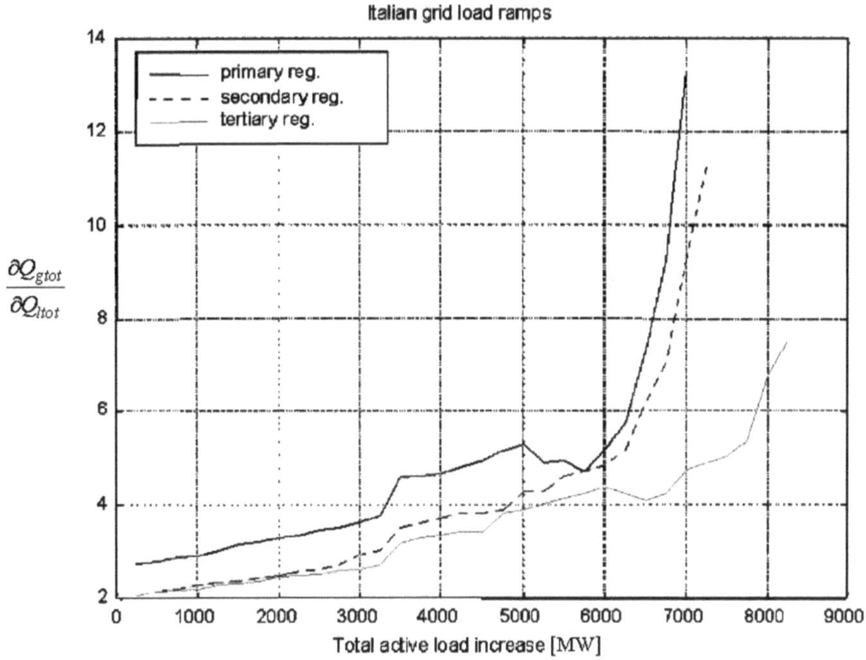

Fig. 5.9 Italian grid load ramps stability margins (static evaluation): $\partial Q_{g\,\text{tot}} / \partial Q_{c\,\text{tot}}$ voltage stability indicator represents sensitivity of total reactive production with respect to total reactive consumption

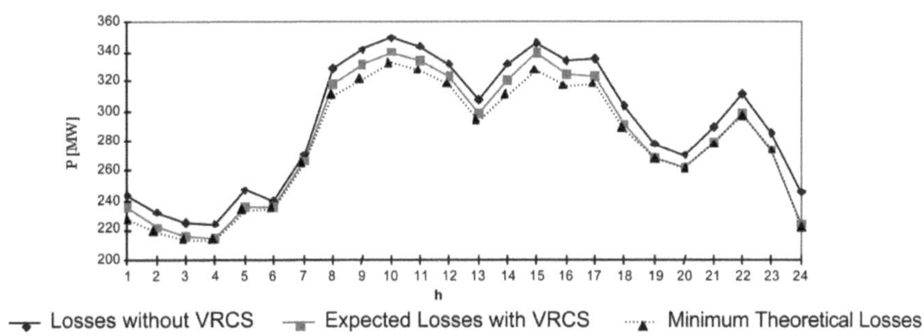

Fig. 5.10 Expected loss reduction in the Italian grid with SVR and TVR

application of the multilevel control system for grid voltage and reactive power regulation (VRCS = SVR + TVR) allows a reduction in transmission losses of about 4–6 %. Such a control system also achieves better service to the final user in terms of operation quality, security and reliability.

5.2.3 Final Remarks on Italian Hierarchical Voltage Control System

The hierarchical voltage control system has been proven to operate with success in the Italian power system since 1985, contributing to simplification and improvement of network voltage operation. Secondary and tertiary voltage regulators and the TVR allow grid operators to achieve a full exploitation of transmission network transfer capabilities, as required by restructured and liberalised energy markets. In the framework of the ancillary services market, data made available by the proposed control system also allow simple, correct recognition of the real contribution of each generator to the voltage service [14].

The Italian experience started with experimental applications in the Florence Area (1985) and in Sicily (1989), which revealed significant benefits. The control system grew step by step, with plants first operating by REPORT high side voltage control, then with RVR participating in SVR pilot node control. The very satisfactory results inspired TERNA (the Italian TSO) to promote widespread application of SVR and the development of TVR-LMC [5–10].

Further developments in control apparatus technology and disputable decisions made on SVR and TVR integration into the TSO's SCADA/EMS control systems caused interruptions and delays in the overall control system's activation and operation. Moreover, forcing SVR and TVR into the SCADA\EMS architecture, which was not designed to properly integrate such real-time automatic voltage regulation, runs a highly probable risk of impoverished implementation of SVR and TVR control loop dynamics.

5.3 Brazilian Hierarchical Voltage Control System

5.3.1 General Overview

In as speedily growing a contest as the Brazilian power system is, the general implementation of a hierarchical voltage control system (HVCS) has undergone many studies to check its efficacy. Since 1990, preliminary investigations related to the prospective application of hierarchical coordinated voltage control to parts of the Brazilian EHV network have been done. The satisfactory results of these are documented [15].

One analysed system is the Rio de Janeiro ("Rio") grid, which is an energy-importing area. This area, which is a part of the Brazilian Southeastern system, has a peak load of approximately 5000 MW in summer time (from January to March). The Rio Area equivalent system model consists of 387 AC buses, 678 AC transmission lines and transformers, 30 power plants and 5 synchronous compensator units. Figure 5.11 illustrates the main transfer corridors that lead to this area. Power that flows into the Rio Area comes through four transmission corridors, identified as

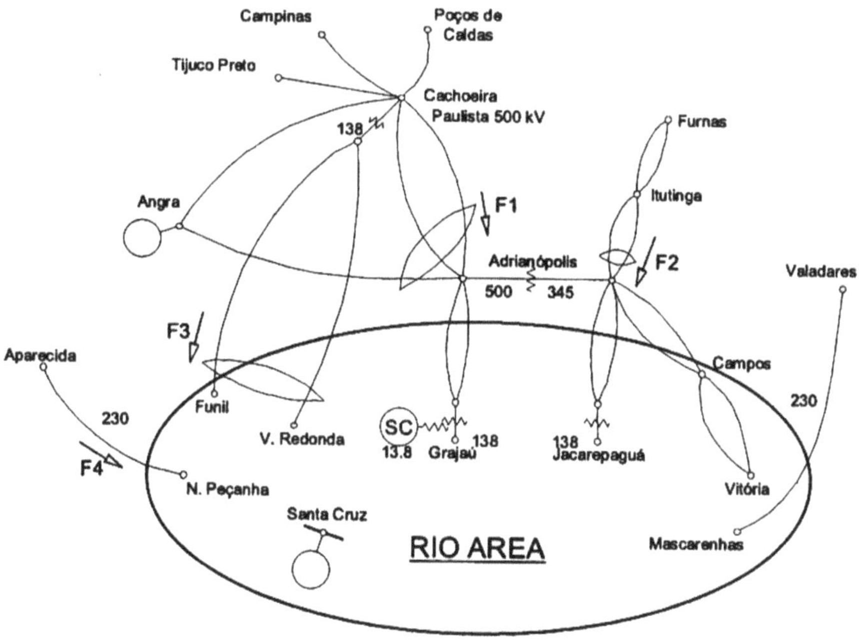

Fig. 5.11 Main transmission corridors leading to the Rio Area (numbers denote voltage levels in kV)

F1, F2, F3 and F4, seen in Fig. 5.11. The main sources of reactive support within the Rio Area are a 2 × 200-MVAR synchronous condenser (SC) at Grajau station and Santa Cruz thermal station. The other reactive sources of interest are located in the transmission system around the Rio Area: the Marimbondo, Furnas and L.C. Barreto power stations and Ibiuna SC. The Rio Area is subdivided into four subsystems: Furnas (124 buses), Light (127 buses), Cerj (57 buses) and Escela (79 buses).

As regards the SVR set up, the pilot bus set-point error is sent to the generator and synchronous condenser units that participate in the SVR scheme. At each unit the error is weighted by a participation factor K_i and integrated. The integrated output signal modulates AVR set-point to remotely regulate pilot bus voltage. Referencing § 4.1.4, the first relevant difference is the direct computing of the AVR set-point updates of the generator units, skipping the generator reactive power control loop. As said before, this direct control of AVR set-points seems obvious from an engineering viewpoint, a simplification, but it is less precise than reactive power control because of AVR set-point offsets and the small range of voltage controllability (± 5 %) with respect to the large reactive power range between the over- and under-excitation limits.

Therefore, generator alignment in a power station is more critical in practice, and there is an increase in the complexity of generator coordination and risk of reactive power recirculation among the generators. Besides this, generator over- and under-excitation limits are indirectly and approximately considered by participation

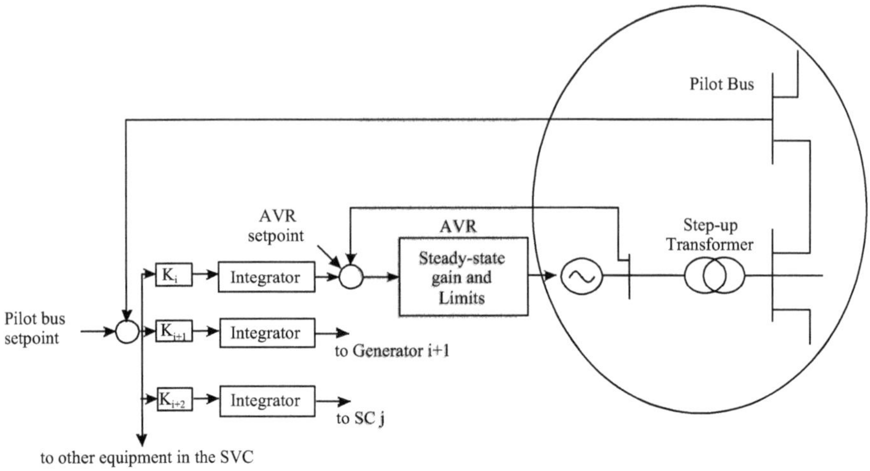

Fig. 5.12 SVR set up in the Rio system studies

factors that also affect dynamic interaction among control generators in the area. Figure 5.12 shows inner (PVR) and outer (SVR) voltage control loop set up; this scheme is useful in studies rather than in practical applications.

5.3.2 *Results of Study Simulations*

The Rio Area is fully represented together with the transmission corridors through which power flows into this area. The remaining parts of the Brazilian south-eastern system are modelled with static equivalents. All participating factors from the control loops are set equal to unity, and integrator time constants are set to 100 s. The AVR's steady-state gains are set equal to 50 p.u./p.u. in all generators and SCs.

SVR Step Response

A step increase of 5% in the reactive load, applied in the Light subsystem that contains the majority of the loads in the Rio Area, shows the SVR closed-loop time response. The pilot bus is the Jacarepagua 138-kV bus (see Fig. 5.11). The power stations participating in the SVR scheme are Furnas, Marimbondo and Santa Cruz, while Grajau and Ibiuna contribute by SCs. In Fig. 5.13 the voltage with and without the SVR scheme for the step disturbance is shown. As expected, in the presence of SVR the voltage of the pilot bus returns to its initial value (the value before the step increase is applied) according to an overdamped closed-loop time response with a time constant of about 100 s.

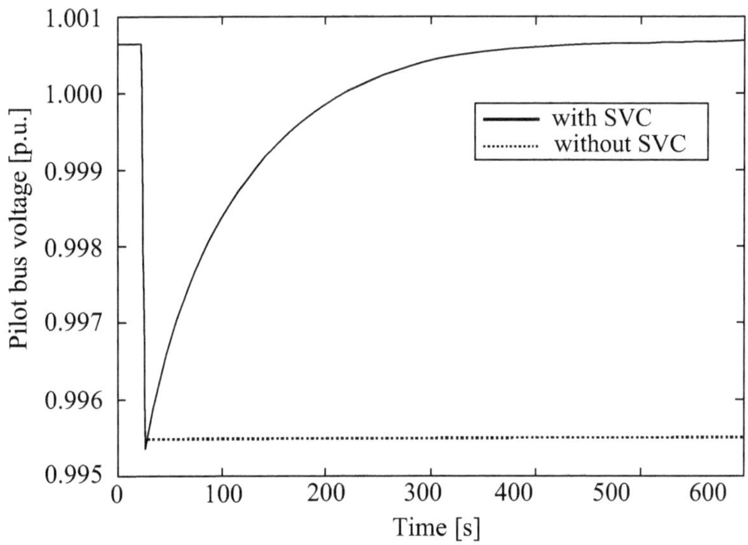

Fig. 5.13 Jacarepagua bus voltages following a step increase in the Light subsystem reactive load (with and without SVR)

Load Variation

Starting from the base case (heavy load condition), a load variation simulation is performed as follows: (a) from 0 to 300 s: 5 % ramp increase in the active and reactive loads of the Light subsystem; (b) from 300 to 900 s: loads remain constant at the final values of the previous step; (c) from 900 to 1200 s: loads are reduced to the initial values through a ramp. Load variation is shown in Fig. 5.14a, together with voltage at the pilot bus (Jacarepagua 138 kV) with and without the SVR scheme, while

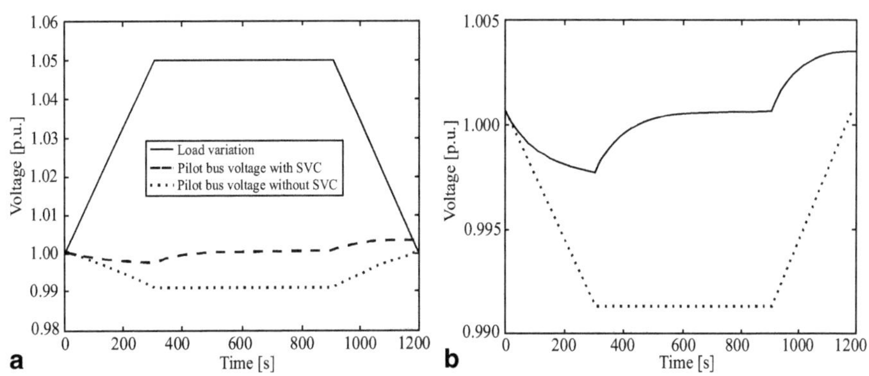

Fig. 5.14 Pilot bus voltage response for Jacarepagua 138 kV at a trapezoidal load variation. **a** pilot bus voltages and trapezoidal-shaped load variation. **b** detailed variation of the pilot bus voltages

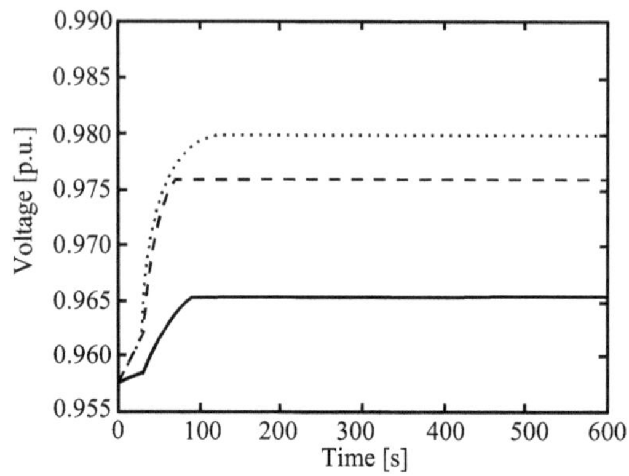

Fig. 5.15 Pilot bus voltage. **a** *solid-line* case (Jacarepagua OLTC alone). **b** *dashed-line* case (Jacarepagua OLTC and Grajau SC). **c** *dotted-line* case (Jacarepagua OLTC, Grajau SC and Santa Cruz thermal units)

Fig. 5.14b only shows pilot bus voltage. The SVR voltage transient clearly shows its lower speed with respect to load ramp slope. When the ramp trend stops, SVR control recovers pilot node voltage to set-point value. As expected, when no SVR scheme is implemented, system voltages have an upside down trapezoidal shape.

Single Contingency Case

SVR performance in the case of an outage of the 500-kV Angra–Adrianopolis transmission line (TL) is shown in Fig. 5.15. After the contingency, a 1.0-p.u. voltage at the pilot bus decreases to 0.958 p.u. In the absence of an SVR system, the voltage clearly does not recover from this value.

Further, three SVR schemes having different control equipment are compared in Fig. 5.15: (a) Jacarepagua OLTC alone; (b) Jacarepagua OLTC and Grajau SC; and (c) Jacarepagua OLTC, Grajau SC and Santa Cruz thermal units. Figure 5.15 shows voltage at the pilot bus for all the three cases and makes evident that the control objective of recovering to 1.0 p.u. is not achieved in the three cases, though with differences. In the first two cases, the control system steady-state errors are due to the OLTC reaching its maximum tap limit (first case) and to the Grajau SC reaching its over-excitation limit (second case). In the third case, a lower steady-state error allows the best result to be achieved, though still in a saturated condition. Through coordination of local resources, as shown in § 4.1.4, saturation can be avoided and a voltage recovery to set-point easily achieved.

Loading the Light Subsystem

Here we show the gain on loading margins when an SVR scheme is used in the Rio Area. A load ramp is applied to the Light subsystem consisting of a 30 % load increase in 1000 s at a constant power factor. Loads at the remaining subsystems are held constant. Only the first and the third cases described earlier (Jacarepagua

Fig. 5.16 Pilot bus voltage

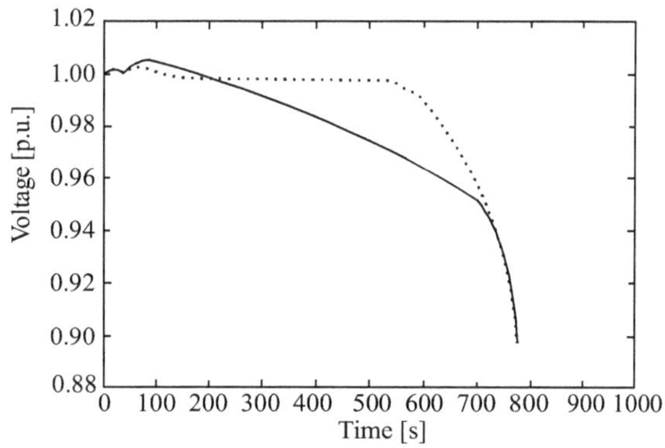

Fig. 5.17 Pilot bus voltage

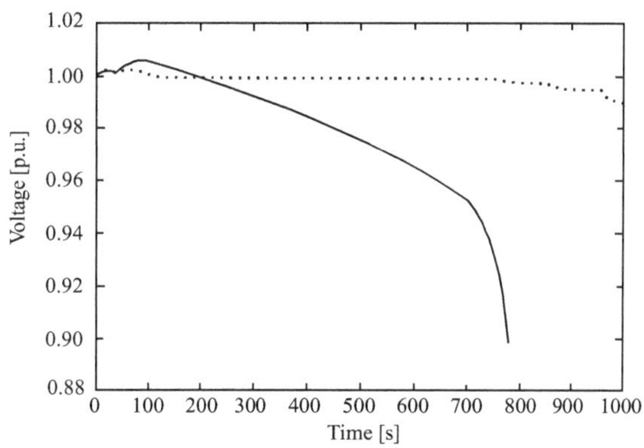

OLTC alone—the solid-line case, and Jacarepagua OLTC, Grajau SC and Santa Cruz thermal units—the dotted-line case) are analysed. The objective of the SVR is to regulate the pilot bus voltage at 1 p.u.

Figure 5.16 compares pilot bus voltage for the two cases. Voltage instability is seen to occur shortly before 800 s of simulation. In fact, in the first case (solid line) voltage deteriorates as the system is loaded. On the other hand, in the third case (dotted line), the voltage at the pilot bus is held constant as long as a reserve of reactive power generation exists.

In order to investigate the capability of the SVR system to increase maximum loadability of a power system, the following case is presented: the same equipment that is considered in the dotted-line case plus the Furnas and Marimbondo power plants and Ibiuna SC. Figure 5.17 shows pilot bus voltage for simulations of the first case (solid line) and this last case (dotted line). The increase in the maximum loadability limit enlarging the SVR control limits can be clearly seen.

To show the voltage instability also occurring in a fourth case, the system was stressed to a 40 % constant power factor load increase. Figure 5.18 shows the pilot bus voltage for this higher loading condition.

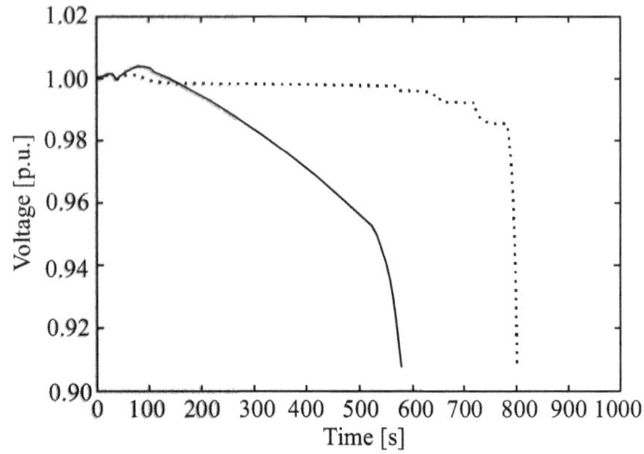

Fig. 5.18 Pilot bus voltage

We note that in the first case (solid line), instability occurs after 600 s and voltage deteriorates as load increases, while in the fourth case (dotted line) the voltage is held constant until after 800 s, when the instability phenomenon occurs. This test shows a voltage stability improvement in the presence of SVR.

5.3.3 Conclusions on the Brazilian Voltage Control System

The control scheme in Fig. 5.12, even if it achieves results similar to those shown in the § 4.1.4 simulations, presents a reason for some critical remarks concerning the system in practice; a few were previously noted for the French CSVC solution:

- No reactive power control loop, but direct control of the primary set-points: In practice, this choice could create dynamic interaction and reactive power recirculation among generators;
- The pilot node control law is purely integral: therefore, it is too slow in the first part of the transient in front of a large contingency;
- The pilot node voltage control is based on a very large number of parallel integral loops (Fig. 5.12), as many as the number of generators and SCs controlling that voltage: This determines a high risk of improper operation in practice because of generator limits that are not properly taken into account and also because of the different dynamics and offsets the considered AVRs can have, with consequent strong interaction among them and with overlapped integral loops.

Looking at the simulations, the preliminary results involving the use of SVR in the Rio Area show the benefits gained regarding voltage profile and security. The results also make evident the importance of properly selecting the pilot buses and reactive power sources participating in the SVR scheme. The simulations also indicate that regulation at higher voltage levels through coordination of all area reactive power resources leads to a better overall control performance and increases the loading margin.

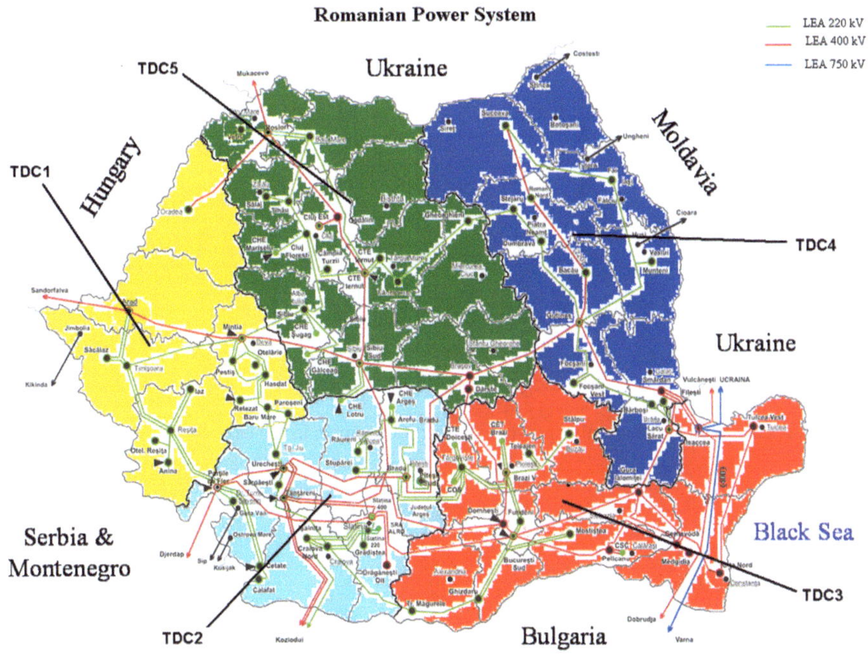

Fig. 5.19 Romanian power grid partitioning into five dispatching regions

5.4 Romanian Hierarchical Voltage Control System

5.4.1 Characteristics of the Studied System

The proposed methodology on the SVR pilot node and control area selection has been applied to the Romanian power system in order to evaluate secondary voltage control structure suitability [16–18]. The case study considers the 220-kV and 400-kV transmission system composed of 254 buses, including some 110-kV buses, 280 lines and 63 generating groups with a total installed power of 20,000 MW. Reference is made to the 2008 winter period having a peak load of 7900 MW.

The Romanian power system is operated and coordinated by the National Dispatching Centre with the support of five territorial dispatching centres (TDCs), chosen according to geographical and administrative criteria (Fig. 5.19).

5.4.2 SVR Area Selection

As far as SVR area partitioning is concerned, three scenarios (5 or 6 or 7 areas) are analysed to link the different values for the electrical distance thresholds among the pilot node buses (values that still guarantee enough electrical distance) with the

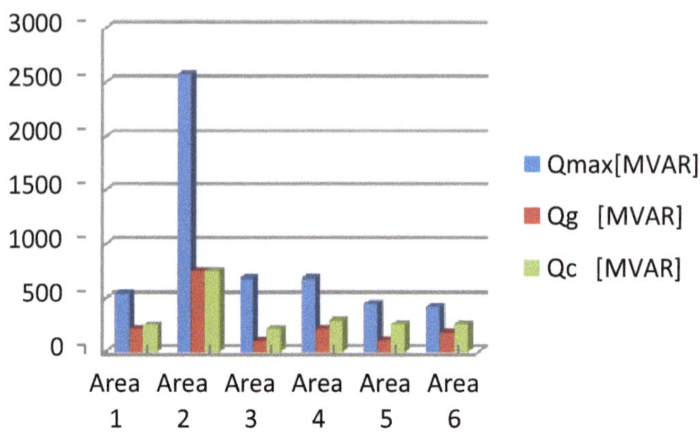

Fig. 5.20 Comparisons of reactive powers in partitioned areas in the case of 6 SVR areas

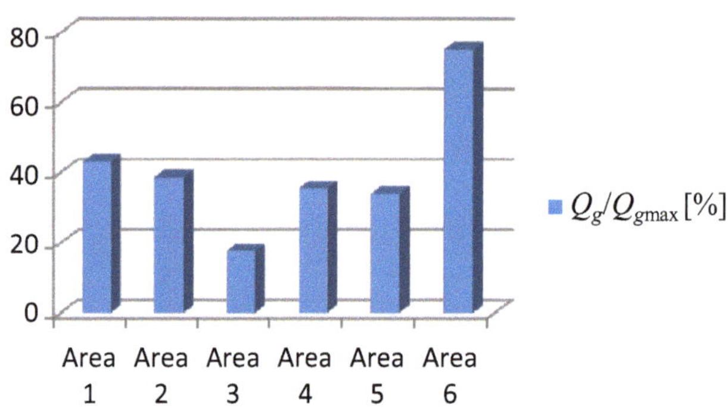

Fig. 5.21 Area ratio Q_g/Q_{max} in the case of six SVR areas

achievable control margin at each area. From each area reactive power involved, it turns out that the six-area case shows the largest control reserve at each area when the already on-field reactive power resources are considered.

Figure 5.20 shows the values of reactive power Q_c required by each area load, reactive power Q_g produced by each area's generators and the maximum area reactive power reserve Q_{max} for the 6-SVR area case.

Figure 5.21 provides the ratio Q_g/Q_{max} [%] for the six-area case, from where the reserve margin of about 60 % or more appears as a general result with the exception of Area 6 having a margin of about 25 %.

Figure 5.22 refers to the Romanian power system subdivided into six SVR areas, with the following 400-kV pilot buses: Mintia (Area 1), Tantareni (Area 2), Domnesti (Area 3), Lacu Sarat (Area 4), Gutinas (Area 5) and Iernut (Area 6). The pilot node set-point values, defined for the considered LF case, assume the following values: 1 p.u. at Area 1; 1 p.u. at Area 2; 1 p.u. at Area 3; 1.02 p.u. at Area 4; 1.03 p.u. at Area 5 and 1.05 p.u. at Area 6.

Some considerations on the impact of operation policy: After 1990, the peak load of the Romanian system decreased constantly from 11,500 to 7900 MW. In

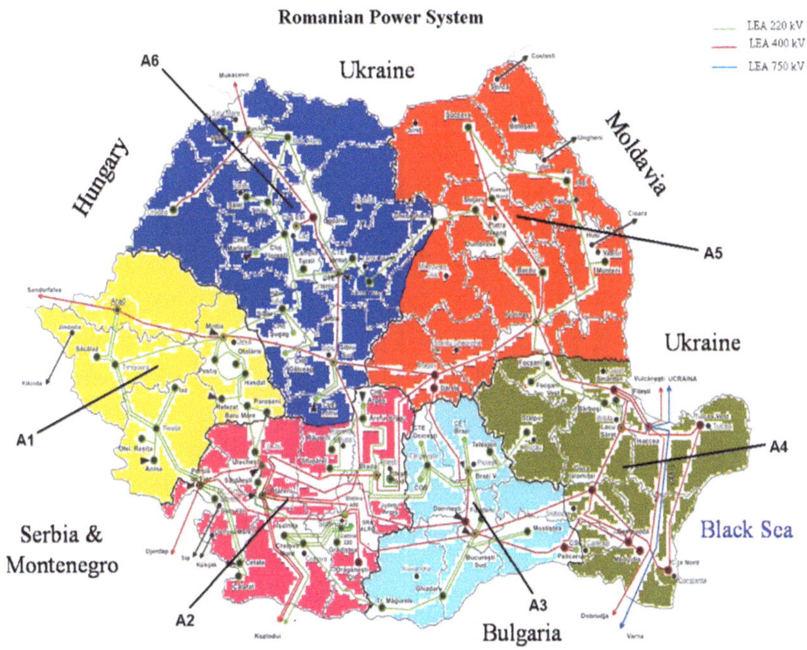

Fig. 5.22 Romanian power grid divided into six controlling areas

this not particularly stressed condition, with lines loaded under the natural power value, usually the lines generate a large amount of reactive power, and the control effort by the generators to support voltages is low in over-excitation but consistent in under-excitation.

Checking the efficiency and robustness of the selected secondary voltage control scheme (six areas), the behaviour of the Romanian power system is tested in front of line outage and generator tripping, and steady-state results after contingency are shown.

Line Outage

In the middle of the blue area (Area 6), the tripping of the 220 kV North–South line: Cluj Floreşti–Tihau is considered. Interest in this area is due to the fact that it is directly interconnected with the ETNSO-E grid, from which it receives 102.8 MVAR.

Figure 5.23 shows the voltage profile (in p.u.) of the most representative buses in Area 6. Here and later, SVR as applied to the Romanian power system, is also called SVCS. Four scenarios are analysed:

- LF: load flow base case;
- $(n-1)$ LF: load flow with the chosen 220-kV line tripping;
- SVR: load flow with the secondary voltage control system included;
- $(n-1)$ SVR: load flow with contingency in the presence of the secondary voltage control system.

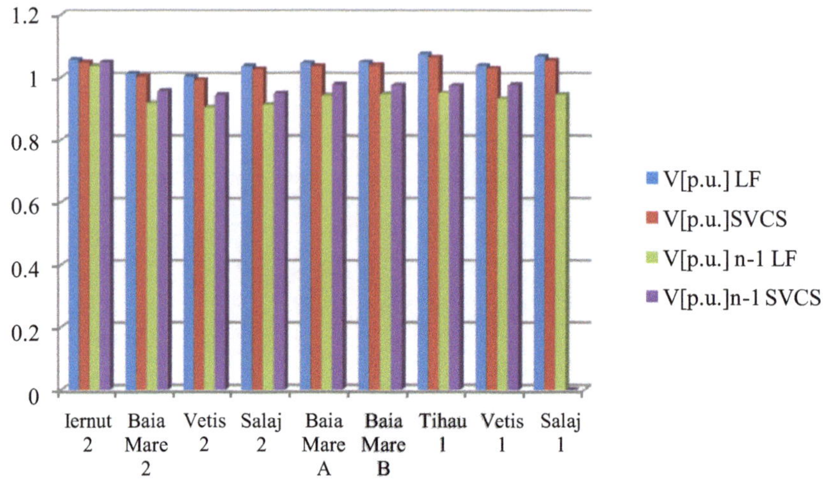

Fig. 5.23 Area 6 voltage profiles with/without SVR and in the presence of a 220-kV line outage

The fact that voltages in the SVR case are lower than in the LF case is simply due to the reduced voltage chosen for the Iernut pilot node with respect to the original LF case value. Usually, the opposite happens because SVR can provide higher voltage values than basic LF. In any event, this voltage difference is not relevant for the objective of the considered test. As can be seen, line outage significantly lowers area voltages, Vetis2 being the most affected bus, with a voltage reduction to 0.9 p.u. Figure 5.23 also shows SVR succeeds in maintaining the set-point voltage at the Iernut pilot node, even in the presence of the contingency.

Under SVR the two controlling generators of Area 6 (Mărişel, Iernut2) have the same reactive output (in p.u.) with respect to their maximum capability limits, either with or without contingency (loaded at 47 % without contingency and at 85 % with contingency). Therefore, generator units under SVR maintain, as expected, the same distance from their capability limits (Fig. 5.24).

This figure also shows very different generator behaviour in the Area 6 considered scenarios: in the base case, without SVR and without contingency, the reactive output of the two generators is already unbalanced and higher than with SVR. The situation is much worse when a contingency occurs: generator Iernut2 reaches its maximum reactive capability, reducing the reactive margin of the area.

Conversely, SVR shows the same reactive loading (in p.u.), and in case of contingency the lowered reactive output by generator Iernut2 results in a larger reactive margin (15 % in the case with secondary voltage control and contingency as compared with 1 % without SVR) due to a better reactive power coordination in the area.

The control generators of the other system areas are not significantly affected by the 220-kV line Cluj Floreşti–Tihau tripping (reactive power outputs of these generators do not appreciably change).

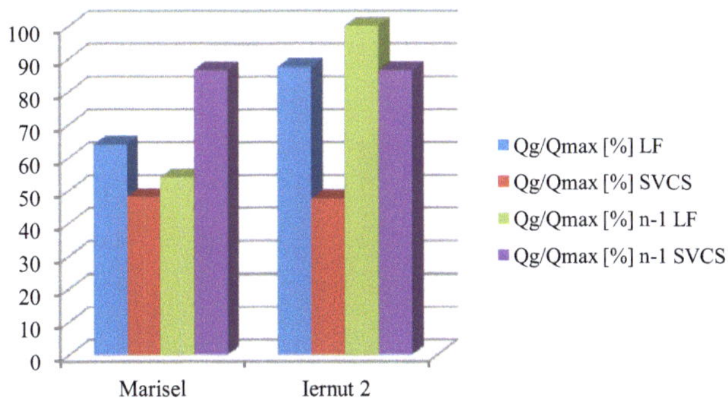

Fig. 5.24 Area 6 reactive power delivery by control generators before and after a line contingence

Generator Outage

In the second scenario, the tripping of the largest generating unit (700 MW) of the Romanian power grid is simulated. Figure 5.25 shows the voltage profile (p.u.) of Area 4's most representative buses (Lacu Sarat is the pilot node) before and after generator tripping as well as with and without SVR control.

In this same case, the SVR significantly contains voltage-lowering in all buses with respect to the $(n-1)$ LF case, so achieving a more robust system with respect to the perturbation. Without SVR, the buses closest to the area with contingency (Medgidia, Cernavoda) are more influenced than the others, and their voltages become lower (of 0.02 p.u.) than in the case with SVR.

Figure 5.26 shows the reactive power outputs of the Area 4 control generators before and after a contingency. The control generator to be tripped has a reactive power output equal to 86.1 MVAR in the LF case and 97.7 MVAR in the SVCS case. Without SVCS we can see that the generating units (Palas, Braila) close to the area most influenced by the contingency provide more reactive power than the others.

The test results show the importance of SVR under steady-state conditions, but mostly at the contingency occurrence. By comparing results of power flow simulations without and in the presence of SVR, we can state that secondary voltage control recovers at best voltages in those areas affected by contingencies, whereas it improves the reduction of active power losses both under steady-state conditions and during transients.

Figure 5.27 shows an SVR loss reduction result of 2.5 % without contingency and 3 % upon a contingency occurrence. Better results in terms of voltage value and losses should be achievable with an SVCS voltage set-point value maintained at the original steady-state value of the SVCS case.

In conclusion, the results of the performed static analysis of the Romanian power system show the importance of proper selection of pilot buses and reactive power control sources defining a secondary voltage control scheme. In the tests considered, the improvements gained by SVR (increase of reactive power control margins,

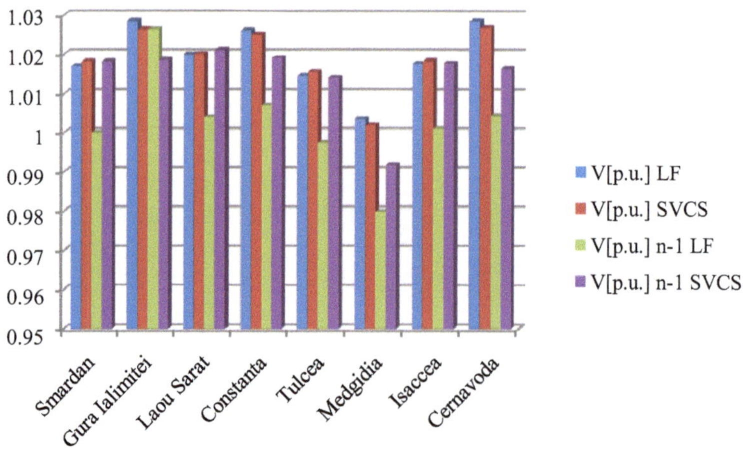

Fig. 5.25 Voltage profile in the case of generator outage

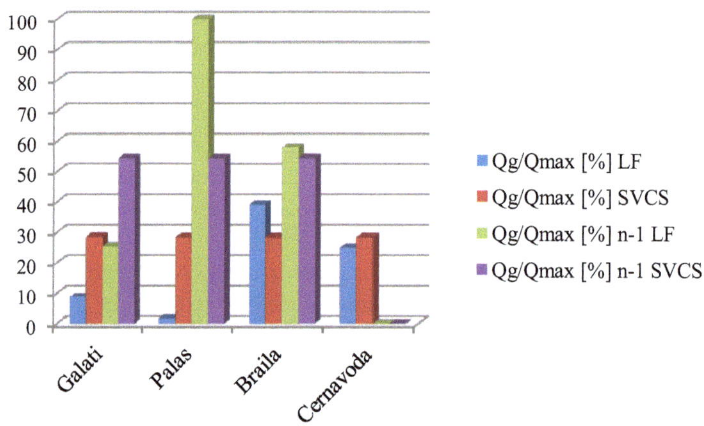

Fig. 5.26 Area 4 reactive power delivery by control generators before and after generator tripping

voltage sustained and loss reduction) are appreciable and allow extrapolation of their more evident impact in front of critical network situations.

5.5 Chinese Hierarchical Voltage Control System

China's hierarchical automatic voltage control system (called AVC) looks mainly to the French solution [19, 21], even when it is distinguished from the French and solved for its limitations. Among these solutions: a reduction of the large number of zones to prevent the risk of increased interaction; a minimisation, at the CSVC level, of voltage control alone, instead of meaningless minimisation of combined

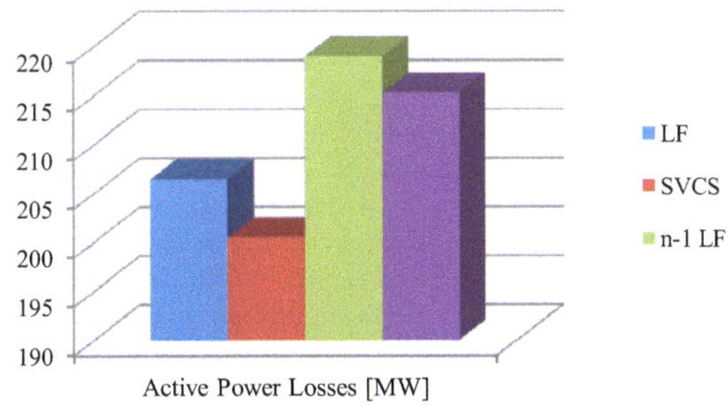

Fig. 5.27 Active power losses of the Romanian system

voltage and reactive power controls. If we take a close look, we see the Chinese solution is linked more to the Italian solution, where pilot node selection methodology (available since 1983) is robust enough to maintain sufficient decoupling over the years and where an adaptive SVR control scheme is considered, too.

As already said, CSVC is based on a quadratic programming model that considers voltage and reactive power constraints in the zone. Generally, there is more than one goal in the objective function. In China [20, 21, 22] the first goal of this model is to control the voltages at the pilot nodes so they follow the optimal set-points, which are updated by the TVR module. The second goal, one with a lower priority, is to equilibrate MVAR reserve distribution among the generators in each control zone so as to enhance the security of the power system.

AVC has been implemented in the Jiangsu provincial power system since November 2002 through the introduction of improvements based on adaptive zone division. Moreover, the application of this control solution has been extended to more than a dozen control centres in China, such as the North China Power Grid, with a generating capacity of 119 GW.

The same SVR control solution has been implemented and tested on-line in the United States at the PJM control centre, with a generating capacity of 164 GW, since February 2010 [19].

References

1. Corsi S, Martins N (eds) (2005) Coordinated voltage control in transmission systems. CIGRE Technical brochure, Task Force 38.02.23
2. Paul JP, Leost JY, Tesseron JM (1987) Survey of secondary voltage control in France: present realization and investigations. IEEE Trans Power Syst 2:505–511
3. Lagonotte P, Sabonnadiere JC, Leost JY, Paul JP (1989) Structural analysis of the electrical system: application to the secondary voltage control in France. IEEE Trans Power Syst 4(2):479–486
4. Lefebvre H, Fragnier D, Boussion JY, Mallet P, Bulot M (2000) Secondary coordinated voltage control system: feedback of EdF. IEEE/PES Summer Meeting, Seattle
5. Corsi S (2000) The secondary voltage regulation in Italy, panel session on power plant (high side) voltage control. IEEE/PES Summer Meeting, Seattle

6. Arcidiacono V, Corsi S, Natale A, Raffaelli C, Menditto V (1990) New developments in the application of Enel transmission system voltage and reactive power automatic control. CIGRE Paper 38/39-06
7. Corsi S, Pozzi M, Sabelli C, Serrani A (2004) The coordinated automatic voltage control of the Italian transmission grid, part I: reasons of the choice and overview of the consolidated hierarchical system. IEEE Trans Power Syst 19(4):1723–1732
8. Corsi S, Pozzi M, Sforna M, Dell'Olio G (2004) The coordinated automatic voltage control of the Italian transmission grid, Part II: Control apparatus and field performance of the consolidated hierarchical system. IEEE Trans Power Syst 19(4):1733–1741
9. Corsi S, Chinnici R, Lena R, Bazzi U et al (1998) General application to the main Enel's power plants of an advanced voltage and reactive power regulator for EHV network support. CIGRE Conference
10. Corsi S, Arcidiacono V, Bazzi U, Chinnici R, Mocenigo M, Moreschini G (1996) The regional voltage regulator for Enel's dispatchers. CIGRE Conference, Paris
11. Corsi S, Marannino P, Losignore N, Moreschini G, Piccini G (1995) Coordination between the reactive power scheduling and the hierarchical voltage control of the EHV Enel system. IEEE Trans Power Syst 10(2):686–694
12. Corsi S, Pozzi M, Marannino P, Zanellini F, Merlo M, Dell'Olio G (2001) Evaluation of load margins with respect to voltage collapse in presence of secondary and tertiary voltage regulation. Bulk Power System Dynamics & Control, IREP-V, Onomichi, Japan
13. Marannino P, Zanellini F, Berizzi A, Medina D, Merlo M, Pozzi M (2002) Steady state and dynamic approaches for the evaluation of the loadability margins in the presence of the secondary voltage regulation. MedPower Conference, Athens, November
14. Corsi S, Pozzi M, Biscaglia V, Dell'Olio G (2002) Fiscal measure of the generators support to the network voltage and frequency control in the ancillary service market environment. CIGRE Meeting, Paris
15. Taranto G, Martins N, Martins ACB, Falcao DM, Dos Santos MG (2000) Benefits of applying secondary voltage control schemes to the Brazilian system. IEEE/PES Summer Meeting, Seattle
16. Eremia M, Petricica D, Simon P, Gheorghiu D (2001) Some aspects of hierarchical voltage-reactive power control. IEEE/PES Summer Meeting, Vancouver, Canada
17. Erbasu A, Berizzi A, Eremia M, Bulac C (2005) Implementation studies of secondary voltage control on the Romanian power grid. IEEE PowerTech Conference, St. Petersburg, Russia
18. Ilea V, Berizzi A, Eremia M (2007) Optimal reactive power flow methodology in power systems with secondary voltage control. 3rd international conference on energy and environment, Bucharest, Romania, November
19. Guo Q, Sun H, Tong J, Zhang M, Wang B, Zhang B (2010) Study of system-wide automatic voltage control on PJM system. IEEE/PES General Meeting, Minneapolis
20. Guo Q, Sun H, Zhang B, Wu W (2005) Power network partitioning based on clustering analysis in MVAR control space. Automat Elec Power Syst 29(10):36–40 (in Chinese)
21. Guo Q, Sun H, Zhang B, Wu W, Li Q (2005) Study on coordinated secondary voltage control. Automat Elec Power Syst 29(23):19–24 (in Chinese)
22. Sun H, Guo Q, Zhang B, Wu W, Tong J (2009) Development and applications of system-wide automatic voltage control system in China. IEEE/PES General Meeting, Calgary, Alberta, Canada

Chapter 6
SVR Dynamic Tests with Contingencies

Examples of SVR dynamic performance in some large power systems are briefly presented according to available data. Most of the traces presented come from very detailed simulations, including all the dynamic aspects of interest. A few data also refer to real systems, confirming the results of studies and the correctness of the modelling used.

The first part of this chapter refers to dynamic performance under normal operating conditions, following load changes and voltage set-point step variations. The second part, on the other hand, considers severe perturbations, determining a large SVR control effort that can also reach control saturation.

6.1 Tests Without Contingencies in Large Power Systems

Under normal operating conditions and in the absence of perturbations, the dynamic performance of an SVR-TVR control system is characterised by continuous and successful attempts made on pilot node voltages for recovery to set-point reference values. Voltage changes are usually determined by load or production plan variations of active and reactive powers as well as by voltage set-point control updates. In the examples ahead, we note mainly the great difference between area reactive power control that is coordinated by SVR and the control case of primary voltage regulation when it operates alone. The simulation mainly shows:

- How the system operator is actually able through SVR to impose the voltage plan in the transmission grid by simple commands from the dispatching control room;
- How fast and concordant is automatic SVR control of the reactive power resources, at the amount and location where it is actually required by the power system voltage needs;
- How the SVR control system is able to maintain unchanged the grid voltage plan or its timely ability to track the new voltage requirements by a higher control level (TVR or system operator).

S. Corsi, *Voltage Control and Protection in Electrical Power Systems*,
Advances in Industrial Control, DOI 10.1007/978-1-4471-6636-8_6

6.1.1 Tests on Italian Hierarchical Voltage Control System

The transients here refer to tests performed in the actual Italian power system, with traces directly collected by the RVR of the Lombardy Region. Obviously, the RVR also collects data when the SVR is out of operation; therefore, a useful comparison on voltages can be made between two very similar days in terms of load and operating condition, one with and the other without SVR [1, 2].

Figures 6.1 and 6.2 illustrate: the large voltage variation at the Baggio bus during a normal day under PVR; the poor effect of the generators' AVR and OLTCs on

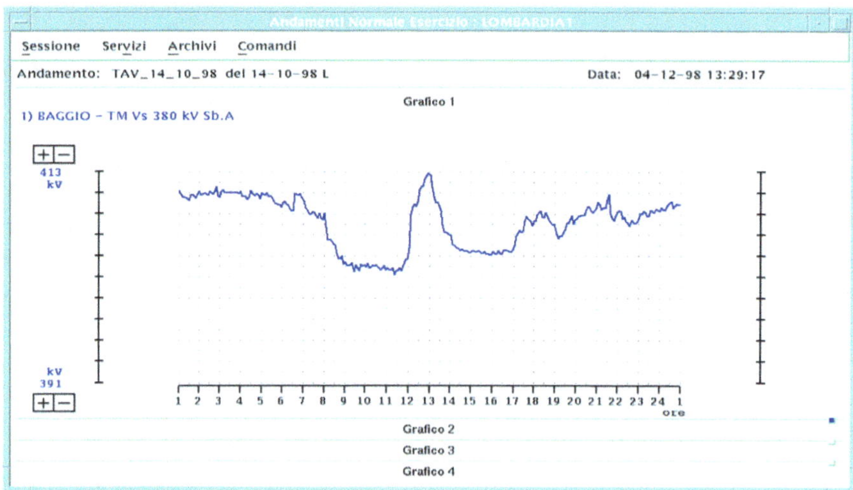

Fig. 6.1 Daily trend under PVR alone of Baggio 400-kV bus voltage due to load variation

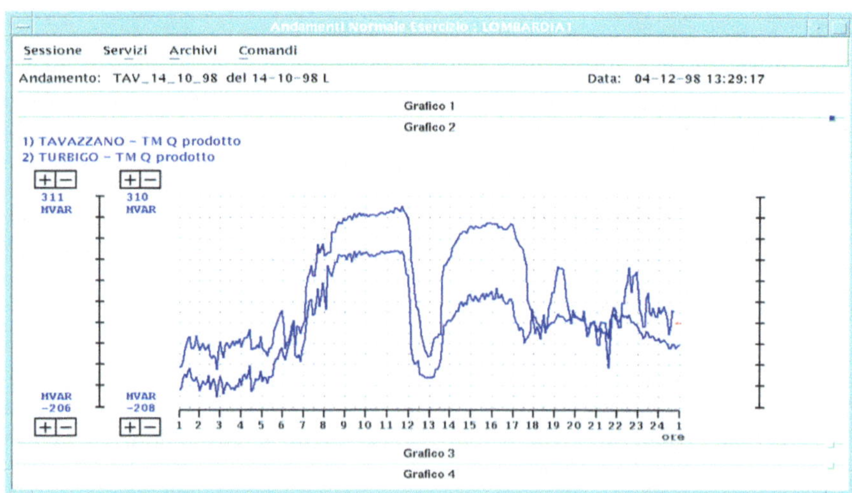

Fig. 6.2 Daily trend under PVR alone of Tavazzano and Turbigo generated reactive powers near Baggio 400-kV bus the same day as in Fig. 6.1

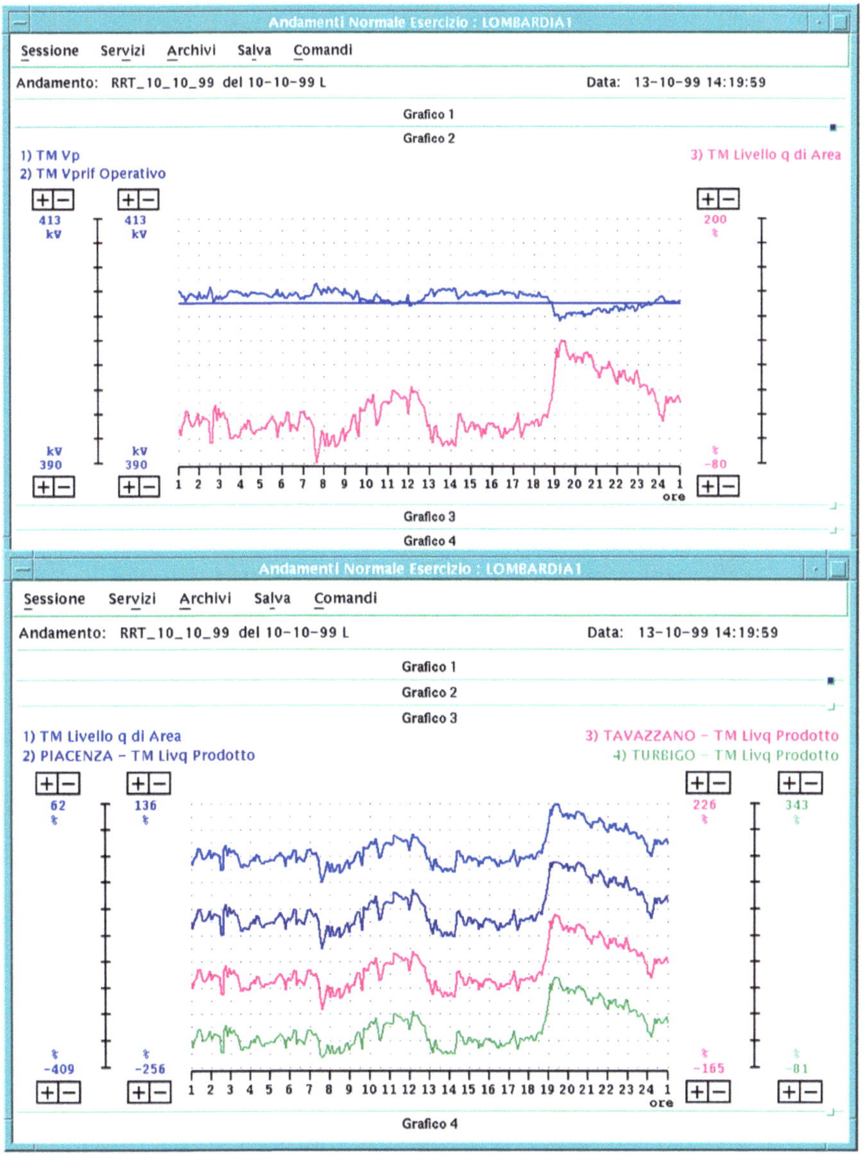

Fig. 6.3 Daily trend under SVR of Baggio 400-kV pilot node voltage and area reactive power level (*top*) and reactive powers of Piacenza, Tavazzano and Turbigo power stations on a day similar to that considered in Figs. 6.1, 6.2

transmission network voltage control; and the poor concordance of the generators' control efforts. Under SVR, Fig. 6.3 (top) shows continuous control by the area reactive power level (red) to maintain pilot node voltage (blue) at set-point value (the constant, black, value). The small voltage differences (between blue and black

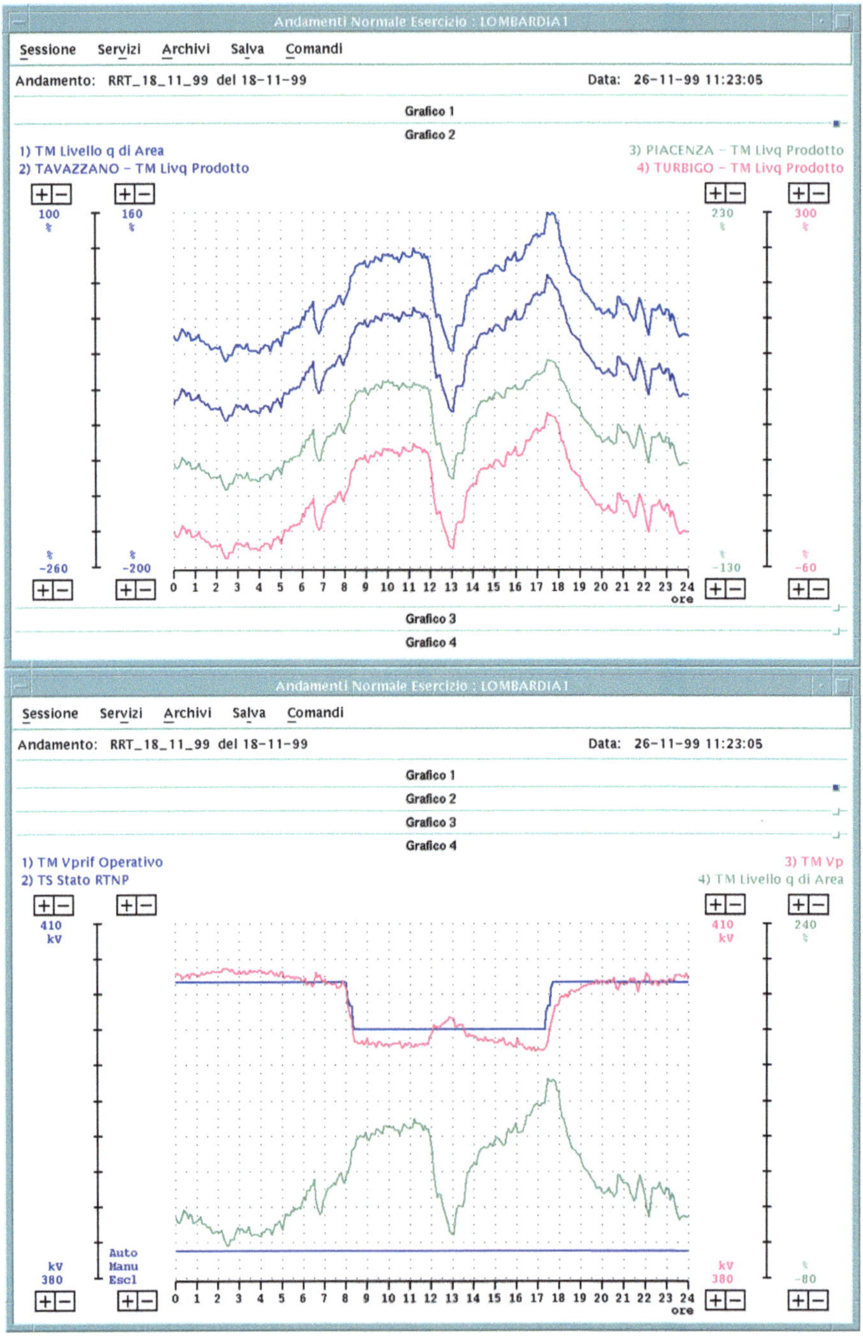

Fig. 6.4 Daily trend under SVR of Baggio 400-kV pilot node voltage and area reactive power level (*top*) and reactive power of Piacenza, Tavazzano and Turbigo power stations, with changes in the pilot node voltage set-point values

traces) are due to the *line drop compensation* effect on the SVR voltage control law. Moreover, Fig. 6.3 (bottom) gives evidence of SVR operating control on a power station's reactive powers by changing them in unison with the tracking of the area reactive power level reference (Fig. 6.3 (top, red)).

Again, in Fig. 6.4 (bottom) we can see the reactive power alignment of the area generators to support the pilot node voltage according to the set-point trend. This performance shows an extreme simplification in grid voltage control and wide area controllability, recognisable by the reactive power level values inside −80% and +60%. A higher control effort in over-excitation in the November 99 test is recognisable at around h.18.00 in Fig. 6.4-top, while a large control effort in under-excitation happens from h.12.00 to 14.00, due to both the voltage peak in Fig. 6.1 and a lowering of the pilot node voltage set-point value. Obviously, the operation in under-excitation is also evident in the first hours of the day due to the low load. Again, differences with the voltage set-point values are due to the line drop compensation effect in the pilot node voltage control loop.

6.1.2 Tests on South Korean Hierarchical Voltage Control System

The transients shown here refer to tests performed on the South Korean power system by a very detailed dynamic simulation model. With the power system under normal operating conditions, the test in Figs. 6.5–6.7 shows the dynamic response of Area 2 SVR control in front of pilot node voltage set-point stepping down and up variations (5 kV). P-I control effects and an absence of the line drop compensation are evident in the pilot node voltage trend seen in Fig. 6.5. The P contribution to the control is mainly evident in Fig. 6.6 in the two sudden but small peaks of the area reactive power level, coincident with the voltage set-point steps.

Figure 6.7 gives evidence of the concordant reaction of the Area 2 generators, first reducing and then increasing their reactive power production according to the dynamics imposed by the pilot node voltage control law. We note how controlled generators react in a timely fashion to variations in area reactive level without any oscillating reactive power exchange among them. Furthermore, an area control effort below 0.1 p.u. (Fig. 6.6) allows for moving the area pilot node about 5 kV.

6.1.3 Tests on South African Hierarchical Voltage Control System

Tests are aimed to show how voltage control is affected differently when a South African transmission system operates with and without SVR. Some of the tests show SVR's timely coordination of all system resources aimed to maintain pilot node voltages (in 10 SVR areas; see Fig. 4.18) at their voltage set-point values while system loads change. Other tests refer to pilot node voltage set-point step variation [3, 4].

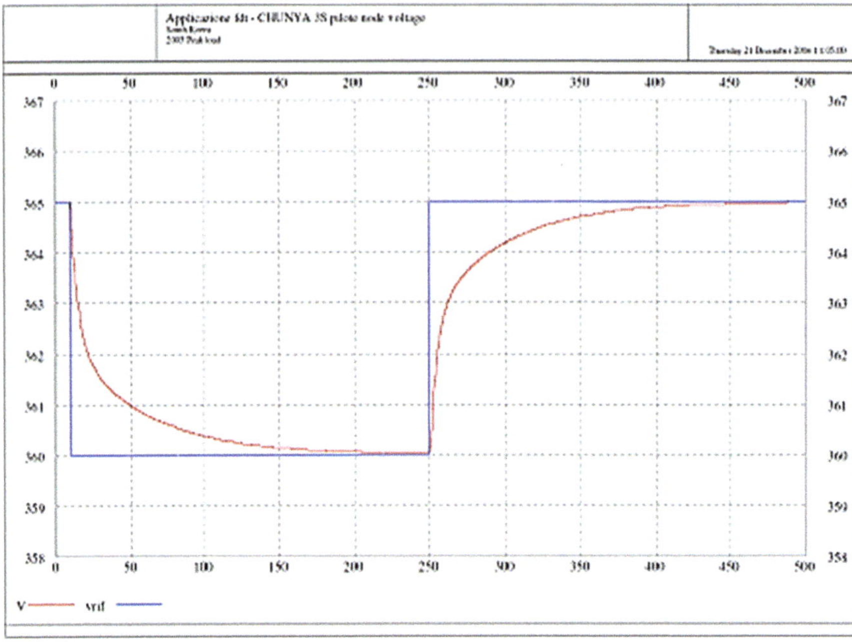

Fig. 6.5 South Korean 2006 power system, Area 2: dynamic response of CHONGYANG pilot node voltage control loop

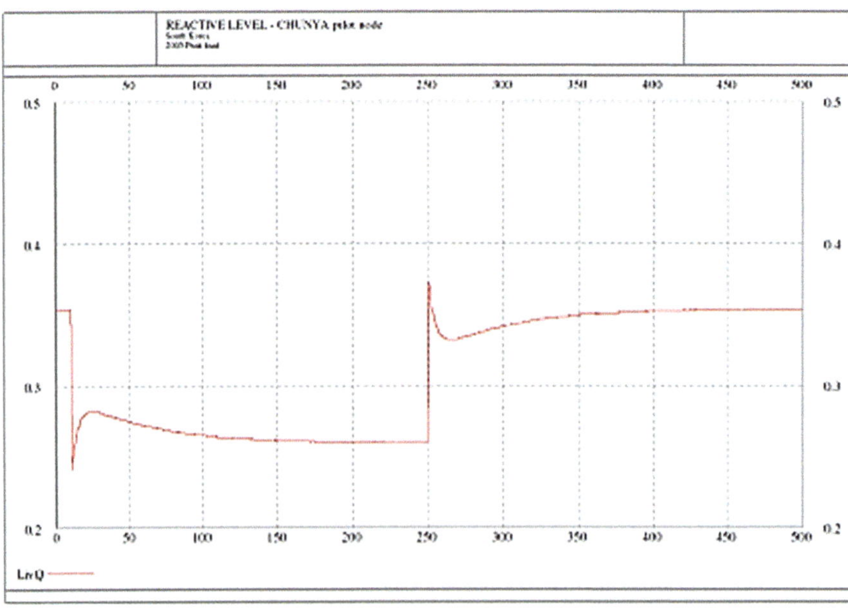

Fig. 6.6 South Korean 2006 power system, Area 2: dynamic response of reactive power level of CHONGYANG pilot node voltage control loop

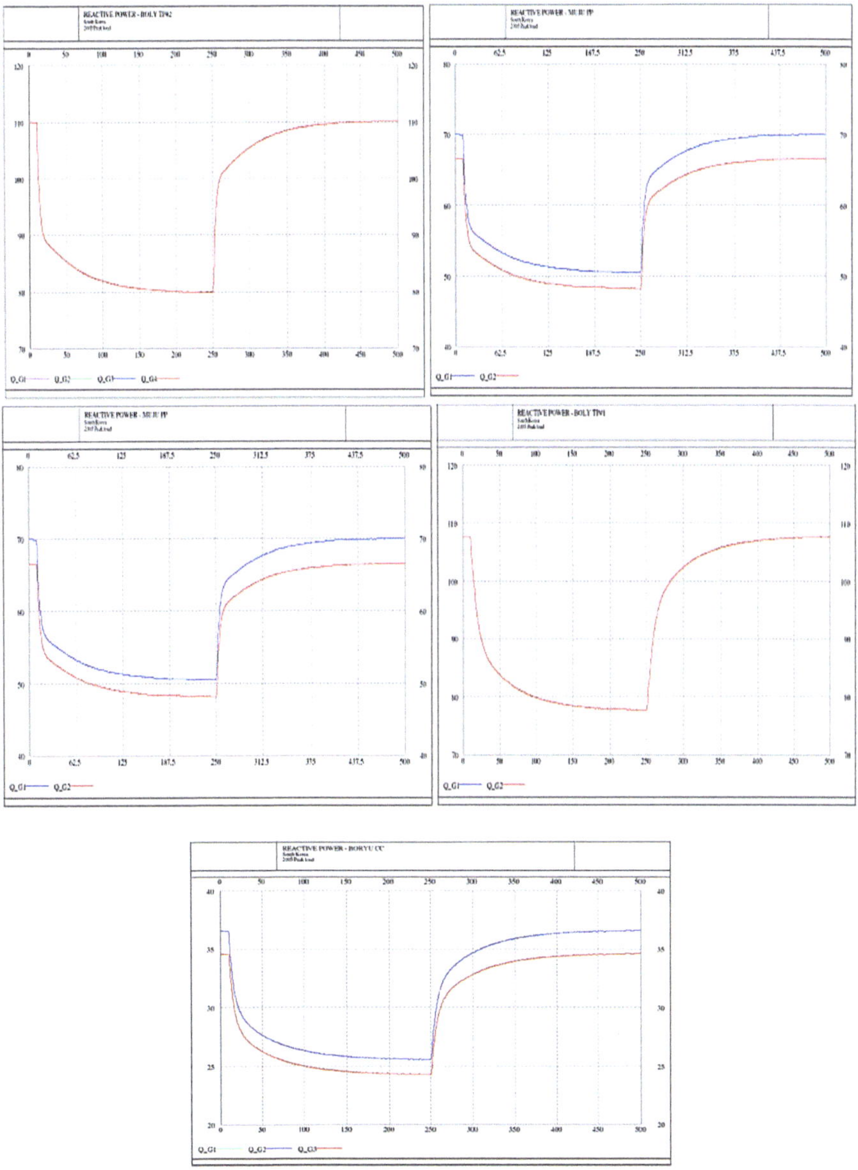

Fig. 6.7 South Korean 2006 power system, Area 2: dynamic response of pilot node voltage control loops: reactive power in the controlled power stations

The transients shown below refer to tests performed by a very detailed dynamic simulation model. They point out improvements provided by SVR in terms of system stability, voltage quality, and control effort minimisation, thereby determining losses as well as OLTC commutation manoeuver reduction. In all the tests shown, each reactive power control resource operates in accordance with its area needs, in unison with the other resources and at the expected time (any delay worsens the operating conditions). Reactive powers provided by the generators for voltage support favour an SVR that shows a lower control effort and higher voltages with respect to cases of PVR (see also Chap. 7). OLTC's superior tap position controllability and fewer commutation manoeuvres favour SVR over PVR (see Chap. 7).

Referring to tests with load step variations, each 2 % step is performed by the use of two subsequent 1 % steps. The results provided give evidence that SVR allows a larger control margin, even at high voltages (+5 %) with very low reactive power control levels (that is, with low reactive powers delivered by the generators and SVCs). The only exceptions are the Apollo and Vulcan areas (Fig. 6.9a). Even in under-excitation, system voltage controllability is wide. We also note:

1. The clear effectiveness of SVR on damping electromechanical oscillations (compare Fig. 6.11 with Fig. 6.14).
2. SVR avoids reactive power recirculation among generators in a power station or among power stations of a given area. These effects and the ability to continuously maintain a high voltage profile in the overall grid allow significant reduction of system losses.
3. SVR clearly reduces the reactive power control effort and so enlarges generator controllability in front of both normal and perturbed system operations.

In Fig. 6.8, load step variations are performed at the Pegasus area only. The other pilot node voltages are sensitive to far perturbations but with very small transients with respect to the Pegasus area charging the larger control effort. This is more evident in Fig. 6.9a, where the Pegasus reactive level shows the largest variations. Figure 6.9b offers a useful comparison in the case of the same load changes but at all the SVR areas: pilot node voltages and reactive control levels move together in unison but with differences in control effort due to the different area loads.

Test on SVR with Load Variation in One Area Only

Figure 6.10 shows how the other SVR areas have small transients and steady-state variations on generator reactive powers and voltages in comparison with the Pegasus ones in Fig. 6.11 where, with the exception of the peripheral the Tutuka power station, all other area power station electrical variables show significant changes.

Figure 6.12 gives evidence of SVC participation in SVR voltage control, mainly from those SVCs that have higher electrical coupling because inside the Pegasus area where the loads change. We notice how the Impala and Athens SVCs work concordantly (notwithstanding their electrical vicinity to one another), without dynamic interaction between their reactive power control loops.

Under primary voltage control alone, the same test on load variation as seen in this §6.1.3.1, but without SVR, shows transmission voltage degradation occurring mainly where the load increases.

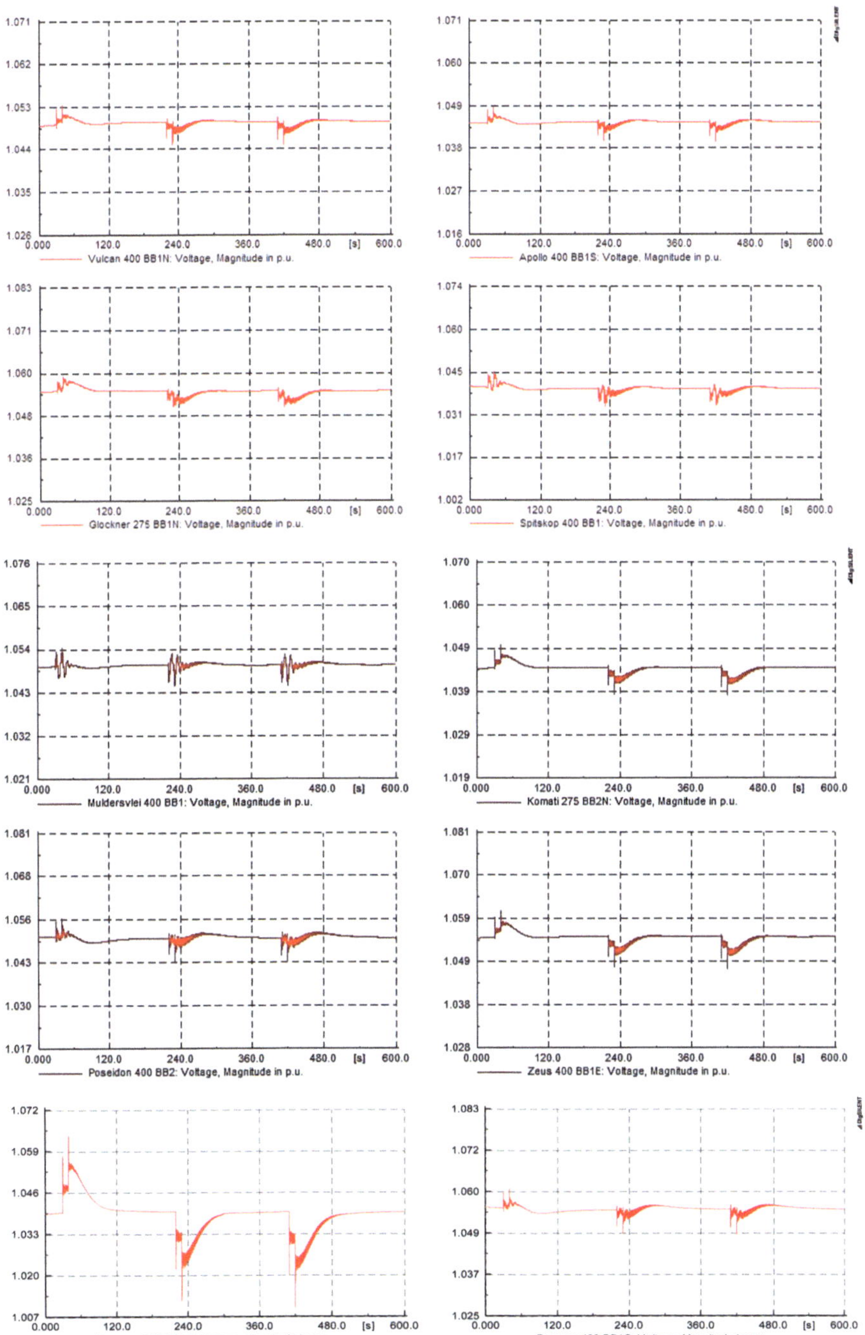

Fig. 6.8 South African 2007 winter peak: SVR pilot node voltage following −2 %, +2 %, +2 % step variations in Pegasus area loads

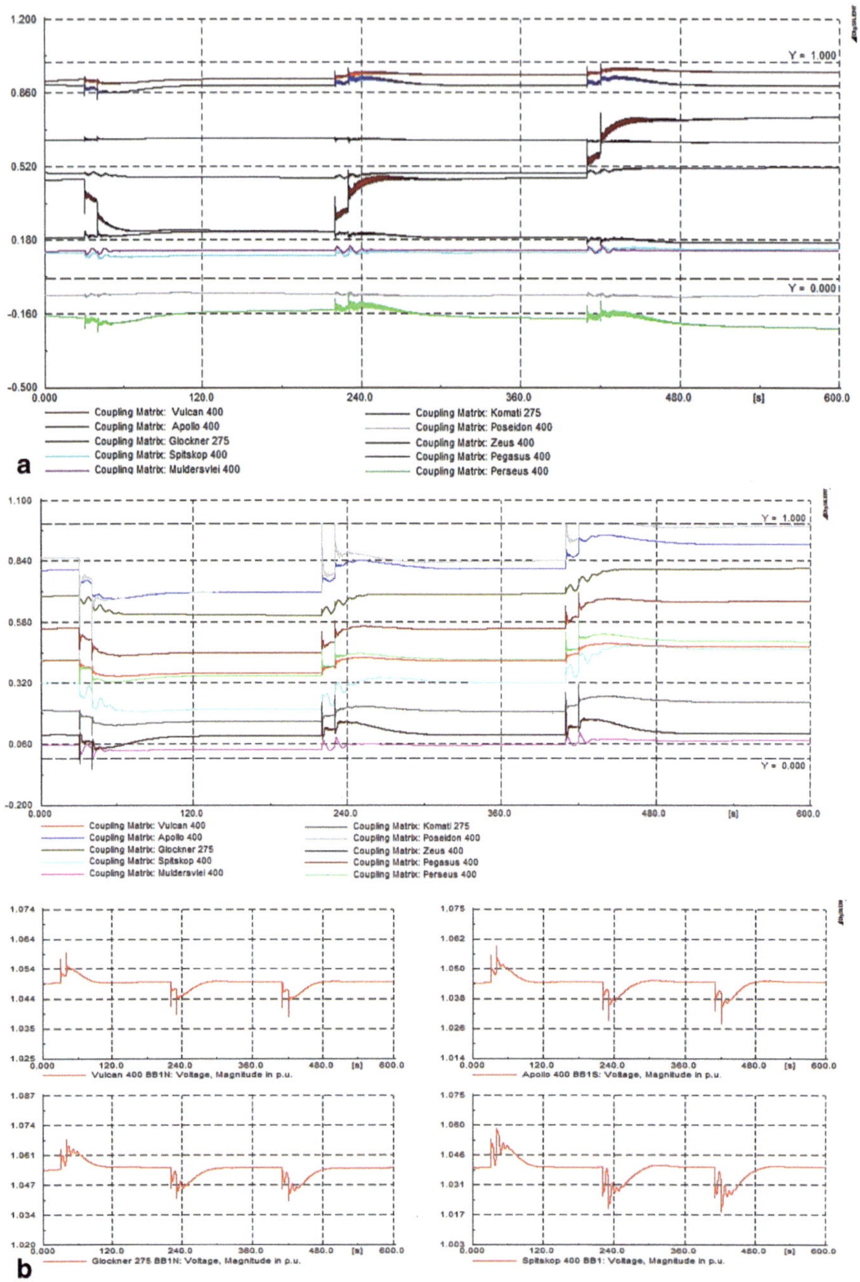

Fig. 6.9 **a** South African 2007 winter peak: SVR area reactive power levels following −2 %, +2 %, +2 % step variations in Pegasus area loads. **b** South African 2007 winter peak: SVR area reactive power levels (*top*), several pilot node voltages (*bottom*) following −2 %, +2 %, +2 % step variations in overall system loads

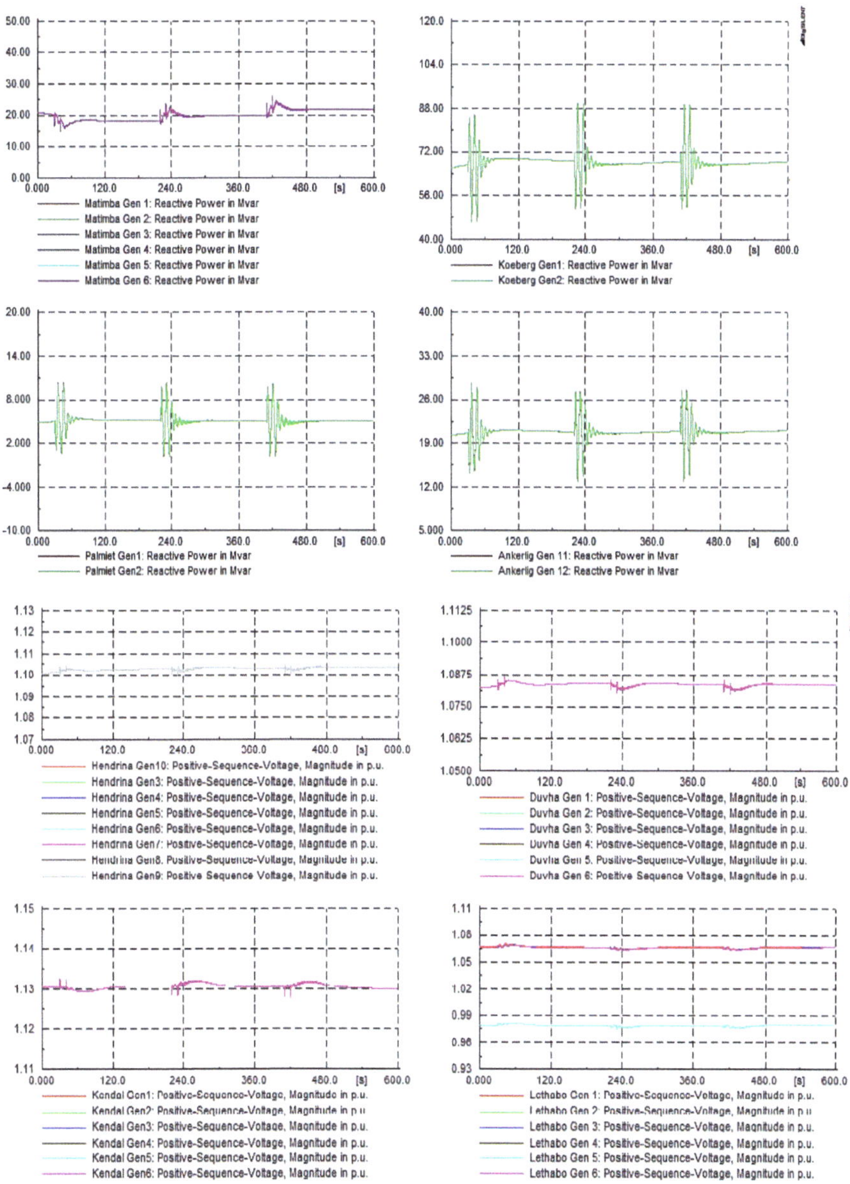

Fig. 6.10 South African 2007 winter peak with SVR: generator reactive power and voltage outside Pegasus area following −2 %, +2 %, +2 % step variations in Pegasus area loads

Figure 6.13 has to be compared with the Fig. 6.8 results related to the Pegasus area load variation. Under PVR only the Pegasus voltage significantly changes, which also offers evidence that the South African pilot nodes were properly selected

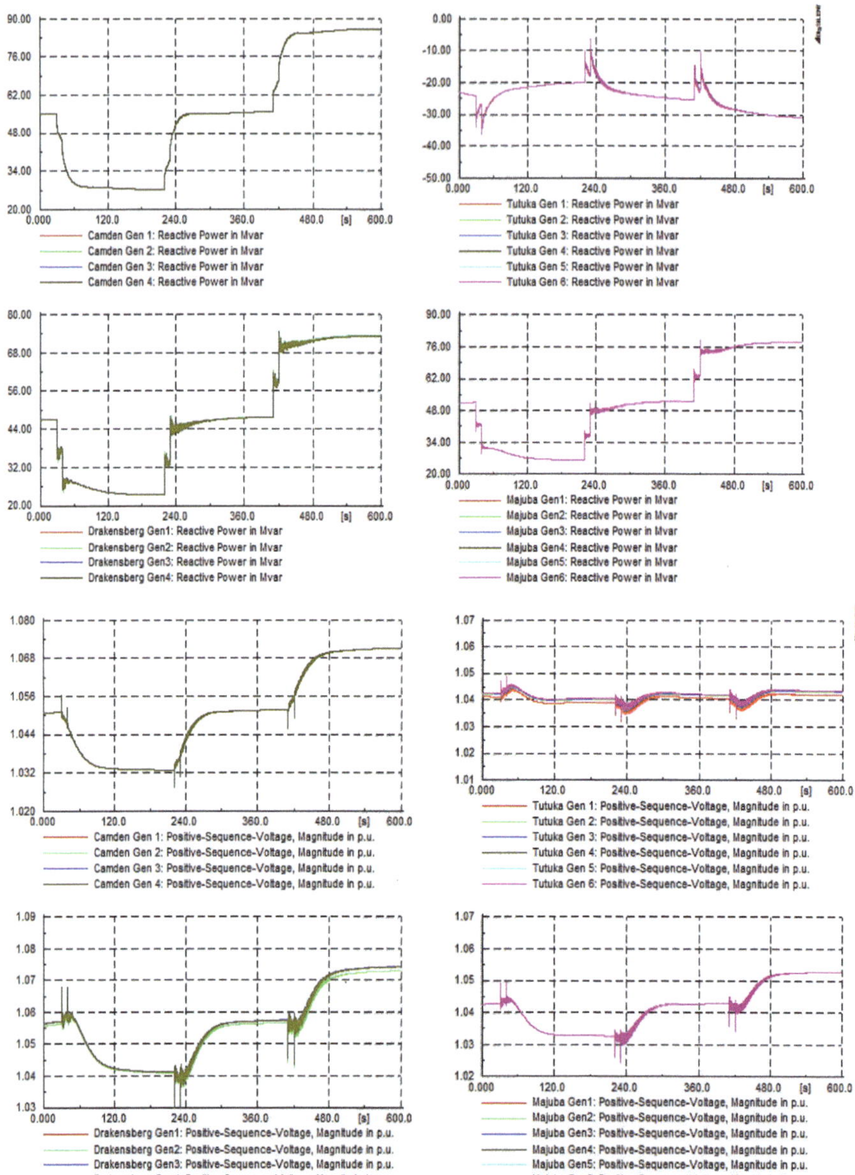

Fig. 6.11 South African 2007 winter peak with SVR: generator reactive power and voltage inside Pegasus area following −2 %, +2 %, +2 % step variations in Pegasus area loads

in amount and grid position. Moreover, thanks to the real-time TVR optimisation, SVR operates automatically at higher steady-state voltages with lower control effort than PVR.

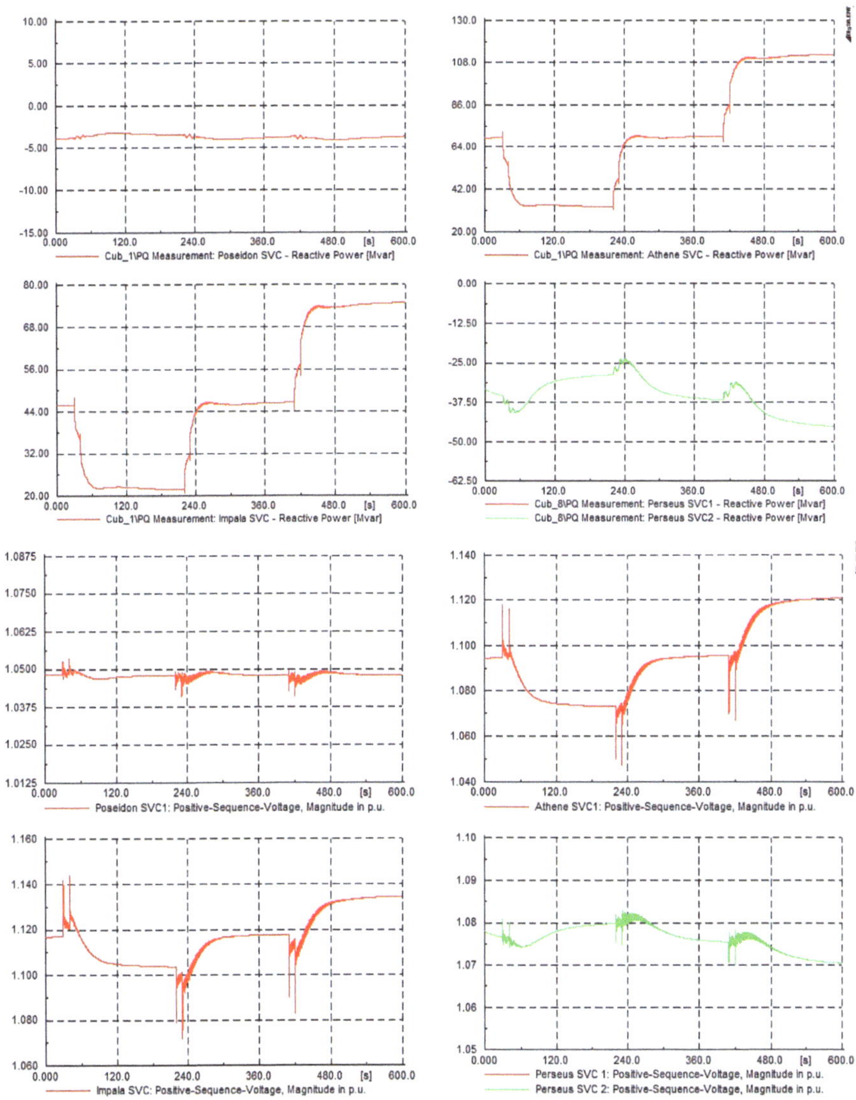

Fig. 6.12 South African 2007 winter peak with SVR: reactive power and voltage of SVCs inside/outside Pegasus area following −2 %, +2 %, +2 % step variations in Pegasus area loads

Test Like § 6.1.3.1, But with PVR Alone. Load Steps at Pegasus Area

Figure 6.14 must be compared with the Fig. 6.11 (SVR) results related to the Pegasus area load variation. With PVR alone, load variations at the Pegasus area determine local electromechanical oscillations at the Drakensberg power station. This happens when local generator AVRs are not more capable of maintaining stator voltages at the set-point value (OEL operation). Moreover, even if the generated

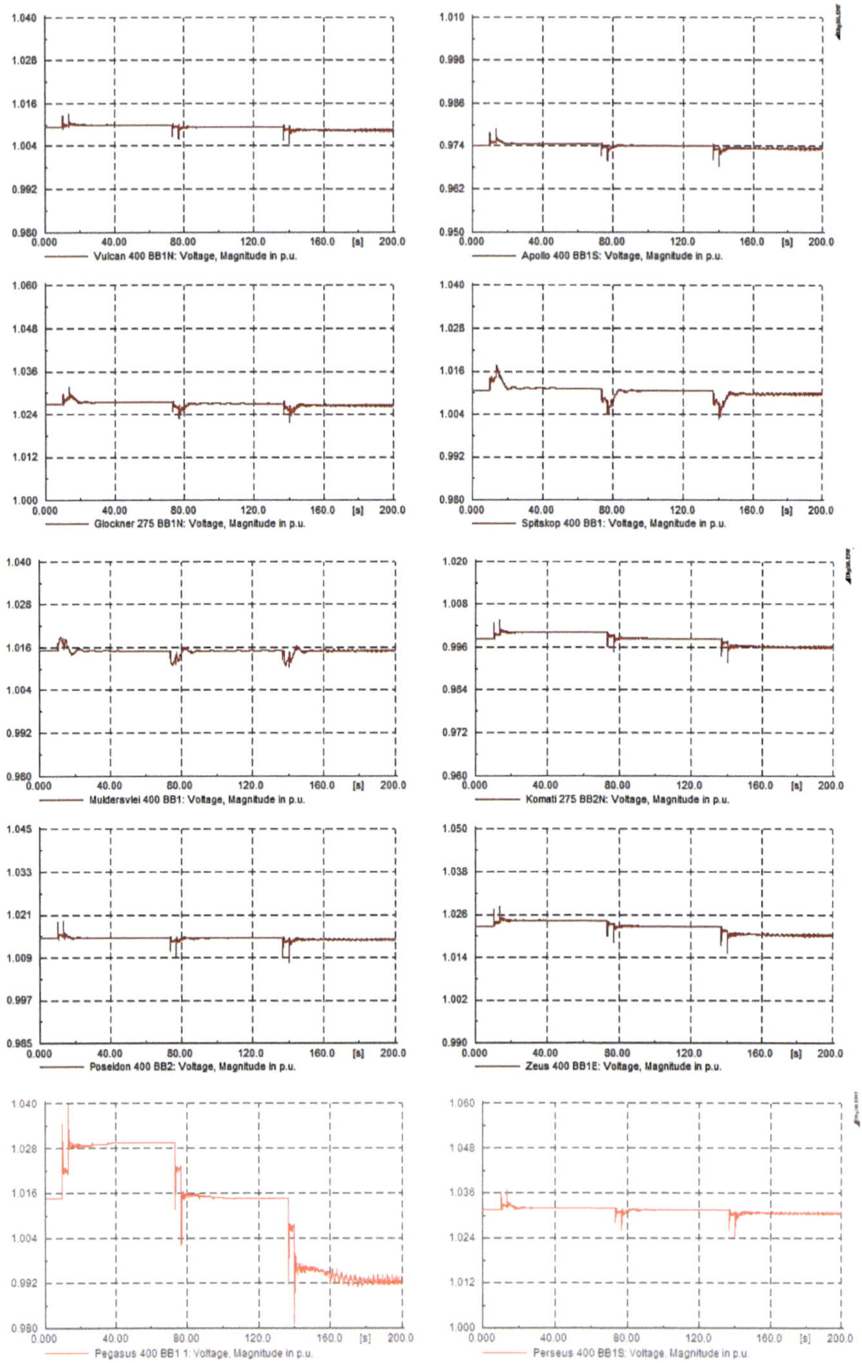

Fig. 6.13 South African 2007 winter peak: PVR voltage at buses called pilot nodes under SVR (see Fig. 6.8) following $-2\%, +2\%, +2\%$ step variations at the Pegasus area loads

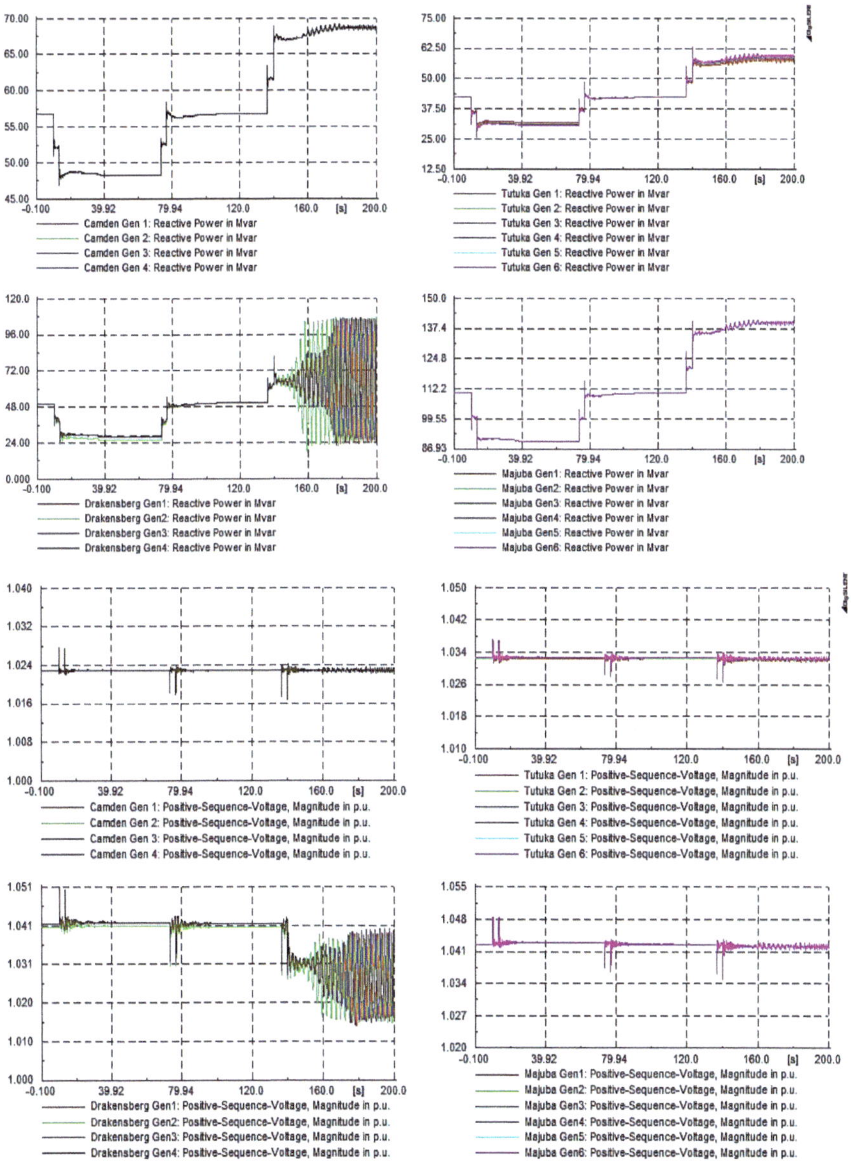

Fig. 6.14 South African 2007 winter peak without SVR: reactive power and voltage of generators inside Pegasus area following −2 %, +2 %, +2 % step variations in the Pegasus area loads

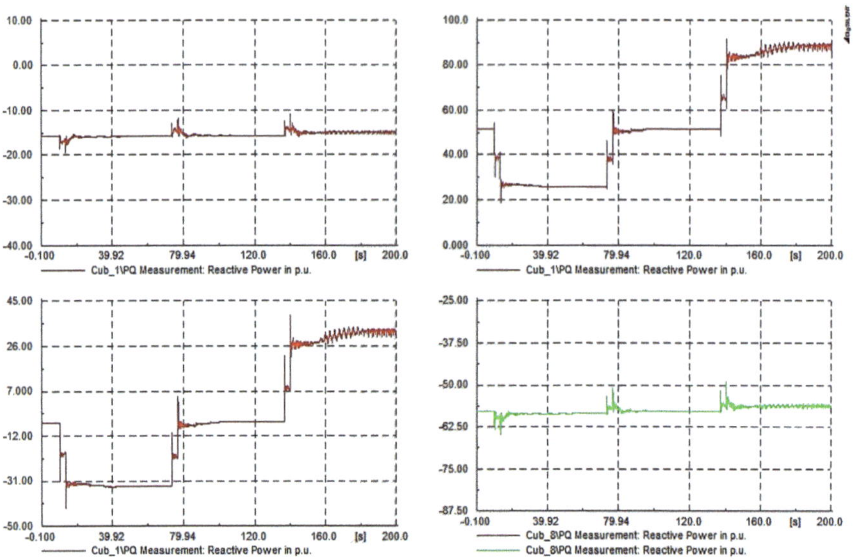

Fig. 6.15 South African 2007 winter peak without SVR: reactive power of SVC located inside/outside the Pegasus area following −2 %, +2 %, +2 % step variations in Pegasus area loads

reactive powers change under PVR (see also the SVCs on Fig. 6.15), they are not sufficient to sustain EHV voltages at the operated values.

Test on SVR with Step Variation at All the Pilot Nodes Voltage Set-Points

This §6.1.3.3 test is fully dedicated to SVR and gives a clear, impressive view of the might of SVR control and the opportunity, never before had by the system operator, to easily and actually move the voltages in the transmission buses as well as to impose a chosen voltage plan on the overall grid. The test consists of simultaneous and concordant change, under SVR control, of the pilot node voltage set-points.

Figure 6.16 shows the dynamic transients of pilot node voltages following set-point step variations. The fact that each pilot node easily tracks the set-point value without appreciable dynamic interaction among the SVR voltage control loops gives evidence of an adequate control law design and a proper pilot node selection. The result is also confirmed by Fig. 6.18, showing the voltages of the other main buses in the power system during the described test: all are strongly affected by their pilot nodes, instantaneously following their voltage trends.

In Fig. 6.17a, all area reactive power level trends are shown. From the area control signals it is evident that each area works during the test as if it were autonomous by facing local needs. Figure 6.17b offers a useful comparison in the case of voltage set-point step variations operated at the Zeus pilot node only. We notice that the Zeus reactive power level significantly changes during the test, as does

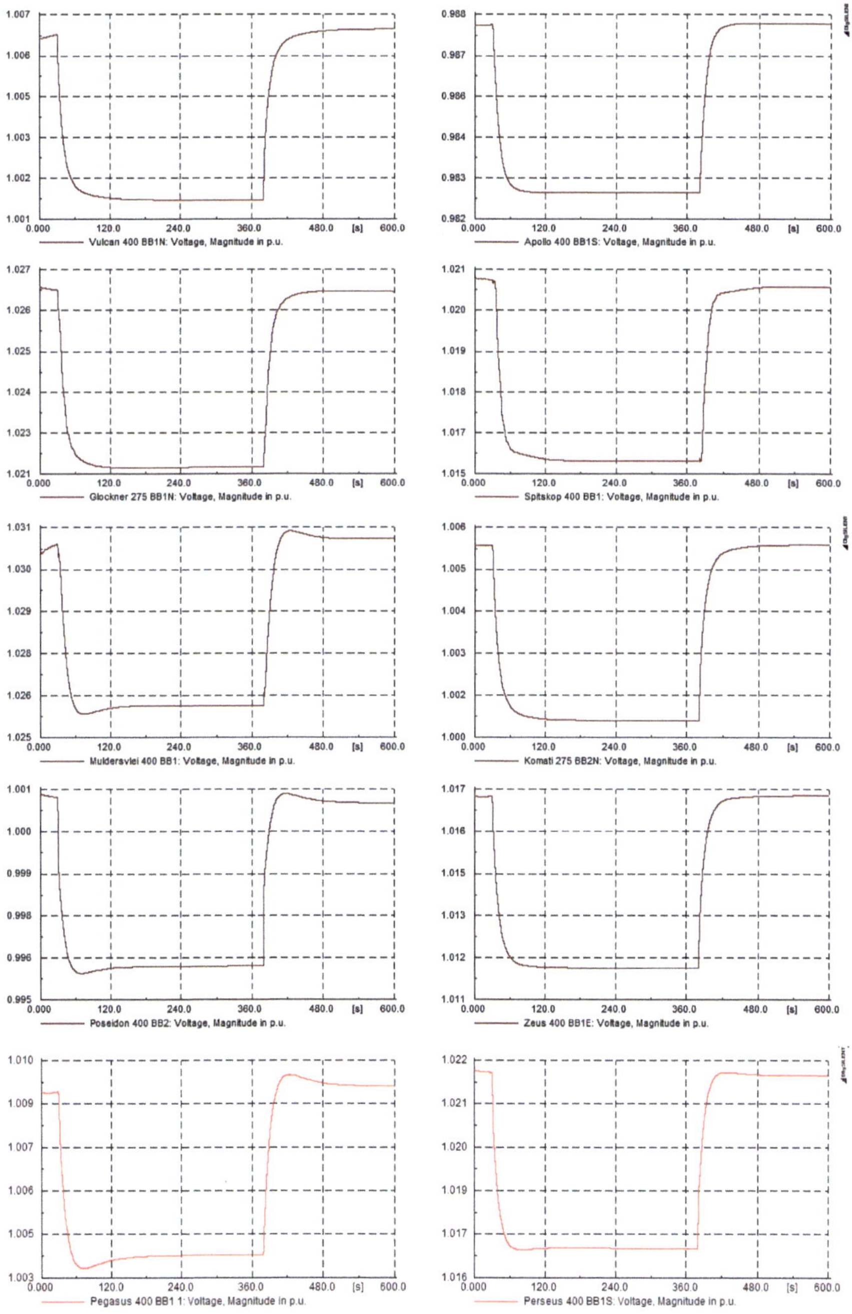

Fig. 6.16 South African 2008 winter peak: SVR pilot node voltage following concordant step variations at all pilot node voltage set-points

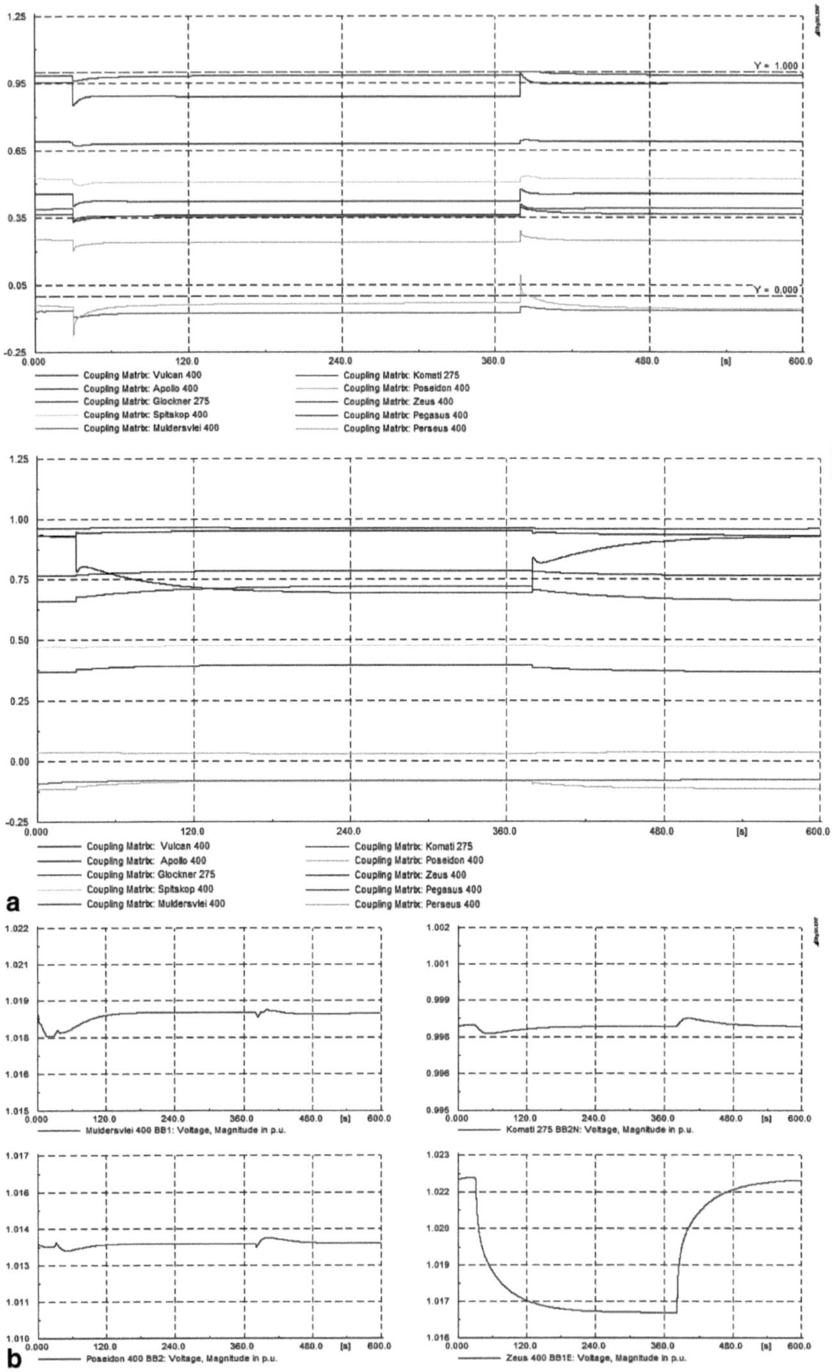

Fig. 6.17 **a** South African 2008 winter peak: SVR area reactive power levels following concordant step variations in all pilot node voltage set-points. **b** South African 2008 winter peak: SVR area reactive power levels (*top*) and several pilot node voltages following step variations at Zeus pilot node voltage set-point

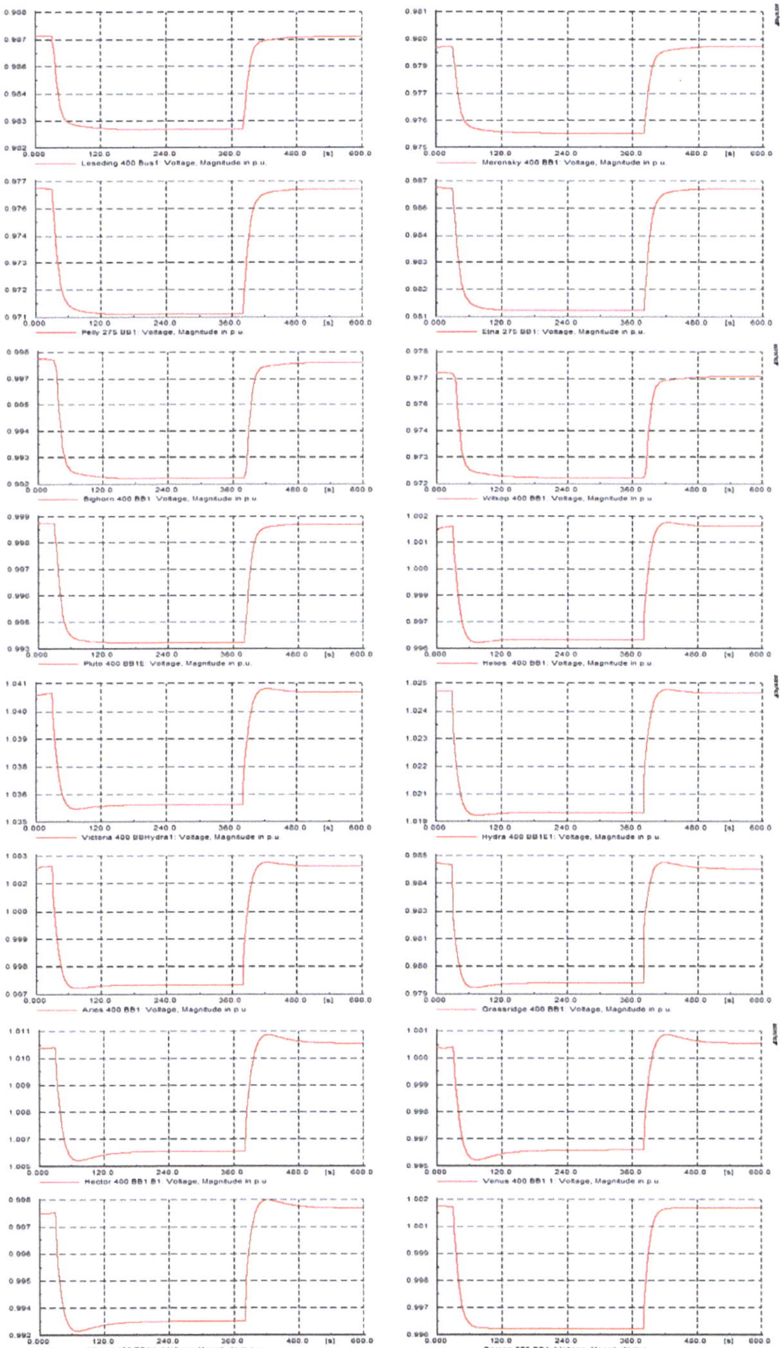

Fig. 6.18 South African 2008 winter peak: voltage trend at system buses imposed by their pilot node voltages following concordant step variations at SVR voltage set-points

the correspondent pilot node voltage. The other pilot nodes are lightly affected by Zeus area changes, again showing how good the operated pilot node selection is. Figure 6.18 confirms SVR strength, showing how all system bus voltages follow their pilot node voltage trends.

6.2 Tests with Contingencies in Large Power Systems

The next examples of SVR control scheme performances consider large power systems under perturbed conditions ($n-1$ or $n-2$) determined by line opening and generator tripping. We also notice the great difference in this case between area reactive power control when automatically coordinated by SVR and the case of primary voltage regulation alone. The simulation mainly shows:

- The timely manner in which SVR begins sustaining voltages, mainly those needing more support after a local contingency, despite system operator manual control;
- The fast and concordant automatic control of reactive power resources in the amount and at the location where it is actually needed by a power system;
- The way the control system uses all available control resources up to their limits, reached at the same time, unless the operating TVR re-optimises in real time the transmission network voltage plan actuated by SVR.

6.2.1 Tests on Line-Opening

Performance Comparison Between SVR with PVR in South Korean Hierarchical Voltage Control System

$n-2$ contingency: Opening two lines of Sinanseong–Sinseosan 765 kV

Figures 6.19–6.23 refer to this contingency. Fig. 6.19 illustrates SVR's ability to recover voltage at the edge buses of two opened lines.

We note that a *special protection scheme* (SPS) operating in the South Korean power system contributes either in the SVR case or when PVR is alone to sustain the otherwise compromised system angle stability. In particular:

- SVR without SPS is stable but with low damping; PVR is very unstable without SPS;
- SVR with SPS is stable with acceptable damping as shown by the Figs. 6.19, 6.21 and 6.22;
- PVR with SPS is stable but with unacceptable damping, as shown in Figs. 6.20, 6.23;

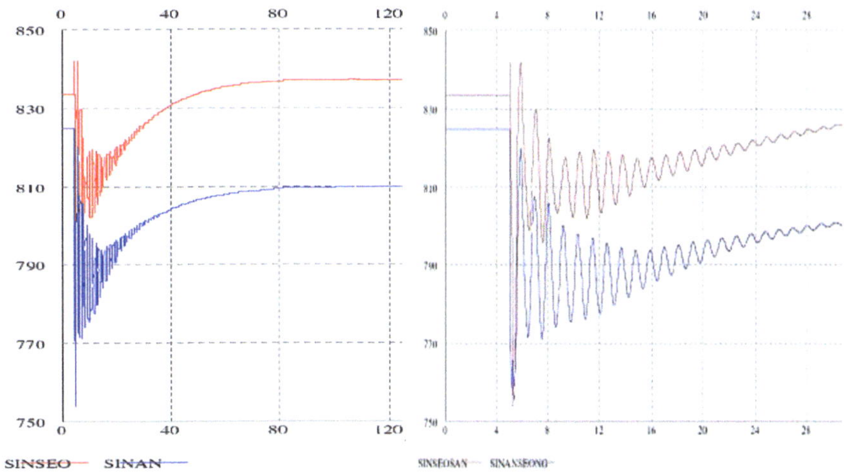

Fig. 6.19 South Korean 2009 peak load: voltage trends under SVR at Sinan and Sinseo system buses, imposed by their pilot node voltage control after the two lines open between them

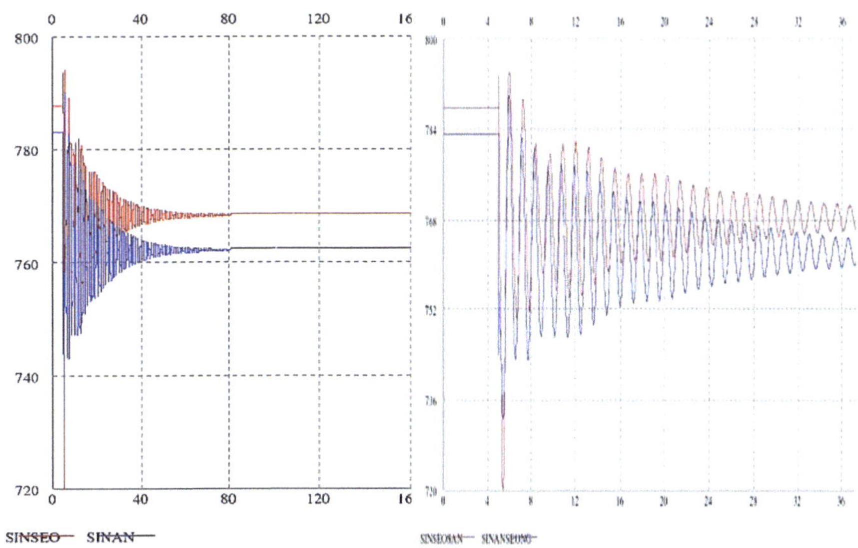

Fig. 6.20 South Korean 2009 peak load: voltage trends at Sinan and Sinseo system buses imposed by PVR alone, after the two lines open between them

- With SVR, voltages are recovered near the values before the contingency after a 60-s transient, with a big control effort by the weakest areas: Area 9, Area 10 (Fig. 6.22). Sinan bus voltage decays, but from 825 to only 810 kV.
- Without SVR, voltages go down to about 8–20 kV without recovering.
- The nearby 1-s oscillation of an electromechanical nature shows an improved damping by SVR, even if this is not the main SVR task.

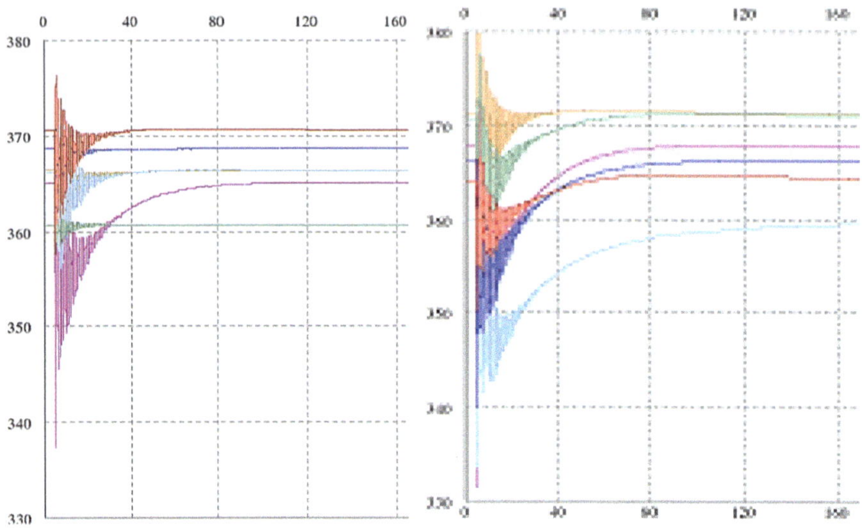

Fig. 6.21 South Korean 2009 peak load: pilot node voltage trends imposed by SVR, after the two lines open between them

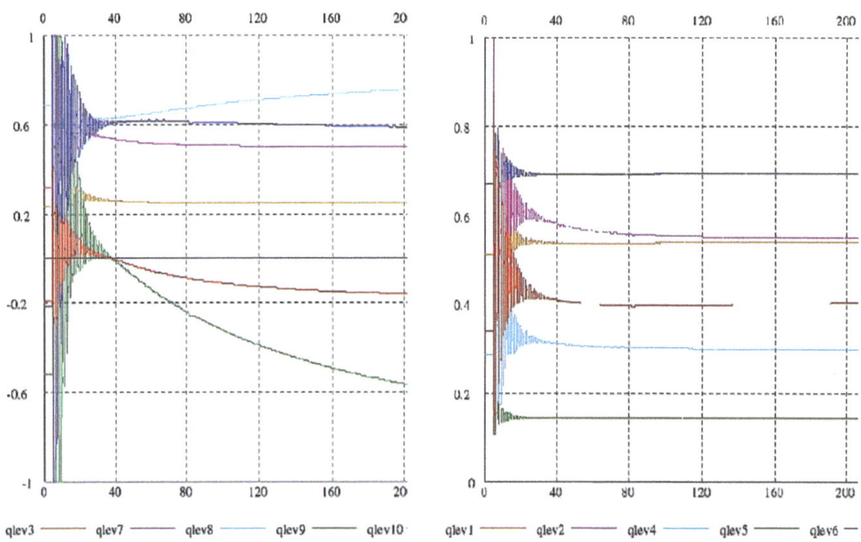

Fig. 6.22 South Korean 2009 peak load: area reactive power level trends imposed by SVR after the two lines open between them

Performance Comparison Between SVR with PVR in Taiwan Hierarchical Voltage Control System

n−2 contingency test: Opening two lines, Luntane (LUNA11) to Chnliaos (CH6B11) and Omeie (OMEB11) to Wufene (WUFB11)

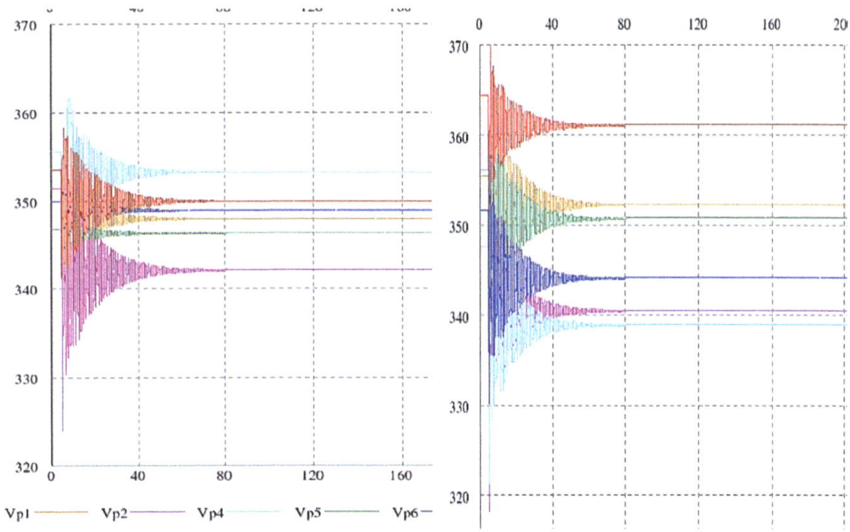

Fig. 6.23 South Korean 2009 peak load: pilot node voltage trends imposed by PVR alone after the two lines open between them

Line opening affects the whole transmission network (see Figs. 6.24 and 6.25). The SVR response correctly recovers pilot node voltages to voltage reference values.

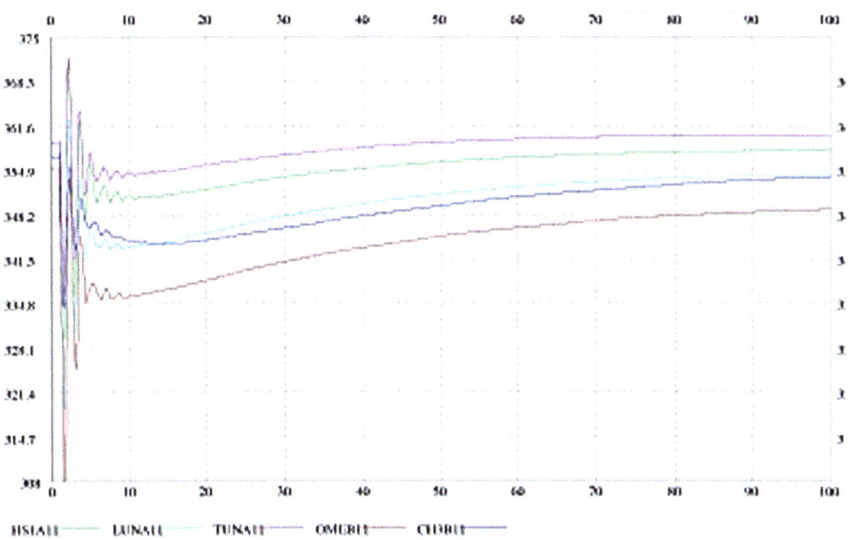

Fig. 6.24 Taiwan 2009 peak load: pilot node voltage trends imposed by SVR after the two lines open between them

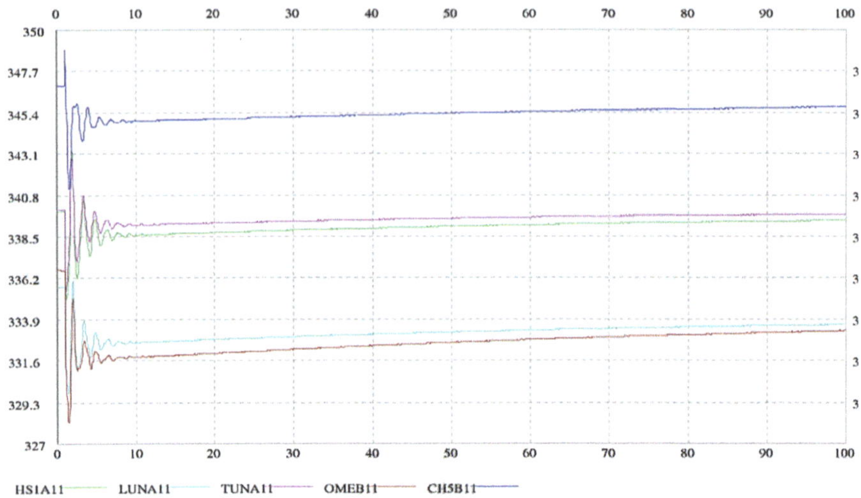

Fig. 6.25 Taiwan 2009 peak load: pilot node voltage trends imposed by PVR alone after the two lines open between them

Performance Comparison Between SVR with PVR in South African Hierarchical Voltage Control System

$n-2$ contingency test: Kendal–Minerva and Duvha–Vulcan 400-kV lines opening in 2008 winter peak load scenario

The opening of Kendal–Minerva and Duvha–Vulcan 400-kV lines causes the redistribution of more than 2500 MW in the Mpumalanga geographical area.

This severe test determines the control saturation of the Vulcan area. This can be easily seen in the way the Vulcan reactive power level suddenly reaches the 1-p.u. value (Fig. 6.27) and by how the Vulcan voltage does not fully recover to set-point value (Fig. 6.26, top left). Figure 6.28 shows the voltage recovery in the system buses (not pilot nodes) due to their pilot node voltage regulation by SVR. With an SVR control system, the grid is more stable and not affected by the persistent electromechanical oscillation of the PVR test.

Considering the case of PVR control working alone (Figs. 6.29 and 6.30), some bus bars, such as Minerva and Merensky 400 kV, have rather low voltage values (Vulcan: −5% with PVR; −1.3% with SVR). The loss increment difference is relevant, being lower with SVR than with PVR due to SVR's ability to concentrate control effort mainly in those areas involved in the contingency (see Chap. 7).

The generator trip can determine the loss of some PSS stabilising controls. Due to the stabilising effect provided by SVR, this contingency carries a minor risk under SVR control. This test comparison also shows that SVR reduces the reactive power control effort, thereby enlarging generator controllability and grid reliability as well.

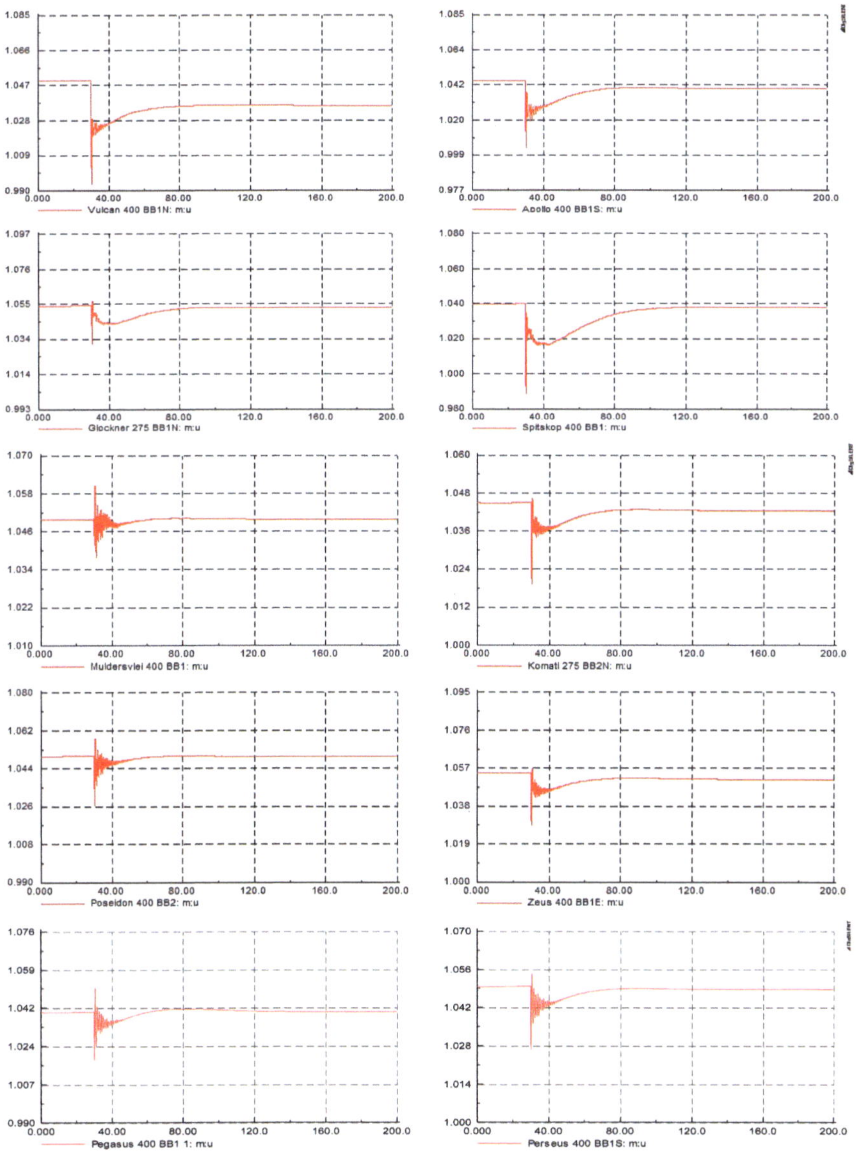

Fig. 6.26 South African 2008 peak load: pilot node voltage trends imposed by SVR after the two lines open between them

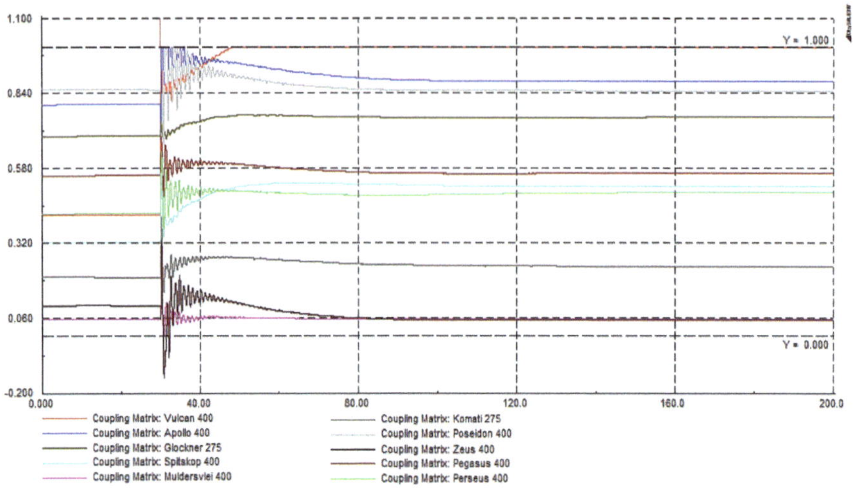

Fig. 6.27 South African 2008 peak load: area reactive power level trends imposed by SVR after the two lines open between them

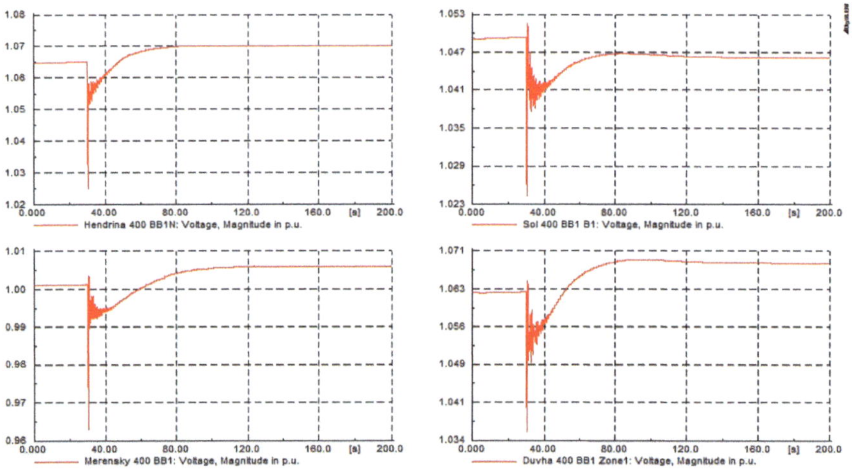

Fig. 6.28 South African 2008 winter peak: voltage trend at system buses imposed by their pilot node voltages after the two lines open under SVR

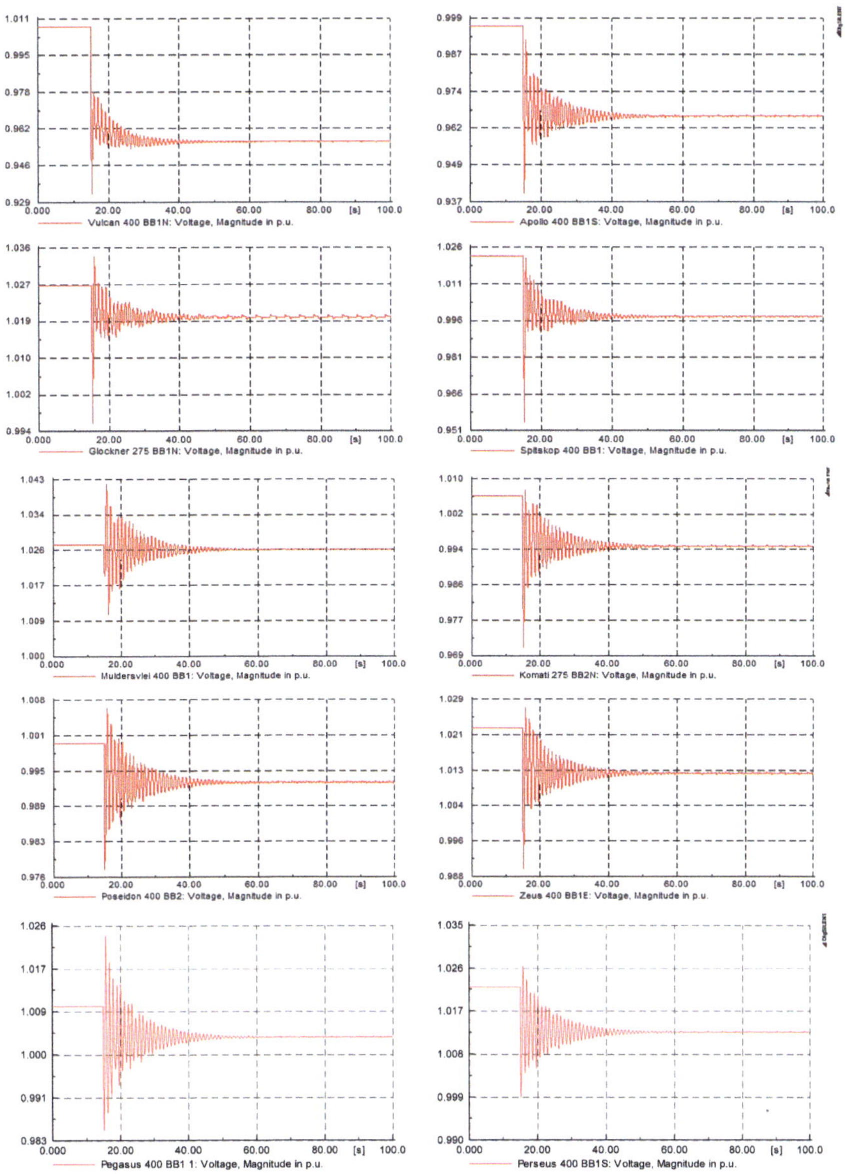

Fig. 6.29 South African 2008 peak load: pilot node voltage trends imposed by PVR alone after the two lines open between them

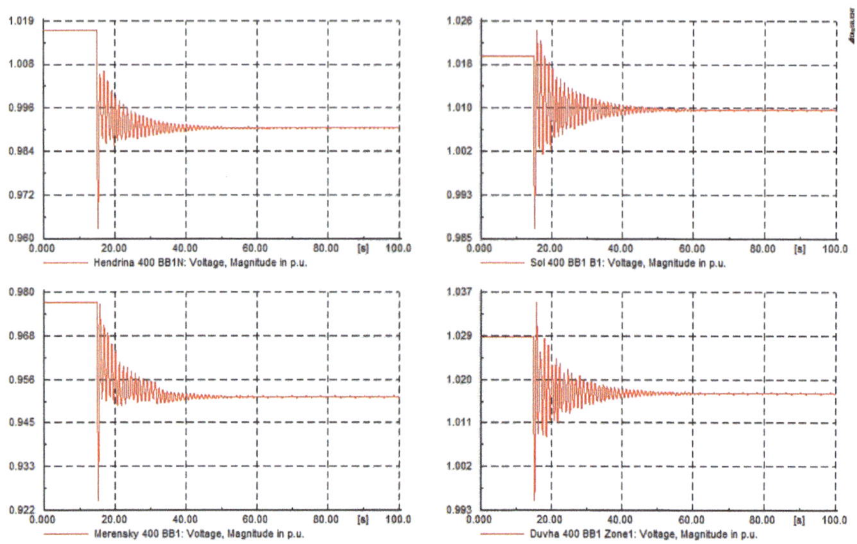

Fig. 6.30 South African 2008 winter peak: voltage trend at system buses without SVR after the two lines open between them (compare to Fig. 6.28)

6.2.2 Tests on Generator Tripping

Performance Comparison Between SVR with PVR in South African Hierarchical Voltage Control System

$n-1$ contingency test: 2008 winter peak load, Koeberg generator trip

The Koeberg generator trip is an important contingency because it causes instantaneous loss at the Muldersvlei area of both 897-MW production and one PSS device, thus increasing system vulnerability by reducing angle stability.

The Muldersvlei area with SVR control does not reach its saturation limits, whereas the Perseus area reaches over-excitation limits with a voltage value slightly lower ($-0.2\,\%$) than the set-point. Under PVR alone, voltage lowering is largely evident at Muldersvlei, Poseidon and Perseus areas (Hydra: $-3\,\%$).

The Poseidon area leaves saturation, whereas Perseus approaches its limits; Perseus area saturation is also evident in the values of reactive production in the area by SVCs and the Tutuka power plant (Figs. 6.31 and 6.32).

In Fig. 6.33, the impact of the generator trip on the overall power system for the Perseus and Poseidon areas mostly involved is evident. Figure 6.34 also shows how Perseus voltage recovery by Tutuka contribution removes the SVC in the Poseidon area from the maximum reactive power limit (Figs. 6.35, 6.36, and 6.37).

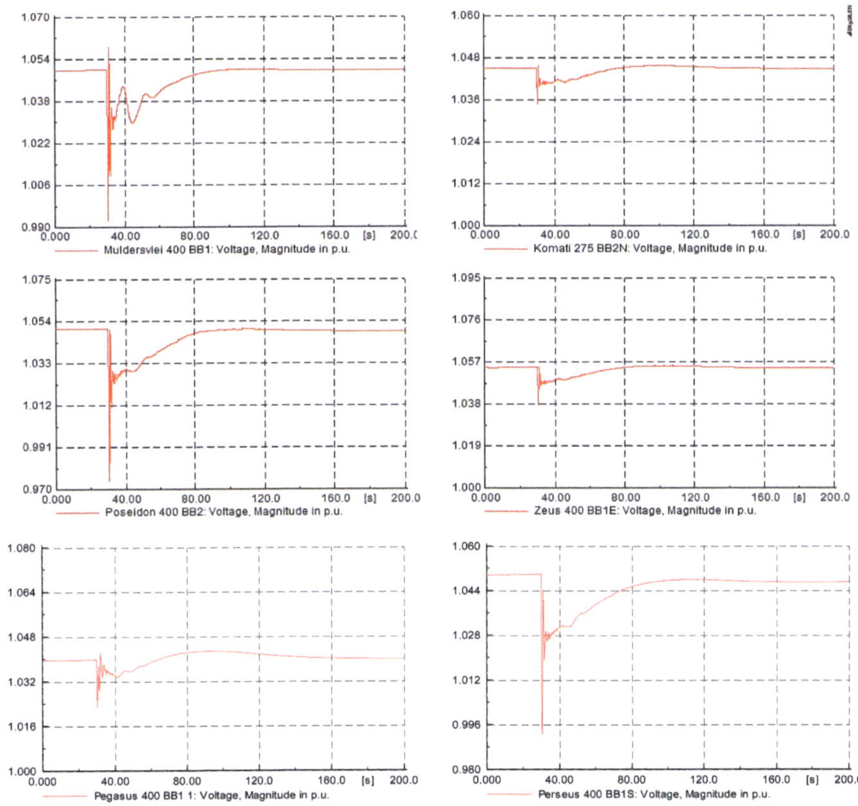

Fig. 6.31 South African 2008 peak load: pilot node voltage trends imposed by SVR following Koeberg generator trip

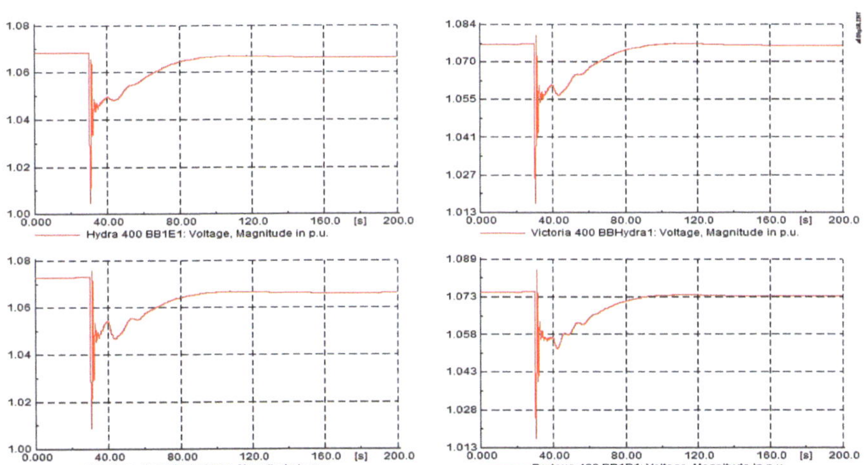

Fig. 6.32 South African 2008 winter peak: voltage trend at system buses imposed by their pilot node voltages following Koeberg generator trip under SVR

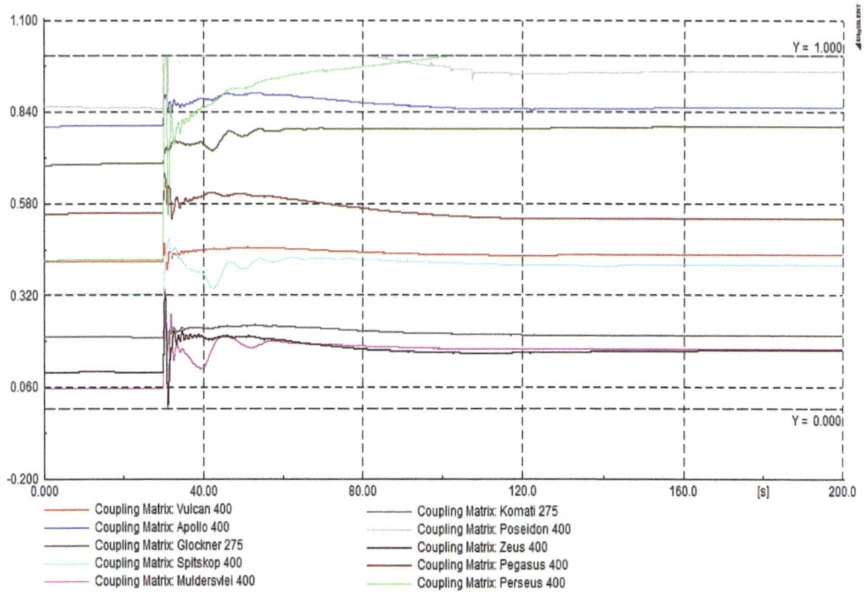

Fig. 6.33 South African 2008 peak load: area reactive power level trends imposed by SVR, following Koeberg generator trip

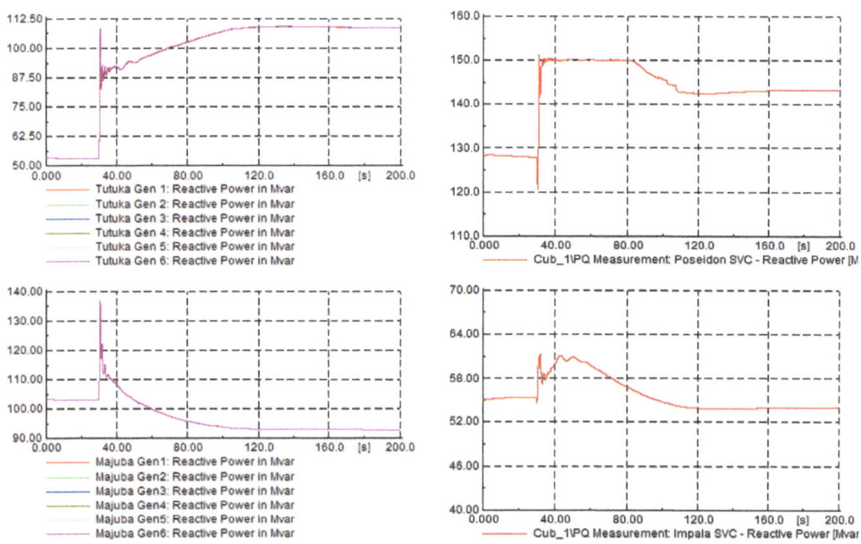

Fig. 6.34 South African 2008 winter peak: generator (*left*) and SVC (*right*) reactive powers trend imposed by SVR following Koeberg generator trip under SVR

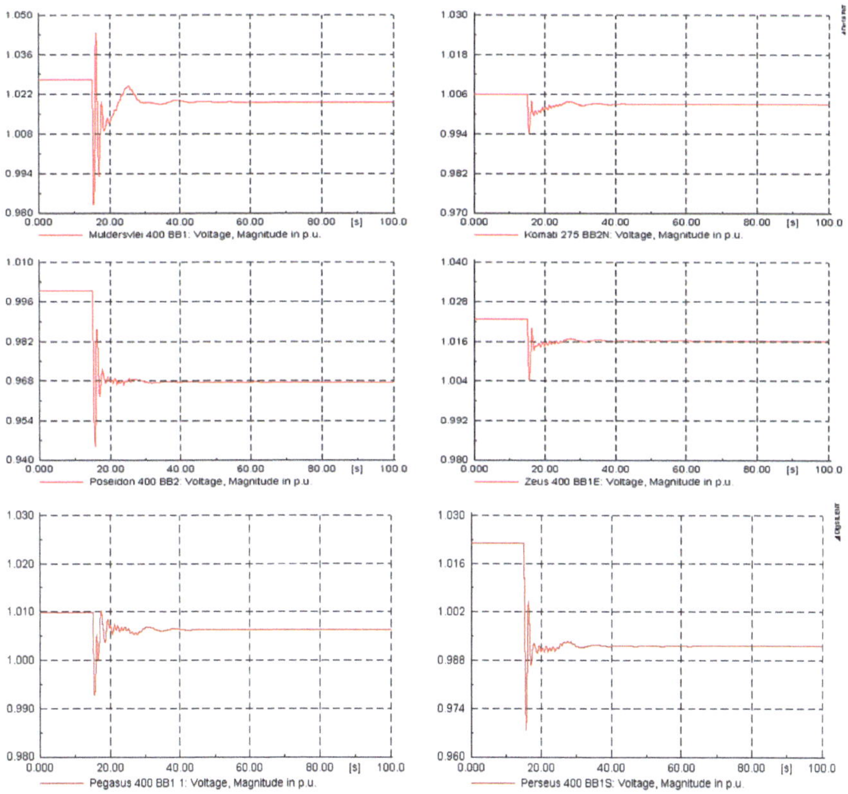

Fig. 6.35 South African 2008 peak load: pilot node voltage trends imposed by PVR working alone following Koeberg generator trip

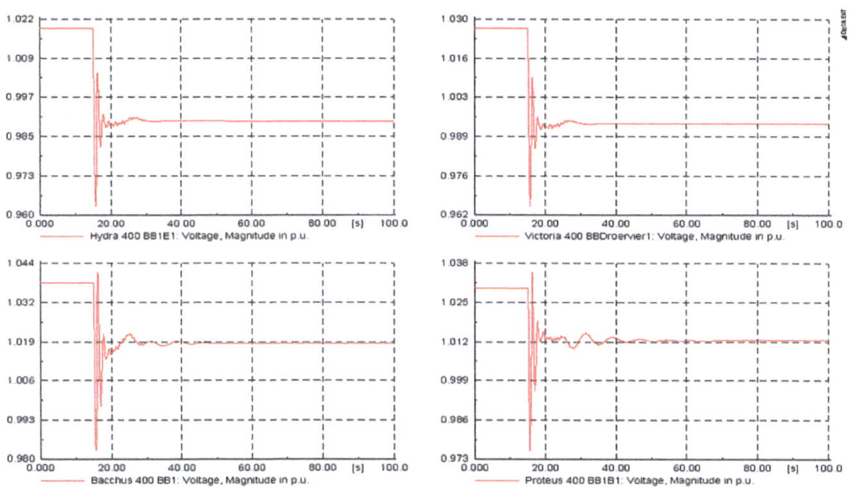

Fig. 6.36 South African 2008 winter peak: voltage trend at system buses under PVR alone following Koeberg generator trip (compare to Fig. 6.32)

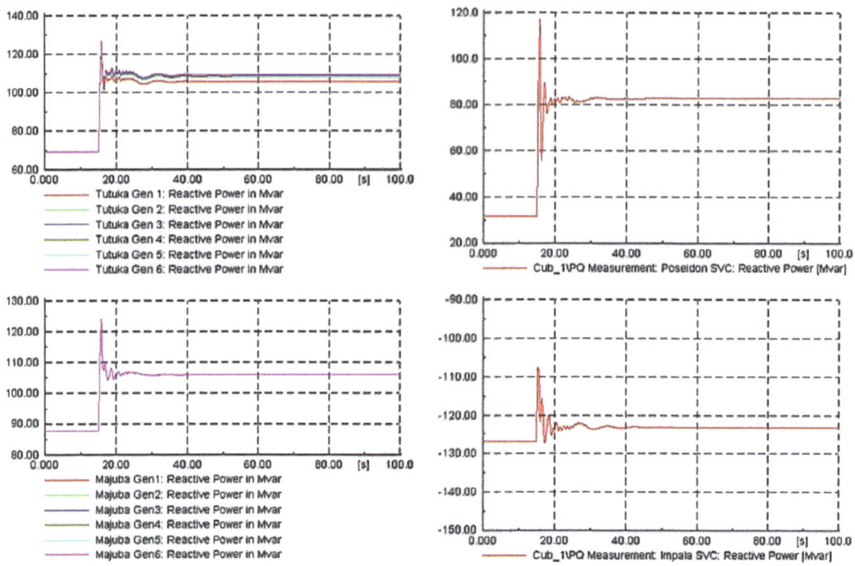

Fig. 6.37 South African 2008 winter peak: generator (*left*) and SVC (*right*) reactive power trend under PVR alone following Koeberg generator trip under SVR (compare to Fig. 6.34)

Performance Comparison Between SVR with PVR in Taiwan Hierarchical Voltage Control System

$n-2$ contingency test: 2009 peak load, two groups opening at the 2NDA power station

The opening of two groups at the 2NDA power plant affects the entire transmission network (see Figs. 6.38 and 6.39). The SVR response correctly recovers pilot node voltages to their set-point values (see Fig. 6.38): voltages do not go transiently down 339 kV; after 10 s they recover over 346 kV.

Without SVR, voltages go down 327 kV and after 10 s they recover over 331 kV (15 kV less than with SVR). It is also useful to offer evidence of the different voltage starting point conditions between SVR-TVR and PVR (due to TVR), which significantly contribute to SVR's higher performance. With SVR, the considered test also presents an advantage of lower circulating currents: the increase of current in the line from TINHUE (TINA11) to SANLUE (SANA11) is 31 % (from 0.326 kA to 0.482 kA); without SVR it is 55 % (from 0.325 kA to 0.504 kA).

The amounts of control power plant reactive power turn out to be smaller than corresponding values under PVR alone.

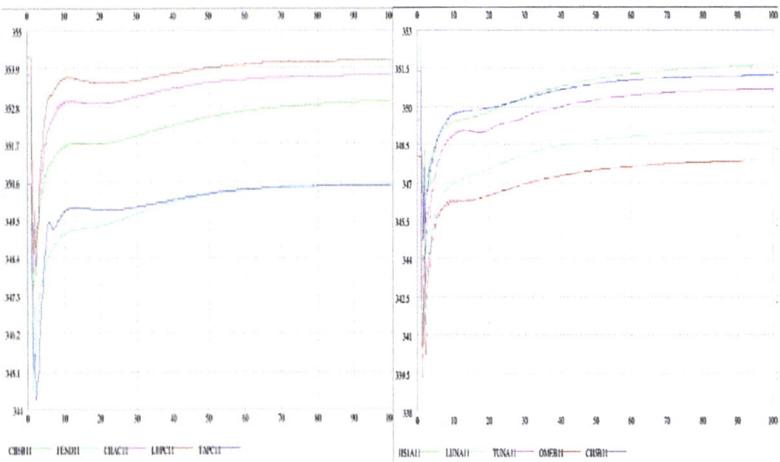

Fig. 6.38 Taiwan 2009 peak load: pilot node voltage trends with SVR following two generators' tripping

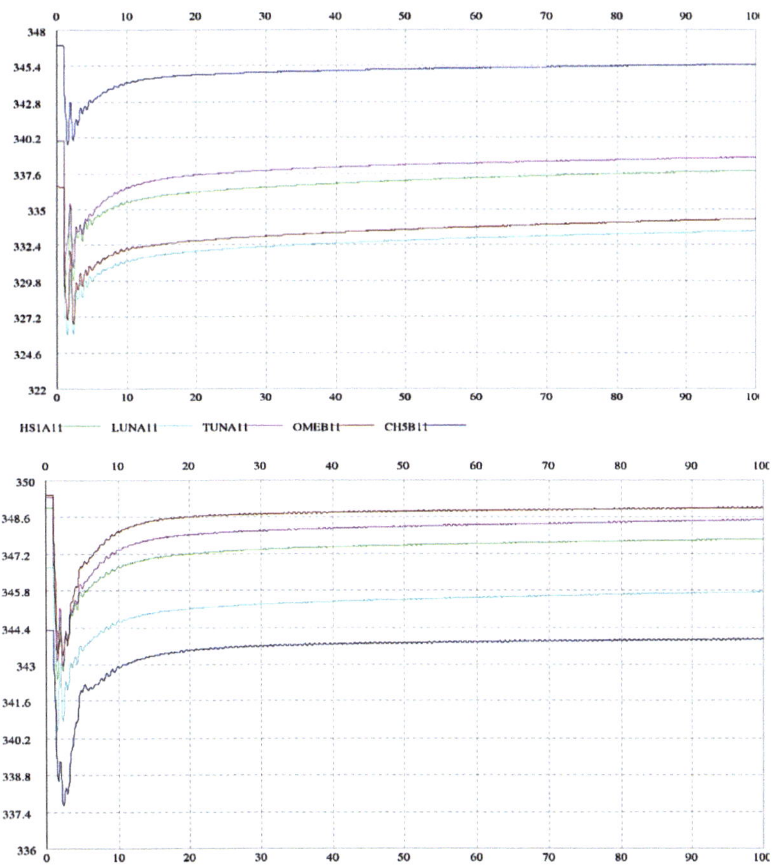

Fig. 6.39 Taiwan 2009 peak load: pilot node voltage trends without SVR following the two generators' tripping

References

1. Corsi S, Pozzi M, Sabelli C, Serrani A (2004) The coordinated automatic voltage control of the Italian transmission grid. Part I: reasons of the choice and overview of the consolidated hierarchical system. IEEE Trans Power Syst 19(4):1723–1732
2. Corsi S, Pozzi M, Sabelli C, Serrani A (2004) The coordinated automatic voltage control of the Italian transmission grid. Part II: control apparatus and field performance of the consolidated hierarchical system. IEEE Trans Power Syst 19(4):1733–1741
3. Corsi S, De Villiers F, Vajeth R (2010) Secondary voltage regulation applied to South Africa transmission grid. IEEE/PES General Meeting, Minneapolis, July 2010
4. Corsi S, De Villiers F, Vajeth R (2010) Power system stability increase by secondary voltage regulation applied to the South Africa transmission grid. Bulk Power System Dynamics & Control, IREP-VIII, Buzios, RJ, Brazil, August 2010

Chapter 7
Economics of Voltage Ancillary Service

Voltage service is one of the major *ancillary services* supporting the operation of a power system. Generators are involved, but so is some grid equipment (as considered in this book). Action occurs mainly on power plant operating controls and at the grid level as well, as made possible by V-WAR.

Here, the complexity of voltage service (VS) cost/benefit analysis is preliminarily introduced to illustrate the significant contributions of SVR-TVR to VS. The cost/benefit examples in the chapter are meant simply to introduce the reader to the relative economic advantages of VS features. Moreover, the correct measure and computing of a power plant's contribution/performance to network voltage regulation is considered so we may gain a proper recognition of its economic value.

The starting point of this chapter's thesis is a criticism of the present generalised situation, where generator effort in terms of reactive power delivery/absorption is generally considered as representative of the concrete network voltage support which generators normally provide. Leaving this unreliable criterion, we know it is difficult to distinguish and compute the various contributions of the aspects and subjects of voltage service. There is a need to see how and under what conditions the measurement of VS can be undoubtedly simplified and recognised. This chapter offers a solution. Indeed, Chap. 7 also demonstrates the simplicity, correctness and indubitableness of the *indicators* of the generator contribution to network voltage support when generators operate under secondary voltage regulation (SVR). We present a new fiscal measure and meter of this contribution to voltage service based on these indicators.

7.1 General Overview

The restructuring of the electric utility industry and the related growing competition in the electricity market have seen the emergence of newly defined power plant services such as those for network voltage and frequency support. The ancillary

S. Corsi, *Voltage Control and Protection in Electrical Power Systems*,
Advances in Industrial Control, DOI 10.1007/978-1-4471-6636-8_7

services market requires services to be measured, evaluated and paid for the same way that electrical energy itself is. Voltage and frequency services are strongly linked with the performance of generator control systems. Therefore, the quest for a correct way to measure on-field power plant automatic regulator performance and effective support of network operation is unavoidable; a concrete solution is needed. Only when one is found will there be correct and indubitable recognition of the economics of generator contribution to ancillary services.

This problem of economic recognition is linked to the individualisation and specification of practical control/monitoring solutions, which must be consistent with national laws and the instructions of the local energy authority. A network system operator (ISO or TSO) manages these services according to his needs in a way that guarantees a secure and reliable network operation, warranting an adequate balance between supply and demand. This not-simple objective can be achieved by the offering of additional powered services to customers.

There needs to be proper recognition of the economic contribution of the different players involved in ancillary services, primarily, the generators. Such recognition confirms that these services are justified by the secure, reliable, efficient and quality improvement they can offer to an electrical energy supply. Complementary to this objective is a cost/benefit analysis of each individual service, one that is aimed at identifying the technical aspects of the generating unit involved and the characteristics of the related controls, which mostly impact a given service's strengths and any economic benefits associated with it.

A crucial ancillary service in the electricity market is generator support of grid voltages, called simply "voltage service". As is well known, voltage service significantly contributes to power system operation. In fact, in a competitive electricity market, system operators need to achieve a high standard of voltage control, so they must manage the voltage service in a way that warrants an adequate covering of system demand, including under perturbed operating conditions, offering customers improvements in the following areas:

- *Voltage quality*: That is, voltage variation reduction around nominal/desired values in front of continuous load changing and network perturbation.
- *Voltage security*: That is, enlargement of system voltage stability margins to reduce the risk of voltage collapse, increase of power transfer capability and containment of dangerous over-voltages, which stress machinery insulation, by regulating and protecting controls. These objectives are linked to real-time availability of generator reactive power reserves in over- and under-excitation not only during normal operation but in emergency conditions, too.
- *Voltage efficiency*: That is, transmission grid loss reduction by maintenance of a suitable high voltage profile and containment of harmful reactive power circulation.

Achievement of these ambitious objectives produces significant economic benefits in power system operation, more or less so according to the effectiveness of the control system used. In practice, the only way to obtain significant results is through the use of a timely, coordinated automatic control of network reactive power resources, one which is able to maintain with continuity a suitable high voltage profile in the transmission grid and contain long distance reactive power circulation against load variation and contingencies. Examples of economic benefits due to SVR-TVR are shown in this chapter, providing useful references and guidelines for directing economic estimates.

7.2 Cost/Benefit Analysis of Voltage Service

Voltage service costs can be subdivided into the *capital* and *operation costs* met by power stations and dispatchers [1–3]. We consider:

- Generator characteristics for operating in over- and under-excitation up to the (more or less large) capability curves, according to generator sizing and design and cooling characteristics, as well;
- Generator transformer sizing and transformation ratio stepping control;
- Sizing of the unit exciter;
- Exciter controller, including AVR, OEL and UEL, line drop compensation, PSS and interface to SQR;
- Control apparatuses supporting the service, like SQR, SVR, TVR and their installation and commissioning costs;
- Special meters for fiscal measurement of voltage service;
- Telecommunications, monitoring systems and SCADA/EMS interfaces linked to the voltage service;
- Voltage service operation costs;
- Voltage service administration and maintenance costs.

7.2.1 Generation Costs

In order for generators connected to an electrical grid to operate, each has to satisfy a given connection prescription imposed by the TSO or equivalent national authority. These rules generally include aspects related to unit controls and their local stable performance. Moreover, the unit contribution to grid voltage controllability, such as minimum generator capability margins or step-up transformer size and transformation ratio settlement, is a mandatory aspect of mainly large power stations. In other words, there are some technical characteristics the production unit must be provided with in order to gain grid connection authorisation. These basic design aspects and control characteristics must be outside any generation cost analysis of voltage service, which conversely should include margins overcoming the basic minimum requisites, discussed next.

Capital Costs

Capital costs refer to:

a) Oversizing of the generator for reactive power production/absorption;
b) Oversizing of the generator transformer according to generator capability oversize;
c) Oversizing of the unit exciter to control generator capability oversize;
d) The exciter controller, including unit voltage regulator with digital calibrator, OEL and UEL, line drop compensation, PSS, SQR interfacing;
e) SQR apparatus (cost shared among power station units). For reasons of homogeneity, this cost is joined to the overall network voltage control system, which is considered later.

Different methodological approaches and traditions refer to these costs. In terms of per cent of total cost, for points (a) and (b) the additional cost could range from 5 to 7 % for thermal units, but it is certainly higher for hydraulic ones.

Considering points (c) and (d) together, the estimated per cent of exciter and control capital costs related to voltage and VAR control is about 65 %. A numerically greater result is shown by FERC (U.S. Federal Energy Regulatory Commission) analysis, estimating the investments for reactive power supply and voltage control from a complete unit equal to 41 % of the total cost: 23 % for voltage and 18 % for VARs. This estimate probably includes the costs for basic generator performance in voltage support.

Going on to point (e), 100 % of the SQR apparatus capital costs must be linked to voltage and VAR control.

Annual voltage service for a 370-MVA generator's capital cost can therefore be preliminarily estimated to be about 0.1–0.4 M€/year.

Operation Costs

Operation costs refer to:

a) System losses;
b) Operation and maintenance (O&M).

Increased losses are due to enlarged generator rating, such as of fan losses and core losses, but also to higher stator copper losses when reactive power is produced or absorbed. Rotor copper and exciter losses are increased or reduced by following the over- and under-excitation conditions. More precisely the increased losses to be considered are those which are:

- In the iron of the alternator and step-up transformer;
- In the copper of the alternator and step-up transformer;
- Mechanical (for the synchronous compensator only);
- In the exciter;
- In the auxiliary services for alternator and transformer cooling;

- For the starting-up of the synchronous compensators.

A reasonable estimate of iron and mechanical losses is 0.2% of each generator oversize. O&M fixed and variable costs also include hardware and software maintenance and operator training. A FERC analysis of the subject attributes to generator O&M costs for VAR and voltage control a percentage of 41% of the total O&M costs for generator, exciter and controls of each.

7.2.2 Transmission Costs

Transmission grid voltages to be controlled require the installation of compensating equipment and local and remote controls of generator reactive powers. Therefore, voltage service costs should include the following.

Capital Costs

These costs refer to:

a) Reactors and capacitors and their switching equipment;
b) OLTCs in transmission and distribution transformers;
c) SVCs and STATCOMs;
d) RVRs for the secondary voltage regulation;
e) National/network voltage regulator (NVR) for the tertiary voltage regulation;
f) Optimal voltage and reactive power forecasting.

Points (d) and (e) will also include SQR costs, for homogeneity reasons (see §7.2.1.1 - Capital costs-). Focusing the analysis on SVR-TVR economic impact, the only cost of interest is that of points (d), (e) and (f), while components of points (a), (b) and (c) already exist on-field.

Considering an SVR-TVR application life of 25 years, the annual cost of SQR, RVR and NVR, adjusted 8% for inflation, is about 1.5 M€/year for a power system of an about 50-GW peak.

Point (f) considers hardware and software tools and their upgrade, with an annual rate estimated at 0.2 M€/year.

Operation Costs

These costs refer to:

a) Losses in network compensating equipment;
b) Losses in electric lines;
c) Operation, maintenance and upgrade of control tools/apparatuses and telecommunication apparatuses related to voltage -VAR control;
d) Load shedding for voltage service.

Points (a) and (b) refer to the very expensive losses corresponding to the some hundred MW continuously lost by a power system and linked to an operated voltage plan. For a power system of 50-GW peak, total transmission losses amount to about 150 M€/year.

Point (c) refers to O&M of:

- Optimal voltage and reactive power forecasting tools;
- TVR regulator apparatus;
- RVR and SQR regulator apparatuses;
- Telecommunication system for voltage-VAR control;
- Voltage and VAR operation in dispatching control centres.

These costs can be estimated in the range 1.5–2 M€/year.

The load shedding costs mentioned in (d) are linked to reactive power unavailability and/or related automatic controls leading to customer disconnection in order to avoid system blackout as a response to voltage problems. Other load losses linked to the previous one are determined by voltage instability up to voltage collapse mainly due to the lack of stabilising controls that are able to prevent line protection interventions.

Lastly, the time during which contractual voltage quality at the customer's supply end is not guaranteed determines a loss of income for the operators involved, too. Of these last three points, the cost reduction estimate can be better analysed and is here after considered an SVR-TVR benefit.

7.2.3 Voltage-VAR Control Benefits

The benefits coming from the general application of secondary and tertiary voltage regulations to a wide area power system are now considered.

The technical benefits include:

- Loss reduction of an amount not lower than 5 % with continuity;
- Increment of reactive power reserves available during transients following large perturbations;
- Widening voltage stability margins: voltage collapse delayed;
- Angle stability improvement: SVR high speed voltage control significantly contributes to damping electromechanical oscillations too;
- Active power transfer capability increase and power flow congestion reduction;
- Increase in lifetime of equipment (OLTC, capacitor banks, shunt reactors) by reduction of switching manoeuvres: 30–70 % reduction in OLTC stepping manoeuvres;
- Voltage support improvement: flat voltages at HV and EHV levels against normal load variation and transient amplitude reduction in front of contingencies.
- Higher voltages with reduced reactive power control effort.

The economic benefits include:

- Production and transmission cost recovery due to loss reduction;
- Investment reduction for VAR compensating equipment (SVC, STATCOM, capacitor banks, shunt reactors);
- Improved management of more economical production under security constraints due to increase in line transfer capacity and reduction in node congestion;
- Reduction of time during which contractual quality of voltage at customer end is not guaranteed;
- Reduction of not-fed loads for security reasons;
- Reduction of heavy costs and negative social impact due to power system blackouts linked to voltage collapse.

Example of Economic Benefits on South African Transmission Grid

Table 7.1 lists the improvement on loss reduction achievable in the South African grid when network voltage control moves from PVR to SVR. SVR-TVR continuously reduces losses during the day as shown.

Another relevant benefit of SVR-TVR, as confirmed by tests on the South African power system, is generator control effort reduction (notwithstanding the higher voltage plan of around 1.04 p.u.) as compared to working conditions with PVR operating alone. Tables 7.2 gives evidence of this important result in terms of absolute value and average percentage with respect to excitation limits, as well. Therefore, with the SVR-TVR system, controllability is increased, having at its disposal a larger amount of reactive power resources for use in front of further load increase or possible system contingencies.

The minimum load case with SVR-TVR, showing a larger number of generators operating in under-excitation, confirms the increased margins on controllability to sustain voltage lowering. This result is due to the line capacitive effect and the increasing reactive power production of the generators when operated at higher voltages without risk (as SVR-TVR is able to do). Table 7.3 provides a complementary view of SVR-TVR control margin enlargement, showing area reactive power level

Table 7.1 South African power system 2007–2008 losses with and without SVR-TVR

2007 Case			
	Losses [MW]		Variation [%]
	PVR	SVR	SVR-PVR
Winter peak	884.16	773.20	−8.41 %
2008 Case			
	Losses [MW]		Variation [%]
	PVR	SVR	SVR-PVR
Winter peak	805.69	737.70	−8.44 %
Summer minimum	381.00	347.30	−8.85 %

Table 7.2 Comparisons of reactive power control efforts with and without SVR-TVR

2007 Winter peak load				
	SVR optimized voltage		PVR provided LF	
	Over-excitation	Under-excitation	Over-excitation	Under-excitation
Reactive power [MVAR]	6042	−222	6663	−105
Reactive power limit [MVAR]	12446	−858	11256	−736
Index [%]	49%	26%	59%	15%
Total reactive power [MVAR]	5820		6558	
2008 Winter peak load				
	SVR optimized voltage		PVR provided LF	
	Over-excitation	Under-excitation	Over-excitation	Under-excitation
Reactive power [MVAR]	5188	–	5879	–
Reactive power limit [MVAR]	13817	–	13817	–
Index [%]	38%	–	43%	–
Total reactive power [MVAR]	5188		5879	
2008 Summer minimum load				
	SVR optimized voltage		PVR provided LF	
	Over-excitation	Under-excitation	Over-excitation	Under-excitation
Reactive power [MVAR]	914	−1559	1380	−751
Reactive power limit [MVAR]	3307	−4672	5245	−3262
Index [%]	28%	33%	26%	23%
Total reactive power [MVAR]	−645		629	

"q" values; that is, the area generators' control effort with respect to their leading or lagging limits (±1.0 p.u.) at the operating conditions of the considered study. It follows from Table 7.3 that only a few areas reach a control effort close to the limit.

To illustrate, we note that for peak 2008 Poseidon operates close to the control limit (the weakest SVR area in the South African grid) and so does Apollo (high area load). For summer 2008 Poseidon operates close to the under-excitation limit, as does Komati (few control generators in the area).

Reduction of Switching Manoeuvres by SVR-TVR in South African Grid

SVR control makes it possible to keep network voltage constant at a predefined profile normally set to 1.04 p.u. This has the following benefits:

Table 7.3 South Africa SVR area control efforts for different case studies

SVR area	2008 summer reactive power level	2008 peak reactive power level	2007 peak reactive power level
Vulcan	−0.235	0.416	0.924
Apollo	0.274	0.802	0.894
Glockner	0.115	0.693	0.490
Spitskop	−0.672	0.324	0.124
Muldersvlei	−0.544	0.056	0.131
Komati	−0.963	0.203	0.641
Poseidon	−0.877	0.855	−0.075
Zeus	0.053	0.100	0.193
Pegasus	−0.133	0.553	0.460
Perseus	0.525	0.423	−0.177

- OLTCs operate at a lower tap. Therefore, during low voltage conditions in the HV grid, there are more taps available to control the secondary voltages.
- Switching operations required at regulating shunt devices and in stepping manoeuvres at the OLTCs are reduced.

To illustrate OLTC manoeuvre reduction, step changes are applied to the load in the South African grid and the number of OLTC manoeuvres recorded. The results are as follows:

- Following a 2 % reduction in load:
  Each stepped OLTC moved one step irrespective of PVR or SVR control;
  59 OLTCs moved under PVR, and only 14 OLTCs moved under SVR.
- Following a 2 % increase in load:
  Each stepped OLTC moved one step irrespective of PVR or SVR control;
  64 OLTCs moved under PVR, and only 47 OLTCs moved under SVR.
- Following a further 4 % increase in load:
  Each stepped OLTC moved one step irrespective of PVR or SVR control;
  10 OLTCs moved under PVR, and only 2 under SVR.

These results clearly show the 30–70 % switching manoeuvre reduction, and therefore component life increases due to SVR-TVR.

Loss Reduction Example Due to SVR benefits in South Korean Grid

An SVR control system's ability to achieve, in the physical process, the optimal voltage plan defined in real time by TVR is shown next. In its turn, TVR also takes into account off-line OPF results provided by LMC.

Accordingly, the main TVR-LMC control issue is system loss reduction, starting from the South Korean operating points in 2006 and 2010, when generators were under primary voltage regulation only, and moving to operating conditions in the presence of SVR and TVR.

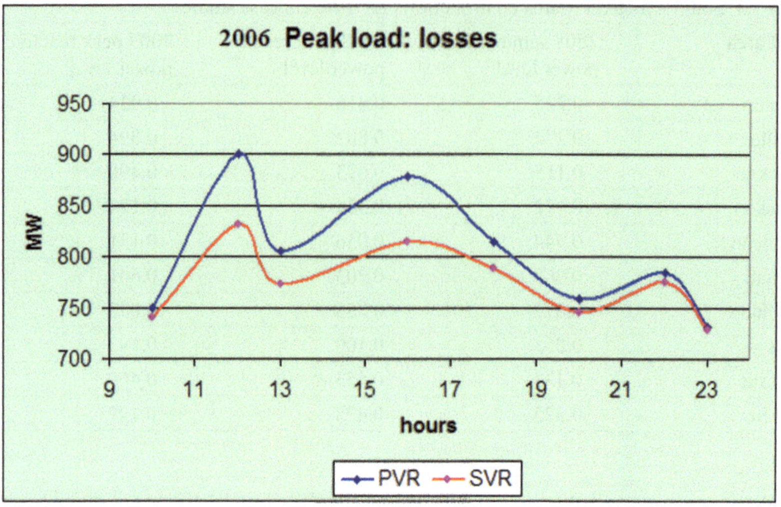

Fig. 7.1 South Korea 2006 peak load losses with (*orange*) and without (*blue*) SVR

The results of the SVR and TVR analysis are in terms of:

- Real losses in the transmission system;
- Global margin of reactive power at disposal on the generation units;
- Distribution of reactive power in the different areas;
- Network voltage profile.

The solution point (called "SVR") is obtained by an OPF taking into account the SVR additional constraints, i.e., those relevant to both the pilot bus voltage and the reactive power production "alignment" of the controlling generators up to their capability limits.

Another step to evaluate system performance is the computing of losses without SVR at the same operating points. This result is called "AVR".

In Fig. 7.1, the orange and blue curves respectively represent system losses with and without SVR at the South Korean 2006 peak load case (54600 MW). In Fig. 7.2, the orange and blue curves respectively represent system losses with and without SVR at the South Korean 2010 peak load case (62327 MW). We see in both cases the relevant impact on losses of an automatic voltage control system able to maintain moment by moment the network voltage plan at the optimised value, as defined by tertiary voltage regulation and, in general, by a predefined optimal voltage plan.

The achievable total loss reduction is, on average, more or less between 4 and 5 % for the 2006 case, while it is about 5–6 % for the 2010 case.

Fig. 7.2 South Korea 2010 peak load losses with (*orange*) and without (*blue*) SVR

Example of economic benefit evaluation for a large power system (50-GW peak load)

The terms referred to are:

- Production cost recovery due to loss reduction: 1.5–2.0 M€/year;
- Investment reduction for VAR compensating equipment (SVC, STATCOM, capacitor banks, shunt reactors): 2.0–2.5 M€/year;
- Production saving due to increased capacity transfer: 2.0–2.5 M€/year;
- Reduction of penalties for voltage contractual quality violation: 0.5–2.0 M€/year;
- Reduction of income for load shedding: 2.0–6.0 M€/year;
- Reduction of both penalties and income for long-time not-fed loads during blackout: 4.0–12.0 M€/year.

The total benefit amount ranges from about 12.0 M€/year up to 27.0 M€/year. This evaluation refers to the value of the euro (€) in 2013.

Obviously, this preliminary estimation of the main SVR-TVR benefits must be examined more deeply and refined to properly and fully consider it as a contribution provided by voltage service in the context of energy tariffs.

7.2.4 SVR-TVR Cost/Benefit Illustrative Case

A very preliminary analysis of the cost/benefit ratio can be now done with regard to the case of existing 50-GW peak power systems. In this case, there is no reason to take into account capital and operating costs of already existing generators and compensating equipment; rather, we consider simply:

- Transmission capital costs (§ 7.2.2.1, (d), (e), (f)) estimated amount: 1.7 M€/year;
- Transmission operating costs (§ 7.2.2.2, (c)) estimated amount: 1.5–2 M€/year.

These cost results amount to 3.2–3.7 M€/year. They must be compared with the SVR-TVR benefit result, for the same system size, in the range 12.0–27.0 M€/year. Therefore, we can conclude without doubt that, even if the performed preliminary and partial economic analysis required deeper, more detailed, checks, it would undoubtedly reveal a large economic advantage to a power system's operation, one that would be obtainable through implementation of wide area automatic voltage regulation based on the SVR-TVR control system.

Hereafter, the synthesis of a preliminary cost/benefit analysis for an SVR-TVR 3-year project on a large power system having a peak load of about 50–60 GW confirms the very attractive results mentioned above. They are summarised in the following:

Duration of project	3 years
Life of results	20 years
Duration of overall project	20 years
Appreciation rate	3%
Failure risk	5%
Appreciation rate with risk	8%
Net present value	NPV = 180,664.00 k€
Internal rate of return	IRR = 52.5%
Payback period	<2 years

7.3 Economic Performance Recognition of Voltage Service

The main objective of this section is to show the simplicity and correctness of the indicators of generator contribution to network voltage support as well as the differences of their performances with respect to contractually agreed-upon control availability when generators operate under a secondary voltage regulation of the transmission grid. In fact, SVR, originally designed for the management of previously vertically integrated power systems, shows high compatibility with the new structure of liberalised electric industries and open energy markets. It gives not only an effective solution to network voltage operation but also a strong simplification and clarity on the subject of fiscal measure of generator support of network voltages.

The SVR power plant regulator (SQR) is the key to the proposal's feasibility. The real contribution of each generator to network voltage support is not simple to define or measure; it generally becomes questionable when it relates to recognition of its economic aspect. The subject under consideration is a well-known complex one for many reasons:

- Generator contribution is, in principle, related to the remote control (on-line/off-line, manual/automatic, etc.) operated by the network dispatcher;
- Correctness at the appropriate time of power plant operator manoeuvres according to dispatcher requests cannot always be monitored and demonstrated;
- A real-time measurement of the coordination of generator reactive power delivery/absorption with respect to the other generators in the power station or in neighbouring plants is a burdensome and complex task, and so, in practice, information is not available;
- Generator over- and under-excitation available capability is more restrictive than nominal values due to limitations imposed by stability problems, protection intervention, a plant operator's unjustified caution and incorrect change in control parameter setting. Therefore, its real contribution is usually far from that shown in nominal curves, which are commonly used instead as the approximate reference by dispatchers.

Moreover, the effectiveness of generator voltage support is strongly linked to:

- AVR stability performance and correct dynamics when operating far from or at over- or under-excitation generator limits;
- Correct coordination between generator protection settings and the generator available continuous-time operation field.

Lastly, national rules/regulations in force on the generator contribution to network voltage and reactive power control often oblige generators, in the absence of a reliable and indubitable monitoring system, to follow apparently simple service rules based on power factor values that, in practice, often determine generator operation against real grid needs: excess of over- or under-excitation with a consequent exchange of reactive power among a plant's internal generators or among plants themselves in a region.

It is evident that the amount of reactive energy itself is not fully representative of the real generator contribution to the voltage service, which, conversely, is substantially related to available reactive power (Q) around the generator's operating point and to its continuous control at the correct rate, simultaneous and consonant with the controls of the other generators and FACTS in the same plant and in neighbouring plants.

Primary voltage regulation, through which AVR maintains generator terminal voltage at set-point value, is not enough for a satisfactory grid voltage service because the AVRs regulate stator edge voltages only, and their set-points are manually controlled by power plant operators, with consequent unsatisfactory performance in terms of timely and coordinated control actions.

For many reasons, the prospect of the proposed new power plant high side voltage regulator (SQR) required by SVR to control generator reactive power offers a true, concrete possibility of achieving an effective voltage service:

- Automatic control in real time and in a coordinated way of the effectively available reactive power resources by generators can be achieved;
- Generator available capability can be continuously and automatically monitored and therefore defined in detail within the contractual agreements;
- AVR stability, dynamics and coordination with generator protections are implicitly guaranteed by SQR, which continuously monitors them as necessary conditions for its correct performance;
- A simple but correct and indubitable monitoring system of the effective contribution of each generator to network voltage support can be achieved through SQR facilities for possible comparison with the contractually expected performance and the correct computing of the generator's economic value.

7.3.1 Voltage Service with SVR: Role Played by Power Plant Voltage and Reactive Power Regulator (SQR)

Aimed at transmission network voltage support, SQR simultaneously controls the reactive powers of station generators at the same percentage value of each one's real over- and under-excitation limits. These dynamic limits, which can also be computed by SQR according to generator thermal prescriptions, are continuously compared with those operating inside the AVR. They represent the generator maximum available capability field, which can be fully or partially assigned to the voltage service by SQR according to contractual agreement as defined between production company and TSO/ISO.

Another constraint on generator support of network voltage is the maximum allowed voltage range at the generator stator edges, which could restrict OEL and UEL limit exploitation. The SQR should also modulate this generator-allowed voltage field so it becomes a contractual tie within a commercial agreement between power plant and TSO/ISO. Besides, the two generator performance fields mentioned can be enlarged or restricted through SQR-setting and therefore more or less remunerated when confirmed by an indubitable and independent special on-line meter.

The SQR also needs to be continuously checked for possible alternative control functionalities:

a) Power plant under SVR (SQR tele-controlled by RVR);
b) Power plant regulation of local HV bus voltage (SQR as HSVR).

These two SQR alternative control modalities, both connected to network voltage support, require a differentiated accounting to distinguish the more valuable (a) performance from the less valuable (b), since they are defective of coordination with other area power stations.

According to the SQR functionalities and limits mentioned, any stated reference to power plant performance in a voltage service contract between TSO/ISO and power station should name the following, under the assumption of SQR mandatory acquisition and activation by the power station:

1. SQR continuous control of all power plant operating units;
2. Unit reactive power availability up to contractually agreed upon OEL and UEL excitation limits;
3. Generator terminal voltage field availability up to contractually agreed upon edges around the nominal value;
4. Generator operation under SQR control, complying with the limits and field mentioned in points 2 and 3, for a yearly amount of hours not below the minimum contractually agreed upon value;
5. Generator operation under (a) or (b) SQR control functionalities, in agreement with yearly percentages defined in the commercial agreement related to the total yearly amount of hours established in point 4.

Obviously, the wider the contractually agreed upon performance (also depending on the generator capital oversize) the more substantial will be the power plant support and the related economic recognition when confirmed by a special, independent on-line meter of on-field performance measurement. Apart from a fiscal meter, generator operation costs (§ 7.2.1) can also be recognised and included in the contract.

From a commercial point of view, a real generator contribution to network voltage support must therefore be correctly monitored and compared with the contractually agreed upon performance, making an impartial economic valuation possible, or for application/pursuance of penalty clauses. This objective could easily be met with a new *fiscal meter* of the power plant generator's contribution to network voltage regulation; one which collects and elaborates (applying cross-correlation and autodiagnostic criteria) the available SQR data related exclusively to the generator's operating state and main SQR control signal measurements. It would not allow any illegal intrusion; only through authorised procedure would it be possible for one to modify its stored measurements. This fiscal meter would also be able to recognise and inform remote terminals as to each generator reaching its contractual performance percentage and the amount of differences present [4].

7.3.2 Voltage Service Indicators

Voltage service evaluation is strongly linked with those generators (the largest on-field controllable reactive power reserve) devoted to the achievement of improvements in voltage *quality*, *security* and *efficiency* in a power system's operation:

- "Quality" means reduction of:
  - Voltage variation in network buses due to load changes and system perturbations;
  - Amplitude and duration of voltage transients following network perturbations;
  - Load disconnections in response to voltage problems.
- "Security" means:
  - Improved stability and therefore reduction of protection interventions;
  - Reduction of voltage collapse risk;

- Reduction of machinery voltage stress and therefore fatigue of generator insulation.
- "Efficiency" means reduction of:
  - Losses in electrical lines and machinery;
  - Investment for additional MVARs, new electrical lines and generating/compensating plants;
  - Warm reserve costs for voltage support;
  - Time recovery, following load rejection due to voltage problems.

As previously said and worth repeating, the achievement of these benefits is strictly linked with the automation level of a network voltage control system as well as with reactive power resource availability and its effective and timely coordination facing real-time power system changes.

In the absence of an automatic transmission network voltage control system, the voltage service cannot be significantly improved with respect to classical off-line reactive power dispatching, which cannot effectively follow real-time continuous changing of grid voltage. Moreover, such an absence penalises the reliability of the monitoring and control of each reactive power resource's real contribution:

- Even when the amount of delivered or absorbed plant reactive power is considered, their real contribution is the net of offsets due to possible compensation with neighbouring plants;
- The amount of reactive energy is simply representative of the average operating condition in a given network area, while reactive power variations around normal operating conditions are most effective in terms of system voltage regulation when they are coordinated in an adequate and timely fashion up to the limits according to real network need.

The presence of a coordinated voltage control system (SVR and TVR) therefore overcomes the above limitations and allows an effective voltage service with significant operation improvement. In addition, it strongly simplifies proper recognition of each generator contribution to the voltage service.

Starting from this perspective and considering just those really accredited generators under SVR performing a recognisable voltage service, we can understand the following indicator proposal based on the single generator contribution to the voltage service. This contribution relates to a real-time checkable availability and concrete usability of the contractual capability limits. Regarding this, points 1–5 of the previous section represent the basis of a commercial agreement between TSO/ISO and power station, dictating generator contribution to the voltage service.

With our aim being the proper checking of the generator's real performance via comparison with contractual obligations, we propose the following indicators.

Index of Generator Available Capability

For each generator under its control, SQR computes at each instant, by taking into account the voltage and active power measurements at stator terminal edges, the corresponding over- and under-excitation limit values to be imposed on generator reactive power. Moreover, SQR checks the limits' consistency with those operating in parallel on the AVR. Obviously, generator limits in the SQR can be selected/changed by a power plant operator in accordance with generator operation/security needs until they are maintained inside the AVR's operating limits. So, changes cannot overcome AVR limits; reduction is allowed. The measure of the dynamic generator reactive power field is sent by SQR to the new *voltage service meter* for a real-time comparison with corresponding limit values, continuously computed by the meter on the basis of contractual OEL and UEL curve parameters.

Therefore, until the generator reactive power limit Q_{lim} value corresponds to the contractual $Q_{\text{lim\_contr}}$, both in over- and under-excitation, generator available capability is 100 %; otherwise it goes down. The availability value is assumed to be 0 % when the generator is outside SQR control.

According to this, a significant index of generator contribution to the voltage service in terms of available reactive power capability is the following:

$$I_q(t) = \frac{100}{T} \int_0^T \frac{Q_{\text{lim}}(t)}{Q_{\text{lim\_contr}}(t)} dt. \tag{7.1}$$

I_q represents the per cent average value in interval T of the correspondence between real and the contractual generator capability. This means $I_q = 100\,\%$ only when $Q_{\text{lim}} = Q_{\text{lim\_contr}}$ along the overall period T, during which the generator is continuously operated under SQR control. We note that I_q could also be below 100 % when voltage limits at the generator's terminal edges prevent reactive power limits from being reached.

Index of Generator Available Voltage Field

For each generator under its control, SQR generally allows the fixing of the available stator voltage field around the nominal value as well as to check its consistency with generator protections and with the control signal forcing the voltage limits. Obviously, generator voltage limits imposed by an SQR can be changed by the power plant operator according to the plant's control requirements and to machinery insulation safety that, for many reasons, could become more restrictive compared to the nominal values indicated by the alternator manufacturer. Such a possible generator voltage field reduction could penalise the exploitation of the allowed generator capability, and for this reason its measurement is necessary.

Because real voltage limiting values cannot a priori be known, they are recognised and memorised by the new voltage service meter when these values are reached and signalled by the SQR. Analogously, their changes are updated when recognised. Therefore, at the SQR the generator operating voltage limits are those most recently measured on-field and their values can be used for a real-time comparison with the corresponding limit values agreed to in a generator voltage service contract.

Accordingly, until the generator voltage field $|V_n - V_{\lim}(t)|$ corresponds to the contractual value of ΔV_{contr}, the available voltage field is 100 %; otherwise it goes down. The availability value is considered to be 0 % when the generator is outside SQR control. Therefore, a significant index of generator contribution to voltage service in terms of available voltage field is as follows:

$$I_v(t) = \frac{100}{T} \int_0^T \frac{|V_n - V_{\lim}(t)|}{\Delta V_{contr}} dt. \quad (7.2)$$

I_v represents per cent average value in interval T of the correspondence between the real and contractual generator voltage field. This means $I_v = 100\%$ only when $V_{\lim}(t) = V_{\lim\_contr}$ along the overall period T during which the generator operates under SQR control.

Furthermore, when possible impediments to the reaching of contractual values of capability limits and stator voltage field are checked, during generator operation, an improper transformation ratio setting of the unit step-up transformer can be more easily recognised.

New Power Plant Meter For Voltage Service

The on-field, actually available generator strength to be used for a network voltage service cannot be known a priori, and what a power plant declares this should be could frequently change or turn out to be unfeasible, without there being any possibility of a reliable check or guarantee.

To avoid this from happening, a new voltage service meter for the power plant, one which fully depends on the presence in the plant of the new voltage and reactive power regulator (SQR), can compute in real time either the previously mentioned I_q and I_v indices or the intervals inside period T (which could last 1 year) during which:

- $I_q = 100\%$
- $0\% < I_q < 100\%$
- $I_q = 0\%$
- $I_v = 100\%$
- $0\% < I_v < 100\%$
- $I_v = 0\%$.

The above SQR computing is done separately for each of the two possible SQR regulating alternatives (a) and (b), mentioned in § 7.3.1. Moreover, the new meter would compute in period T for each generator in the power station the amounts of reactive energy, both in over- and under-excitation operating conditions.

It must be pointed out that the progressive values of all generator performance indicators, calculations and measurements can be monitored (but not modified) by power plant operators and TSO/ISO at any time through telecommunications. The TSO/ISO during contract definition/renewal only is authorised to remotely set into the fiscal meter the contractual parameters, which represent generator performance references for the immediate future.

On the basis of the values reached along period T by the aforementioned generator performance indicators, it will be possible on the one hand to compare, in a precise way, generator availability and contribution to voltage support with the elements of the voltage service contract and on the other hand to attribute to each generator a highly correct and indubitable recognition of the economic value of its network voltage support as already provided.

7.3.3 Simplicity, Correctness and Indubitableness of Proposed Indicators

Simplicity The proposed methodology to compute and monitor indicators based on real-time measurements of generator contribution to voltage service is very simple to apply when a power plant is operating under SQR. The main reasons are:

- The analytical functions demanded are simply integrals and counters of the present values of measurements and logic states available at the SQR apparatus interface;
- The proposed new meter exclusively interfaces with the SQR and monitors in real-time the computed indices to the power plant and remote (TSO/ISO) operators by ordinary local and remote telecommunications.

Correctness Until power station generators are under SQR control, the following performances are continuously monitored and checked by SQR:

- On-line, real coordination of the generator's available operation fields with its protections;
- Correct operation of the AVR.

If a generator has problems with primary voltage regulation, it is automatically shut down by SQR control and a signal requiring AVR maintenance is sent by the SQR to plant operators.

Operation under SQR allows the correct use of the generator reactive power up to the contractual capability limits, avoiding wasted reactive power dispersions or overprotective interventions by plant operators. A possible lack of coordination between the AVR and SQR capability curves is signalled by the SQR to operators, requiring maintenance to update it to a correct setting. Under SQR all units move in a coordinated way using just the minimum reactive power reserves to maintain the grid at the desired voltage value. No useless reactive power exchanges among generators will be possible under SQR, especially in the presence of SVR, which coordinates all the SQRs.

The generator reactive power must be used in a timely fashion for that which is necessary and without limitation within the capability agreed upon. For this reason, proper recognition of the generator's economic value is necessarily based on the availability of continuously checked generator capability limits, which are correctly computed and ensured during generator operation by SQR, and because possible differences with contractual values are timely and automatically taken into account by the fiscal meter mentioned above.

Indubitableness The proposed methodology is fully based on real-time field measurements of the generator performance, the same measurements used by SQR with a careful and deep surveillance on data congruity. Therefore, real data on generator availability for the voltage service can be continuously monitored and stored. The proposed fiscal meter takes all the information it needs, including the modifications employed at any time on the SQR and its out-of-service or maintenance periods. Furthermore, the fiscal meter would apply the necessary mechanisms for checking the congruity of the acquired data and elaborating cross-correlations. Lastly, the new meter, besides using autodiagnostic algorithms, would actuate protective mechanisms against an intrusion also sending remote alarm signals.

Final Remarks On New Voltage Service Meter

For each generator, the fiscal computing of the progressive time period during which the available capability field corresponds to the contractual ties as well as the comparison with the agreed-upon yearly per cent allows the proposal of an indubitable performance index providing both:

- Generator reactive power reserve, available in real time and usable without limitation for network voltage support;
- The coherence of this result with the contractual tie.

This and other indicators inferred from the proposed method and related to the generator's different possible operating conditions and their duration are computed through counters and their usefulness is shown.

The simplicity, correctness and indubitableness of the proposed method encourage a practical, extensive application of a new meter, fully dedicated to voltage service monitoring and accounting. The chapter has aimed to show how such a method can recognize a proper contribution and make the related payment to a given generator on the basis of its real and indubitable support to voltage service.

References

1. Corsi S, Arcidiacono V, Cambi M, Salvaderi L (1998) Impact of the restructuring process at Enel on the network voltage control service. Bulk power system dynamics and control. IREP-IV, Santorini

2. Berizzi A, Sardella S, Tortello F, Marannino P, Pozzi M, Dell'Olio G (2001) The hierarchical voltage control to face market uncertainties. Bulk power system dynamics & control. IREP-V, Onomichi
3. Corsi S (2001) The indubitable recognition of the generators contribution to the network voltage support in a market environment. Bulk power system dynamics & control. IREP-V, Onomichi
4. Corsi S, Folcini L (1999) Criteri di rilevamento e contabilizzazione delle prestazioni dei sistemi di controllo del sistema elettrico SISCORE project. Internal Report, Enel AT-UCR No. 99/692

Chapter 8
Voltage Stability

Distinguishing voltage stability from the classic power system angle stability problem, as it is generally understood and classified, is the chapter starting point [18, 25, 26]. Evidence is also given to the significant contribution of power system voltage control loops (AVR and SVR) to electromechanical oscillation stability [14, 16], to counteract the tendency to associate voltage control with voltage stability alone. These preliminary clarifications help us differentiate the voltage instability phenomenon as substantially linked to maximum line loadability while increasing the load. The classic voltage-power (*V-P*) curves of the Thevenin equivalent circuit are introduced as the main evidence in support of the voltage instability process [8, 20, 23, 24, 29]. The nose tip of such a curve gives the correct information on maximum loadability when a power system's detailed dynamic model of a considered large or equivalent scheme is used. The dependence of the nose shape on the on-load tap changer (OLTC) and the over-excitation limit (OEL) dynamics, and the load characteristics and differences with or without SVR are clearly evidenced. The fact that voltage instability appears to be strongly influenced by power system dynamics is widely demonstrated with comparisons of different operating conditions of the power system control loops [2, 4, 6, 7, 9, 10, 13, 15, 17, 21, 22, 27, 30]. System voltage collapse as the terminal event of an instability process's deterioration leading to blackout is described as an irreversible process [3, 5, 12, 19, 28]. Examples of large power system voltage instability followed by voltage collapse are provided. A brief mention of the voltage instability Hopf–saddle-node bifurcation method is also made [1, 12].

8.1 General Overview on Stability

The stability of a dynamic system is a fundamental, vital characteristic of its proper performance. Therefore, any regulating control should have, as a mandatory, implicit peculiarity, robust stability of the controlled system dynamics. This means inside its allowed operating field any dynamic process can operate only when its stability is guaranteed there.

S. Corsi, *Voltage Control and Protection in Electrical Power Systems*,
Advances in Industrial Control, DOI 10.1007/978-1-4471-6636-8_8

The stabilising task in a process is usually charged, it is said, by a control solution, but that notwithstanding, instability could also result from the extreme conditions under which a process is required to operate or because of system structural changes or also due to not well-tuned control parameters/switching logic as well as untimely protection operation.

The classic analysis on system stability refers to the linearised dynamic model of the process represented by transfer functions or state variable matrices. As is well-known, all system characteristic polynomial poles or system state matrix eigenvalues must have negative real parts to guarantee system asymptotic stability.

This linearised model provides an important help in the framework of small variation stability analysis around the selected system operating point. Nevertheless, it does not properly cover performances linked to large perturbations or when the process approaching significant structural and operating point changes.

Despite the linearised model, where the same kind of analysis is used for all the physical processes, when instead we approach nonlinear stability analysis, we find each system has its own peculiarities linked to physical aspects, design characteristics and limits, as well as to the types of perturbations possible. In this case the amplitude and length of specified perturbations significantly change the stability results, with sudden or slow instability trend possible up to a general collapse of the process.

Regarding electrical power systems, "stability" basically refers to two kinds of phenomena linked to the electromechanical and electrical nature of the process. The two stability aspects (*angle stability* and *voltage stability*) interact, but they also can be distinguished well from one another.

Electromechanical instability is linked to the rotating masses of the operating alternators and the differences of rotor speeds either among themselves or with grid frequency. Analogously to a mass hanging from a solid ceiling by a spring, the prevailing dynamics of such a system is of an oscillating nature. More precisely, typical transients of the generator electromechanical loop when operating in a large power system put in evidence:

- Oscillating inter-area rotor angle and active power exchange (lack or unsatisfactory value of damping torque);
- Monotonic rotor acceleration leading to loss of synchronism (lack of synchronising torque);
- Long term frequency control of generation-load imbalance.

The voltage instability phenomenon is basically linked to voltage-lowering in the main system buses due to the local load increase and lack of system reactive powers to sustain the bus voltages. Typical transients of this phenomenon put in evidence:

- Monotonic load bus voltage-lowering while increasing the power transfer to that bus;
- Operation of generator OELs and grid transformer OLTCs leading to the containment of the generators delivered reactive powers in one side and OLTCs inverse operation on the other side;

- After reaching the load maximum power transfer, the voltage at the load bus suddenly falls, thereby reducing the loading and determining the loss of synchronism by the local generators. This irreversible process leads to the so-called *voltage collapse.*

Voltage instability transients can be long term in the case of slow load increase or short term after a large contingency suddenly determining a voltage collapse. Usually, when talking about voltage stability, we mainly refer to the grid long term phenomenon (minutes) or to transient stability in front of a large contingency (ms). This notwithstanding, the case of dynamic load components operating in the short-term time scale (100 ms-1 s) has to be considered, too. Such components include electronically controlled load (FACTS), induction motors and HVDCs. Moreover, local voltage control of rotating generators falls within this time frame. In other words, load-driven and generator-driven stability problems come together in the short term voltage stability time scale.

In short, power system voltage stability classification includes:

- Long term voltage stability;
- Short term voltage stability;
- Transient voltage stability (following large contingency during an operating condition with low stability margin).

Hereafter, the analysis begins considering the stability of:

- System components provided with a voltage control loop;
- Wide area voltage control loops (basically HSVR or SVR).

This analysis helps clarify and distinguish the role operating voltage control loops can play on angle instability with respect to their other contributions linked to power system voltage instability, basically due to line overloading and saturated controls. System voltage collapse following voltage instability is deeply investigated, too.

Before getting into detail on the subject of voltage instability, first we offer a preliminary introduction to the link between the voltage and angle stabilities.

8.2 Electrical Power System Stability

In the context of an electrical power system, *stability* essentially means the way rotating synchronous generators tend to run in synchronism. The stability concept is therefore widely associated with the equilibrium on transmittable power from generators to loads. This moment by moment equilibrium obviously depends on the balance between the system's mechanical and electrical equivalent torques but also on the safeguard of acceptable parameter values by the operating points of the generators, lines, substation components and loads. Among others, two very important control solutions widely help support the operating equilibrium within safe limits: voltage and speed control at the generators and voltage and frequency control at the grid side.

It should also be said that the considered voltage control loops only marginally influence the dynamic performance of the overall electromechanical process, but they do so enough to contribute to its stability/instability. This is confirmed by the block scheme in Fig. A.3 (of Appendix A), allowing the statements:

- The electromechanical control loop is responsible for a low damped oscillation mode. In fact, it consists of a cascade of two integrators having in feedback a block with a transfer function (TF) representable, in the first approximation, by a proportional coefficient;
- The speed regulator loop can significantly contribute to the phase margin increase of the electromechanical loop only in the case of its high cut-off frequency. When that happens, the equivalent block substituting the integrator with the speed regulator is a first order TF;
- The voltage control loop moderate effect can lead the overall system, under given conditions, toward an oscillatory electromechanical instability;
- Nevertheless, voltage control loop high speed can significantly contribute to electromechanical oscillation damping by additional power system stabilising (PSS) feedback, as will be shown later.

Moreover, we will also demonstrate in what follows how the SVR, which definitely operates through generator AVRs, contributes to overall power system angle stability.

From these preliminary considerations it is possible to state that generator or grid voltage control loop stability is not solely for its own purpose; voltage stability also contributes to the stability of the electromechanical loop with which the voltage loop interacts. The proper understanding of these points opens a window on a concrete view of power system stability.

A helpful approach toward understanding the subject is to reduce the complexity of the power system process to the simplest equivalent model, composed of:

- A generator feeding by an equivalent line a large, dominant power system represented by an equivalent infinite bus. This is the classic equivalent model represented in Fig. 8.1, generally used to analyse generator electromechanical dynamics with respect to the grid.
- An equivalent generator representing the overall power system feeding a given load bus by an equivalent line. This is the classic Thevenin equivalent model, represented in Fig. 8.2, generally used to check the load change effect on grid voltages.

These simple processes contain all the basic elements for a general understanding of stability in power systems. They will be frequently discussed hereafter.

8.2.1 Transient Stability

The term "transient stability" refers to a power system's ability to recover to normal operation following a major disturbance (e.g., loss of generation, line opening, load shed, and faults in general).

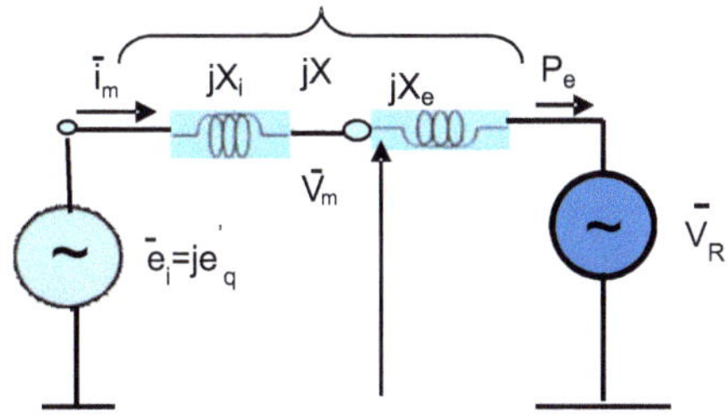

Fig. 8.1 Generator feeding an infinite bus

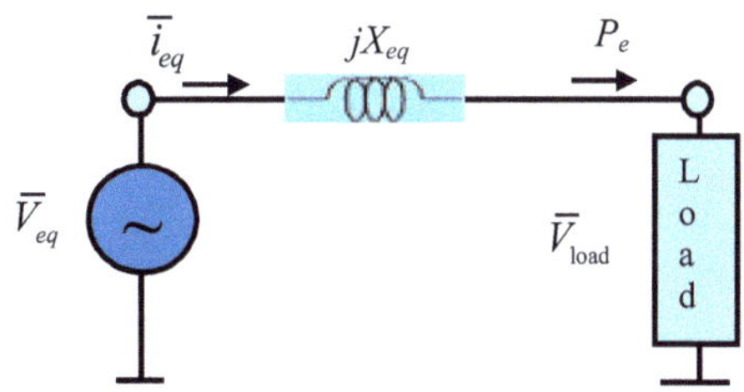

Fig. 8.2 Equivalent generator feeding a load

Considering the simple model in Fig. 8.1, the transmitted active power P is given by

$$P = V^2 \frac{\sin(\delta)}{X}, \tag{8.1}$$

where V is the magnitude of both the generator and infinite bus voltages that are assumed under voltage control (the infinite bus by definition and the generator through its AVR). Power angle δ lies between the sending end machine internal voltage and the infinite bus voltage. The relationship between P and δ is shown in Fig. 8.3 for both the uncompensated and compensated cases.

From the uncompensated case, the theoretical maximum transmittable power defining the steady state stability limit is reached with $\delta = \pi/2$: $P_{max} = V^2/X$.

It is interesting to observe that connecting a synchronous compensator or an SVC at the midpoint (V_m) of the transmission line (Figs. 8.1 and 8.3 compensated case), both being able to regulate voltage at that point at the same value of the sending or receiving ends, Eq. (8.1) applied to each half of the line becomes:

$$P = V^2 \frac{\sin(\delta / 2)}{(X / 2)}.$$

This shows how voltage control at the line midpoint doubles the maximum power transfer at $\delta = \pi$. This also shows how the steady state maximum transmittable power limit of the uncompensated case can be increased up to the maximum line capacity transfer.

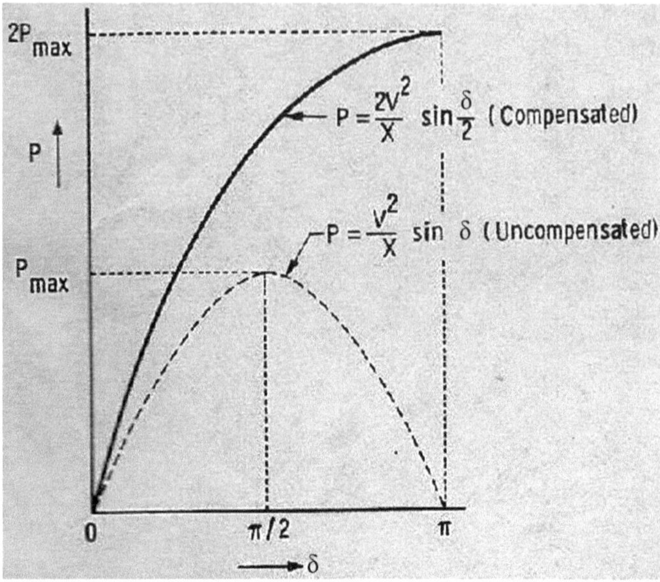

Fig. 8.3 Power transmission characteristic of the Fig. 8.1 two-generator system, both in the uncompensated and compensated cases

The considered compensation also increases the steady state stability limit as shown in Fig. 8.4b with respect to the Fig. 8.3a uncompensated case. More precisely, assuming that in both the uncompensated and compensated systems the transmitted power and the fault are the same, the steady state operating points move along the power characteristics as shown in Fig. 8.4:

- Prior to the fault, each system transmits electrical power P_0 at angle δ_1 (case (a)) and δ_{c1} (case (b)); mechanical power $P_M=P_0$;
- During the fault the transmitted electrical power becomes zero, while the mechanical input power to the generator remains constant. Therefore, the generator accelerates from the steady state angles δ_1 (case (a)) and δ_{c1} (case (b)) to δ_2 and δ_{c2}, respectively, at which the fault clears;
- The accelerating energies in the two cases are represented by the A_1 areas;
- After fault clearing, transmitted electrical power exceeds mechanical input power and the machine decelerates, but its angle further increases due to the kinetic energy accumulated in the rotor;
- The maximum rotor angles δ_3 (case (a)) and δ_{c3} (case (b)) are reached when the decelerating energies defined by the A_2 areas are equal to the corresponding accelerating energies defined by the A_1 areas;
- If the maximum rotor angles δ_3 (case (a)) and δ_{c3} (case (b)) are below the critical value δ_{crit} and δ_{ccrit}, respectively, the system will remain transiently stable;
- The critical rotor angle represents the rotor angular swing beyond which rotor deceleration cannot be maintained;
- The difference between maximum angular swing and the critical angle determines the margin of transient stability, that is, the still available decelerating energy represented by areas A_{margin} and $A_{\text{c margin}}$.

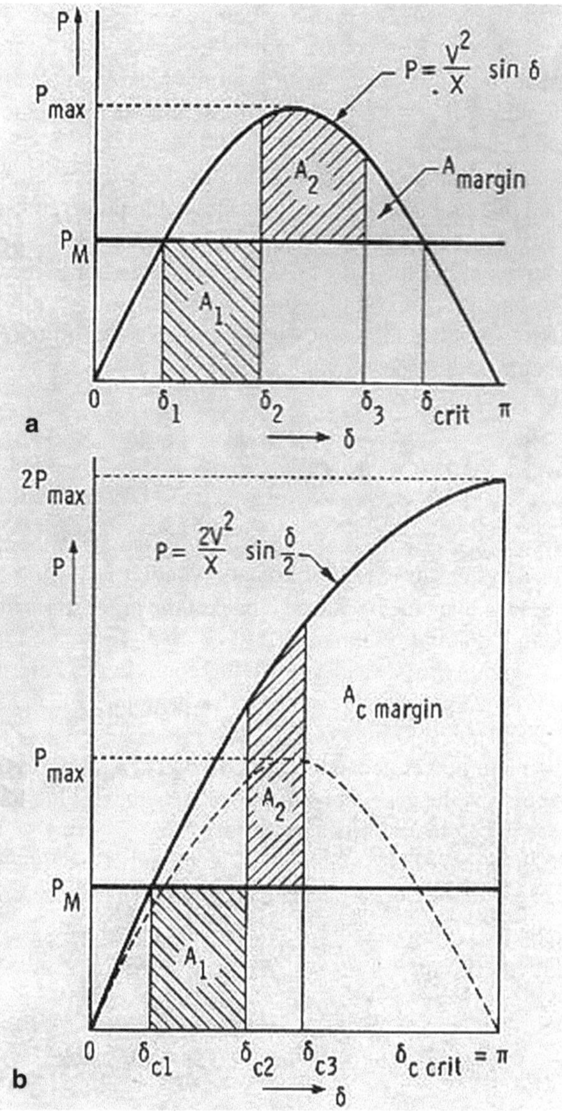

Fig. 8.4 Equal area transient stability margin for a two-generator power system without (**a**) and with (**b**) compensation

Comparing cases (a) and (b) in Fig. 8.4: a substantial increase in the transient stability margin of the compensated ideal case is evident—that is, when a line's midpoint is voltage-regulated by an automatic control and the voltage "phasor" remains constant module throughout the process, except possibly during the fault, while its phase angle follows the generator (rotor) angle swings (no real power exchange with the compensating system). We note this compensation significantly increases the transmittable power, too.

An ideal synchronised voltage source in the above elementary stability analysis is also assumed to provide reactive power as needed, without limitation.

In the considered midpoint compensating scheme, the reactive power demand at constant midpoint voltage rapidly increases with the transmitted power, reaching a maximum value equal to $4P_{max}$ at the maximum steady state real power transmission limit of $2P_{max}$ (P_{max} being the maximum transmittable power of the uncompensated case).

In real applications the rating of the compensating static equipment is lower than that required for maximum attainable power transfer. For this reason, a practical synchronous compensator or SVC approximates an ideal synchronised voltage source only as long as MVAR demand does not exceed its capacitive rating. When this happens, the compensation at the midpoint is equivalent to a fixed capacitor unable to regulate midpoint voltage and consequently provides less transient stability margin with respect to the ideal solution.

8.2.2 Steady-State Stability

System stability in front of small perturbations is the classic dynamic requirement for a system recovery to normal operation following a minor disturbance. Therefore, the dynamic linearised model of the process has to exhibit stability by positive damping of the system modes.

There exist particular operating conditions where the power system may have very small positive or even negative damping, which could result in sustained voltage and power swings with the consequent high risk of loss of synchronism between the main power generators. The two-generator equivalent system in Fig. 8.1 also represents the generator-infinite bus scheme in Fig. A.1, Appendix A, with the corresponding linearised model in Fig. A.6.

Preliminarily assuming nil interaction between the electromechanical and voltage loops, the electromechanical transients can be described by the so-called "swing equation", linking generator mechanical (P_m) and electrical (P_e) powers to unit angular momentum M and rotor angular position δ with respect to the infinite bus:

$$M\frac{d^2\delta}{dt^2} = P_m - P_e.$$

The difference $P_m - P_e$ is the accelerating power.

At small variations:

$$M\frac{d^2\Delta\delta}{dt^2} = \Delta P_m - \Delta P_e.$$

Assuming constant mechanical power and with

$$P_e = \frac{VV_R \sin\delta}{X},$$

we get

$$\Delta P_e = \frac{\partial \Delta P_e}{\partial V}\Delta V + \frac{\partial \Delta P_e}{\partial V_R}\Delta V_R + \frac{\partial \Delta P_e}{\partial \delta}\Delta \delta.$$

Due to the fact that $\Delta V = \Delta V_R = 0$ because of controlled voltages, the swing equation becomes

$$M\frac{d^2\Delta\delta}{dt^2} + \frac{\partial \Delta P_e}{\partial \delta}\Delta\delta = 0.$$

The corresponding characteristic equation indicates the angle δ undamped oscillation mode, the roots being on the imaginary axis of the s-plane, with an angular frequency of

$$\Omega_0 = \sqrt{\frac{1}{M}\cdot\left.\frac{\partial \Delta P_e}{\partial \delta}\right|_0}.$$

A similar result is achieved when considering the voltage regulation of an intermediate bus along reactance X by static compensating equipment. This means a compensator that maintains constant the terminal voltage is not effective in damping power oscillations.

In order to make damped power oscillations, midpoint voltage V_m must be controlled according to $d(\Delta\delta)/dt$. In this case in fact, the swing equation yields a damping effect:

$$M\frac{d^2\Delta\delta}{dt^2} + K\left(\left.\frac{\partial \Delta P_e}{\partial V_m}\right|_0\right)\frac{d(\Delta\delta)}{dt} + \frac{\partial \Delta P_e}{\partial \delta}\Delta\delta = 0.$$

In contrast to the previous case of compensation supporting voltage and transient stability improvement, now oscillation damping is achieved by compensator MVAR output controlled with the terminal voltage following the power system frequency variations.

Apart from the local compensator is the case of secondary voltage regulation on the considered line midpoint, this because SVR regulates mid-point V_m in closed loop by the equivalent generator reactive power. In this case, $\Delta V \neq 0$ and SVR control, overlapping the generator voltage loop, modifies the interaction between the equivalent generator electromechanical and the voltage control loops. This modifies the transfer function of the electromechanical loop feedback, so contributing to oscillation damping (details follow).

In addition, when the preliminary assumption of nil interaction between the electromechanical and voltage loops is removed, the previous elementary equations become more complex, and stabilising/destabilising effects has to be considered. One must remember that stabilising additional feedbacks (provided by PSS) applied to the generator AVR (see Fig. A.5) play the fundamental role of increasing the interaction between voltage and electromechanical loops, thus damping the electromechanical oscillation by linking generator voltage control to rotor speed/frequency variations (see next § 8.2.3).

8.2.3 Generator AVR Contribution to Steady-State Stability

Referring to Fig. A.5, the voltage control loop impact on the electromechanical loop can also be represented by a simplified control scheme (Fig. 8.5):

The transfer function $K'(s)$ is obtained by elaborating the block diagram in Fig. A.5:

$$K'(s) = -hh_1 \cdot \frac{\frac{\mu(s)}{sT'_{d0}}}{1+h_2\frac{\mu(s)}{sT'_{d0}}} = -\frac{hh_1}{h_2} \cdot \frac{\frac{h_2\mu(s)}{sT'_{d0}}}{1+h_2\frac{\mu(s)}{sT'_{d0}}} = -\frac{hh_1}{h_2} \cdot G_v(s).$$

The phase margin γ gained by $K'(s)$ gives the contribution to the electromechanical loop stability provided by the voltage control loop. By itself, the second order undamped electromechanical loop has zero control margins: $\gamma = 0$.

With $K'(s)$, two possible effects can be obtained at cutting frequency ω_0 (see Fig. 8.6):

1. $0 < \gamma(K'(j\omega_0)) < \pi$ (stabilising effect);
2. $-\pi < \gamma(K'(j\omega_0)) < 0$ (destabilising effect).

Therefore, with $\gamma > 0$ the system is stable, whereas with $\gamma < 0$ the system is unstable. Because $G_v(j\omega)$ has a negative phase,

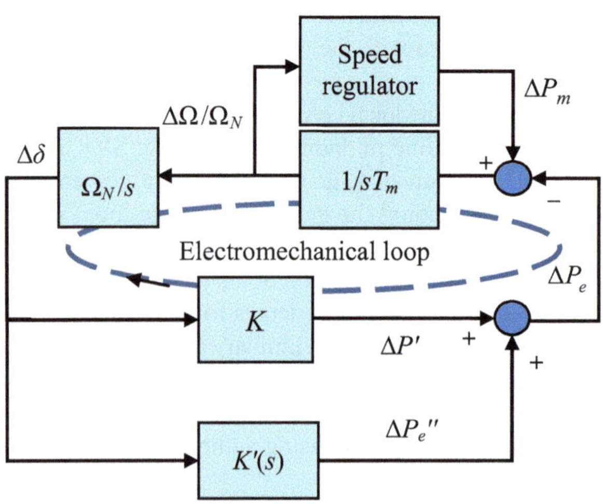

Fig. 8.5 Generator linear model of the electromechanical loop

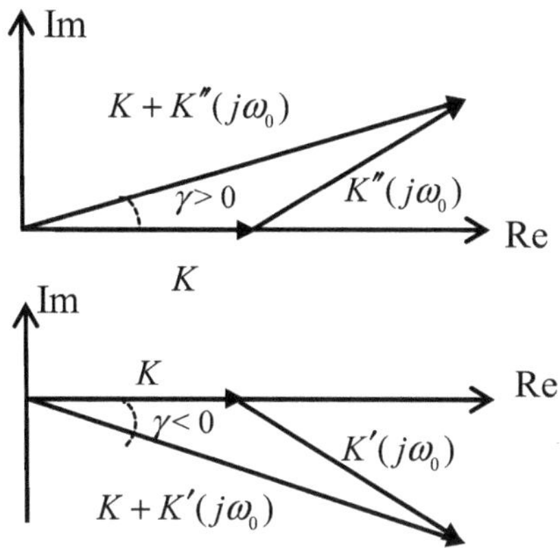

Fig. 8.6 Phase margin contribution to the electromechanical loop by the voltage control loop

$$G_v(j\omega)=\frac{h_2\cdot\dfrac{\mu(j\omega)}{j\omega T'_{d0}}}{1+h_2\cdot\dfrac{\mu(j\omega)}{j\omega T'_{d0}}}=\frac{1}{1+\dfrac{j\omega T'_{d0}}{h_2\mu(j\omega)}}.$$

The $K'(j\omega)$ phase is positive (stabilising) if the sign of hh_1/h_2 is negative, otherwise $K'(j\omega)$ determines a destabilising effect.

The sign of h, h_1 and h_2 here after are evaluated by referring to Fig. 8.1: Since $P_e = e_i V_R \sin\delta / (X_i + X_e)$, then:

$$h=\left.\frac{\partial \Delta P_e}{\partial e'_q}\right|_{P_{e0}}=\left.\frac{V_R\sin\delta}{X_i+X_e}\right|_0=\left(\frac{P_e}{e_i}\right)_0.$$

h has the same sign as $P_{e,}$ therefore:

- $h>0$ if $P_{e0}>0$, that is, a generating unit;
- $h<0$ if $P_{e0}<0$, that is, a pumping unit.

$$h_1=\left[\frac{\partial V_m}{\Delta\delta}\right]_{V_m=V_{m0}}=-\frac{X_eX_i}{X_i+X_e}\left(\frac{P_e}{V_m}\right)_0.$$

h_1 has the opposite P_e sign, therefore:

- $h_1>0$ if $P_{e0}<0$, that is, a pumping unit;
- $h_1<0$ if $P_{e0}>0$, that is, a generating unit.

$$h_2=\left[\frac{\partial V_m}{\Delta e'_q}\right]_{\delta=\delta^0}=\frac{X_e\cos\delta^0}{X_i+X_e}>0.$$

h_2 is always positive due the small value δ assumes.

According to the above evaluations, a generating unit is characterised by $hh_1/h_2<0$. Therefore, the voltage regulator (AVR), $-\frac{hh_1}{h_2}G_v(s)$, always determines a destabilising effect on the electromechanical loop. At no load ($P_{e0}=0$) the destabilising effect becomes nil. In general, reducing the delivered active power makes the AVR destabilising effect progressively reduced.

In conclusion, the generator voltage regulator mainly impacts angle stability. This allows an understanding of why the control solutions aimed to damp the electromechanical oscillations operate through the voltage control loop, as will soon be introduced.

Electromechanical Oscillation Damping Through Additional Feedbacks on Generator Voltage Control Loop

At the input summing junction of the generator voltage control loop, additional feedbacks to stabilise rotor angle transients can be reclosed, thereby obtaining positive damping of electromechanical oscillating phenomena. As already introduced, voltage dependence on rotor speed contributes to the introduction of stabilising terms into the model equations. This corresponds to a consideration of a signal proportional to rotor speed entering into the AVR input junction through a given filter $K_{PSS}(s)$, summing it to the voltage set-point value (see Fig. A.5, dashed lines).

According to this feedback the electromechanical loop becomes as seen in the following Fig. 8.7 block scheme, where

$$\hat{K}_{\Omega}(s)=\frac{K_{PSS}(s)G_v(s)}{h_2}.$$

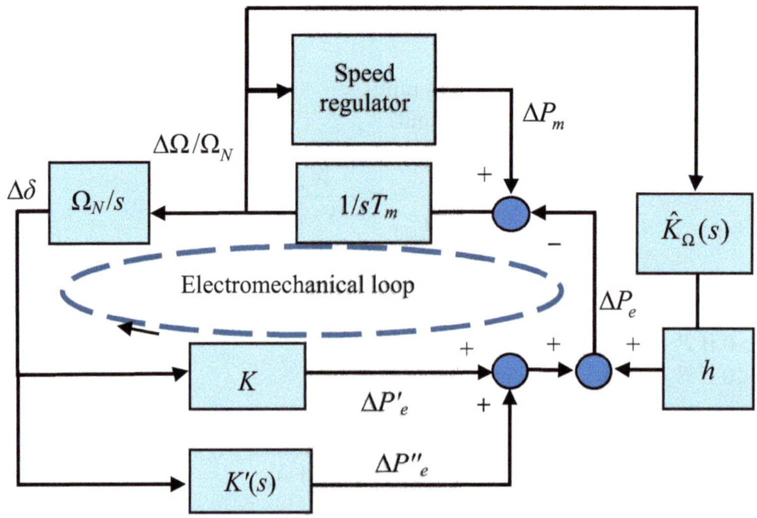

Fig. 8.7 Feedback from the rotor speed to the AVR input

$\hat{K}_{\Omega}(s)$ represents the link between the internal transient e.m.f. and rotor speed variations.

$K_{PSS}(s)$ is chosen in a way that determines (around the electromechanical oscillation frequency ω_0) the phase alignment between the internal transient e.m.f. and the rotor speed variations. In other words, if the resulting $\hat{K}_{\Omega}(j\omega)$ is a real positive number, then its feedback around the block $1/sT_m$ partially compensates the integrator phase delay. Accordingly, the electromechanical loop phase margin γ can be moved to positive values. This objective also means $K_{PSS}(j\omega)$ has to compensate $G_v(j\omega)$ by a PD transfer function. We note that the proportional term has the major effect if ω_0 has a very low value with respect to the voltage control loop cut-off frequency ω_c. The opposite occurs if $\omega_0 >> \omega_c$ is the derivative term having a more stabilising role.

In practice, to minimise noise in additional feedbacks, the $K_{PSS}(s)$ output can be obtained by a linear combination of available signals from the electromechanical loop: rotor speed $\Delta\Omega/\Omega_N$ and its derivative $\Delta P_e = -sT_m\Delta\Omega/\Omega_N$. Therefore, stabilising PSS are usually achieved through the use of the following feedbacks:

$$K_{PSS}(s)\frac{\Delta\Omega}{\Omega_N} = -K_c\Delta P_e + K_{\Omega}\frac{\Delta\Omega}{\Omega_N}.$$

Usually, $K_c \approx 0.3$, whereas $1 < K_{\Omega} < 10$, depending on the ω_0 value. When $P_e = 0$, stabilising signals are ineffective.

In conclusion, even if only a small damping of the electromechanical oscillations is gained, the use of generator AVR to achieve angle stability between generators and inter-areas in a power system is a very common and effective practice.

Results of a simulation study of an actual, critical event and a related, exciting on-field experiment are seen in Fig. 8.8a and b, a demonstration of the stabilising power of PSS additional feedbacks. The event occurred at Porto Tolle, a large power station in northern Italy, where angle instability due to a transmission line opening caused the generators to trip.

A study to reconstruct the event clarified what took place, with the support of detailed dynamic simulation, as the instability was due to an out-of-service PSS. Figure 8.8a shows how the instability appeared without PSS (left part of transient) and the stabilising effect of PSS when it was switched on at $t = 17$–18 s (right part of transient).

Because it was difficult for the TSO to accept this study's thesis, an on-field test was organised, reconstructing the same event, this time recording the generator transients. After the instability appeared, timely PSS manual switching-on at the generator AVR apparatuses caused electromechanical oscillation damping, which is recorded in the second half of the Fig. 8.8b traces. A very good similarity can be seen between the study and real system results.

This on-field test, which was trusted by the Italian TSO and power station, was highly convincing proof of the relevant importance of the PSS. After this event, the power station observed a rigid control on PSS active operation.

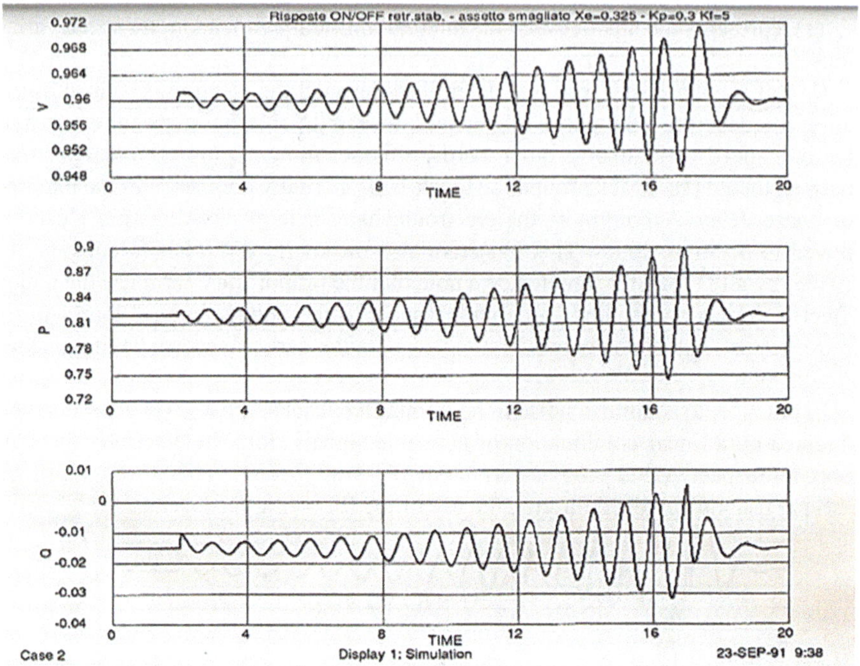

Fig. 8.8 **a** Test at Porto Tolle power station: manual switching-on of PSS feedbacks a few seconds after the rise of unstable electromechanical oscillation due to HV lines opening. Results of simulation study are shown.

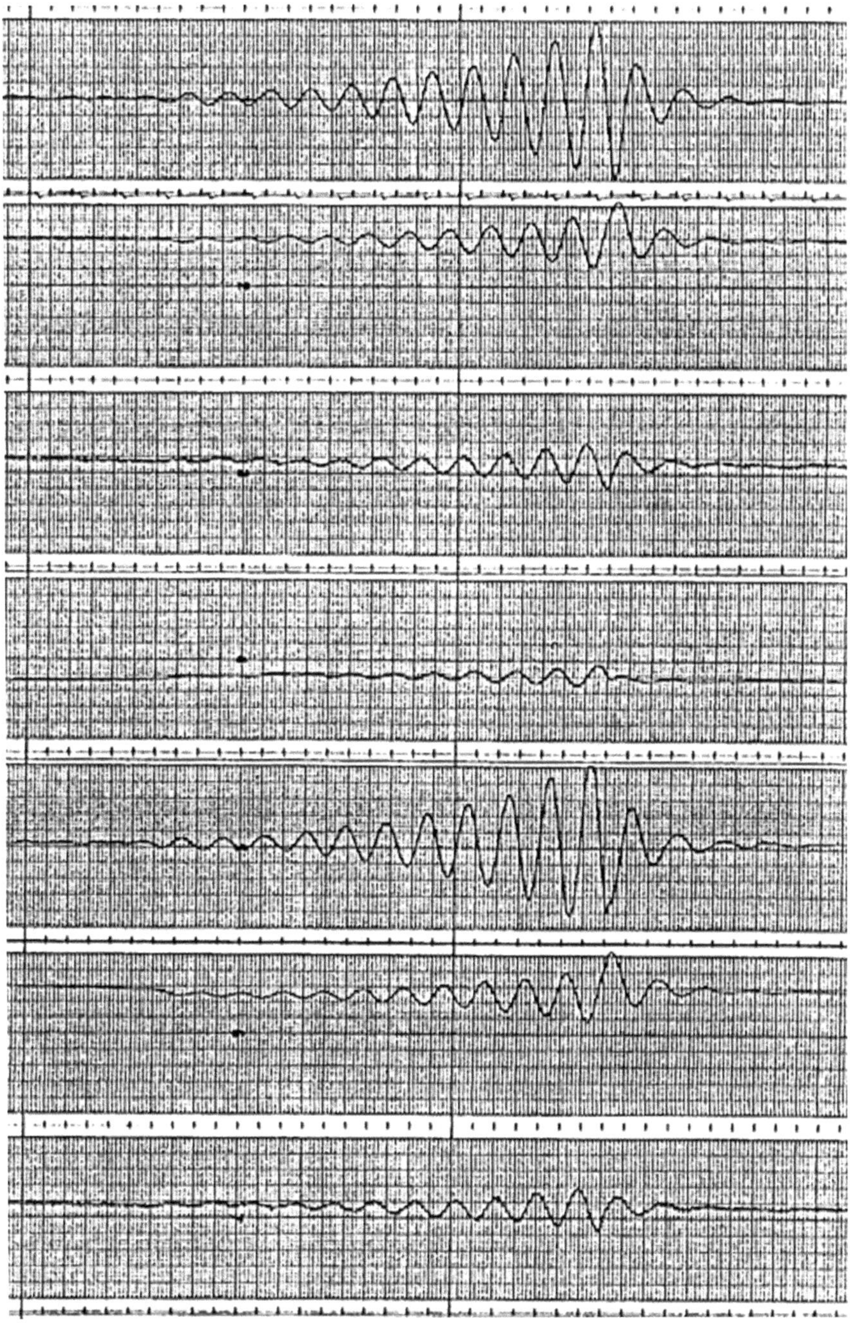

Fig. 8.8 b Test at Porto Tolle power station: manual switching-on of PSS feedbacks a few seconds after the rise of unstable electromechanical oscillation due to HV lines opening. Results of two generators' on-field test recording of an actual event at Porto Tolle power station. Traces show the manual switching-on of PSS feedbacks a few seconds after the rise of an unstable electromechanical oscillation due to the HV lines opening. From the top: P_2, V_2, Ω_2, Q_2, P_3, V_3, Ω_3. Time scale: 1 s=1 cm; P: 2%=1 cm; V: 1%=1 cm; Ω: 0.05 Hz=1 cm; Q: 4%=1 cm

8.2.4 SVR Contribution to Angle Stability

Reference is made to the equivalent system in Fig. 8.1, now representing an SVR area equivalent generator regulating its pilot node voltage $V_p = V_m$, while the remaining part of the power system is described by an infinite bus having sufficient electrical distance from the considered pilot node.

Combining the linearised models of both the alternator in Fig. A.5 and the secondary voltage control loops in Fig. 3.24, the equivalent SVR area generator is described by the block scheme in Fig. 8.9.

The model clearly shows the three overlapped control loops: the inner and faster due to the generator stator edge voltage control loop; the intermediate and less fast loop, of integral type, representing the reactive power control required by the SVR to sustain pilot node voltage; the outer and slower PI control loop regulating V_p at the values required by the V_{pref} set-point.

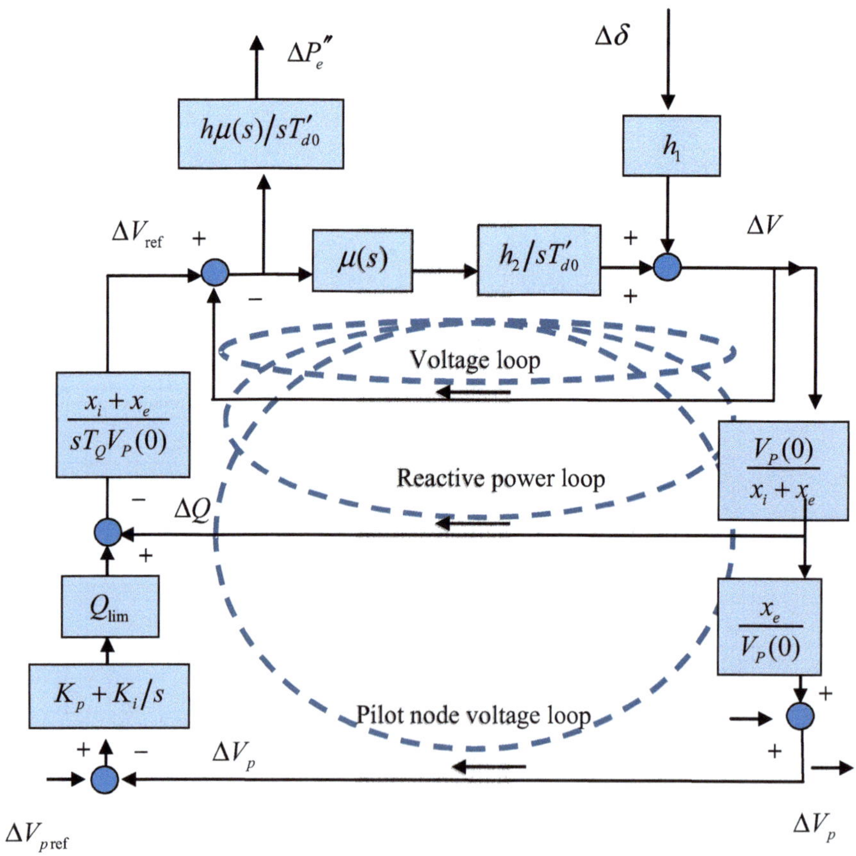

Fig. 8.9 Linear model of pilot node voltage control loop showing links with equivalent generator voltage and electromechanical loop

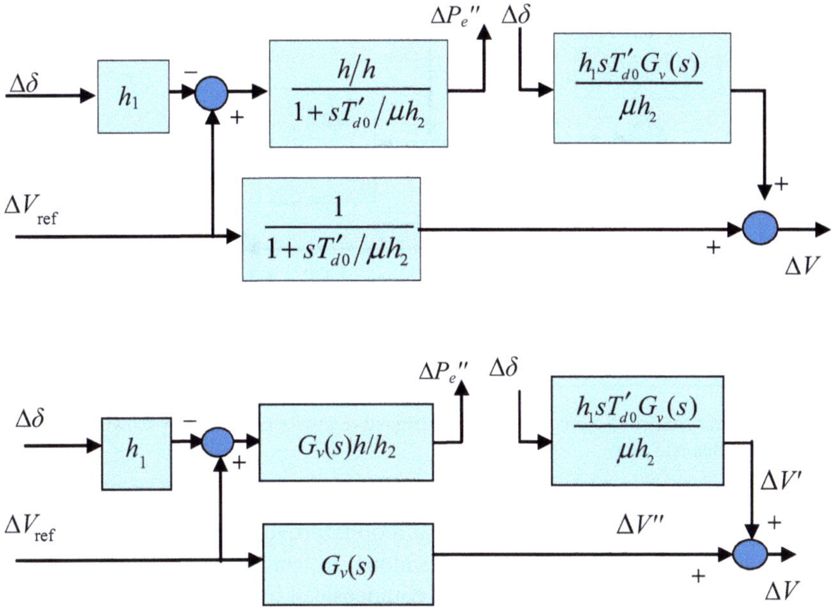

Fig. 8.10 Elaborated linear model of equivalent generator voltage control loop showing links with electromechanical loop

From Fig. 8.9 and confining the analysis to the inner generator voltage control loop, the following elaborated block diagram can be achieved (Fig. 8.10) by assuming no further simplification is introduced: $\mu = \mu(s)$.

Until now, nothing new has been shown other than what was seen before regarding AVR control; that is:

Dependence $\frac{\Delta P_e''}{\Delta\delta} = -\frac{hh_1}{h_2} G_v(s) = K'(s)$ is the contribution of the generator AVR alone.

Now the generator's slower reactive power control loop is reclosed. We notice in the enlarged model that the voltage control loop dynamics can be properly neglected ($G_v(s) \approx 1$) mainly at those blocks interacting with the slower reactive power control loop, as shown in Fig. 8.11.

Elaborating this scheme the same as done before with the generator voltage control loop (Fig. 8.10), the result achieved in Fig. 8.12 puts in evidence how the reactive power control loop adds a new contribution to the link with the electromechanical loop.

Defining $\tau = T'_{d0}/\mu h_2$:

$$\frac{\Delta P_e''}{\Delta\delta} = -\frac{hh_1}{h_2} \cdot G_v(s) - \frac{hh_1}{h_2} \cdot G_v(s) \cdot \frac{s\tau}{1 + sT_Q(x_i + x_e)},$$

$$\frac{\Delta P_e''}{\Delta\delta} = K''(s) + K''(s) = -\frac{hh_1}{h_2} \cdot G_v(s) \cdot \frac{1 + s[T_Q(x_i + x_e) + \tau]}{1 + sT_Q(x_i + x_e)}$$

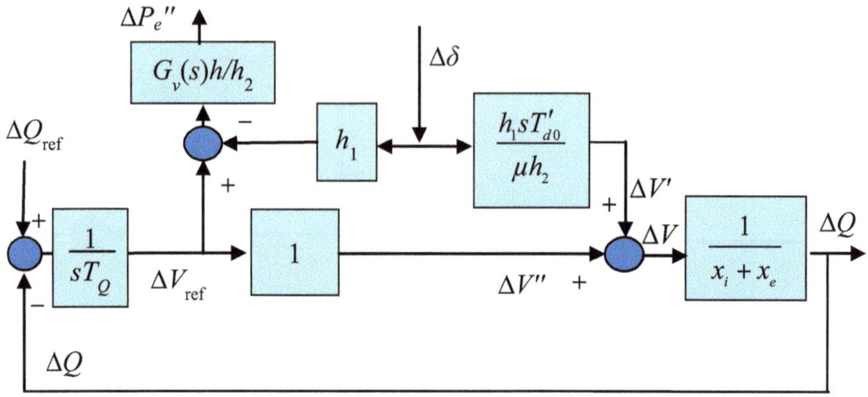

Fig. 8.11 Linear model of equivalent generator reactive power control loop showing links with the electromechanical loop

We notice that the reactive power control loop improves the unstable condition of the voltage control loop when it operates alone. In fact, the negative phase margin due to $G_v(s)$ is now certainly reduced or compensated by the new term:

$$\varphi(1+sT_Q(x_i+x_e)+\tau) > \varphi(1+sT_Q(x_i+x_e)).$$

Lastly, adding the external pilot node voltage control loop, the block scheme in Fig. 8.13 shows the further and slower closed loop having an impact on the electromechanical transients.

Elaborating this scheme the same way as before, with the generator reactive power control loop (Fig. 8.12), the achieved result in Fig. 8.14 puts in evidence how the secondary voltage regulation adds a further contribution to the feedback of the equivalent generator electromechanical loop.

The additional contribution is:

$$\frac{\Delta P_e'''}{\Delta\delta} = -\frac{Q_{\lim}k_I\left(x_i+x_e\right)\left(1+s\frac{k_p}{k_I}\right)}{s\left(1+sT_Q\left(x_i+x_e\right)\right)}\cdot\frac{s^2T_Q\tau h_1 x_e}{V_P(0)\left(1+sT_Q\left(x_i+x_e\right)\right)}\cdot\frac{1+s\left(T_Q\left(x_i+x_e\right)+\tau\right)}{\left[1+\dfrac{Q_{\lim}k_I x_e\left(1+s\frac{k_p}{k_I}\right)}{sV_P(0)\left(1+sT_Q\left(x_i+x_e\right)\right)}\right]},$$

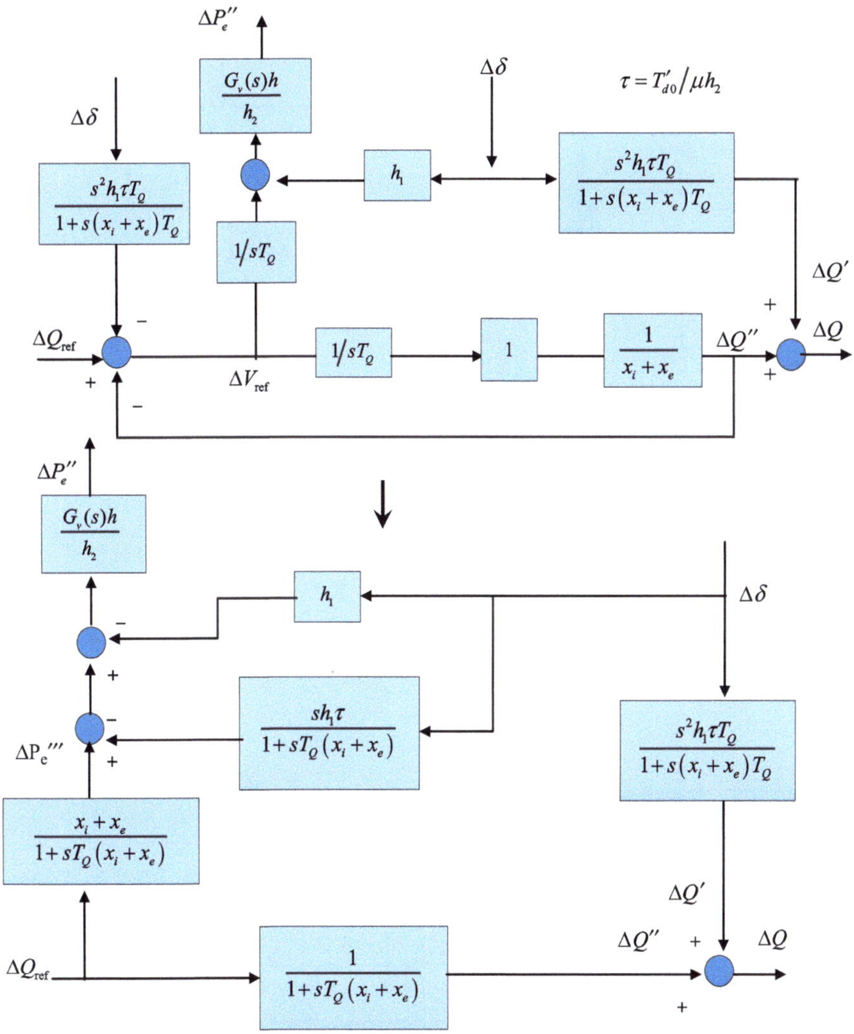

Fig. 8.12 Elaborated linear model of equivalent generator voltage and reactive power control loops showing the two corresponding links with the electromechanical loop

$$\frac{\Delta P_e'''}{\Delta\delta} = \frac{-Q_{\lim} T_Q \tau h_1 x_e k_I \left(x_i + x_e\right)}{\left[V_P(0)s\left(1+sT_Q\left(x_i+x_e\right)\right)+Q_{\lim} k_I x_e\left(1+s\dfrac{k_p}{k_I}\right)\right]} \cdot \frac{s^2\left(1+s\dfrac{k_p}{k_I}\right)}{\left(1+sT_Q\left(x_i+x_e\right)\right)},$$

$$\frac{\Delta P_e'''}{\Delta\delta} = \frac{-h_1 T_Q \tau(x_i + x_e)}{\left[s^2\dfrac{\left(V_P(0)T_Q\left(x_i+x_e\right)\right)}{Q_{\lim} k_I x_e}+s\dfrac{V_P(0)+Q_{\lim} k_P x_e}{Q_{\lim} k_I x_e}+1\right]} \cdot \frac{s^2\left(1+s\dfrac{k_p}{k_I}\right)}{\left(1+sT_Q\left(x_i+x_e\right)\right)}.$$

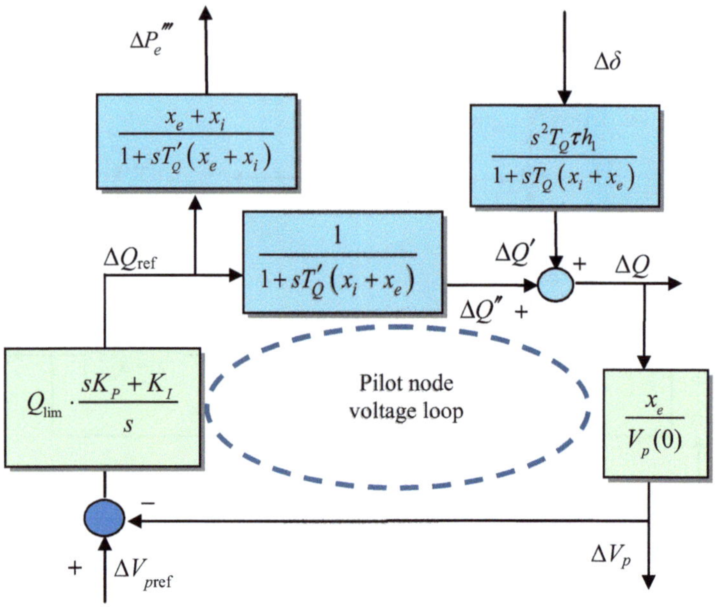

Fig. 8.13 Linear model of the pilot node voltage control loop showing a further link with the equivalent generator electromechanical phenomena

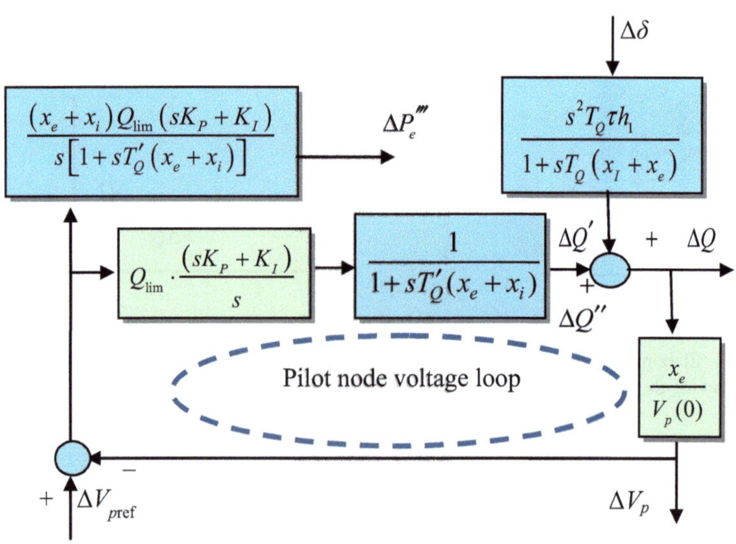

Fig. 8.14 Elaborated linear model of pilot node voltage control loop showing its further link between $\Delta\delta$ and $\Delta P_e''$ in addition to the two links described before on the same quantities (Figs. 8.12 and 8.13)

Commenting on this equation:

- h_1 is a positive term: it contributes no phase;
- s^2 is a term contributing π positive phase;
- $[\cdot]$ is a second order term with negative phase $\varphi([\cdot])$: $0 > \varphi([\cdot]) > -\pi$;
- $k_P/k_I \approx 40 > T_Q(x_i + x_e) \approx 5$: combining the phases of the two real singularities contributes a positive phase.

Therefore, the phase contribution θ to the electromechanical loop due to the SVR feedback ($\Delta P_e'''/\Delta\delta$) is positive, providing a stabilising effect.

In summary, the Fig. 8.15 block scheme represents, coming from the equivalent process in Fig. 8.1, the additional contributions to the electromechanical loop feedback K provided by the three overlapped electromagnetic closed-loop controls considered here: the generator AVR and the SVR with generator reactive power and pilot node voltage controls. In Fig. 8.15, T_P corresponds to:

$$T_P = T_Q(x_i + x_e)\tau.$$

The measure of the three contributions affecting system angle stability is given by the algebraic sum of their phasor angles amounting to θ ($\Delta P_e''/\Delta\delta$), which is added to the zero phase contribution of the feedback K. According to the sign of the resultant feedback vector phase, positive or negative, the electromechanical loop may respectively result as stable or unstable.

More precisely: The additional feedbacks providing $\Delta P_e''$ from $\Delta\delta$ as the combination of the generator AVR link ($K'(s)$), the reactive power control loop link ($K''(s)$) and the pilot node voltage control loop link ($K'''(s)$) contributions is now shown in Figs. 8.15a and 8.15b.

Because the h_1 parameter is negative, as was already seen, all additional feedbacks mentioned have positive gain, therefore, their phases depend on the contributions of the singularities alone.

Therefore, at the electromechanical oscillation frequency (ω_0), the feedback phase contributions are:

$$\begin{aligned}
&\varphi(K'(\omega_0)) < 0,\\
&\varphi(K''(\omega_0)) > 0,\\
&\varphi(K'''(\omega_0)) > 0.
\end{aligned}$$

In fact, at ω_0, $G_v(j\omega_0)$ provides a very small phase delay, easily recovered by the other positive contributions provided by the SVR reactive power (Q) and pilot node (V_p) control loops, as shown in Fig. 8.16.

In conclusion, even though achieved through a simplified analysis based on an equivalent model of the process, the result undoubtedly confirms how the SVR provides a stabilising effect on the electromechanical oscillations, mainly those between grid areas at low frequency. This happens notwithstanding that the SVR primary objective is voltage support and stability. Figure 8.17 provides an example of such a stabilising SVR contribution.

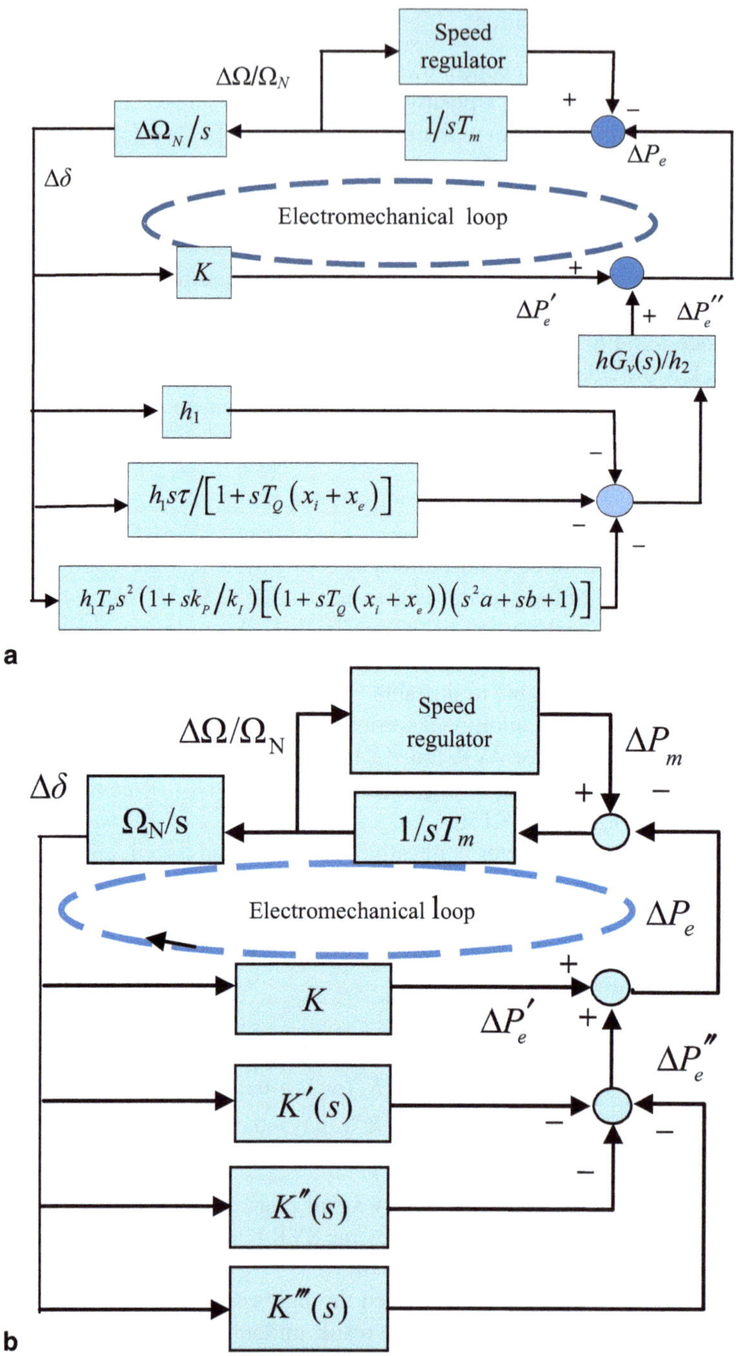

Fig. 8.15 **a** Elaborated linear model of the equivalent generator voltage, reactive power and pilot node voltage control loops showing the three corresponding feedbacks to the electromechanical loop. **b** Elaborated linear model of the equivalent generator voltage, reactive power and pilot node voltage control loops showing the three corresponding feedbacks to the electromechanical loop

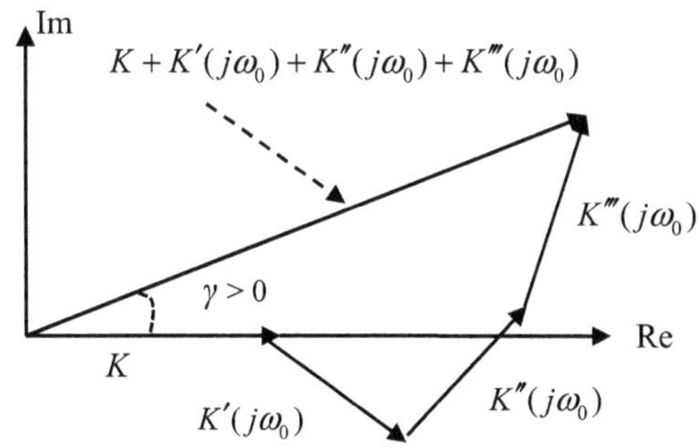

Fig. 8.16 Phasor diagram of the different contributions to the electromechanical control loop feedback and their resultant vector with positive (stabilising) phase angle ($\gamma>0$)

The Drakensberg generators show unstable electromechanical oscillations that become damped in the presence of SVR.

Up until now, the power system stability we have considered offers evidence of a relevant link with the generator electromechanical loop, and the simple equivalent model with two generators (Fig. 8.1) has been used to understand the angle stability phenomenon. From now on, conversely, the proper equivalent model to refer to when considering voltage instability and voltage collapse phenomena is that of a generator feeding a load, as in Fig. 8.2.

8.3 Voltage Stability: Introduction

We have clarified the impact of already operating voltage control loops on angle stability, as well as the potential applicability of such loops to the electrical process. With this done, the subject of voltage stability can be now introduced without ambiguity.

To this end, we now leave the equivalent scheme of Fig. 8.1 that was largely utilised previously and substitute the Thevenin's equivalent in Fig. 8.2, which represents a generator feeding a load. This is the more proper scheme for describing the voltage instability phenomenon.

Voltage instability arises from long-term or sudden-increase of load, followed by local OLTC and OEL operations and maximum power transfer reaching local loads. After the dynamic maximum line loadability has been overcome, the system operating condition gives evidence, under a load-restoring attempt, of a progressive or sudden reduction of local voltages up to the point of system collapse.

More precisely:

- Load increase combined with its electrical distance from the generators and the approaching power transfer limits are the driving forces of the voltage instability phenomenon;

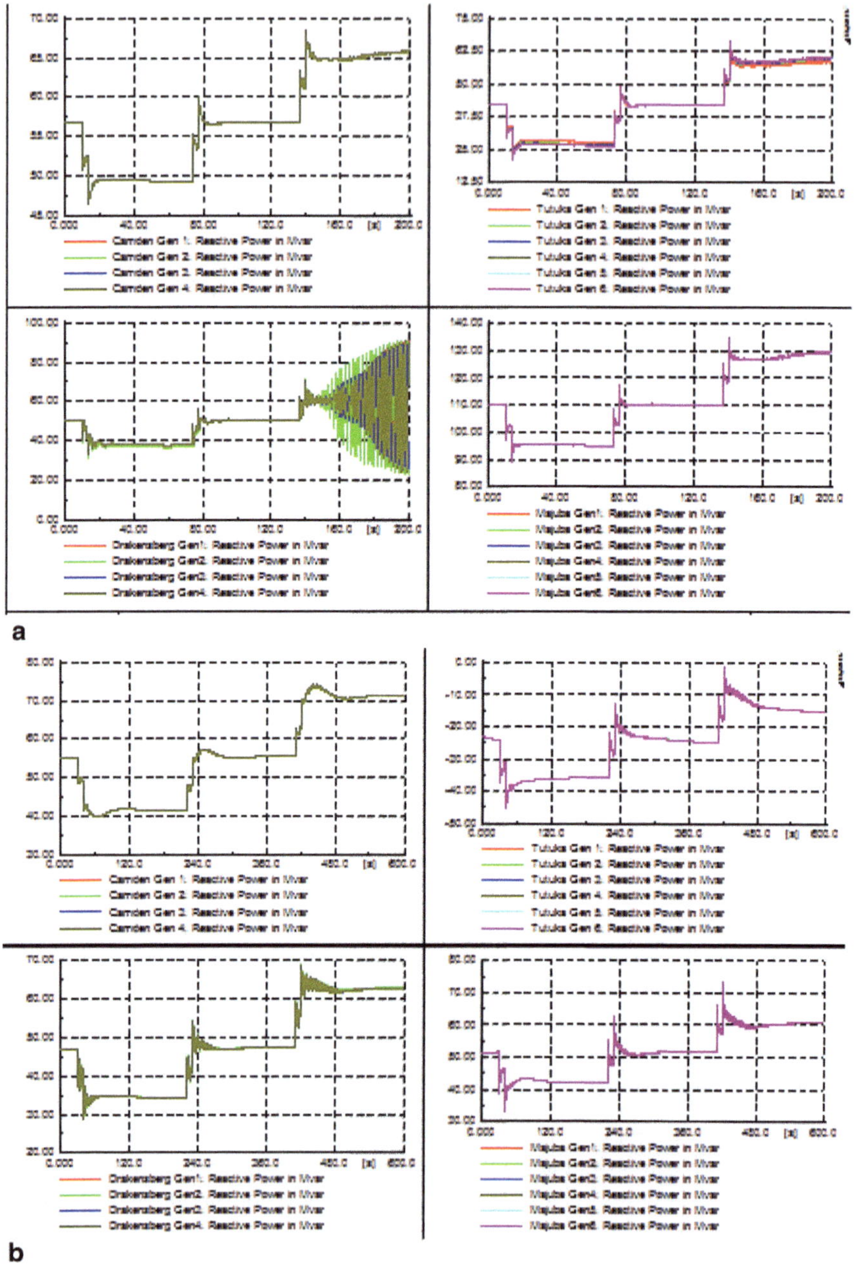

Fig. 8.17 South Africa 2007 generators reactive power following load variation without (**a**) and with (**b**) SVR

- System "static limits" coming from the static equations are less restrictive than the "dynamic limits" imposed by the control system operating on the generators (mainly the OELs) and the transformers (OLTCs in closed loop). Therefore, a detailed dynamic model of the system is needed to analyse the voltage instability otherwise unrealistically delayed by a simple static and optimistic analysis;
- The *V-P* dynamic curve linked to the OEL and OLTC closed-loop operation shows the correct voltage degradation at load increase;
- The voltage dynamic dependence from the power transfer increase, while the OELs and OLTCs are operating, significantly contributes to both voltage-lowering and the progressive reduction of maximum power transfer loadability, indicated by the tip of the "nose" in the *V-P* curve (see later explanation);
- The reactive power provided by generators, compensating equipment and lines contributes to sustaining the voltages up to their dynamic maximum delivery, limited by generator OELs and voltage-lowering on lines and compensating capacitors;
- OLTC stepping-up increases the load seen by the transformer HV side, therefore anticipating OEL limiting and voltage-lowering, up to reaching the extreme "inverse operating condition", unless saturated or locked before.

What the above anticipates with regard to the voltage instability phenomenon is here after analysed in detail and largely justified.

8.3.1 Relationship Between Load Power and Network Voltage

The conventional system equivalent structure, to which a large variety of books and papers refer when introducing the basics of voltage stability, is represented in Fig. 8.18 with the obvious meaning of the symbols: an equivalent generator feeds the load impedance through the equivalent line.

According to what has been said before the appearance of this scheme, assuming a constant value for generator voltage allows analysing the power system phenomena at steady state sinusoidal operating condition through a static equation linking the electrical variables evident in the figure. This is a preliminary but useful analysis describing the basic relationship linking load power transfer to grid voltage.

First, we compute under proper constraints the maximum power that can be delivered to the load.

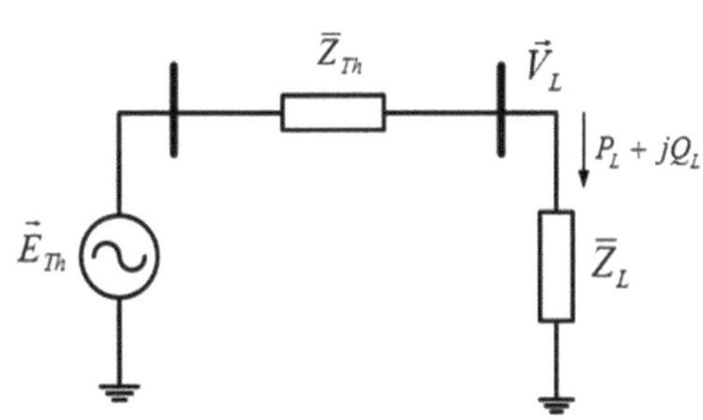

Fig. 8.18 Two-bus Thevenin equivalent circuit

Without constraints:

$$\bar{I} = \frac{\bar{E}_{Th}}{(R_{Th}+R_L)+j(X_{Th}+X_L)},$$

$$P_L = R_L I^2 = \frac{R_L E_{Th}^2}{(R_{Th}+R_L)^2+(X_{Th}+X_L)^2}.$$

Maximising P_L with respect to the load variables R_L and X_L:

$$\frac{\partial P_L}{\partial R_L} = E_{Th}^2[(R_{Th}+R_L)^2+(X_{Th}+X_L)^2]-2R_L E_{Th}^2(R_{Th}+R_L)=0,$$

$$\frac{\partial P_L}{\partial X_L} = -2R_L E_{Th}^2(X_{Th}+X_L)=0.$$

Under the assumption $R_L>0$, the solution is unique:

$$R_L = R_{Th}, \qquad X_L = -X_{Th}, \qquad Z_L = Z_{Th}^*.$$

Therefore, power is maximised when the load impedance is the complex conjugate of line impedance. This means the impedance seen by the generator is purely resistive:

$$R_L + R_{Th} = 2R_{Th}, \qquad P_{L\max} = \frac{E_{Th}^2}{4R_{Th}}.$$

Corresponding to

$$V_L = \frac{E_{Th}}{2}.$$

In transmission lines, resistance value is negligible with respect to reactance value, so $P_L$max is reached at very low R_L, that is, as $P_L$max tends to infinity. Therefore, current I goes to infinity while X_L+X_{Th} tends to zero. This is an unrealistic situation where a highly capacitive load would be required to compensate the inductive nature of the line impedance.

An analysis closer to the real power system operating condition is now presented by specifying the load power factor value ϑ, here after assumed constant.

With constraints:

$$X_L = R_L \tan\vartheta,$$

$$\bar{I} = \frac{\bar{E}_{Th}}{(R_{Th}+R_L)+j(X_{Th}+R_L\tan\vartheta)},$$

$$P_L = R_L I^2 = \frac{R_L E_{Th}^2}{(R_{Th}+R_L)^2+(X_{Th}+R_L\tan\vartheta)^2}.$$

Maximising P_L with respect to the unique load variables R_L:

$$\frac{\partial P_L}{\partial R_L} = E_{Th}^2\left[\left(R_{Th}+R_L\right)^2+\left(X_{Th}+R_L\tan\vartheta\right)^2\right]$$
$$-2R_L E_{Th}^2\left[\left(R_{Th}+R_L\right)+\left(X_{Th}+R_L\tan\vartheta\right)\tan\vartheta\right]=0,$$
$$\frac{\partial P_L}{\partial R_L} = E_{Th}^2\left[\left(X_{Th}^2+R_{Th}^2\right)-R_L^2\left(1+\tan^2\vartheta\right)\right]=0\Rightarrow\left|Z_{Th}\right|=\left|Z_L\right|,$$
$$\frac{\partial^2 P_L}{\partial {R_L}^2}=-2R_L\left(1+\tan^2\vartheta\right).$$

It turns out under the assumption $R_L > 0$ that the solution to the second derivative is negative, so equality between load and line impedances corresponds to maximum power transfer while under load change.

Therefore, under constant power factor, the power transfer to the load is maximised when the magnitudes of the load and line impedances are equal. That is:

$$R_L\big|_{P\max}=\left|Z_{Th}\right|\cos\vartheta,$$
$$X_L\big|_{P\max}=\left|Z_{Th}\right|\sin\vartheta.$$

Under the assumption of $R_{Th}=0$ (this simplification is usually assumed in transmission line analysis), the load parameters maximising the power transfer at constant power factor are:

$$R_L|_{P\max}=|X_{Th}|\cos\vartheta,$$
$$X_L|_{P\max}=|X_{Th}|\sin\vartheta,$$
$$R_L|_{P\max}+jX_L|_{P\max}=X_{Th}(\cos\vartheta+j\sin\vartheta),$$
$$(R_L|_{P\max})^2+(X_L|_{P\max})^2=X_{Th}^2\Rightarrow|Z_L|_{P\max}|=|X_{Th}|=|Z_{Th}|.$$

Therefore, the previous maximum power transfer result is confirmed: load power is maximised when the magnitudes of the load and line impedances are equal. Moreover, in this operating condition electrical values corresponding to maximum power transfer are:

$$\overline{I}=\frac{\overline{E}_{Th}}{(R_L)+j(X_{Th}+R_L\tan\vartheta)},$$

$$P_{L\max}=R_L\big|_{P\max}I^2=\frac{R_L\big|_{P\max}E_{Th}^2}{(R_L\big|_{P\max})^2+(X_{Th}+R_L\big|_{P\max}\tan\vartheta)^2}=\frac{\cos(\vartheta)E_{Th}^2}{2X_{Th}(1+\sin\vartheta)},$$

$$Q_{L\max}=X_L\big|_{P\max}I^2=\frac{X_L\big|_{P\max}E_{Th}^2}{(R_L\big|_{P\max})^2+(X_{Th}+R_L\big|_{P\max}\tan\vartheta)^2}=\frac{\sin(\vartheta)E_{Th}^2}{2X_{Th}(1+\sin\vartheta)},$$

$$V_L\big|_{P\max}=\left|\overline{V}_L\big|_{P\max}\right|=\left|\overline{I}X_{Th}(\cos\vartheta+j\sin\vartheta)\right|=\frac{E_{Th}}{\sqrt{2(1+\sin\vartheta)}}.$$

Assuming a constant power factor, the result depends on the value of line reactance only.

V-P Curve Basics

Replacing, for simplicity, the Thevenin impedance in Fig. 8.18 by a pure reactance X_{Th} and assuming the ideal voltage source as the phase reference with constant amplitude E_{Th}, placed on the real axis, t the voltage at the load bus varies with respect to load active power P_L and load reactive power Q_L according to (8.2) can be shown to be:

$$V_L = \sqrt{\frac{{E_{Th}}^2}{2} - Q_L X_{Th} \pm \sqrt{\frac{{E_{Th}}^4}{4} - {X_{Th}}^2 {P_L}^2 - X_{Th} E_{Th} Q_L}} \tag{8.2}$$

In fact, from the scheme in Fig. 8.18,

$$\overline{V}_L = \overline{E}_{Th} - jX_{Th}\overline{I} = E_{Th} - jX_{Th}\overline{I}, \qquad \overline{I} = \frac{\left(E_{Th} - \overline{V}_L\right)}{jX_{Th}},$$

$$\begin{aligned} P_L + jQ_L &= \overline{V}_L \overline{I}^* \\ &= \frac{\overline{V}_L\left(E_{Th} - \overline{V}_L^*\right)}{-jX_{Th}} \\ &= \frac{\left[\left(V_L \cos\vartheta + jV_L \sin\vartheta\right)E_{Th} - V_L^2\right]}{-jX_{Th}} \\ &= \frac{-V_L E_{Th} \sin\vartheta}{X_{Th}} - \frac{j\left(V_L^2 - V_L E_{Th} \cos\vartheta\right)}{X_{Th}}. \end{aligned}$$

These are the well-known power flow equations at fixed load powers.

Eliminating ϑ, the following equation is obtained:

$$\left(\frac{V_L E_{Th}}{X_{Th}}\right)^2 = P_L^2 + Q_L^2 + \left(\frac{V_L^2}{X_{Th}}\right)^2 + \frac{2Q_L V_L^2}{X_{Th}}, \tag{8.3}$$

And from (8.3) comes the following:

$$\left(V_L^2\right)^2 + \left(2Q_L X_{Th} - E_{Th}^2\right)V_L^2 + \left(P_L^2 + Q_L^2\right)X_{Th}^2 = 0.$$

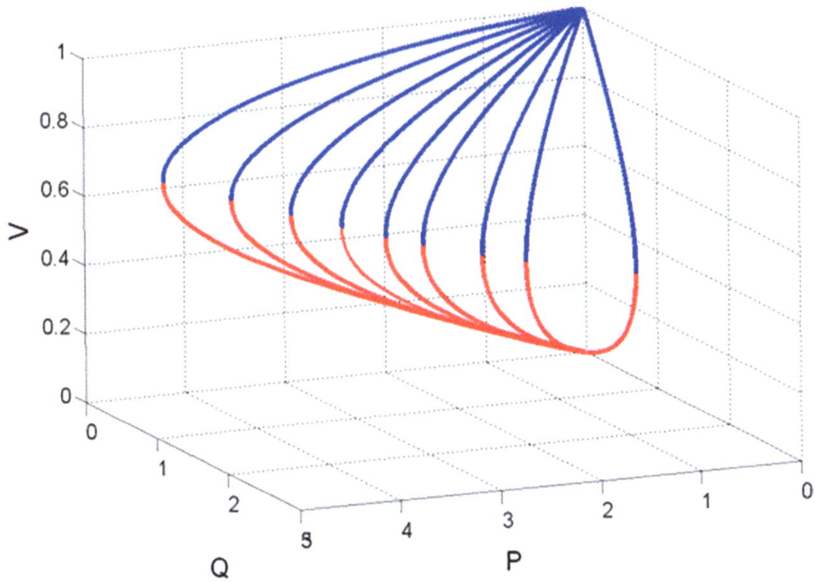

Fig. 8.19 Three-dimension nose curves

This second order equation in V_L^2 has a solution when

$$(2Q_L X_{Th} - E_{Th}^2)^2 - 4\left(P_L^2 + Q_L^2\right)X_{Th}^2 \geq 0.$$

All equation solutions respecting this condition are of interest, satisfying possible operating conditions.

In (P_L-Q_L-V_L)-space, Eq. (8.3) and its solution (8.2) define a surface represented in Fig. 8.19. The surface upper part of the surface, at the highest voltage values, corresponds to the Eq. (8.2) solution with the plus sign, while the lower part related to the low voltage solutions, corresponds to the minus sign.

The points of the surface with two coincident solutions describe the curve of the maximum power ($P_{L\max}$, $Q_{L\max}$) whose points are located at the edges between the blue and red half-curves.

Each of the curves shown refers to a fixed load tan(ϑ) (constant power factor). Projecting one of these on the *V-P* plane provides the curve of load voltage as a function of the active power represented in Fig. 8.20. This "nose curve" is largely used to interpret the voltage instability phenomenon and to identify the maximum loadability limit of the line.

Obviously, increasing the inductive part of the load contributes to anticipating voltage-lowering and the approach of the nose tip at lower active power. Vice-versa, increasing load capacitive effect or increasing compensation contributes to sustaining voltage at high values, thus delaying the nose tip that comes from higher active power values.

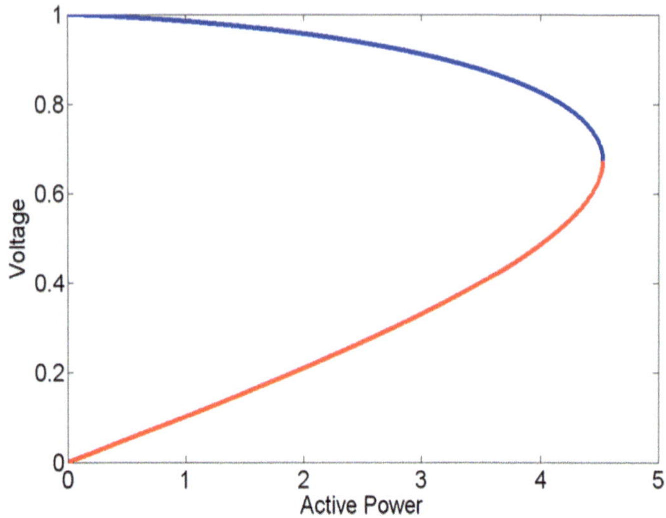

Fig. 8.20 The well-known nose curve at constant load power factor

We note:

- In the equivalent scheme, at a fixed active power in the load there are two possible operating conditions below the maximum. They differ according to voltage value; the higher voltage point corresponds to the normal operating condition;
- Maximum power transfer corresponds to a unique available operating condition representing the system limit in terms of power transfer and maximum power the load can absorb;
- In general, the intersection between the system *V-P* curve in Fig. 8.20 and any possible *V-P* load curve determines the operating point that becomes critical when maximum power transfer is approached, or worse if located at low voltage.
- Under a smooth load increase, the operating point moves along the *V-P* nose curve from high voltage values to points of low voltage, after overcoming the nose tip. The operating point is lost after the load (represented as constant *P*) increases over the nose tip value. The operating point can be also lost after a large system perturbation, causing a sudden retreat in the *V-P* curve toward the axes.

The pure static equivalent scheme considered does not allow a correct description of both the system transients that occur as a result of equilibrium loss and the impact of control systems on the shape of the *V-P* curve. Represented in Fig. 8.20 are the equilibrium characteristics of the system dynamics before the operation of generator OELs and with open loop on OLTCs. Moreover, when the operating point is run along the *V-P* curve, the speed values it moves are not correctly provided by the static model used.

Therefore, simply referring to the static *V-P* curve, the dynamical aspects linked to the voltage instability phenomenon seem to be frequently overlooked in the choice of models used to study the voltage stability problem.

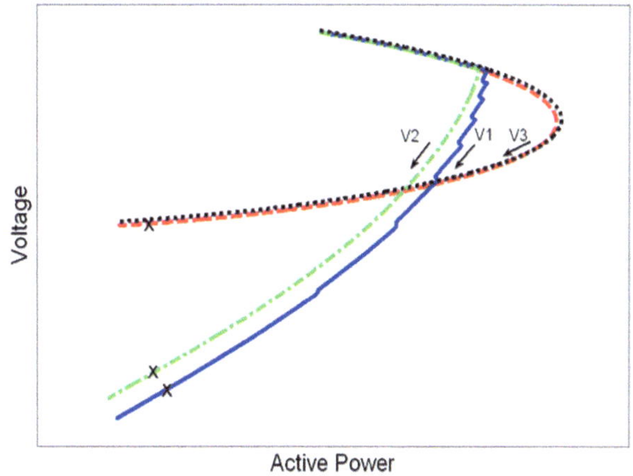

Fig. 8.21 "Nose" curves: One obtained from static models (*dotted*), the other three obtained from dynamic models. With AVR only (*dashed*), AVR+OEL (*dash-dotted*) and AVR+OEL+OLTC (*solid*)

It is conversely relevant to perform the analysis via a credible equivalent network model showing loadability conditions that are often significantly different from those indicated by a simple static model. In fact, the "nose" assumes diverse forms according to the dynamic aspects and load characteristics considered. Moreover, for a given load increase there is a variety of possible "noses", consisting (even in some small portions) of unstable points and run by diverse speeds driven by different closed-loop control configurations. Therefore, the analysis framework is very interesting and complex.

In summary, evidence is given to the results qualitatively depicted in Fig. 8.21, not only referring to the simple system based on the Thevenin equivalent but also to the high-voltage Italian electrical system, described by a detailed dynamic model. More specifically, confirmation is given to the following statements:

a. The generic static *V-P* curve of a load bus, based on the assumption the equivalent Thevenin generator seen from the bus (Fig. 8.18), is ideal; with neglected dynamics it describes operating points often not coherent with the actual physical system. Referring to the common way real loads are represented, the static *V-P* curve of an EHV bus indicates the maximum loadability point with active power values higher than what is seen in real situations. Furthermore, all the points of static *V-P* curves are stable equilibria. This fact is not confirmed when a more representative dynamic model is used.
b. Considering the same equivalent model of item a), but with a nonideal generator under AVR control with excitation ceiling limit and with a governor under speed regulation: the dynamic *V-P* curve of the load bus approaches the static curve referred to in item a) as the AVR gain is much higher, and until the ceiling limit is not reached. After excitation saturation, the *V-P* curve lies inside the static nose curve. According to the characteristic of the load, some points on the *V-P* curve can be unstable equilibria (in Fig. 8.21, the part of the curves on the left of the "X" mark) with or without ceiling saturation.
c. Considering the model of item b) but with a realistic AVR with OEL: increasing the load, the OEL significantly modifies the *V-P* nose shape, lowering maximum

loadability and contributing to instability of low voltage *V-P* curve equilibrium points. This happens for most of the loads (Fig. 8.21, dash-dotted line).

d. Last, we cannot ignore OLTC transformers and the importance of their dynamic model on determining the *V-P* curve's "nose" shape and on the speed of the operating point running along the curve when the load increases, particularly when the OLTC in the local load bus works in the vicinity of the maximum power transfer limit (Fig. 8.21, solid line).

The "X" mark in Fig. 8.21 *represents the point whose left V-P curve points are unstable, and the symbol "V" indicates the speed at which, at a specific voltage level, the equilibrium point moves according to the load increase. Usually, these speed values increase with voltage-lowering and are such that V1 > V2 > V3.*

The objective of the next section is to show how the results of *V-P* curves obtained from the proposed equivalent model of the grid seen by a given EHV bus (§ 8.3.1.2 (Proposed Equivalent System)–8.3.1.4 (V-P curve analysis for a more realistic generic load representation)) are coherent with those of a detailed model representing a very large grid, including the detailed dynamics of various OELs and OLTCs (§ 8.3.1.5 (V-P curve analysis for the Italian system)).

Proposed Equivalent System

This section proposes a one-machine dynamic equivalent model that has two objectives: first to aggregate dynamic aspects to the static equivalent model (as depicted in Fig. 8.18) for analyses of voltage stability based on Thevenin equivalents; and second, to serve as a test system for simulations of *V-P* curves.

Figure 8.22 shows the one-line diagram of the proposed dynamic equivalent system. It consists of a 370-MVA/20-kV round-rotor synchronous machine (six-order model), one 380-MVA–20-kV/400-kV step-up transformer, one 460-MVA–400 kV/132-kV step-down transformer, six parallel 64-MVA–132-kV/20-kV OLTC distribution transformers, and two parallel 400-kV/100-km overhead transmission lines. Data for the network and for the dynamic components, as well as the block diagrams of the voltage and speed regulators, are given below. Data utilised in numerical simulations represent actual components taken from the Italian electrical system.

Data Used in the Dynamic Model

Figure 8.23 shows a block diagram of the automatic voltage regulator and the over-excitation limiter (OEL). The OEL model is of the summing type with soft limiting, which retains the normal voltage regulator loop [16].

AVR and OEL $K_A = 500$ p.u./p.u., $T_A = 0.03$ s, $T_B = 1.0$ s, $T_C = 10$ s, $K_{OEL} = 1.0$ p.u./p.u., $T_{OEL} = 10$ s, $I_{fd}^{ref} = 2.5$ p.u.,

$V^{ref} = 1.03$ p.u., $E_{max} = 5$ p.u., $E_{min} = -1$ p.u., $I_{min} = 0.0$ p.u.

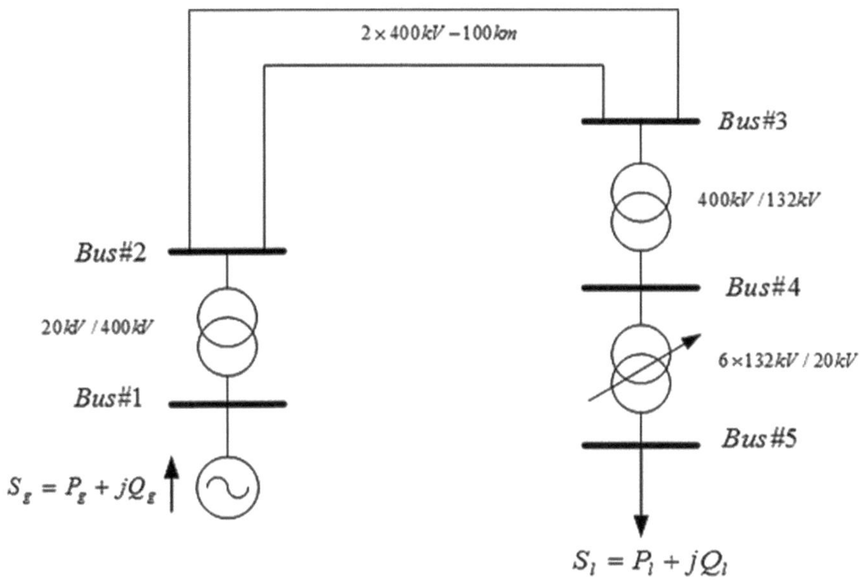

Fig. 8.22 One-line diagram of the test system

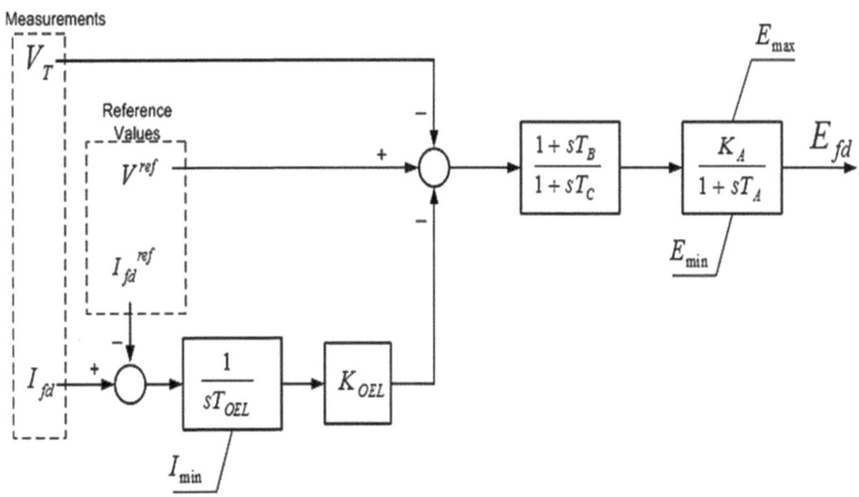

Fig. 8.23 AVR and OEL block diagram

Figure 8.24 shows the block diagram of the speed regulator.

Speed regulator $K_R = 20$ p.u./p.u., $T_R = 0.04$ s, $T_D = 3.0$ s, $T_E = 10$ s, $P_{\max} = 0.9$ p.u., $P_{\min} = 0.0$ p.u.,

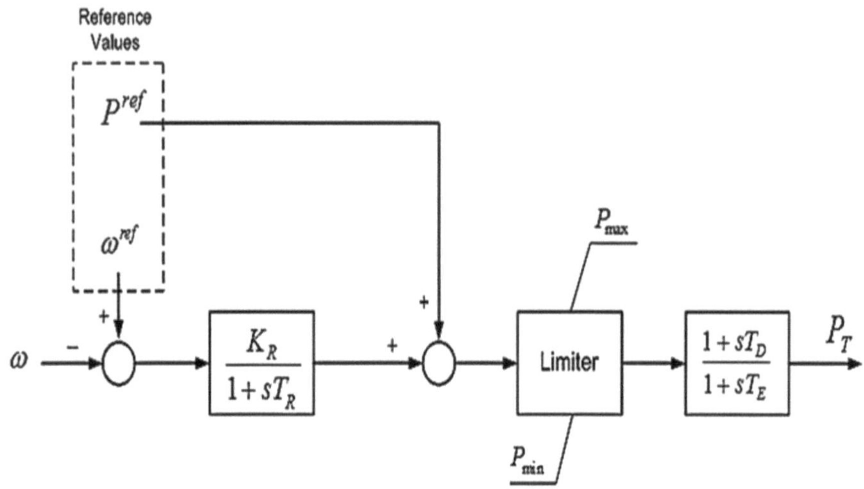

Fig. 8.24 Speed regulator block diagram

P^{ref} is either 160/370 p.u. or 280/370 p.u., depending on the operating condition under analysis.

Generator $H = 9.26$ s, $D = 0.0$ p.u./p.u., $r_a = 0.0014$ p.u., $x_l = 0.193$ p.u., $x_d = 1.9$ p.u., $x_q = 1.7$ p.u., $T_d' = 1.27$ s,

$T_q' = 0.235$ s, $X_d' = 0.302$ p.u., $X_q' = 0.5$ p.u., $T_d'' = 0.027$ s, $T_q'' = 0.012$ s, $X_d'' = 0.204$ p.u., $X_q'' = 0.3$ p.u.

Overhead transmission line Length = 100 km, rated voltage = 400 kV, rated current = 1 kA, nominal frequency = 50 Hz, resistance = 0.029 Ω/km, reactance = 0.3833 Ω/km, susceptance = 2.859 μS/km.

Step-up transformer Rated power = 380 MVA, HV-side = 400 kV, LV-side = 20 kV, r = 0.1896 %, xl = 12.68 %,

xm = 0.18 %, rf = 0.0.

Step-down transformer Rated power = 460 MVA, HV-side = 400 kV, LV-side = 132 kV, r = 0.1638 %, xl = 14.66 %, xm = 0.26 %, rf = 0.0.

Distribution OLTC transformer Rated power = 64 MVA, HV-side = 132 kV, LV-side = 20 kV, r = 0.4406 %, xl = 22.5 %, xm = 0.5 %, rf = 0.078 %, additional voltage per tap = 1.25 %.

OLTC mechanism Senses LV-side, minimum tap changer delay = 5 s (initial and subsequent), minimum voltage = 0.9 p.u.,
maximum voltage = 1.1 p.u., minimum tap position = −10, maximum tap position = 10.

Analysis of V-P Curves of the Test System

The analysis performed is fully oriented to recognise, for each given network structure, the voltage stability margin (seen from a given bus) with respect to the operating point. Therefore, the real limit is clearly indicated through load ramp increase, instead of through severe system contingencies with transients dominated by a large variety of phenomena, including voltage degradation/instability. Clearly, the steady state condition after a contingency is a new starting point to be checked against voltage instability by repeating slow load increase. According to this view, the most significant approach to discovering the voltage stability limit is to define those system structures and load characteristics to be analysed by load ramp increase.

A comprehensive analysis of nose curves for the test system shown in Fig. 8.22 follows. The presentation of the results is organised according to the closed-loop controls that are active. Three cases are defined as follows:

- Case 1: OLTC and OEL in service;
- Case 2: OLTC out of service and OEL in service;
- Case 3: OLTC and OEL out of service.

Cases 1 and 2 represent realistic operating conditions of a power system, where OELs are always in service while the OLTCs may be locked. Case 3 is a reference case since it sufficiently represents the static *V-P* curve (actually, the real static curve corresponds to Case 3 with the AVR having an integral control law without ceiling limit). In all cases, the automatic voltage regulator and the speed regulator are in service. Moreover, these three cases are compared to each other according to:

- Load type characteristic;
- Initial operating condition;
- Voltage bus level.

Two initial operating conditions are considered: Condition 1, when initial load is 160 MW (0.43 p.u.) and 0 MVAR, and Condition 2, when initial load is 280 MW (0.76 p.u.) and 20 MVAR. Also, two buses are considered: the EHV bus (Bus #3) and the load bus (Bus #5). Finally, with respect to load type, the common ZIP model is used: 100 % P-constant, 100 % I-constant and 100 % Z-constant, for both active and reactive power, are considered to analyse their specific effects.

For all the cases studied, the load increase rate is equal to $\Delta P_L = 0.5$ MW/s and $\Delta Q_L = 0.5$ MVAR/s according to:

$$P_L = (P_o + \Delta P_L) \times \left(\frac{V_L}{V_o} \right)^{\alpha}, \qquad Q_L = (Q_o + \Delta Q_L) \times \left(\frac{V_L}{V_o} \right)^{\beta},$$

where P_o, Q_o and V_o are nominal values, and α and β are constants to model the load type characteristic.

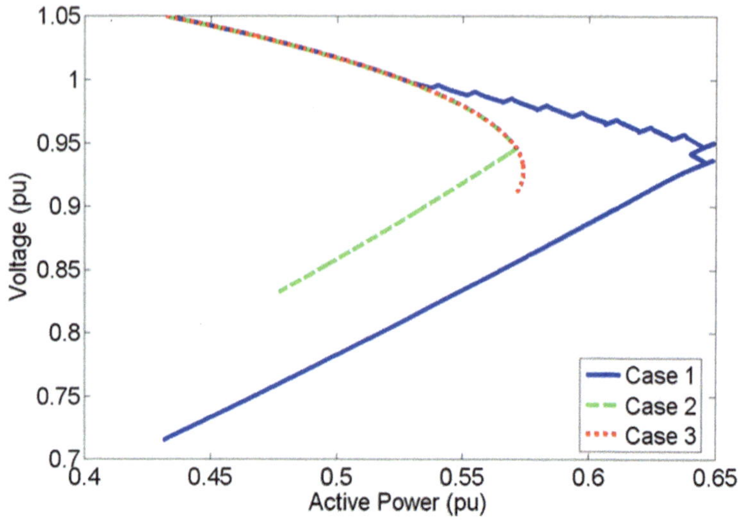

Fig. 8.25 Nose curves at Bus #3 for Condition 1 and load as Z-constant

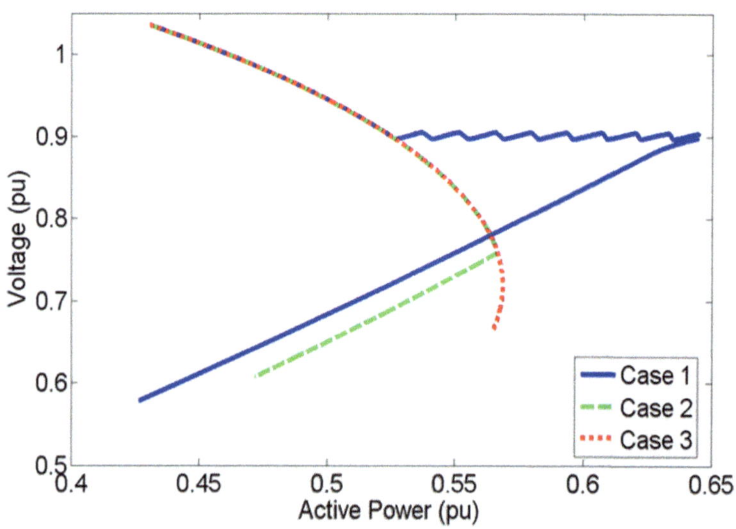

Fig. 8.26 Nose curves at Bus#5 for Condition 1 and load as Z-constant

Z-constant load type ($\alpha = \beta = 2.0$)

This subsection's results are obtained with the load represented as a *Z*-constant type. Figures 8.25 and 8.26 show the nose curves when the system is in Condition 1 for Bus#3 and Bus#5, respectively. Figures 8.27 and 8.28 show the nose curves when the system is in Condition 2 for Bus#3 and Bus#5, respectively. Large differences

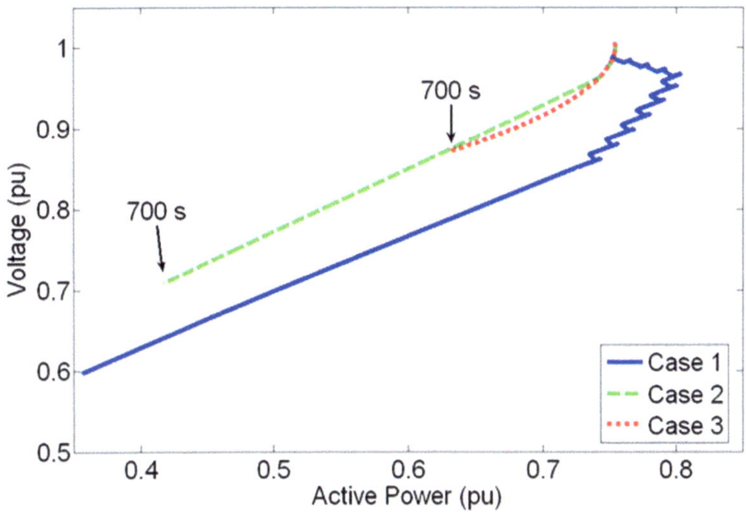

Fig. 8.27 Nose curves at Bus#3 for Condition 2 and load as Z-constant

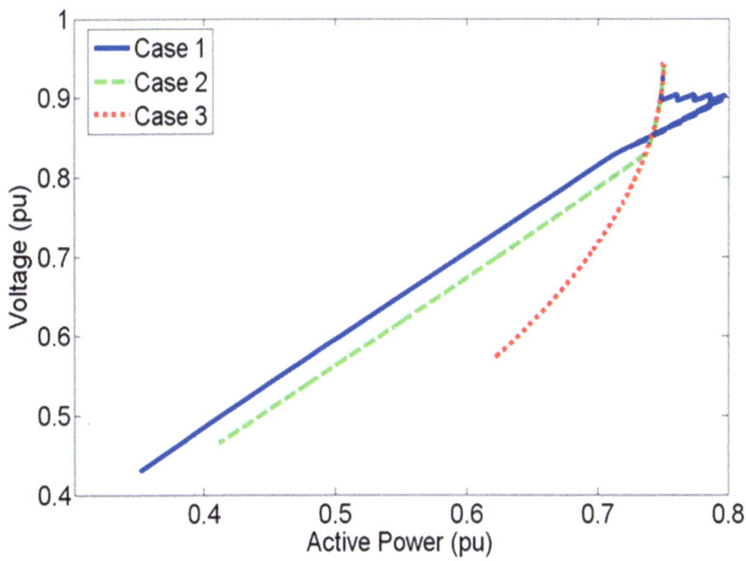

Fig. 8.28 Nose curves at Bus#5 for Condition 2 and load as Z-constant

in the shapes of these curves, particularly at the nose when the control loops start to operate, can be clearly seen.

The simulation time for the results presented in Figs. 8.25–8.28 is 700 s. Due to the type of load, the system does not present voltage stability problems. Thus, nose curves shown in these figures are drawn up to this simulation time.

Table 8.1 Key recorded time (s) with respect to Fig. 8.25

Case	T_{max}	$T_{0.85}$	T_{coll}	$T_{s\&g}$
1	311	424	∞	∞
2	464	656	∞	∞
3	574	>700	∞	∞

Table 8.1 shows the time, in seconds, to reach maximum power transfer (T_{max}), the time to reach 0.85 p.u. of voltage in the EHV bus ($T_{0.85}$), the computing collapse time (T_{coll}) and the actual time of instability ($T_{s\&g}$). A record of $T_{0.85}$ is of particular importance because, in general, at this voltage level under-voltage relays normally start to operate. T_{coll} is defined as the time in which the simulation program has numerical difficulties to converge. Often, the system crosses the voltage stability limit some time before it. This led to the introduction of $T_{s\&g}$, the minimum simulation time for the system to autonomously evolve to instability after load increase stops and simulation continues. Values in Table 8.1 are time values with respect to the nose curves presented in Fig. 8.25.

The follow are comments on Fig. 8.25 and Table 8.1:

1. Significant increasing of OLTC affects power transfer limit.
2. Voltage deterioration acceleration by the OLTC. Even though the OLTC tries to maintain voltage, it drives the system into operating points where dynamic mechanisms quickly depress voltage.
3. T_{coll} and $T_{s\&g}$ are infinity because there is no voltage instability problem when loads are represented as *Z*-constant type: the load characteristic always intersects the *V-P* curve at a stable point.
4. With respect to the curve of Case 2, one can see that the system changes immediately from a situation of increase in power with increase in load admittance (or decrease in load impedance, if you will) to a situation of decrease in power with increase in load admittance. This is explained by the fact that when the generator reaches its over-excitation limit, the equivalent Thevenin impedance seen from the bus near the load increases. This occurs because the synchronous reactance now can be seen as part (and a big part) of the equivalent Thevenin impedance. At that point, the impedance of the load is higher than the equivalent Thevenin impedance, and the system is beyond the maximum power transfer limit.
5. The time to reach T_{max} and $T_{0.85}$ is very different among Cases 1, 2 and 3, and the speed to run along the curve of Case 3 is considerably slower with respect to the other two cases.

From Fig. 8.26 we observe the actuation of the OLTC in trying to maintain the voltage on Bus#5 at the minimum deadband value of 0.9 p.u. The effects of this actuation are clear in the first 8 tap changes. After that, the generator OEL starts to operate and mainly drives the dynamics of the system, thus mitigating the effect of the last 2 tap changes of the OLTC.

Table 8.2 Key recorded time (s) with respect to Fig. 8.27

Case	T_{max}	$T_{0.85}$	T_{coll}	$T_{s\&g}$
1	90	126	∞	∞
2	4	325	∞	∞
3	4	>700	∞	∞

Bus#5 shows the relevant difference in terms of voltage-lowering with respect to EHV Bus#3. At Bus#5, voltage decay is more evident. Moreover, differences in the nose shapes are markedly evident at Bus#5, as well.

Table 8.2 shows time values for the key points of the nose curves presented in Fig. 8.27 related to the Z-constant load case.

The following comments refer to Fig. 8.27 and Table 8.2:

6. Comments 1, 2 and 3, made before, are still valid.
7. Under this heavier-load initial operating condition, the system is very close to the maximum power transfer limit. In this case, OEL actuation (Case 2) occurs when the nose curve is already at its lower part. In this situation, the voltage rate of change in Case 2 is higher (e.g., at 700 s it is roughly double) with respect that of Case 3.
8. The speed of voltage-lowering is markedly different in all three cases. Case 3 resembles the static *V-P* curve.

Figure 8.28 again puts in evidence the very low voltage value seen at Bus#5 with respect to Fig. 8.27, relating to the EHV Bus. Figure 8.29 refers to Case 1 of Table 8.2 and shows the transients of the OEL, which begins to operate at 89 s, and of the OLTC. Under the considered working condition, where the OLTC is stepping down and the OEL is starting to limit, a stop on the active power increase occurs at 90 s together with a fast voltage-lowering. The voltage at Bus#3 changes from 0.98 p.u. to 0.85 p.u. in 36 s.

I-constant load type ($\alpha=\beta=1.0$)

All results in this subsection are obtained with the load represented as I-constant type. Figures 8.30 and 8.31 show the nose curves when the system is in Condition 1 for Bus#3 and Bus#5, respectively. Figures 8.32 and 8.33 show the nose curves when the system is in Condition 2 for Bus#3 and Bus#5, respectively.

Table 8.3 shows time values with respect to the simulation results presented in Fig. 8.30.

The following are comments on Fig. 8.30 and Table 8.3:

9. Comments 1 and 2 are still valid.
10. T_{coll} and $T_{s\&g}$ are finite and different because now the loads are represented as I-constant type and stability problems start to occur with voltages at very low

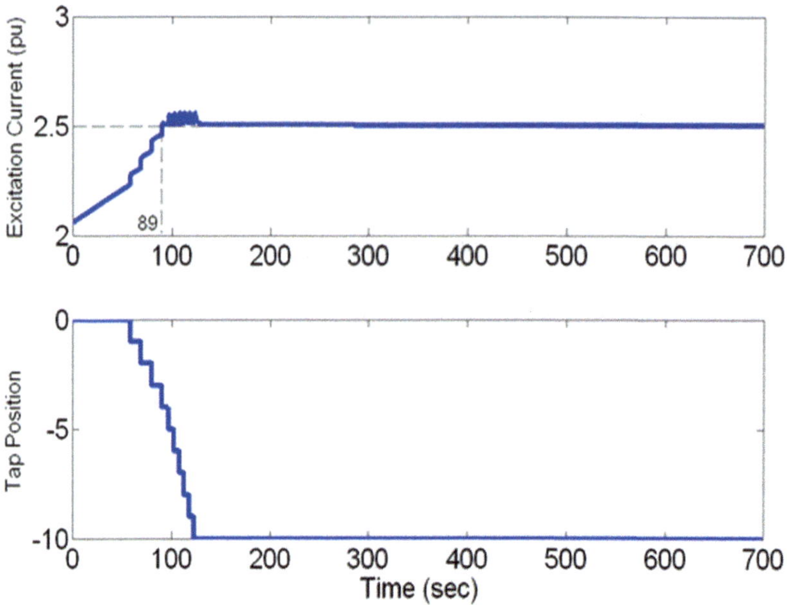

Fig. 8.29 Excitation current and tap position in Case 1 simulation

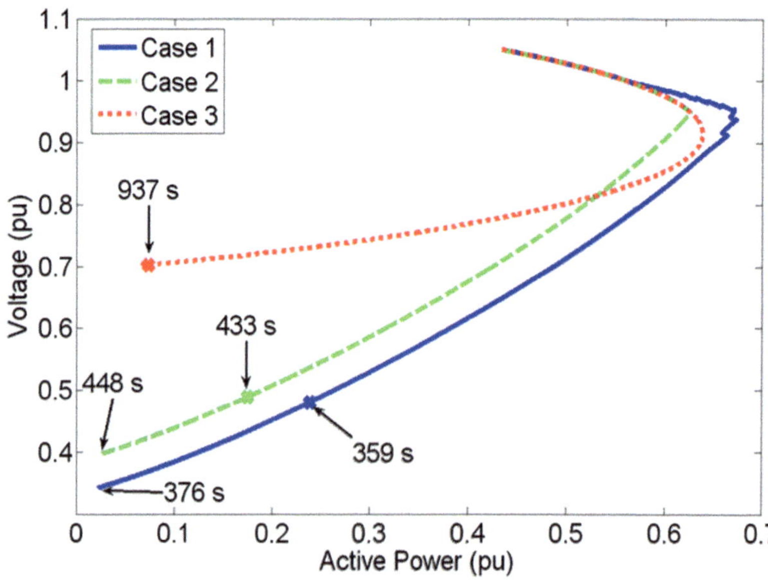

Fig. 8.30 Nose curves at Bus#3 for Condition 1 and load as I-constant

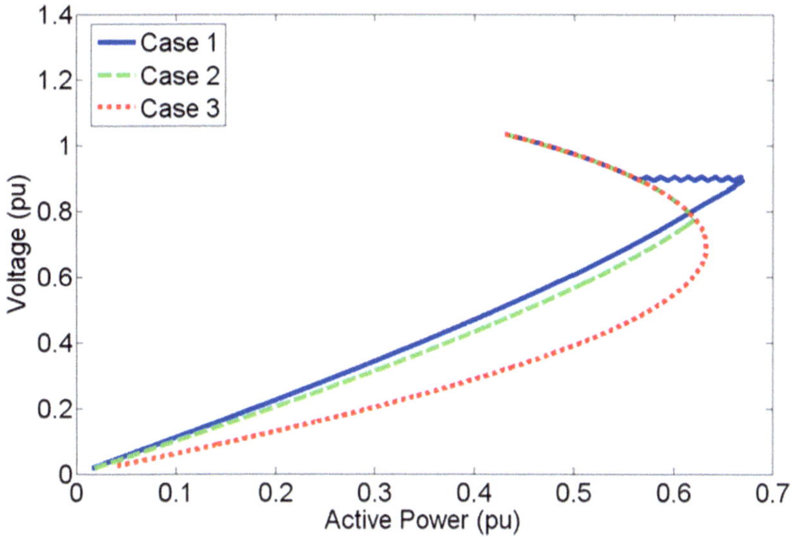

Fig. 8.31 Nose curves at Bus#5 for Condition 1 and load as I-constant

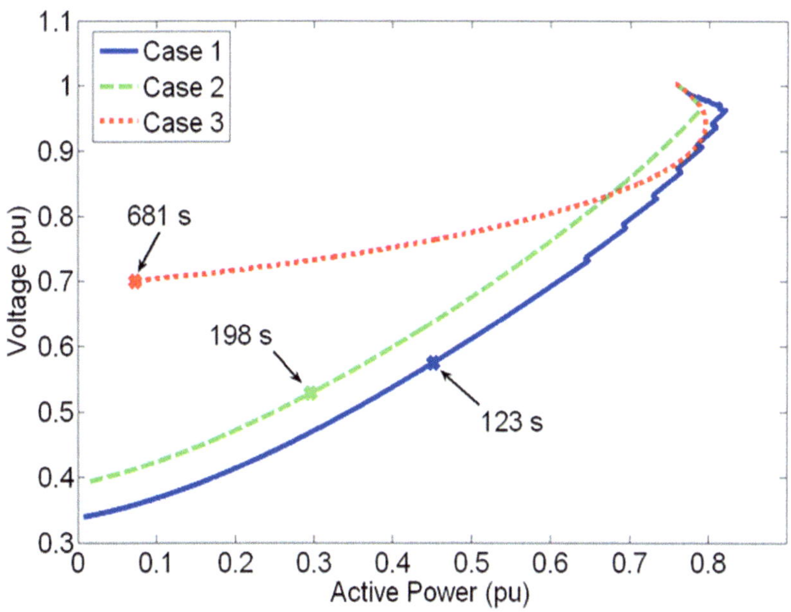

Fig. 8.32 Nose curves at Bus#3 for Condition 2 and load as I-constant

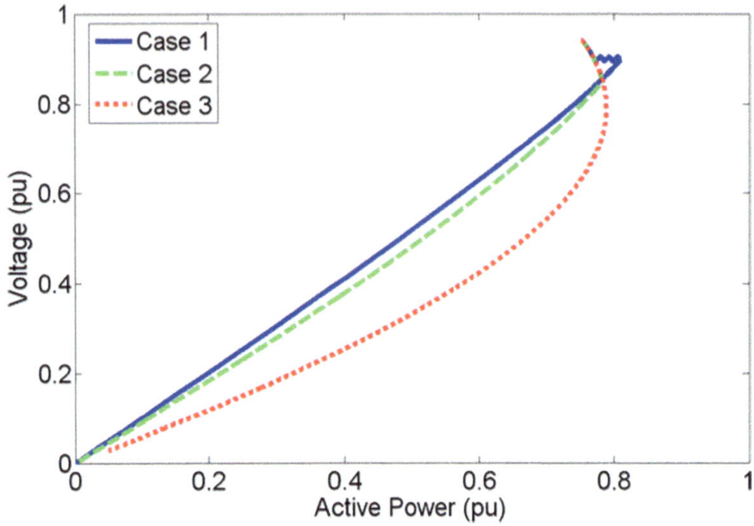

Fig. 8.33 Nose curves at Bus#5 for Condition 2 and load as I-constant

Table 8.3 Key recorded time (s) with respect to Fig. 8.30

Case	T_{max}	$T_{0.85}$	T_{coll}	$T_{s\&g}$
1	255	276	376	359
2	292	334	448	433
3	386	553	940	937

Table 8.4 Key recorded time (sec) with respect to Fig. 8.32

Case	T_{max}	$T_{0.85}$	T_{coll}	$T_{s\&g}$
1	74	91	144	123
2	84	124	214	198
3	144	345	681	681

values. Here and after the relevant table information comes from the column time differences and not from the absolute values linked to the chosen test.

11. The difference between T_{coll} and $T_{s\&g}$ is about the same for Cases 1, 2 (17 s and 15 s, respectively). This difference approaches zero for Case 3 (3 s).
12. In terms of speed to run along the curves, Case 3 (also representative of the static *V-P* curve) is very slow with respect to Cases 1 and 2, again confirming the distance of the static curve from the real process.

Table 8.4 shows the times with respect to the simulation results presented in Fig. 8.32. As is obvious, at the higher load the events are anticipated with respect to Table 8.3. Analogously, the events at Bus#5 come before those at Bus#3.

Figure 8.34 shows the strong impact of the combined actions of OEL and OLTC on nose shape for Case 1 and Condition 2. Three seconds before T_{max} (see Table 8.4) the OEL begins to operate in a closed loop with a dominant time constant of few seconds.

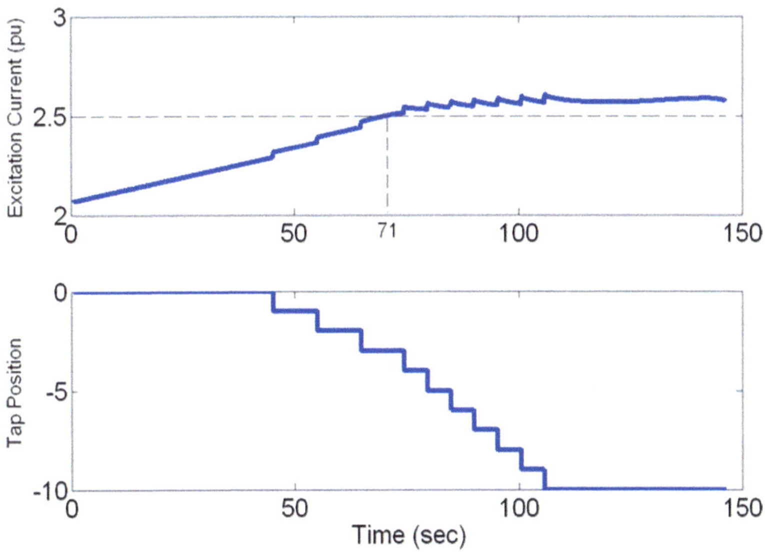

Fig. 8.34 Excitation current and tap position in Case 1 simulation

Again, Case 3 is not able to correctly reconstruct in value and in time the real process, which instead moves according to the Case 1 and 2 dynamics.

P-constant load type (α=β=0.0)

This subsection's results are obtained with the load represented as P-constant type. Figs. 8.35 and 8.36 show the nose curves when the system is in Condition 1 for Bus#3 and Bus#5, respectively. Figs. 8.37 and 8.38 show the nose curves when the system is in Condition 2 for Bus#3 and Bus#5, respectively.

Table 8.5 shows time values with respect to the simulation results presented in Fig. 8.35.

The following are comments on Fig. 8.35 and Table 8.5

13. In all three cases, voltage instability is mainly driven by OEL dynamics.
14. Instability occurs before the maximum power transfer point. Only points on the upper part of the nose curve can be shown.
15. Instability occurs at a high voltage profile. This confirms that relying on voltage profile alone as an indication of proximity to voltage instability is not a safe procedure.
16. With this load type, Case 3 is able to represent with sufficiently good approximation the real process performance.

At Bus#5, voltage-lowering is more evident, as it is in all the tests performed. Table 8.6 shows the times with respect to the simulation results presented in Fig. 8.37.

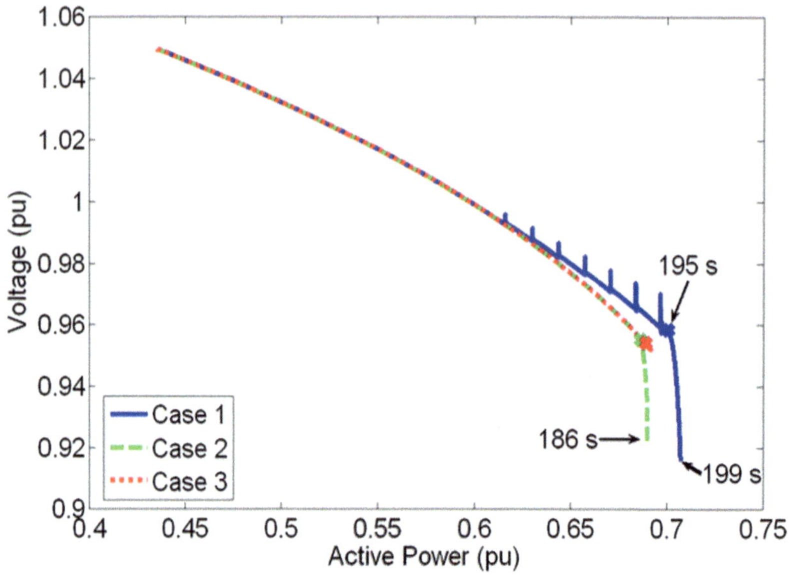

Fig. 8.35 Nose curves at Bus#3 for Condition 1 and load as P-constant

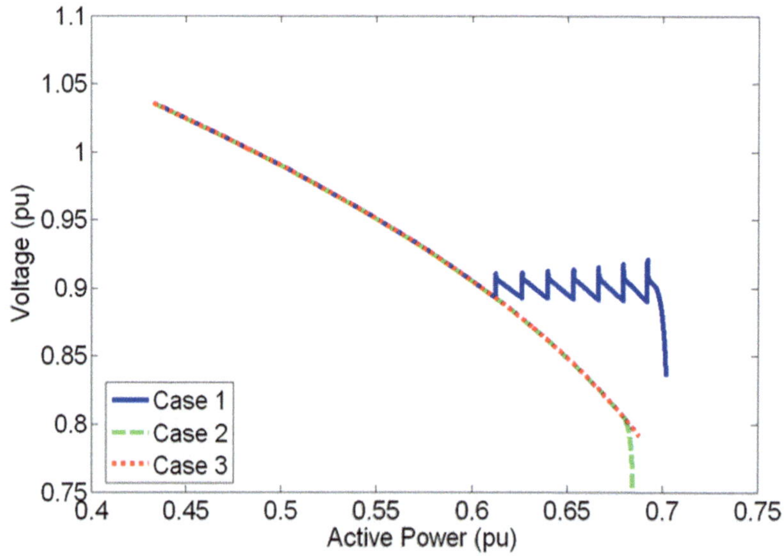

Fig. 8.36 Nose curves at Bus#5 for Condition 1 and load as P-constant

Figure 8.39 shows again the combined actions of the OEL and OLTC in determining (see Table 8.6) the beginning of voltage instability for Case 1 and Condition 2.

The main results of the comprehensive analysis made in this section are:

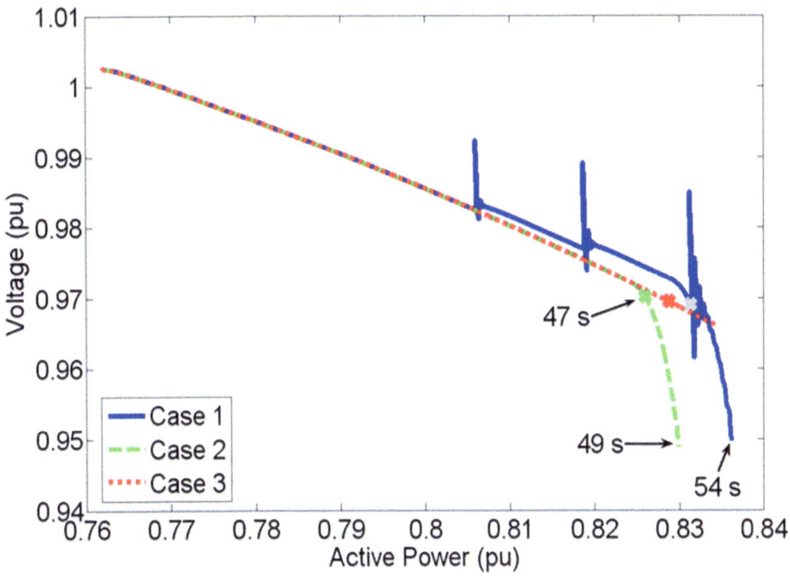

Fig. 8.37 Nose curves at Bus#3 for Condition 2 and load as P-constant

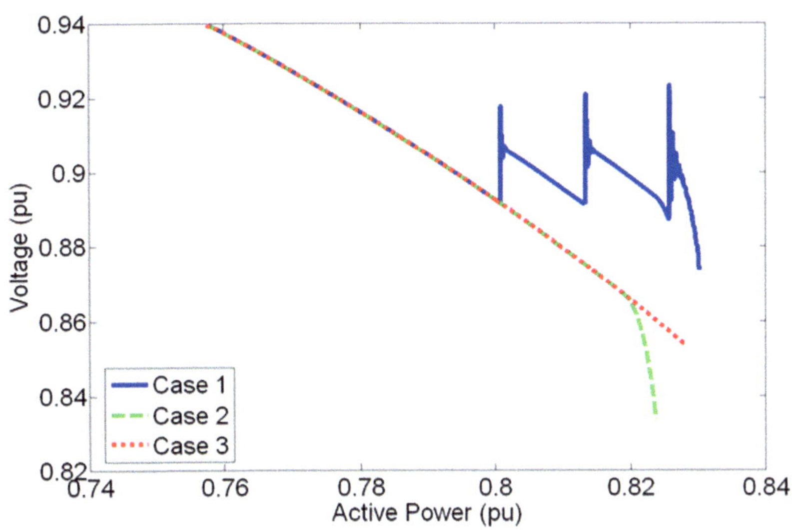

Fig. 8.38 Nose curves at Bus#5 for Condition 2 and load as P-constant

- The static *V-P* curve is very different from the possible dynamic ones. Smaller differences are seen in the P-constant load case;
- Real voltage instability is always determined by the OEL or by the combined action of the OEL and OLTC.

Table 8.5 Key recorded time (s) with respect to Fig. 8.35

Case	T_{max}	$T_{0.85}$	T_{coll}	$T_{s\&g}$
1	199	>199	199	195
2	186	>186	186	184
3	189	>189	189	186

Table 8.6 Key recorded time (s) with respect to Fig. 8.37

Case	T_{max}	$T_{0.85}$	T_{coll}	$T_{s\&g}$
1	54	>54	54	51
2	49	>49	49	47
3	53	>53	53	49

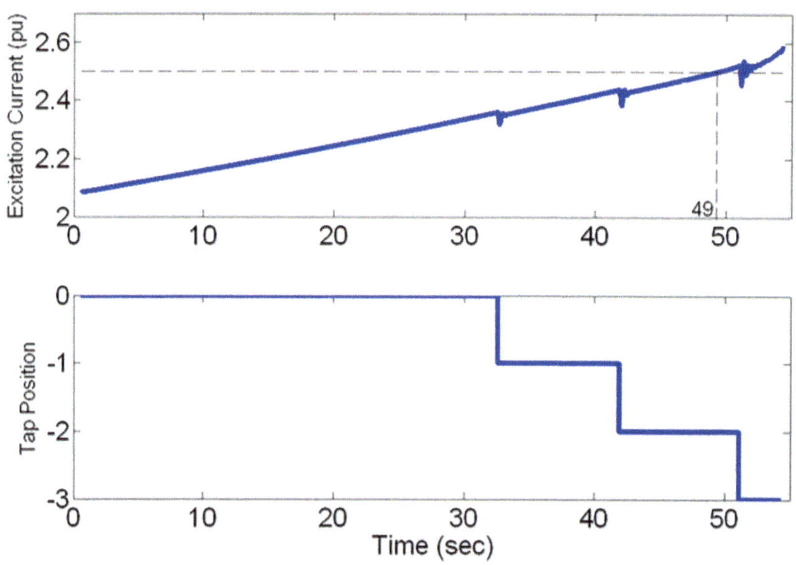

Fig. 8.39 Excitation current approaching the limit and tap position in Case 1 simulation

Now the analysis necessarily moves from the basic, simplified but theoretical approach to a more realistic system contest, where the load seen by EHV bus is generally represented by a combination of the previously considered theoretical typologies, and understanding the simulation results is necessarily assisted by the given schematic and a simplified introduction.

V-P Curve Analysis for a More Realistic Generic Load Representation

In this section, simulations are performed considering the load characteristic with $\alpha=0.7$ and $\beta=2.0$. These coefficients are used in simulations for the Italian system and presented in the next section.

Figures 8.40 and 8.42 show the nose curves at Bus#3 and Bus#5, respectively. Real load analysis confirms relevant differences between the nose of the static *V-P* curve (Case 3) and the dynamic ones (Cases 1 and 2) for both the EHV (Bus#3) and the LV (Bus#5) buses. More precisely:

- Maximum loadability differs by about 7 %;
- Instability begins at very high voltages and at very different times between Case 3 and Cases 1 and 2, as is also shown in Table 8.7, with differences of about 300 s;
- Clearly confirmed, as is also shown in Fig. 8.41, is the relevant effect of the OEL alone or in combined action with the OLTC in triggering real voltage instability.

Table 8.7 shows time values with respect to the simulation results presented in Fig. 8.40.

To give a better understanding of $T_{s\&g}$, Fig. 8.43 shows voltage at Bus#3 for cases where the load increase stops at 67 s (dashed line) and at 68 s (solid line). At 40 s, voltage at Bus#5 reaches 0.9 p.u. and triggers the OLTC clock. At 45 s, the first tap position is changed in order to support the voltage at Bus#5. Subsequent tap changes occur at 55, 65, 74, 82, 88, 93, 98, 103, 108 s. At 66 s, the OEL starts

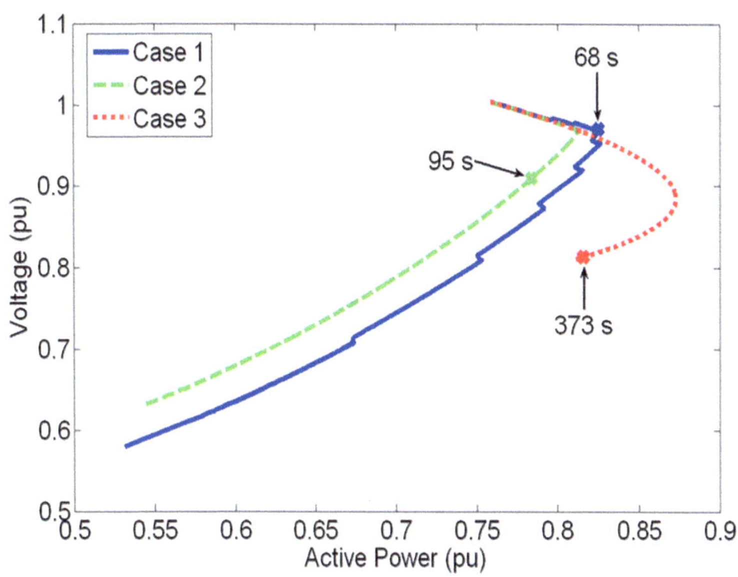

Fig. 8.40 Nose curves at Bus#3 for Condition 2 with the load represented as a generic dynamic model

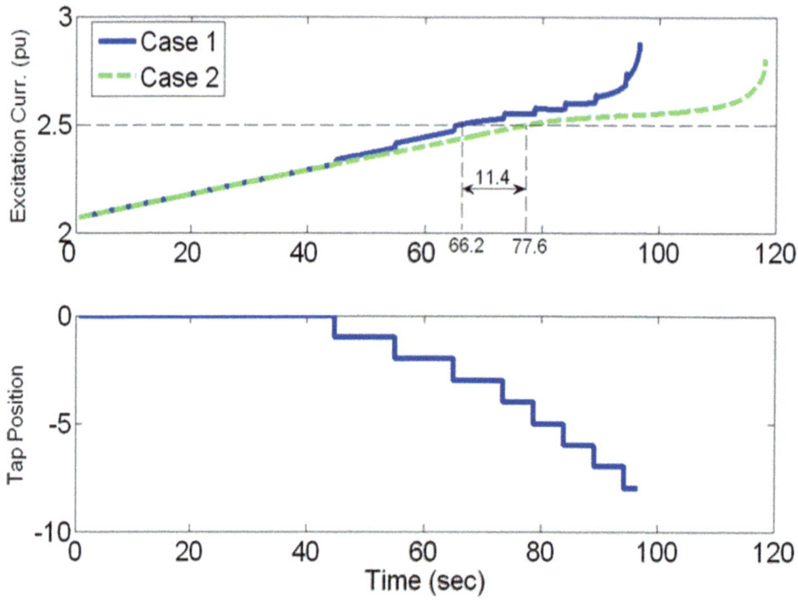

Fig. 8.41 Excitation currents in Cases 1 and 2 and tap position in Case 1

to activate. So, the evolution of OEL and OLTC dynamics combined with the time that load increase stops drives the stability of the system. In other words, when load increase stops at 67 s the stability boundary is not crossed and the system dynamics find a stable equilibrium. On the contrary, when load increase stops at 68 s the stability boundary is crossed and the dynamics of the system do not find stable equilibrium.

All the above results confirm statements a), b), c) and d) anticipated in § 8.3.1.1–(V-P curve basics). Lastly, the 30s difference between T_{coll} and $T_{s\&g}$ in Case 1 clearly shows the difference between voltage instability and loss of convergence in system computing (collapse).

V-P Curve Analysis for The Italian System

This section's objective is to compare the performances of the proposed equivalent dynamic model with the results of a wide, detailed, multivariable dynamic model of a large power system.

The analysed Italian system contains 380-kV and 220-kV networks (Fig. 8.44 depicts the 380-kV network only). The system configuration has 2549 buses, 2258 transmission lines, 134 and 191 groups, respectively, of thermal and hydro generators. System load is approximately 50 GW, fully represented as a static model with $\alpha=0.7$ and $\beta=2.0$ (§ 8.3.1.3 (Analysis of V-P curves of the test system)). Similarities of test results have shown that it was not worth representing loads using a 50 %

Table 8.7 Key recorded time (s) with respect to Fig. 8.40

Case	T_{max}	$T_{0.85}$	T_{coll}	$T_{s\&g}$
1	74	86	97	68
2	79	104	118	95
3	260	322	374	373

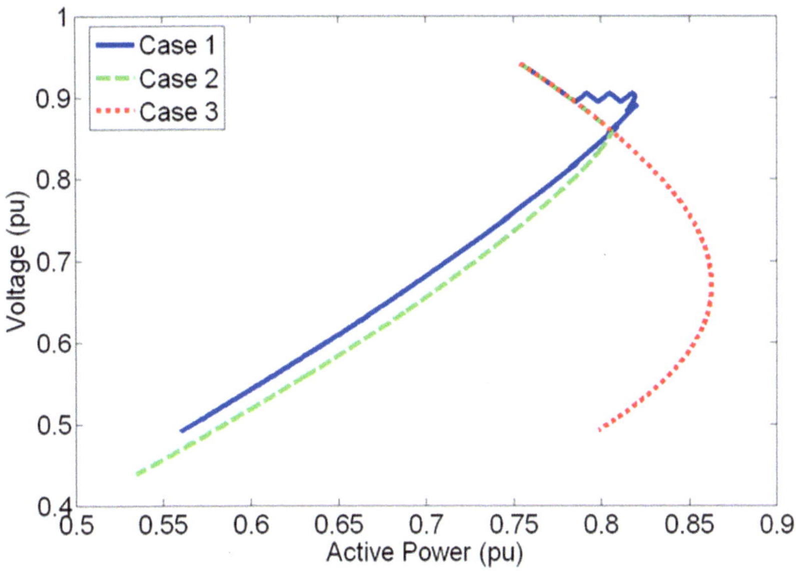

Fig. 8.42 Nose curves at Bus#5 for Condition 2 with the load represented as a generic dynamic model

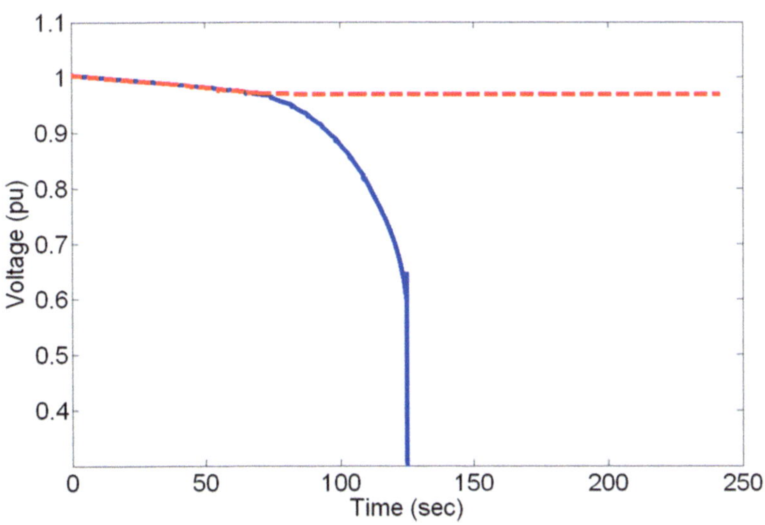

Fig. 8.43 Voltages at Bus#3 when stopping the load increase at 67 s (*dashed*) and at 68 s (solid)

Fig. 8.44 380-kV Italian network

static model and 50 % dynamic model with a power-restoring time constant of 5 s. The considered Italian system is under primary voltage and frequency control only and its initial condition is very stable.

Two set of tests were performed, one at the Milano area (region marked in the north part of Fig. 8.44 and enlarged in Fig. 8.45), and another at the Firenze area (region in the centre marked in Fig. 8.44 and enlarged in Fig. 8.46). The analysis performed in the Milano area consisted of increasing the local area by a rate of

Fig. 8.45 Enlarged detail of the area in the north part, marked in Fig. 8.44

10 %/min, maintaining constant the power factor. This rate is justified because the starting operating point is very far from instability. In order to enforce a voltage instability to a given bus, which allowed the opportunity to monitor the critical bus, we increased the load at Brugherio 380-kV bus at a rate of 20 %/min.

The load at Brugherio is fed through three 380-kV/132-kV OLTC transformers. Figure 8.47 shows the nose curves of the Brugherio 380-kV bus in four cases described as follows:

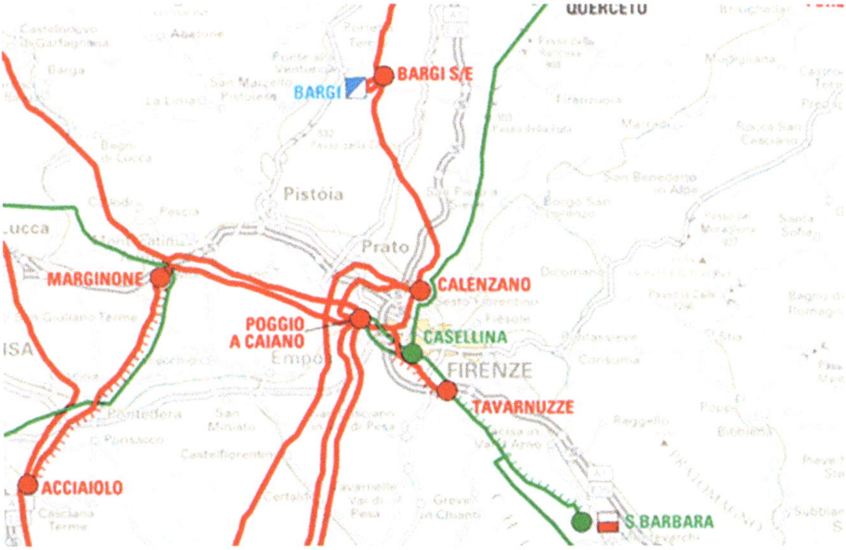

Fig. 8.46 Enlarged detail of the area in the centre part, marked in Fig. 8.44

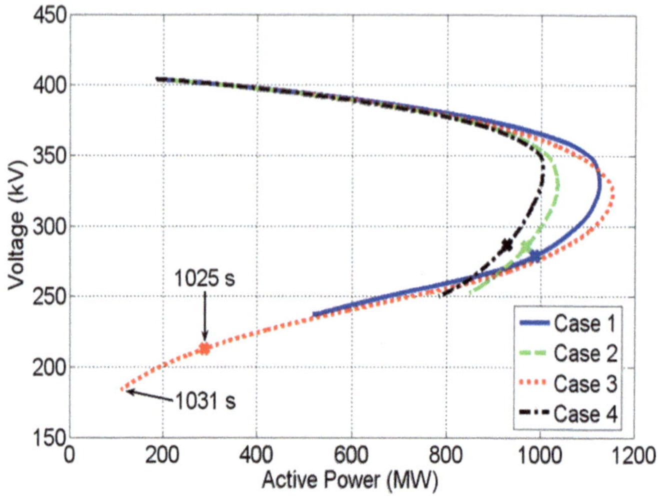

Fig. 8.47 Nose curves at Brugherio 380-kV bus

- Case 1: all OELs in service and only Brugherio OLTCs in service;
- Case 2: all OELs in service and all OLTCs out of service;
- Case 3: all OELs and OLTCs out of service;
- Case 4: all OELs and OLTCs in service.

Case 1 better fits the equivalent dynamic model with OLTC as given in Fig. 8.22. Case 4 represents a more realistic situation, to which Case 1 must be compared. Case 2 is another realistic situation, with all OLTCs locked. Case 3, as before, represents the reference, due to its close approximation to the static *V-P* curve.

Figure 8.48 shows the tap positions of the three Brugherio OLTC transformers for Case 1. It can be seen that a continuous tap variation model is employed in the simulation. OLTCs cease to operate when their tap position reaches the nondimensional value of 0.8 (lower tap limit). Figure 8.49 shows the Case 4 tap position of the three OLTC transformers in Brugherio (the three heavier solid lines) and the other OLTCs working in the same grid area (lighter solid lines). It can be seen the Brugherio OLTCs reach their lower tap limits before the others, due to the load increase profile, which enforced a higher increase rate at the Brugherio bus.

Table 8.8 shows the times with respect to the simulation results presented in Fig. 8.47 for the Brugherio 380-kV bus. $T_{0.85}$ was defined for a base voltage of 400 kV, i.e., the time when voltage reached 340 kV.

Figure 8.50 shows OEL indicators for Case 1 of some of the large generation groups electrically close to the Brugherio 380-kV bus. The closest groups are Tavazzano1, Tavazzano2 and Turbino, shown in heavy solid lines. When this indicator reaches zero the machine over-excitation limiter begins to work in closed loop. Three other groups (Vado, Spezia and La Casella), although electrically farther from Brugherio, reach their over-excitation limits before Tavazzano and Turbigo, due to their higher initial excitation condition.

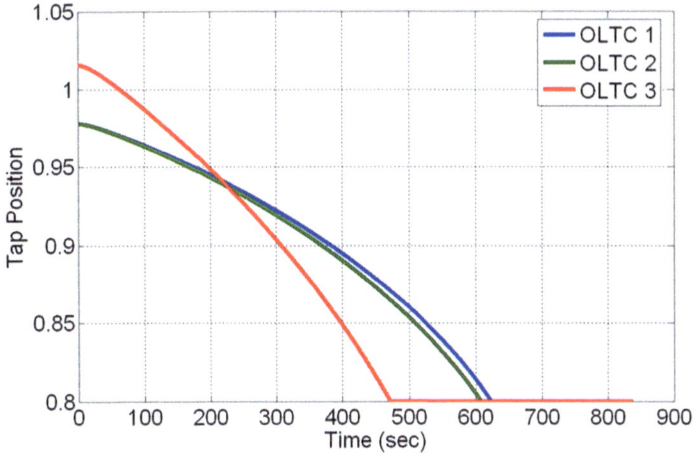

Fig. 8.48 Brugherio tap evolution corresponding to Case 1

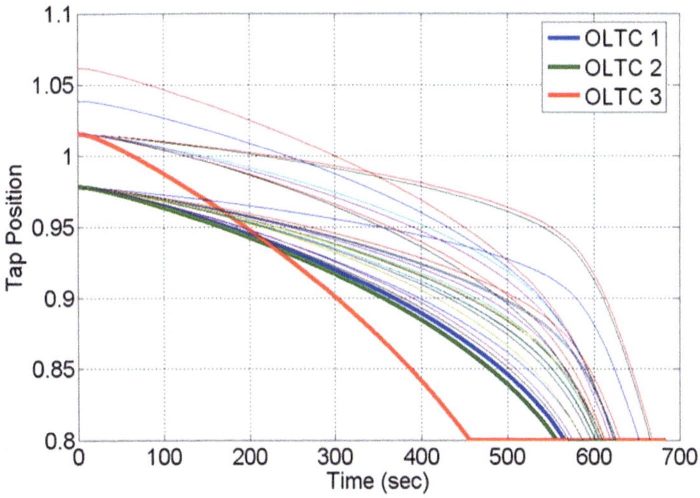

Fig. 8.49 Brugherio tap evolution corresponding to Case 4

Table 8.8 Key recorded time (s) with respect to Fig. 8.47

Case	T_{max}	$T_{0.85}$	T_{coll}	$T_{s\&g}$
1	715	686	838	805
2	740	710	845	815
3	881	800	1031	1025
4	585	587	684	660

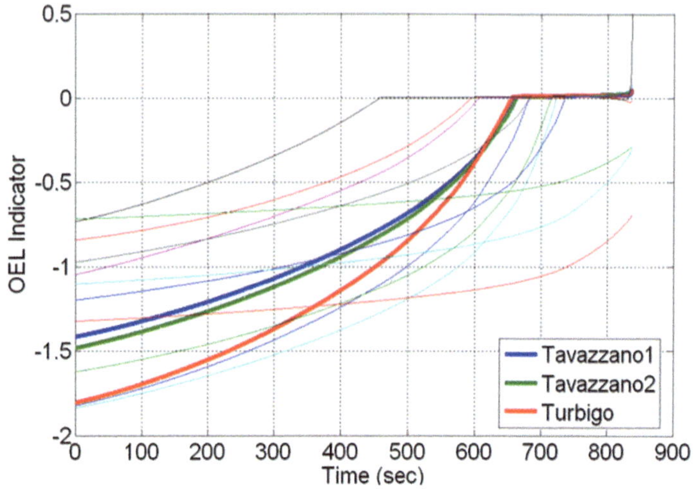

Fig. 8.50 OEL Indicators corresponding to Case 1

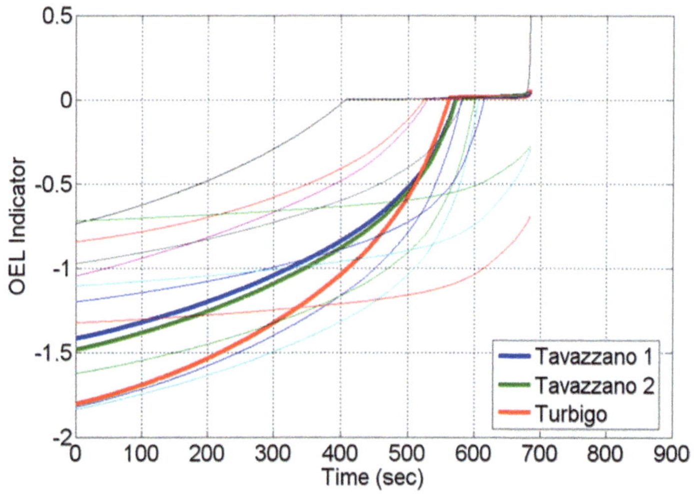

Fig. 8.51 OEL Indicators corresponding to Case 4

Figure 8.51 shows OEL indicators for Case 4. One can see that when all OLTCs are unlocked (Case 4), Tavazzano and Turbigo's OELs start operating 100 s earlier with respect to their operation in Case 1, where only Brugherio's OLTCs are unlocked.

The comparison of the results presented in (§ 8.3.1.4 (V-P curve analysis for a more realistic generic load representation)) with those in (§ 8.3.1.5 (V-P curve analysis for the Italian system)) is of real interest, mainly concerned with the general trends of the nose curves and with the voltage instability point along them. Due to the natural difficulties of comparing the operating conditions of the two systems, the analyses present relevant numerical differences, which will be clarified later. Starting with Case 3, both show instability at low voltage values, particularly in the Italian system, and relevant delay with respect to the other more realistic cases.

For the two Case 1 examples, both nose tips are reached at higher voltage values than for the corresponding Case 3 tips. The main difference is between the two points corresponding to $T_{s\&g}$—the one in Fig. 8.40 is located on the nose tip whereas the one in Fig. 8.47 is at a low voltage, below the nose tip.

The continuous integrator utilised in the Italian system OLTCs does not give rise to the sawtoothed shape seen in Fig. 8.40, where a discrete stepping control law is used in the OLTC of the equivalent model. Comparing Case 2, the nose shape is smoother in Fig. 8.47 due to the different time the many generators in the Milano area reach their OEL reclosure (in contrast to the single generator in Fig. 8.40). This is also the reason the time span between the OEL operation in Tavazzano and Turbigo (Fig. 8.51, Case 4) and the $T_{s\&g}$ value (660 s in Table 8.8) is not small (approximately 100 s).

For the two considered Case 1 examples, the large system shows instability point at low voltage, whereas the small system shows this point at high voltage. The main reason for this difference is the local OLTCs, in the large system, have reached their saturation before the local OELs begin to operate. Therefore, the lack of simultaneous operation of OELs and OLTCs is the main reason for the noticed delay and lower voltages in Fig. 8.47.

Case 4 is the most realistic of the four cases, comparable only with the other curves in Fig. 8.47. It clearly shows that the combined action of the OELs and OLTCs makes the nose shorter with respect to Case 2 and anticipates the time of instability. From the comparison with Case 3, representing the static *V-P* curve, the difference in terms of loadability and voltage instability timing are very large.

The load difference at the tip between Case 3 and Case 2 is of about 10% in Fig. 8.47 and 7% in Fig. 8.40. Moreover, Table 8.8 shows large difference in time between Case 3 and the other cases, again confirming the inadequacy of the *V-P* static curve to correctly describe the voltage-lowering and voltage instability.

Figure 8.52 shows the nose curves of the Poggio a Caiano 380-kV bus (Tuscany) in four cases, described as follows:

- Case 1: all OELs in service and only the Poggio a Caiano OLTC in service;
- Case 2: all OELs in service and all OLTCs out of service;
- Case 3: all OELs and OLTCs out of service;
- Case 4: all OELs and OLTCs in service.

The proximity of the curves for Cases 1 and 2 differs from the Fig. 8.47 results. This is explained by the fact that at Poggio a Caiano 380-kV bus a large part of the local load is fed directly at 380 kV. Only a small part of the local load (25%) is fed through OLTC transformers at the Casellina 220-kV bus (see Fig. 8.46). Accordingly, the Case 1 nose in Fig. 8.47 is closer to Case 3 nose, whereas in Fig. 8.52 it is closer to Case 2 nose.

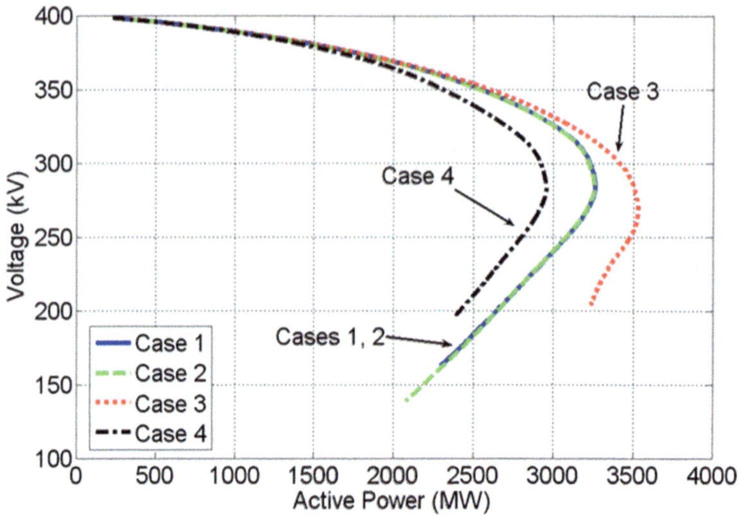

Fig. 8.52 Nose curves for Poggio a Caiano 380-kV bus, Italy

Table 8.9 Key recorded time (s) with respect to Fig. 8.52

Case	T_{max}	$T_{0.85}$	T_{coll}
1	1279	933	1329
2	1287	937	1338
3	1458	966	1602
4	1160	838	1206

Again, Fig. 8.52 and Table 8.9 confirm the relevant differences in maximum loadability and in time to collapse between the static and dynamic *V-P* curves.

In conclusion, evidence is given to the ability of the proposed equivalent dynamic model to represent the real system process of voltage degradation and voltage collapse, even if Case 4 leaves room for possible improvement (see next paragraph).

It is worth noting that in the tests shown, made in the Italian network (starting from a very stable initial condition), maximum load is generally reached at a voltage profile lower than 0.85 p.u. after a heavy load increase. Thus, before reaching the maximum loadability point, a protective relaying not represented in the simulations will actuate. If secondary voltage control [9, 10] were present, the nose curves shown in Fig. 8.47 and Fig. 8.52 would obviously have a flatter shape on the upper part (see § 8.3.1.7-(V-P curve in presence of grid automatic voltage regulation). Under different and more critical operating conditions of the Italian power system, voltage instability can result at higher voltages.

Understanding and Modeling Voltage Instability

We needed the *V-P* curve analysis above to understand what is primarily responsible for the voltage instability phenomenon. We analysed the voltage stability problem by showing the strong impact of the generator OELs and transformer OLTCs on the shape of the *V-P* curve and on the speed at which the equilibrium point runs along the curve. The results showed that the Thevenin equivalent of the grid seen by a given EHV bus cannot be represented simply by a constant-voltage generator feeding the load by a constant reactance. The relevant differences between the static and dynamic equivalent models were made evident vis-à-vis maximum loadability, running speed along the *V-P* curve and instability of a portion of the equilibrium points. A large grid described by a detailed dynamic model confirmed the results of the proposed dynamic equivalent model. This proposed model is the minimum necessary but satisfactory representation of the real power system through which incoming voltage instability is recognised with high reliability as seen by the considered load bus. Again, the most important dynamic aspects linked to voltage instability are equivalent generator OEL and transformer OLTC at the load bus.

The OLTC representation of the remaining power system should require a more complex equivalent model in terms of structure and control parameters, not easily defined by simple equivalent Thevenin's impedance. In fact, the other OLTCs not too far from the load bus impact its nose curve by lowering the maximum loadability and anticipating the instability point. These effects can be reproduced through the use of an equivalent dynamic model by adding, along equivalent transmission lines in Fig. 8.22, a new load bus electrically close to the considered one, having a local equivalent OLTC. Obviously, the increased number of parameters of this more complex equivalent model progressively changes values with the added equivalent OLTC and load mostly influenced by the surrounding network characteristics of the load bus being analysed.

We should point out that in the case of real-time identification of Thevenin equivalent impedance, an OLTC representation of the remaining system is not required, as impedance dynamic identification already includes the effects of system OLTC tap changes. The relevant differences in terms of loadability and time to instability between a static and a dynamic model confirm that only the dynamic model can shows the correct nose shape and instability point. In fact, starting from a common equilibrium point for both static and dynamic models, the process degradation due to load increase up to the maximum power transfer into the considered bus will show two different *V-P* curves. These curves present large differences in time for operating point to run along them, and therefore on the amount of load increase required to reach their respective maximum loadability values.

We confirm that the tips of the nose curves do not correspond to the voltage instability point, even in the case of a P-constant load, when system dynamics are considered, in which case, instability begins a short time before the occurrence of the tip. In the other cases, instability normally occurs after the nose tip, going further along the curve as high as the Z-constant portion. Accordingly, any voltage instability indicator based on a maximum loadability finding would appear unrigorous, mainly in the presence of a real load.

OEL operation is extremely relevant to the triggering of voltage instability under certain operating conditions. Nevertheless, the span of time between local OEL operation and voltage instability could be great. Therefore, to be effective, real-time information about OEL operation should cover a large part of the grid around the considered bus. In any event, the availability of such a large amount of data (still very difficult to access in practice for real-time use) makes the problem of deciding the correct time at which to consider OEL data critical for voltage instability.

The analysis performed also shows the high speed of voltage-lowering after the nose tip, before the onset of real instability. In practice, this still stable but fast-changing operating condition appears to be of concrete interest, more than the first unstable point at very low voltage. Therefore, the start of fast voltage-lowering is the point we should recognise to counteract the instability phenomenon in a timely fashion, and therefor increase system security and reliability. Accordingly, the maximum loadability point really represents the "gate" after which fast voltage-lowering becomes very critical, carrying a risk of protection intervention and system security degradation.

It is worth noting that a dynamic representation of the load (one determining a partial power self-restoration with a time constant of about 5 s) does not qualitatively change the major results presented here. Since all simulations are based on a gradual and continuous increase of load admittance, the slow power recovery taken into account plays, in our experience, a secondary role on the values and stability of *V-P* curve equilibrium points: the primary role is played by voltage control and excitation limiters.

About the load characteristic impact on loadability and voltage instability point: we are driven to think about the usefulness of *V-P* curve analysis on voltage instability when considering loads as P-constant, I-constant and Z-constant (see previous results). Therefore, it is natural we should wonder whether these analyses are simply an academic exercise, finalised toward a better understanding of mixed ZIP load cases, or if conversely they have a real meaning and interest in the practical sense. For example, the case of the P-constant is very far from the correct result, notwithstanding how often it is used in literature! It becomes realistic only when the Thevenin equivalent is redefined at each step by transferring the Z-constant and I-constant portions of the actual load to inside the equivalent Thevenin impedance [20]. In doing so, the new *V-P* curve maximum loadability point should correspond in terms of time and voltage value to the correct voltage instability point, determined by the load-independent Thevenin equivalent.

To indicate voltage instability proximity based on local phasor measurements, the identification of the two parameters of the simple equivalent circuit in Fig. 8.18 should require a very high-speed computation, according to the results shown. In fact, especially during the fast voltage degradation period, determined by the combined actions of the OELs and OLTCs, *V-P* curve equilibrium points quickly change value and must be adequately tracked. This implies that neither off-line parameter identification methods related to a specific normal operating condition nor those which are on-line but slow to identify and update parameters are appropriate for indicating precise, on-time voltage instability risk.

Even though in "highly specialised load-flow programs" the static model can detect voltage instability (building "static equilibria" correspondent to the equilibria of the above underlined dynamic model [12]), perfectly achieving such correspondence is not a trivial task. This concern is exemplified in OLTC dynamics, which should be modelled as a set of algebraic equations governed by time delays and voltage thresholds and with the restriction that the tap is changed only between subsequent positions. Another example of this concern is the OEL effect not precisely described by changing a generator bus from a *PV* to a *PQ* type. The OEL modifies the reactive power during its operation under a load increase; being an active closed-loop control, it has an impact on stability characteristics of the *V-P* curve equilibrium points.

Up until now, the dynamic analysis performed was limited to process component control loops like those of generator OELs and transformer OLTCs.

Other dynamic effects with an impact on voltage instability are those provided by grid automatic voltage regulations such as secondary voltage regulation or SVCs, STATCOMs and UPFCs. This grid voltage continuous controls contribute to the changing shape of the *V-P* curves, delaying a rise in instability, as is shown next.

V-P Curve in Presence of Grid Automatic Voltage Regulation

Grid automatic voltage regulations were introduced in Chaps. 3 and 4. They continuously control HV bus voltage by relatively fast loops provided by local compensating equipment like SVC, STATCOM, UPFC, or by remote control of the generator reactive power through a hierarchical control system integrating SVR and TVR.

Any automatic grid voltage regulation determines continuous voltage support to the buses controlled. This support persists until the full use of available control resources (reactive power) is allowed. Therefore, as power system loads grow, regulated voltages (under a hypothesis of integral control law) remain unchanged at their set-point values, at steady state. Understanding this, up to the point when voltage controllability is guaranteed, is very easy. This also means the shape of *V-P* curves of regulated voltages, and therefore process loadability, differs from previously examined cases that grid voltage controls.

To best characterise the effects of automatic pilot node voltage regulations (SVR) on *V-P* curves (hence on power system voltage stability), Thevenin's equivalent scheme must be slightly modified by distinguishing the generators that contribute to the voltage regulation of the considered load bus (pilot node) from the remaining part of the power system outside this task, which is simply represented by a link to an "infinite bus" (see Fig. 8.53).

In fact, as already seen, SVR is a decentralised voltage control (as any grid voltage regulation should be) having only a few pilot nodes, each independently regulated by the generators operating inside their own grid areas. Accordingly, the tests

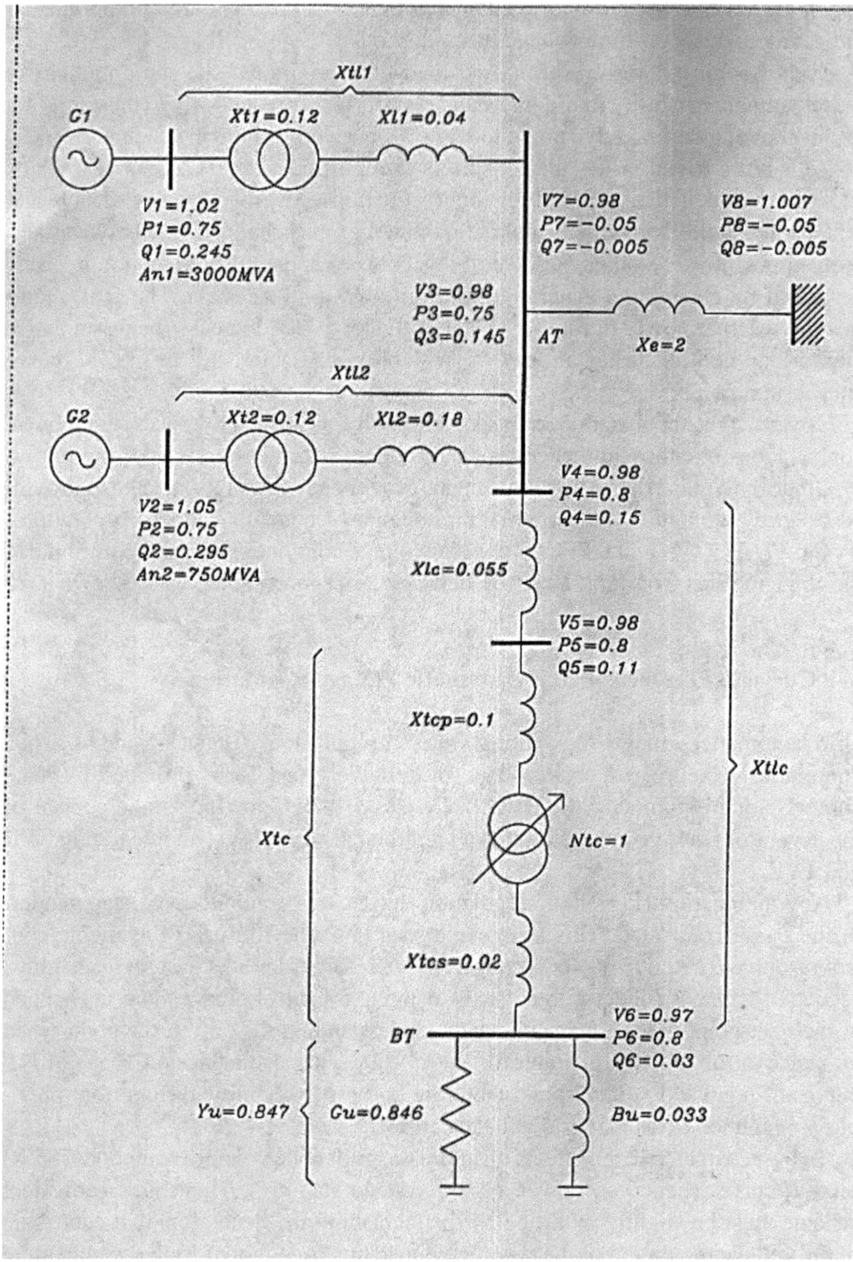

Fig. 8.53 One-line diagram of the equivalent system to be used for tests on voltage stability in the presence of SVR

of further interest for completing our voltage stability analysis are those carried out in the presence of automatic grid voltage regulation at a selected equivalent SVR area. Figure 8.53 is the minimal equivalent scheme appropriate to this check.

By substituting SVR control generators in the proposed equivalent scheme (Fig. 8.53) with compensating equipment provided with a local automatic voltage control loop (SC, SVC, STATCOM, etc.), we note that the performance results achievable are very similar to those under SVR, as is obvious. They differ simply in control saturation limit, depending on the control reactive power that each may contribute.

Test System

To demonstrate how system loadability and voltage stability is influenced by SVR control impact, a proper test system is proposed, taking into account the following:

- Primary voltage regulation;
- Generator over-excitation limits
- Transformer on-load tap changer;
- Secondary voltage regulation.

The test system used is shown in Fig. 8.53, where the HV bus (V7, V3, V4) is the unique pilot node of the SVR area with control generators G1 and G2 and equivalent area load Y_u. The infinite bus represents the remaining part of the power system. In the equivalent power system mathematical model, each equivalent generator is described by a detailed sixth order model.

Unit transformers are simply modelled by their leakage reactance. The equivalent load supplied by the OLTC is assumed linear and is modelled by an equivalent admittance Y_u. Generator voltages are controlled by the AVR including additional stabilising feedback. AVR also includes the over-excitation limit for restraining the field current below its maximum value. The SVR control scheme is the same already presented in Chap. 4, which includes the generator reactive power control loop.

Load voltage is controlled through the OLTC transformation ratio changes (N_{tc}=HV/LV). An OLTC integral-continuous control scheme has been used rather than a discrete-time scheme, being an equivalent one representing a number of OLTCs operating in the considered area.

The equivalent system initial condition has been chosen to represent a proximity to the system voltage stability limit.

Dynamic Analysis

The dynamic model used allows the analysis of the curves corresponding both to steady state equilibrium points while system parameters change and system transients following perturbations.

The static figures shown refer to the pilot node voltage while changing:

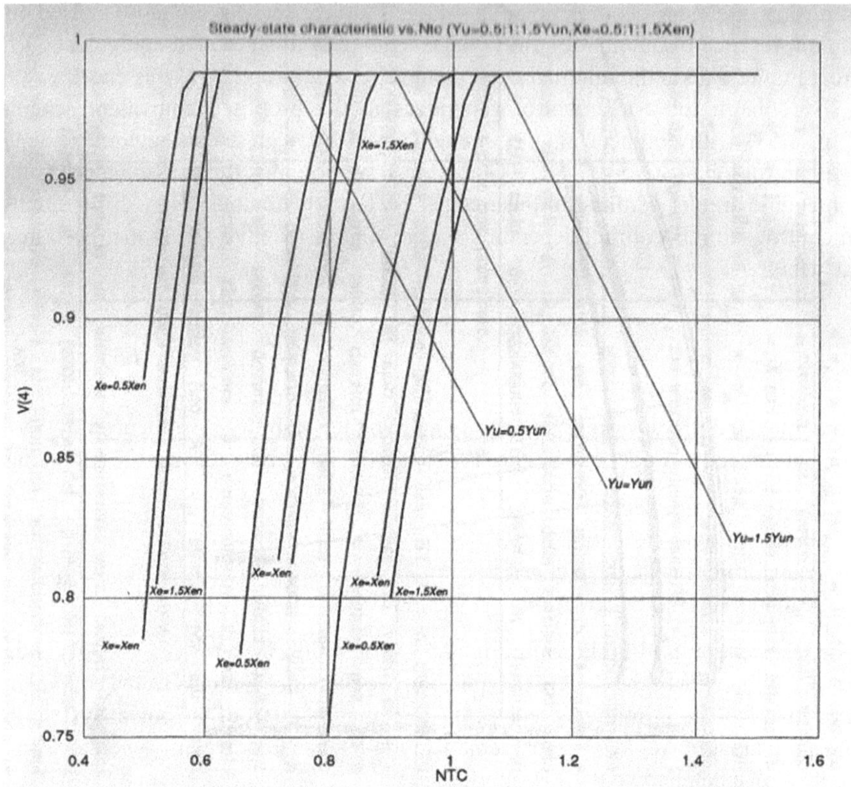

Fig. 8.54 Static characteristics of the pilot node voltage with respect to OLTC transformation ratio N_{tc} in the presence of SVR and with different X_e values

- N_{tc}: $0.5N_{tcn} \leq N_{tc} \leq 1.5N_{tcn}$ ($N_{tcn} = 1$ p.u.);
- Y_u: $0.5Y_{un} \leq Y_u \leq 1.5Y_{un}$ ($Y_{un} = 0.847$ p.u.);
- X_e in the field: $0.5X_{en} \leq X_e \leq 1.5X_{en}$ ($X_{en} = 2$ p.u.).

Figure 8.54 makes evident the effect of N_{tc} manual change when AVR, SVR and OEL operate in a closed loop for three different Y_u values. We notice the pilot node voltage is maintained constant at set-point value until the load increase seen by the generators and SVR, due to the transformation ratio reduction, determines SVR saturation at the beginning of OEL operation. The higher are the loads and the electrical distance from the infinite bus the more anticipated is the voltage drop.

Without SVR, the voltage trend differs from Fig. 8.54, becoming progressively lower by following the N_{tc} reduction as determined by the OLTC operation with the aim of maintaining the voltage value (V(6)) unchanged at load side.

Figure. 8.55 shows pilot node voltage versus load active power increase, when AVR, SVR and OEL operate in closed loop. This represents the classic *V-P* curve having a shape to be compared with those analysed previously without SVR. It is

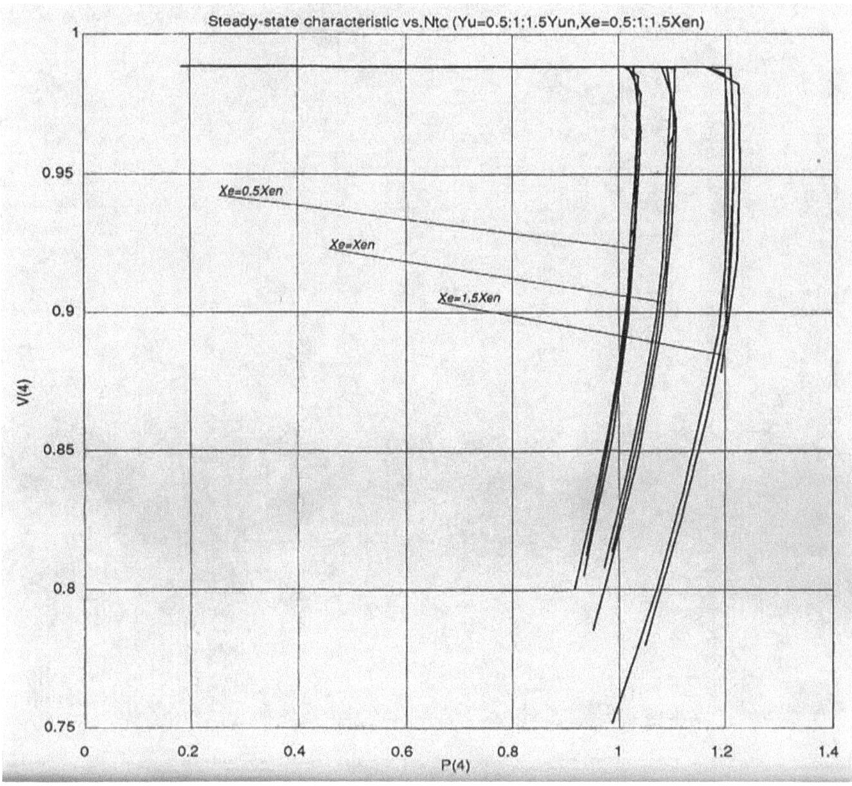

Fig. 8.55 Static characteristics of pilot node voltage with respect to generator OEL operation in the presence of SVR

evident that the nose tip is reached at the time the OELs begin to operate. Before that time, the load transformer OLTC does not need to change N_{tc} due to the constant voltage at the pilot node.

Voltage anticipates its decline while increasing electrical distance from the infinite bus due to the OEL operation. This figure confirms the strong impact on the *V-P* curves by the grid voltage controls that significantly modify the *V-P* curve shape.

Moving from system static curves to transient following load perturbation, from here the voltage instability phenomenon is represented by its traces following a load increase: $\Delta Y_u = 50\,\%$ at 100 s. In this case, the load step is too high for the SVR that suddenly is saturated, losing any opportunity to hinder voltage-lowering. Usually, reaching generator voltage values below 0.85 %, voltage protection commands the unit trip. In the transients shown, this protection is not simulated, therefore the generators continue to stay linked to the grid during voltage collapse, but lose synchronism, as evidenced by the angle oscillations in the last portion of the transients (Fig. 8.56).

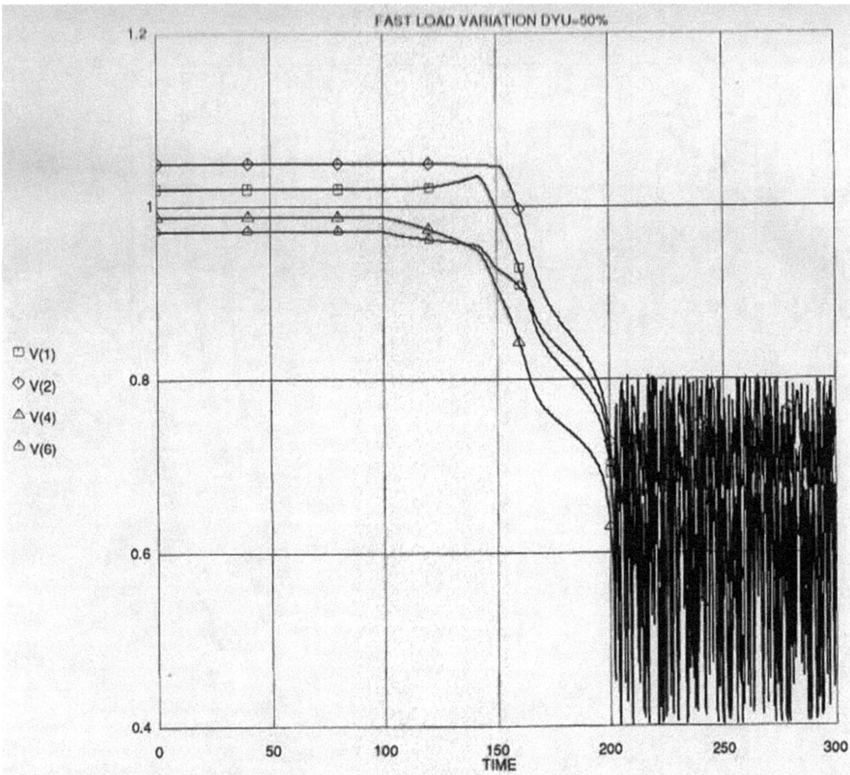

Fig. 8.56 System voltage transients following a sudden load increase in the presence of AVR, SVR, OEL and OLTC closed-loop controls

8.3.2 Distinguishing Voltage Instability from Voltage Collapse

Until now, voltage stability analysis and performance results have allowed us to state some basic and fundamental considerations [25], most of them previously introduced (§ 8.3.1.6: Understanding and modeling.......). Here it is useful to remember them:

- Thevenin's equivalent requires a detailed dynamic model to correctly reconstruct voltage instability. Therefore, its equivalent parameter (Fig. 8.18) values continuously and quickly change (due to the operating controls) while approaching maximum loadability.
- Local area generator OEL and transformer OLTC dynamics strongly impact bus maximum loadability and voltage instability [8].
- P-constant load analysis is not adequate for a correct identification in real systems, where a mixed ZIP load is usually adopted.
- The maximum loadability point (*V-P* nose curve) differs from the real instability point, which in general occurs a short while later, at a lower voltage.

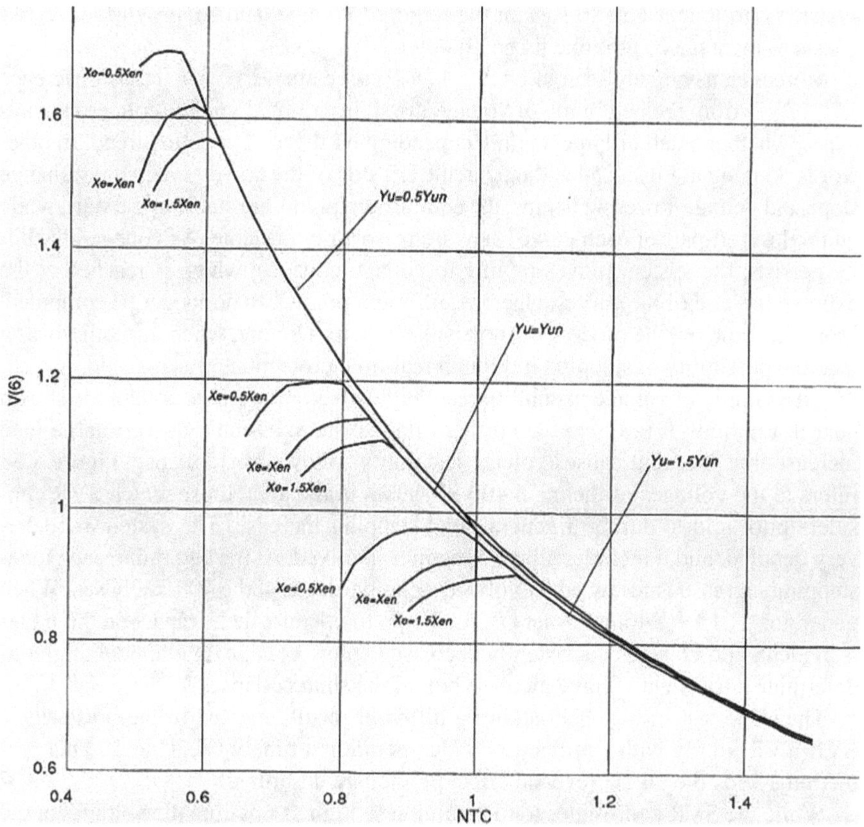

Fig. 8.57 Static characteristics of the load bus, low voltage side, with respect to OLTC transformation ratio N_{tc}, in the presence of SVR

- The last basic consideration refers to the high speed the *V-P* curve equilibrium point moves from the curve's nose tip to the first unstable point at lower voltage.

All these facts are relevant in practice for identifying the nose tip and preventing against voltage instability. In particular the need for a very fast identification procedure when the voltage instability limit is approached is an essential requisite.

Another relevant point is the difference that exists between the beginning of voltage instability that can be stopped with adequate very fast control/protection and irreversible and subsequent voltage collapse: at that time any further control cannot avoid local or wide area blackout.

To justify this statement we first refer to Fig. 8.54, representing system equilibrium points at the load transformer, HV side. Reducing N_{tc}, pilot node voltage begins decreasing until computing cannot discover further equilibrium points. On the LV side, voltage V_6 grows while reducing N_{tc} until both V_6 and N_{tc} go down in parallel before reaching points where the computing does not find more solutions (see Fig. 8.57).

Notwithstanding the SVR strong voltage support, transformer reversal action with respect to N_{tc} reduction begins at a given time. This is because the power system is no longer able to sustain the required voltage, and so any further N_{tc} requests to increase V_6 provoke its going down.

Moreover, as already seen in § 8.3.1.4 (V-P curve analysis ......), a time difference ($T_{coll}-T_{s\&g}$) from the beginning of voltage instability to final voltage collapse always exists, whether small or large, mainly depending on the load variation trend. In other words, looking at Figs. 8.54 and 8.57 at the left side of the curves, where they change slope and voltage-lowering begins, the equilibrium points are stable at the start, while on the last left part of each curve, they are reasonably unstable. As voltage instability persists, the system moves rapidly to voltage collapse, which is reached at the extreme left end of the curves, where equilibrium points can no longer be computed. From this time on, the process is irreversible, whereas before, when instability was a fact, the possibility of stopping it through real-time protections was realistic.

An example of voltage instability recovery that avoids voltage collapse is shown next in Fig 8.59. It refers to the northern Italy system, assuming a very large load increase in a way that causes voltage instability followed by collapse. Figure 8.58 refers to the voltages of the main 400-kV buses in the area, those which SVR considers pilot nodes, during a general load-stepping increase. The system model is very detailed, and it includes all the dynamics involved. At the beginning each load-stepping increase is followed by voltage stepping-down at the 400-kV buses. When generator OELs and transformer OLTCs begin to operate, their impact on the traces is evident, and voltages consistently decrease. At low voltage, load model upgrades determine a transient voltage increase before the final collapse.

The same test in Fig. 8.59 shows a different result, enacted in the presence of SVR-TVR jointly with a protection logic operating a timely OLTC lock. This protection avoids the OLTC reversal effect previously described.

While the SVR contributes to maintaining as high as possible the voltages at the beginning, its anticipated saturation at about 200 s leaves to the protection the over-

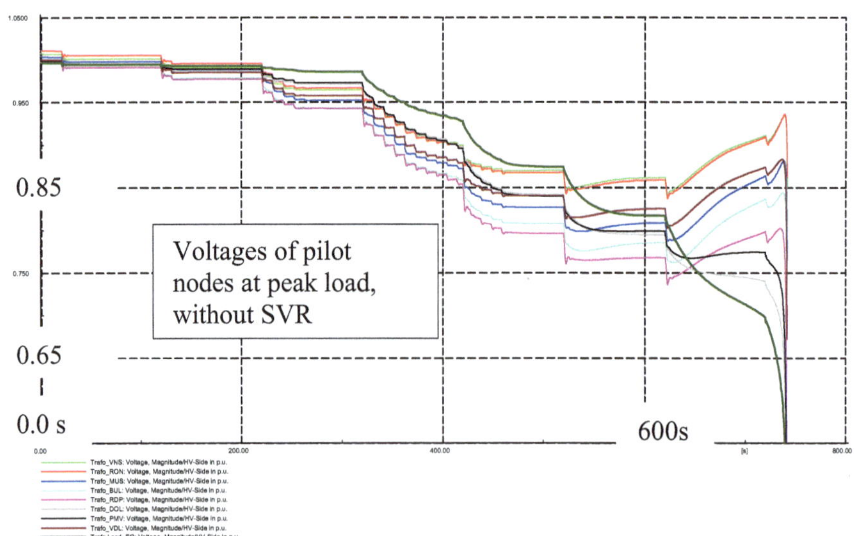

Fig. 8.58 Northern Italy: load increase, primary voltage regulation only, no protection, voltage instability up to collapse

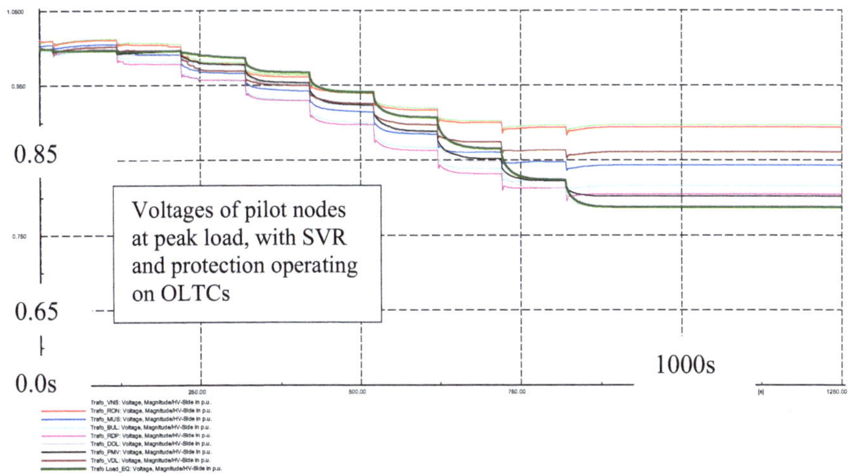

Fig. 8.59 Northern Italy: load increase, SVR-TVR-OLTC lock based on real-time VI indicator; no protection

all onus to stabilise the process. A real-time voltage stability indicator (described in Chap. 9) helps the protection to timely locking the OLTC transformation ratio. The satisfactory result is shown in Fig. 8.59 where, it has to be noticed, the load still increases up to 800 s without determining the voltage collapse as conversely happens in Fig. 8.58 at 740 s.

The voltage collapse can be avoided thanks to the timely identification of the incoming voltage instability and the correspondent protection controls on the OLTC lock/unlock. This control, to be effective, must operate inside the interval (Tcoll-Ts&g) otherwise, after the collapse beginning, it becomes ineffectual.

Further Examples of Voltage Instability and Collapse

Case of South Africa 2008 Winter Peak: "contemporarily increasing of system loads"

Figure 8.60 shows the voltage transients of few important transmission buses under primary voltage regulation (PVR) alone. That is, when operating in closed loop the generators' AVR, OELs and the transformers' OLTCs only. The transients determined by the load ramping increase show the voltage-lowering while the OELs and the OLTCs are active. The instability comes out at about 90 s and the voltages speedily goes down up to reach the collapse condition where the curves computing does not provide further equilibrium points.

Under SVR (Fig. 8.61), the network voltage instability becomes evident when the total loads (active and reactive power) increase reaches +8.5 % at 105s; instead operating PVR by alone the instability raises when the load increase reaches +6.5 %. Considering that the total load is of about 34,500 MW, the 2 % difference corresponds to about 700 MW.

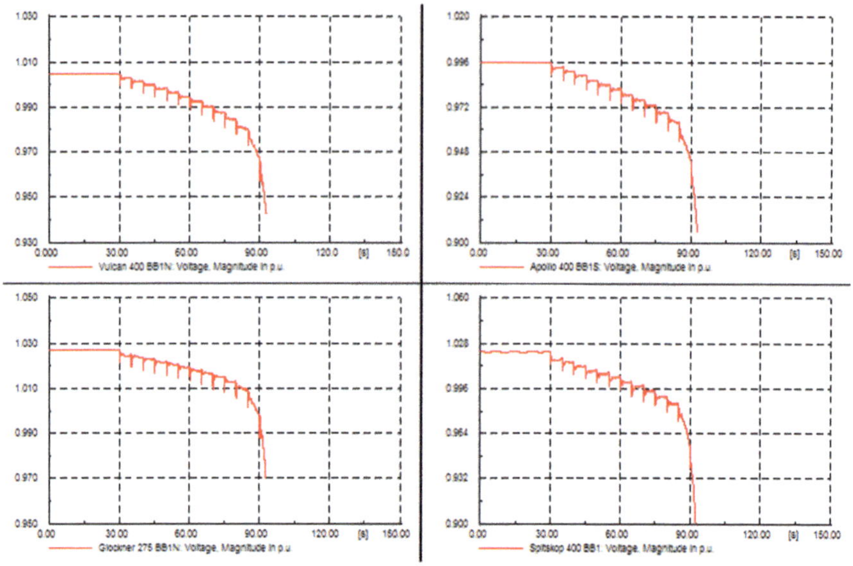

Fig. 8.60 South African transmission grid voltages transients following a progressive load increase in the presence of AVR (primary voltage regulation), OEL and OLTC closed-loop controls

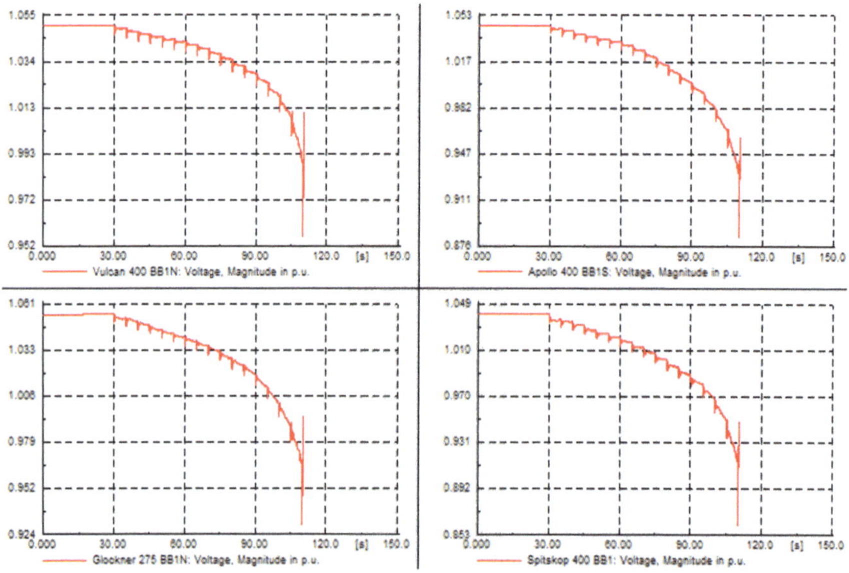

Fig. 8.61 South African transmission grid voltages following a progressive load increase in the presence of AVR and SVR (primary and secondary voltage regulations), OEL and OLTC closed-loop controls

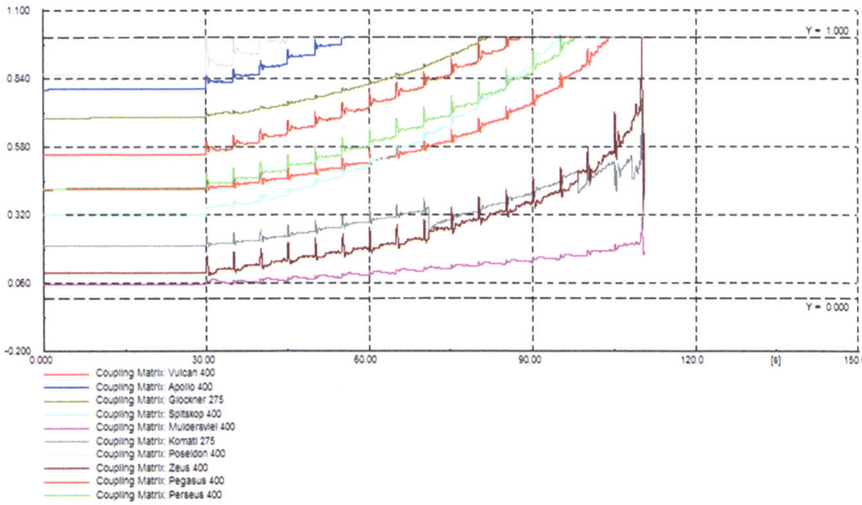

Fig. 8.62 South African transmission grid: transients of SVR area "reactive power levels" following a progressive load increase in the presence of AVR, SVR, OEL and OLTC closed-loop controls

The traces of the SVR pilot nodes (the same busses considered under PVR alone) again show the operation of the generators OELs and transformers OLTCs while increasing (sequence of small steps) the load up to the collapse after 100 s, with small differences among the SVR pilot nodes. Under SVR, the losses increase at 110 s is of about 300 MW (150 MW at 93 s). With PVR alone, the voltage collapse appears at 93 s when the losses increase is already greater than 260 MW.

The traces of the ten area reactive power levels in Fig. 8.62, show the SVR operation while the load increases in the power system, progressively growing up to saturation when reaching the generators' OELs limits in each different SVR area. Some areas begin to saturate before 60 s, most before 90 s and the last at about 100s when the voltage collapse is achieved.

This is a further clear example of the strong link the voltage instability has with the OELs and the OLTCs operation.

The very interesting result in Fig. 8.62 clearly shows the approaching SVR saturation (all area reactive power levels reach their 100 % value) before voltage instability start-up and final black-out. The same results are seen in Fig. 8.63, where collapse (Fig. 8.63a) delayed about 30 s by the SVR (Fig. 8.63b) again demonstrates the SVR stabilising effect.

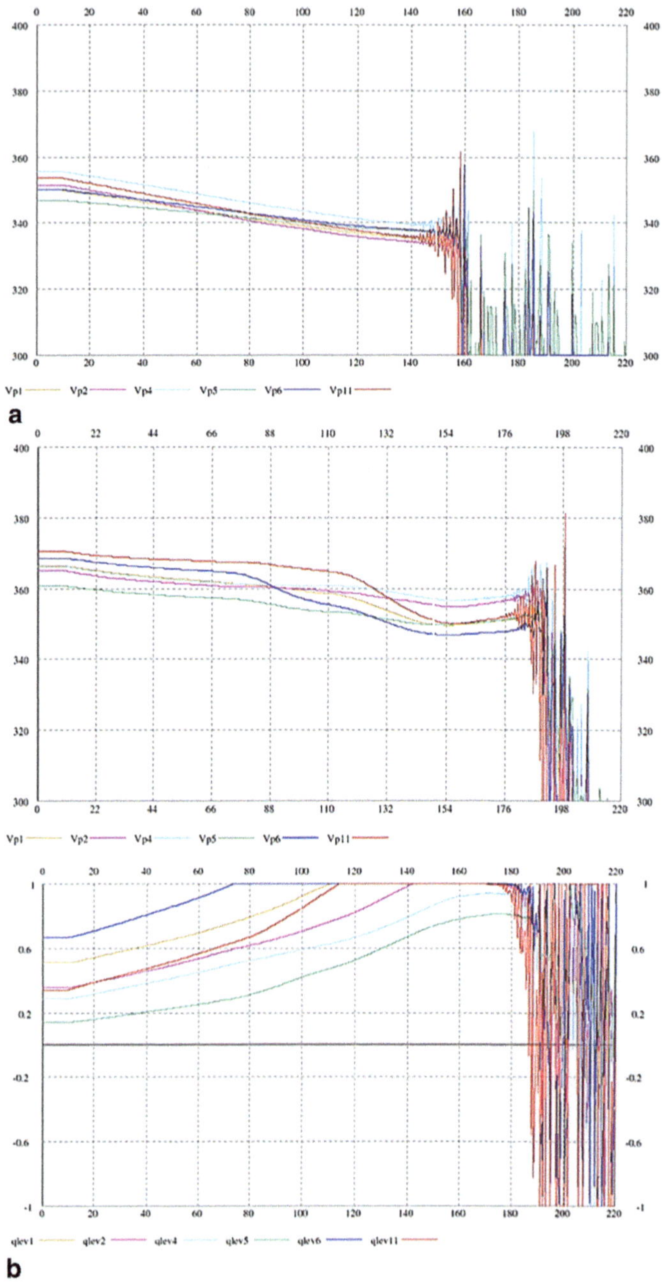

Fig. 8.63 **a** Voltage collapse in a real power system: transients of HV buses without SVR following a progressive load increase in presence of AVR, OEL and OLTC closed-loop controls. **b** Voltage collapse in a real power system: transients of HV buses with SVR (*top*); SVR "reactive power levels" (*bottom*), following a progressive load increase in the presence of AVR, OEL and OLTC closed-loop controls

8.3.3 Voltage Instability and Bifurcation Analysis

Premise Before closing this chapter on voltage instability, the subject of *saddle-node* and *Hopf bifurcation* analysis, usually recalled upon reconstruction and interpretation of voltage collapse phenomena, cannot be circumvented. This is for completeness, but it is also to provide as simple an interpretation as possible of a particular, frequently mentioned voltage stability methodology in literature.

As will be summarised here, saddle-node and Hopf bifurcation theories analyse the trajectories of a generic nonlinear dynamic process around equilibrium states and in front of parametric perturbations. Such methods recognise the process stability/instability around a selected operating point by following criteria based on the Lyapunov's stability theory.

In principle, these investigation methods add nothing to the description of voltage instability and collapse phenomena as already introduced; nevertheless, they are of interest as a different analytical approach, one that is used when we want to refer to the overall power system with a rigorous detailed model. The methodology avoids, in the first stage, the shortcut of equivalents to simplify power system nonlinearities and model complexity.

Obviously, through this approach (i.e., without simplifications) system model complexity is consistent with many consequent computational problems. Therefore, this method provides a possible way around stability analysis that can help us overcome such computational issues and achieve reliable results.

More precisely, the method here described requires the use of acceptable simplification and linearisation to reveal necessary voltage instability conditions. This means hidden in the folds of the computational method of large scale power systems, some uncertainties and approximations of the achieved results are possible, notwithstanding the great computational effort that is needed.

It should also said that the benefit of bifurcation theory for instability recognition lies in a computational method whose objective is the selection of critical performances of nonlinear dynamic processes operating under singular conditions. This exceptional extreme operating condition can therefore be extrapolated rather than rigorously computed/identified.

In conclusion, the complex and burdensome computational bifurcation methods presented are useful for off-line studies and comparison, yet they appear to have come along too late to be useful or supportive in the electrical power system real-time operation.

Equilibrium and Stability of Dynamic Systems

As already introduced, voltage stability analysis is largely performed by considering trend and characteristics of power system trajectories around system equilibrium states. Accordingly, the model of the electrical power system, linearised around possible equilibrium points as well as dynamic model formulation in terms of state variables are useful simplifications for many analyses, including the voltage instability considered here.

State Vector Differential Equations

A linear dynamic system can be described by a set of first order differential equations in system state variables "x" (a linear combination of the x provides an alternative set of state variables). More precisely the linear equations of an "invariant" system may be written as follows:

$$\begin{aligned}
\dot{x}_1 &= a_{11}x_1 + a_{12}x_2 + \cdots + a_{1n}x_n + b_{11}u_1 + b_{12}u_2 + \cdots + b_{1m}u_m \\
\dot{x}_2 &= a_{21}x_1 + a_{22}x_2 + \cdots + a_{2n}x_n + b_{21}u_1 + b_{22}u_2 + \cdots + b_{2m}u_m \\
&\vdots \\
\dot{x}_n &= a_{n1}x_1 + a_{n2}x_2 + \cdots + a_{nn}x_n + b_{n1}u_1 + b_{n2}u_2 + \cdots + b_{nm}u_m,
\end{aligned}$$

where $\dot{x}_i = dx_i/dt$, $i=1, 2,\ldots, n$, and u_j is the jth control variable, $j=1, 2,\ldots, m$; a and b are constant parameters.

This set of instantaneous and simultaneous first order differential equations may be written in a matrix form as

$$\begin{bmatrix} \dot{x}_1 \\ \dot{x}_2 \\ \vdots \\ \dot{x}_n \end{bmatrix} = \begin{bmatrix} a_{11} + a_{12} + \cdots + a_{1n} \\ a_{21} + a_{22} + \cdots + a_{2n} \\ \vdots \\ a_{n1} + a_{n2} + \cdots + a_{nn} \end{bmatrix} \begin{bmatrix} x_1 \\ x_2 \\ \vdots \\ x_n \end{bmatrix} + \begin{bmatrix} b_{11} + b_{12} + \cdots + b_{1m} \\ b_{21} + b_{22} + \cdots + b_{2m} \\ \vdots \\ b_{n1} + b_{n2} + \cdots + b_{nm} \end{bmatrix} \begin{bmatrix} u_1 \\ u_2 \\ \vdots \\ u_m \end{bmatrix}$$

or

$$\dot{X}(t) = AX(t) + BU(t).$$

X and U are respectively the vectors of the state and control variables. $A(_{n \times n})$ and $B(_{n \times m})$ are the state and the control matrices.

The state matrix eigenvalues (n), that is, the (λ_i), $i=1, 2,\ldots, n$, solutions of the following characteristic polynomial equation, are the system poles:

$$\det(\lambda I - A) = 0,$$

where the operator "det $(\cdot)$"computes the determinant of the matrix equation between brackets. I is the unitary square matrix with coefficients "1" in the main diagonal and "0" in the remaining coefficients.

In general, outputs of a linear system are a linear combination of the state and control variables that can again be represented by a vector matrix equation:

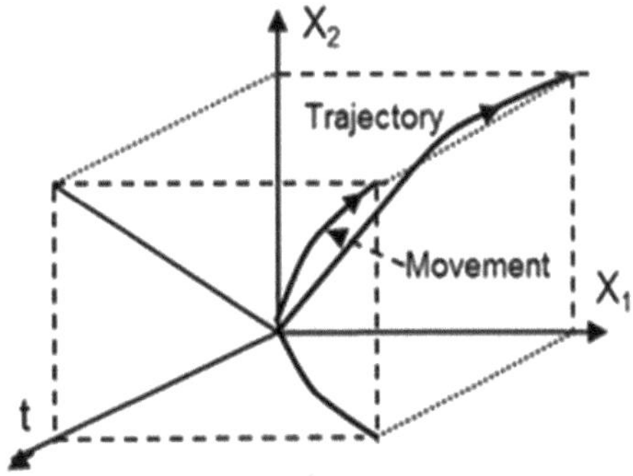

Fig. 8.64 Movement and trajectory of the considered second order dynamic system

$$Y(t) = CX(t) + DU(t),$$

where Y is the "r" order output signal vector, and $C(_{r\times n})$ and $D(_{r\times m})$ are the output matrices linked to the state and control variables, respectively.

The linear system equilibrium state is represented by the equation:

$$\dot{X}(t) = 0$$
$$AX_{eq}(t) + BU(t) = 0 \Rightarrow X_{eq}(t) = -A^{-1}BU(t)$$

Therefore, $X_{eq}=0$ if $U=0$, while $X(0)=X_0$.

This equilibrium equation also gives evidence of the fact that the equilibrium state usually changes with time ($X_{eq}(t) \neq X_{eq}(t+T)$), unless $U(t)=K$. Therefore, the system can have branches of equilibrium points due to the control variable or other changeable parameters hidden inside the A and B matrices (case of variant system).

After these preliminary considerations, the dynamic system "movement" is introduced and represented in the system hyperspace by the trend, with respect to time, of the system state (the sequence of points obtained by combining, at each instant, the state variable values), starting from the equilibrium point at $t=0$ or different initial conditions around it.

A simple example based on a second order linear system clarifies the meaning of the state movement, as shown by the following Fig. 8.64

Consider the dynamic system

$$\dot{x}_1(t) = x_2(t)$$
$$\dot{x}_2(t) = u(t) = k,$$

to which corresponds the movement:

$$x_1(t) = k\frac{t^2}{2}, \quad x_1(0) = 0,$$
$$x_2(t) = kt, \quad x_2(0) = 0.$$

As can be seen, in this case the initial state is always the origin, but this is not a rule. The equilibrium state is given by

$$x_2(t) = 0,$$
$$u(t) = 0.$$

In fact this system, characterised by two series integrators, cannot find equilibrium unless the two integrator inputs are zero. The two system eigenvalues are $\lambda_{1,2}=\pm j$.

This system has a feeble and unique equilibrium point at the origin of the axes, after which any input or disturbance ideally moves the system state indefinitely far.

As can be seen in Fig. 8.64, the trajectory is the projection of the movement on the state plane.

The dynamic system's movements are usually considered to start from the equilibrium state (at $t=0$), usually assumed to be stable, but also dependent on the initial operating conditions (due to system perturbations) as well as on the process inputs.

The trajectories are the system movement projections on the state hyperspace, at $t=0$. As in the Fig. 8.64 example, movement trajectory is represented by an oriented curve. In fact, the trajectories have a direction imposed by the process movement that changes with time. Moreover, trajectories take complex forms even when they refer to linear systems. Therefore, it is easy to presume that they can become intricate in the case of nonlinear dynamic processes.

Trajectory curves, oriented according to movement, progressively move away from or approach the equilibrium point, starting from the initial state. As is well known, the equilibrium points under analysis may be stable or at the stability limit or definitely unstable, and the corresponding trajectories have different shapes accordingly. Along their directional gait, the trajectories may rotate around or intersect the equilibrium points, thus determining bifurcations on them, as considered in the following § 8.3.3.2 (Equilibrium points and Trajectories.....).

Trajectory intersections and bifurcations on equilibrium points give evidence of process stability. This could also mean that, in case system stability can be deduced with the help of the corresponding linearised system, the trajectories' shapes would strongly depend on the Jacobian eigenvalues, as discussed in the following.

Real Nonlinear Dynamic Systems

Most dynamic systems in practice are nonlinear and described in compact form from equations like the following:

$$\dot{z}(t) = \tilde{f}(z(t), u(t)),$$

where $z(t)$ is an $(n \times 1)$ state vector, $z_i(t)$, $i=1, 2,\ldots, n$; $u(t)$ is an $(m \times 1)$ control vector, $u_i(t)$, $i=1, 2,\ldots, m$; and each $\tilde{f}_i$, $i=1, 2,\ldots, n$, is a nonlinear function of all z_i, u_i. In the case of a "*variant*" system, the nonlinear function also depends on system parameters. System time response is linked to an initial condition of the state vector at $t=0$: $z(0)=z_0$.

Assuming the existence and uniqueness of the system solution, its trajectory is a curve in the state hyperspace, linked in same way to the trajectory of the linearised model that starts from the same initial conditions and moves around the equilibrium point z_{eq} as a solution of $\dot{z}(t) = \tilde{f}(z_{eq}(t), u(t)) = 0$.

- According to the stability definition proposed by Lyapunov, the considered equilibrium point can be called *stable* if all equation solutions with an initial condition close to z_{eq} remain near z_{eq} for all t.
- The stability is called *asymptotic* when the trajectories starting from initial conditions of a given neighbourhood around z_{eq} approach z_{eq} as $t \rightarrow \infty$.
- A nonstable equilibrium point is called *unstable*.

Often the stability of a nonlinear system is determined by examining the linearised system around the equilibrium point:

$$Z(t) = z(t) - z_{eq},$$

$$A = \left.\frac{\partial \tilde{f}}{\partial z}\right|_{z_{eq}} = \left|\tilde{f}_z\right|,$$

where A is also defined as the $\tilde{f}$ *Jacobian*.

The equilibrium point stability is often determined by looking at the A eigenvalues of the linearised model:

- If all system eigenvalues have a negative real part, then equilibrium point z_{eq} is asymptotically stable;
- If at least one system eigenvalue has a positive real part, then the equilibrium point z_{eq} is unstable.

This result is independent from the control vector U trend. Therefore, linear system stability can be also analysed without considering control inputs: $U=0$.

We need to clarify that in the linearised model the region of attraction of an asymptotically stable equilibrium is the whole state space. Conversely, in a nonlinear system either no equilibrium point exists or there can be more than one. Moreover, the region of attraction of a stable equilibrium may be limited. Therefore, one stable equilibrium point does not guarantee nonlinear system stability that conversely would require the analysis at any of its equilibrium points to infer its global stability.

Equilibrium Points and Trajectories with Saddle-Node Bifurcation

As already introduced, the main distinction between stable and unstable equilibrium points can be also described by referring to system state trajectories. Consider the following first order dynamic system:

$$\dot{x}(t) = ax(t)$$

.Its equilibrium point is $x_{eq}=0$, while movement is done according to

$$x(t) = x_0 e^{at}.$$

System movement determined by an initial state different from equilibrium is represented in Fig. 8.65. We notice that the system eigenvalue (a) determines the movement trend that asymptotically moves to x_{eq} either starting from $x_0>0$ or $x_0<0$ with $a<0$, and conversely diverging from x_{eq} with $a>0$.

The trajectories of these movements, like their images at $t=0$, are located on the vertical axes together with the equilibrium point and oriented according to the arrows.

We also notice the arrows indicate the trajectories moving toward the equilibrium point in the case of asymptotic stability; conversely, they leave the equilibrium point in the case of instability.

There are other cases in which parts of the trajectories move toward and others far from the equilibrium point: Such is the case of a *saddle-node equilibrium point.*

Referring to the above example, with $a=0$, the movement $x(t)=x_0$ is still stable (Lyapunov) in the linearised system. Conversely, referring to the real nonlinear

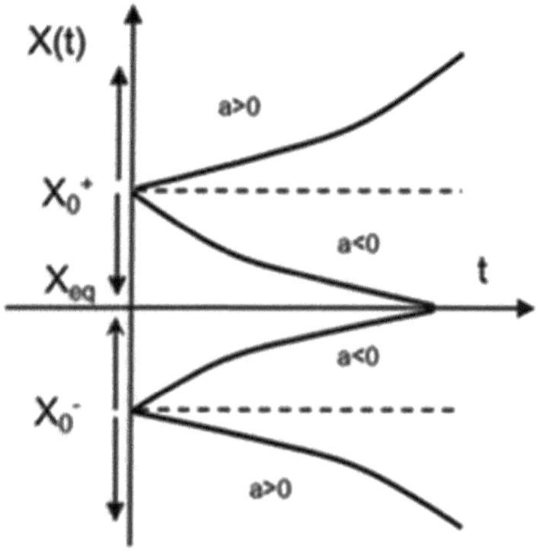

Fig. 8.65 Movements and trajectories of stable ($a<0$) and unstable ($a>0$) dynamic system

system, a knowledge of the Jacobian matrix eigenvalues with zero real part is not enough to reach a conclusion regarding system stability. This is because the real result could be strongly dependent on the system initial condition value and the trend of the input signals as well.

In power system nonlinear dynamic system cases, trajectories with saddle-nodes can be easily found and usually linked to voltage collapse. How the saddle node is linked to voltage instability will be introduced in the following by the concept of "*bifurcation*" due to a sudden change of the nonlinear system's significant parameters (such as system structure or inputs).

Therefore the nonlinear system model has to be represented showing also the parameter vector $\Gamma_{(k\times 1)}$; (ς_i=1,2,...,k) of the possible different system structures:

$$\dot{z}(t) = \tilde{f}(z(t), u(t), \Gamma). \tag{8.4}$$

For every value of the vector couple $(u(t), \Gamma)$, the equilibrium points of the system are given by the solutions of the following equation that defines the $(m+k)$-dimensional equilibrium manifold in the $(m+k+n)$-dimensional space of "states, inputs and structures":

$$\tilde{f}(z_{eq}(t), u(t), \Gamma) = 0.$$

The existence of an equilibrium value $z_{eq}^1(t) = g^1(u^1(t), \Gamma^1)$ corresponding to the couple $(u^1(t), \Gamma^1)$ is based on the assumption of the nonsingular value of the Jacobian:

$$\det\left(\tilde{f}_z^1(z_{eq}^1(t), u^1(t), \Gamma^1)\right) \neq 0.$$

The equation $z_{eq}^1(t) = g^1(u^1(t), \Gamma^1)$ represents a branch of equilibrium points of (8.4). As we know the power system can have a sequence of different equilibrium points for a given structure and load changes. Now we can assume that, with the same couple $(u^1(t), \Gamma^1)$ there exists another equilibrium result:

$$z_{eq}^2(t) = g^2(u^1(t), \Gamma^1)$$

$\tilde{f}_z^2$ being nonsingular.

Analogously, in a power system with starts at a different initial condition, the equilibrium points that are reached differ from one another, even considering the same system structure and load trend, because voltages, currents and losses differ. Then $z_{eq}^2(t)$ represents another branch of equilibrium points as a function of $(u^1(t), \Gamma^1)$.

The term "bifurcation" comes from the intersection of different branches of the nonlinear system equilibrium points. System branches due to the same system parameters may intersect each other, determining a bifurcation point.

Relevant to the steps in the further analysis is the fact that at the bifurcation point *the system Jacobian is singular*, and under a given condition it can be also recognised to be a saddle-node. Therefore, the so-called *saddle-node bifurcation* (SNB) is characterised by the following necessary condition:

$$\det\left(f_{\chi}^{SNB}\left(z_{eq}^{SNB}(t),u(t),\Gamma\right)\right)=0.$$

The Jacobian singularity does not allow eigenvalues to be computed, but in the case of SNB (coming from equilibrium state branches characterised by opposite signs of the real parts) eigenvalues can be extrapolated with zero real parts.

- At a saddle-node bifurcation, two equilibrium points, one stable and the other unstable, coalesce and disappear or emerge simultaneously. One of the equilibrium points has a real positive and the other a real negative eigenvalue, both becoming zero at the bifurcation.

An example of saddle-node bifurcation follows. Consider the dynamic system

$$\dot{x}\ (t)\ = x^2(t)+\mu$$

having equilibrium points at

$$x^2(t)+\mu=0 \Rightarrow x_{eq}(t)=\pm\sqrt{-\mu}.$$

The two equilibrium points assume values depending on μ as follows:

- for $\mu>0$, $x_{eq}(t)=\pm j\sqrt{\mu}$; this result means no single equilibrium point;
- for $\mu<0$, $x_{eq}(t)=\pm\sqrt{\mu}$;
- for $\mu=0$, $x_{eq}(t)=0$.

In the case of $\mu<0$, the system has two real equilibrium states located in the x_{eq} - μ plane along a parabola, trending and symmetric with the μ-axis.

As far as system trajectories are concerned, these converge to a single point as $\mu \rightarrow 0$ by following two half-parabola trends. Therefore, the equilibrium point at $\mu=0$: $x_{eq}=0$, is a saddle-node

$$\left.\frac{\partial f}{\partial x}\right|_{x_{eq}}=0$$

satisfying the necessary condition for the bifurcation.

More precisely, concerning trajectories, with movement $\mu=0$ given by

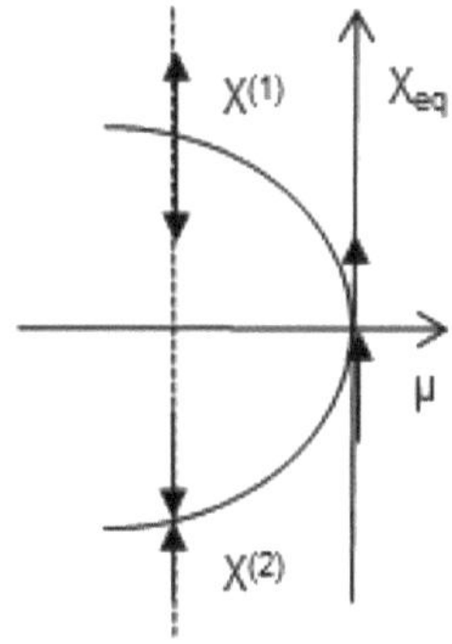

Fig. 8.66 Trajectories of a saddle-node bifurcation at axes' origin coming from stable ($x^{(2)}(t)$) and unstable ($x^{(1)}(t)$) equilibrium points

$$x(t) = \frac{1}{1 - x_0 t},$$

then for negative initial conditions the movement $x(t) \to 0$, and the corresponding trajectory moves to $x_{eq}(t) = 0$, while for positive initial condition the first part of the trajectory diverges to ∞ when $x_0 t = 1$. This result confirms the fact that $x_{eq} = 0$ is a potential saddle-node bifurcation. The following check on the equilibrium branch trend confirms the fact.

In the case of $\mu < 0$, the two possible equilibrium points are towards the positive ($x^{(1)}(t)$) and towards the negative ($x^{(2)}(t)$) arches of the equilibrium parabola, as a function of μ (see Fig. 8.66).

Being the Jacobian of the considered system $\left.\frac{\partial f}{\partial x}\right| = 2x_{eq}$, ($x^{(1)}(t)$) is an unstable equilibrium state because the trajectories move away from it, whereas the negative ($x^{(2)}(t)$) is stable because the trajectories move towards it.

In conclusion, the equilibrium point at $\mu = 0$: $x_{eq} = 0$ is a true saddle-node bifurcation state coming from two branches of stable and unstable equilibrium states.

When system singularities are complex-conjugate, as most power system singularities are, then SNB is called *Hopf bifurcation*, and the above concepts remaining unchanged on equilibrium state bifurcations trajectories and their use for instability analysis.

The emergence of oscillatory instability from a stable equilibrium point comes from a parameter variation forcing a pair of complex eigenvalues to cross the imaginary axis in the complex plane, going toward the positive real axis.

Saddle-node Bifurcation and Power System Loadability

Referring to the electrical power system, any relevant change in inputs (loads or generator production) or in system structure (line opening, generator tripping, etc.)

determines significant changes in power system movements and trajectories as well as in the Jacobian coefficients at any equilibrium point.

Moreover, equations describing the electrical process loadability near the maximum power transfer reach sensitivity and Jacobian extreme (0 or ∞) values, as can be easily imagined from the previous considerations. This confirms the fact that in the case of voltage instability/collapse, the lack of a solution at the considered equilibrium point by the mathematical model process helps us associate the result to a Hopf–saddle-node bifurcation. Some synthetic evidence will be provided on this intuitive statement.

We consider an electrical power system *V-P* curve based on a detailed and complex model approaching a maximum loadability point. We also consider a complete model, including all power system dynamics to which the voltage instability phenomenon pertains as well as showing the vector "p" of the possible power system parameters, including a scalar function γ representing electrical power system loadabilty. The electrical power system model to be considered is represented below:

$$\begin{aligned} \dot{x}(t) &= \tilde{f}(x(t), u(t), \Gamma) \qquad & \dot{x}(t) &= \tilde{f}(x(t), p), \\ 0 &= g(x(t), u(t), \Gamma) \qquad & \Rightarrow 0 &= g(x(t), p). \end{aligned} \tag{8.5}$$

The equilibrium condition is given by

$$\begin{aligned} \tilde{f}(x(t), p) &= 0, \\ g(x(t), p) &= 0. \end{aligned} \tag{8.6}$$

For the purpose of understanding and reconstructing voltage instability via the model properties, having already considered time decomposition and therefore reasonably reduced model order, a simplified system equilibrium model (8.6) must be considered acceptable, though still large in size and not linear.

At equilibrium points, the trajectories depend on the parameter values (p) that include changes in both load and power system structure. Concerning loadability, the independent parameters $u(t)$ correspond to load demand. Because the loadability limit is the operating point where demand reaches the maximum value, when the limit is approached no further solution to Eq. (8.5) is achieved.

In a power system there are many different ways to reach the loadability limit and there is not necessarily one limiting value: each combination of $u(t)$ and Γ yields one specific limit. From the mathematical view point, loadability limit corresponds to the maximum of a scalar function γ of p over all possible solutions of (8.5).

The optimisation problem to be considered is therefore:

$$\max_{p} \{\gamma(p)\}$$

$$\text{subject to: } \begin{cases} \tilde{f}(x(t), p) = 0 \\ g(x(t), p) = 0 \end{cases} \Rightarrow \psi(x(t), p) = 0\,.$$

The optimality necessary conditions are based on a Lagrangian function giving:

$$\psi(x(t), p) = 0,$$
$$\frac{\partial \gamma}{\partial p_i} \neq 0; \quad i = 1, 2, \ldots, \overline{p},$$
$$\det\left(\frac{\partial \psi}{\partial x}\right) = 0.$$

Therefore, at the loadability limit the Jacobian $\partial \psi / \partial x$ is singular. This means the Jacobian $\partial \psi / \partial x$ approaches zero real-part eigenvalue at the loadability limit.

In conclusion, the Jacobian $\partial \psi / \partial x$ with eigenvalue close to the origin can be used to indicate voltage instability and the start of collapse.

Necessary Condition From the above the difficulty of the maximum loadability process is evident, as inferred by the use of a large-scale nonlinear dynamic model. It is also evident that there exists a method based on Hopf–saddle-node bifurcation which entails computation of the Jacobian, that is, linearization of an electrical power system model at the equilibrium point.

Lastly, the instability solution is indirectly deduced by a check that the Jacobian determinant cannot be computed. From this we infer the necessary condition indicating the electrical power system under an operating condition for which the corresponding linearised model shows eigenvalues on the imaginary axis.

Comparisons with other methods' results, more or less rigorous or approximate, provide optimistic results, thus confirming the method's validity. We should say also that this complex and ponderous method is of minor interest for real-time, timely control by the system operator in the face of unexpected voltage instabilities.

This conclusion clearly indicates the need for real-time power system control and protection of real-time voltage instability indicators. These are widely discussed in the next chapter.

References

1. Ajjrapu V, Lee B (1992) Bifurcation theory and its application to nonlinear dynamic phenomena in an electric power system. IEEE Trans Power Syst 7:424–431
2. Ajjarapu V, Lee B (1998) Bibliography on voltage stability. IEEE Trans Power Syst 13(1):115–125
3. Barbier C, Barret JP (1980) An analysis of phenomena of voltage collapse on a transmission system. Revue Generale d'Electricité, CIGRE (Special Issue)
4. Berizzi A, Bresesti P, Marannino P, Granelli GP, Montagna M (1996) System-area operating margin assessment and security enhancement against voltage collapse. IEEE Trans Power Syst 11:1451–1462
5. Cañizares CA, Alvarado F, DeMarco CL, Dobson I, Long WF (1992) Point of collapse method applied to ac/dc power systems. IEEE Trans Power Syst 7:673–683

6. Cima E, Cova B, Marconato R, Salvadori G, Salvati R, Scarpellini P (1996) A powerful simulator for investigating severe dynamic phenomena during system major disturbance. Paper No. 38–305, CIGRE Session, Paris, August 1996
7. Concordia C (1991) Voltage instability. Int J Elec Power Energy Syst 13(1):14–20
8. Corsi S, Taranto GN (2007) Voltage instability The different shapes of the "nose". Bulk Power System Dynamics & Control, IREP-VII Conference, Charleston, South Carolina, August 2007
9. Corsi S, Pozzi M, Sabelli C, Serrani A (2004) The coordinated automatic voltage control of the Italian transmission grid—part I: reasons of the choice and overview of the consolidated hierarchical system. IEEE Trans Power Syst 19(4):1723–1732
10. Corsi S, Pozzi M, Sforna M, Dell'Olio G (2004) The coordinated automatic voltage control of the Italian transmission grid—part II: control apparatuses and field performance of the consolidated hierarchical system. IEEE Trans Power Syst 19(4):1733–1741
11. Deuse J, Stubbe M (1993) Dynamic simulation of voltage collapses. IEEE Trans Power Syst 8(3):894–904
12. Dobson I (1992) "Observations on the geometry of saddle node bifurcation and voltage collapse in electrical power systems" Circuits and Systems I: Fundamental Theory and Applications, IEEE Transactions on (Volume:39 , Issue: 3)
13. Gao B, Morison G K, Kundur P (1996) Towards the development of a systematic approach for voltage stability assessment of large-scale power systems. IEEE Trans Power Syst 11(3):1314–1324
14. IEEE Working Group on Voltage Stability (1990) Voltage stability of power systems: concepts, analytical tools, and industry experience. IEEE Special Publication 90th 0358-2-PWR
15. IEEE Working Group on Voltage Stability (1993) Suggested techniques for voltage stability analysis. IEEE Special Publication 93th 0620-5-PWR
16. IEEE Task Force on Excitation Limiters (1995) Recommended models for overexcitation limiting devices. IEEE Trans Energy Conver 10(4):706–713
17. Koessler RJ, Feltes JW (1993) Time-domain simulation investigates voltage collapse. IEEE Comput Appl Power 6(4):18–22
18. Kundur P (1994) Power system stability and control. McGraw-Hill, New York
19. Lee B, Ajjarapu V (1993) Period-doubling route to chaos in an electric power system. IEEE Proc C 140(6):490–496
20. Milosevic B, Begovic M (2003) Voltage-stability protection and control using a wide-area network of phasor measurements. IEEE Trans Power Syst 18(1):121–127
21. Morison GK, Gao B, Kundur P (1993) Voltage stability analysis using static and dynamic approaches. IEEE Trans Power Syst 8(3):1159–1171
22. Overbye TJ, DeMarco CL (1991) Improved techniques for power system voltage stability assessment using energy methods. IEEE Trans Power Syst 6(4):1446–1452
23. Phadke AG (1993) Synchronised phasor measurements in power systems. IEEE Comput Appl Power 6(2):10–15
24. Smon I, Verbic G, Gubina F (2006) Local voltage-stability index using Tellegen's theorem. IEEE Trans Power Syst 21(3):1267–1275
25. Taylor CW (1994) Power system voltage stability. McGraw-Hill, New York
26. Van Cutsem T (2000) Voltage instability: phenomena, countermeasures, and analysis methods. (Invited Paper). Proc IEEE 88(2):208–227
27. Van Cutsem T Vournas C (1998) Voltage stability of electric power systems. Norwell, Kluwer
28. Verbic G, Gubina F (2004) A new concept of voltage-collapse protection based on local phasors. IEEE Trans Power Delivery 19(2):576–581
29. Vu K, Begovic MM, Novosel D, Saha MM (1999) Use of local measurements to estimate voltage-stability margin. IEEE Trans Power Syst 14(3):1029–1035
30. Yorino Y, Sasaki H, Masuda Y, Tamura Y, Kitagawa M, Oshimo A (1992) An investigation of voltage instability problems. IEEE Trans Power Syst 7(2):600–611

Chapter 9
Voltage Instability Indicators

In the first section of the chapter, we discuss how to recognise voltage instability using reliable indices, mainly those having real-time performance. The introduction provides a general overview of voltage stability indices (VSI) as proposed in the literature, distinguishing the great amount of VSI that fall into the *off-line* category from the few belonging to the *on-line* group that can be properly considered true real-time indicators.

Off-line indices, also called "static indices", are introduced through the basic elements they consider: load flow Jacobian matrix singularity, sensitivity analysis, modal analysis and continuation method on iterative simulations. Those declared by the literature as belonging to the on-line category, but which are based on Prony analysis and power system dynamic model identification, are far from being real-time due to the computing complexity they entail. Therefore, this group is more properly associated with off-line VSI.

True real-time, on-line, short-term voltage instability forecasting requires a strong computing simplification based on real-time, high-speed field measurements, unconventional identification methodologies and very fast telecommunications where wide area power systems are considered. Specifically, on-field availability of an automatic control system for transmission network voltage real-time support (SVR and TVR) allows for a simple definition and the application of a real-time, on-line dynamic indicator of a grid's operating point when voltage instability is near. More precisely, the index needs to be computable in a very simple way (as with a deterministic algorithm) and in a short time (a few seconds), as accomplished from actual measurements and control signals that are already available in SVR and TVR control apparatuses.

The high-speed electrical measurements provided by PMU (phasor measurement units) are a pragmatic alternative for achieving reliable, real-time, on-line indicators. A powerful solution is one based on high speed identification of a power system's Thevenin equivalent as seen by a grid bus provided with PMU; due to its novelty, high speed identification must be seriously considered by the world community to use in developing true, timely, real-time indicators, which will increase power system security. Obviously, telecommunications play an important role in prompt real-time performance; they must be high speed and dedicated/tailored to

S. Corsi, *Voltage Control and Protection in Electrical Power Systems*,
Advances in Industrial Control, DOI 10.1007/978-1-4471-6636-8_9

the task, primarily when real-time indicators are used to protect the power system process.

9.1 Introduction

Voltage instability in transmission networks has led or significantly contributed to a number of major blackouts around the world. Its recognition is therefore crucial towards effective control and protection interventions.

Identifying voltage instability conditions and thus preventing network collapse requires on the one hand evaluation of voltage profile security and on the other timely control actions of the preventive type. Experience has shown that monitoring voltage levels and/or using load flow calculations for checking the consequences of a possible outage of one or more pieces of equipment, once low voltage thresholds are exceeded, may produce an insufficient strategy. In other words, reduced voltage levels or load flow divergences may often lead to inaccurate or late indicators of voltage instability proximity to the limit.

On the problems of voltage stability and collapse the literature is wide: excellent reports [1–4, 43] include the description of basic phenomena, terms and definitions, modelling and simulation requirements, different kinds of voltage stability indices, various scenarios of incidents and an extensive list of references. The literature [5–18] also shows the growing trend in studying the phenomenon by taking into account system dynamics through detailed mathematical models inclusive of unit regulation and limitation loops, OLTCs and network voltage control loops (SVR and TVR). These dynamic aspects contribute to the proper consideration of nonlinear effects when control loops reach their own limits and system protections intervene.

The same literature shows the great interest in a representative indicator of power system or network area proximity to voltage stability or collapse limit. At the same time, system operator demand for effective and reliable proximity indicators to voltage instability limits has grown, mainly for the planning of studies but also for network optimisation and short-term operation up to real-time monitoring and control.

Expanded utilisation of voltage stability indicators for different needs obviously requires specialised characteristics, depending on the bounds on computing time that strongly affect real-time and short-time performance of operation indices; therefore, the use of VSI requires distinguishing the different modalities of voltage stability information as it supports the planning stage with respect to the information needed by real-time operation and control. In particular, computing efforts depend on the complexity of the network model for estimating the value of the index, as well as on the typology of the network data (i.e., past (actual) data, forecasting data and real-time data related to days, hours and minutes up to seconds before). Voltage stability indices entail particular computing procedures and analytical functions and therefore differ greatly from each other. Attempts to classify them have been proposed in literature.

Even though are many ways available to predict voltage instability, the majority of power system blackouts around the world are nonetheless not predicted for a number of reasons (unavailability of tools, insufficient tools, operators who are not well-trained in using predictive tools). It is therefore useful to understand the reasons for the generally poor results seen in voltage instability forecasting.

First, it is evident that off-line studies based on "reference" load flow cases and DSA analyses of given contingencies, although they can help operators

- avoid risky power system operation conditions, and
- gain a clear awareness on the impact of contingencies in certain circumstances,

they cannot cover all real, high-risk events, *even with off-line study support*. Actions by untimely or overly conservative operator control are inevitable, as is operator delay in recognising and correctly facing high risk cascading/high speed voltage instabilities.

Such off-line voltage instability studies are generally linked to a detailed multivariable power system model based on reliable state estimation and related sensitivities and Jacobian computing. Starting from these data and modelling, many *off-line and not real-time* voltage instability indicators have been proposed; they can be effectively used (even with the underlined limits).

Section 9.2 provides information and references on voltage stability/instability indicators [1–35] whose computing time is generally too long with respect to the voltage instability dynamics considered (mainly from the time a real instability is about to commence up to the moment of collapse). These VSI can be properly applied to off-line analyses, useful for programmatic interventions such as power system planning and operation dispatching.

Evidently, indices and related off-line checking do not adequately meet the need for avoiding voltage instability in practice by facing voltage instabilities which are unknown and outside-the-operator. On the other hand, it is well known that if we recognise on time and with high reliability any incoming voltage instability, independently from a priori knowledge of a power system's operating structure and load conditions, the information will surely allow us to block incoming instability in all controllable cases (about 100 % of possible events) through proper controls.

Consequent to these considerations is the relevance that real-time voltage instability indicators must assume. They should be reliable and fast enough to alert operators and, principally, activate on-time power system controls and protections, thus stopping instability before an irreversible voltage collapse.

The "real-time" attribute of the computing method pertains to computing methods and indicators being able, at a given instant, to unequivocally recognise and contextually signal an incoming high-voltage instability risk, doing so according to real-time measurements at the same instant or in a narrow, linkable time span. Risk indication is therefore "real-time" when it is based on actual process data that sound an unequivocal alarm of imminent instability (thus avoiding untimely, too-conservative control and preventing collapse).

The literature provides only a few cases of real-time VSI [29–35]. They include attempts at fast, real-time, neural-based VSI, whose objective is to show how close an operating point is to instability (i.e., the margin of instability). In fact, for an index of this kind to be real-time it must necessarily be based on real-time measurements of the power network and its controls. Moreover, it must be computed in a short time (a few seconds) compared to network voltage instability dynamics and the delay process on component protections. Only a neural index having these characteristics can pull the trigger for automatic network control and grid protective action, whose goal is to delay or prevent the network from overtaking unstable operating conditions.

We have said that moving from a consideration of off-line to real-time (on-line) voltage stability indicators (VSI) diminishes the kinds of VSI available. We name only two possible categories:

1. VSI based on real-time data coming from a wide area control system (when one exists and operates) that regulates transmission grid voltages in real time;
2. VSI based on data coming from PMUs.

In fact, either PMUs or a wide area voltage control system (V-WAR) can provide real-time information on power system performance based on instantaneous measurements of the electrical process (PMU) or of wide area voltage control system performance with a computing time of few tens of milliseconds. Telecommunication speed impacts real-time performance mainly when real-time indicators are used to protect the process.

This chapter refers to the subject introduced, providing details and clarification. Special emphasis is given to real-time VSIs due to their importance. Their computing complexity (PMU-based case) or simplicity and ease (V-WAR–based case) are both shown in detail.

9.2 Off-line Voltage Instability Indicators

Premise In recent years many indices and criteria for voltage stability have been studied and proposed in the literature [10]. The general goal has been to measure the margin between a power system's point of operation and the instability limit, thereby providing an early warning of a potentially critical situation. As is obvious, any voltage stability/instability computing method on its own provides results that can be used to indicate instability, stability or distance from instability.

An examination of the attributes of indices in use or under development shows that their main concern is off-line preventive analysis related to system planning. Their computation usually requires information on system state condition, load flows, sensitivities and algorithms; depending on the model used, these attributes fall in a category of analysis that is either small variation around a given state or large perturbation. Computation of these indices [11–20, 24–26, 28–35] is often burdensome due to the iterative methods used for wide network models, repeated load flows, Jacobian and sensitivity matrix computing, optimisation procedures and so on.

Moreover, many of these voltage stability indices refer to power system models that lack dynamic and system controller nonlinear effects. These aspects of the model cannot be considered in real time because they require overly long computation time and expensive resources if a detailed dynamic simulation is to be accomplished. In addition, the on-line environment increases the risk of uncertainty in a number of network parameters.

The traditional approach to identifying power system saddle-node bifurcation is extremely time-consuming [21]. It involves solving for and tracing equilibrium points along a path defined by a scheduled power system operating scenario, after which the system dynamic Jacobian matrix and its eigenvalues are computed at each equilibrium point until a zero eigenvalue is detected. In the computing of each equilibrium point, an iterative procedure including load flow is applied to take into account voltage-regulating and reactive power-limiting control characteristics of the operating system.

Some simplifications made under certain assumptions and limiting hypotheses have been obtained [7, 22–24]. Another step ahead toward simplification [25] shows an approach that identifies voltage instability by direct equilibrium tracing: the *continuation method* described in [26] is used to directly trace the system equilibrium manifold defined by the power system dynamic model.

There are a variety of bifurcations linked to voltage instability, in particular: Hopf bifurcation (HB); saddle-node bifurcation (SNB) and limit-induced bifurcation (LIB). LIB is introduced to identify approaching dynamic limits of a power system, like generator OELs (over-excitation limiters). Several indices have been proposed to determine the proximity of the power system operating point to the closest bifurcation point [27–29]. However, all these indices suffer one or more of the following defects:

- They are computationally expensive;
- They are based on approximate power system models;
- They do not capture the full dynamic behaviour of the power system.

To overcome these problems, other methods based on the Prony identification technique (which does not require a particular system model) have been considered. The Prony method develops equivalent linear models and determines their modal content and helps in tuning controllers using system measurements [27]. This requires a system linearised model identification (an autoregressive model) and least squares method to compute system dynamic order, which is usually unknown [28]. Furthermore, to provide a correct response the Prony method must be applied after filtering the signal used from both the high frequency component and the mean value. This method, based on direct measurements and on-line model identification, achieves simplifications but is still far from true real-time use. Nevertheless, it indicates a step towards the successful development of on-line indicators presented in § 9.3.

Like the continuation method on equilibrium states, power system security level can be also investigated by an iterative procedure based on subsequent load variation and computation of the load flow solution achieved via a detailed power system model at any iteration. Suitable sensitivity indicators taking into account the impact of nonlinearity and the operating regulator (AVR, SVR, TVR) allow us to define the minimum next-step load increase and therefore the distance from maximum system loadability [35, 36]. Again, these optimisation iterative methods are useful for operational planning studies because of the computing effort they entail.

In conclusion, the wide variety of voltage instability indicators described up until now cannot be used for real-time applications; they can be used only for off-line studies.

9.2.1 Basics of Off-line Indices Based on Jacobian Singular Values

The load flow problem does not find an acceptable solution in the case of the singularity of the power system Jacobian. Even if this circumstance does not directly imply singularity of the state matrix of $\partial\psi / \partial x$ (see § 8.3.3.2 (Equilibrium Points and Trajectories with Saddle-Node Bifurcation)) for the power system model, the use of an index based on Jacobian proximity to this critical situation is clearly linked to the imminent instability of the system.

More precisely, considering the power system dynamic model

$$\begin{cases} \dot{x} = f(x(t), u(t), p) \\ y = g(x(t), u(y), p) \end{cases},$$

with p being the grid parameter vector, the linearised model around the equilibrium point $\alpha_0(x_0, u_0)$ becomes

$$\begin{cases} \partial\dot{x} = F_x(\alpha_0, p)\partial x + F_u(\alpha_0, p)\partial u \\ \partial y = G_x(\alpha_0, p)\partial x + G_u(\alpha_0, p)\partial u \end{cases}.$$

Matrices $F_{x,}$ $F_{u,}$ G_x and G_u represent the partial derivatives in $\alpha_0(x_0, u_0)$ of f and g.

As is well known, power system stability depends on the eigenvalues of F_x computed by the characteristic equation: $\det(sI - F_x(\alpha_{0,} p)) = 0$.

Denoting

$$\dot{z} = \begin{cases} \dot{x} = f(x(t), u(t), p) \\ 0 = \dot{g}(x(t), u(y), p) \end{cases} = \varphi(z, u, p),$$

the linearised model around the equilibrium point becomes

$$\partial\dot{z} = \Phi_z(\alpha_0, p)\partial z + \Phi_u(\alpha_0, p)\partial u,$$

where $\partial z = [\partial x, \partial y]^T$ represents the combined vector of the state and output variations, while ∂u refers to the inputs:

$$\Phi_z(\alpha_0, p) = \begin{bmatrix} F_x(\alpha_0, p) & 0 \\ G_x(\alpha_0, p) & I \end{bmatrix},$$

$$\Phi_u(\alpha_0, p) = \begin{bmatrix} F_u(\alpha_0, p) \\ G_u(\alpha_0, p) \end{bmatrix}.$$

From the above definitions, it clearly appears that $\det(\Phi_z(\alpha_{0,} p))=0$ corresponds to $\det(F_x(\alpha_{0,} p))=0$, and therefore the state matrix of the linearised model is singular under this condition. This also means at least one solution of the characteristic equation $\det(sI-F_x(\alpha_{0,} p))=0$ has zero real part.

The above clarifies that the singularity of the Jacobian of the power system's linearised dynamic model gives credible information on system instability. In particular, through diagonalisation of the Jacobian $\Phi_z(\alpha_0, p)$ matrix:

$$\Phi_z(\alpha_0, p) = T\Lambda T^{-1}, \quad \Lambda = \mathrm{Diag}\{\lambda_i\}.$$

A possible voltage stability index is given by

$$VSI_{\Phi z} = Min\ (\lambda_i).$$

An alternative approach consists of the use of the LF Jacobian, obtained by the power system static model: $J(\alpha_0, p)$. A computing simplification based on the determinant or the eigenvalues of the LF Jacobian module $|J(\alpha_0, p)|$ obviously assumes that singularity of the LF Jacobian $J(\alpha_0, p)$ implies singularity of the Jacobian $\Phi_z(\alpha_0, p)$ of the linearised dynamic model.

Sensitivity Analysis

Other examples of static indicators based on the properties of the Jacobian LF $J(\alpha_0, p)$ reference *sensitivity analysis* and *modal analysis*. A power system algebraic model described by nonlinear equations is

$$\begin{bmatrix} P \\ Q \end{bmatrix} = f\left(\begin{bmatrix} V \\ \vartheta \end{bmatrix}, p\right),$$

where P and Q respectively represent the vectors of the active and reactive power injected into the grid buses, while the V and ϑ vectors refer to the modules and phases of the bus voltages. System parameters are represented by the p vector.

Model linearisation around $\alpha_0(V_0, \vartheta_0)$ gives

$$\begin{bmatrix} \Delta P \\ \Delta Q \end{bmatrix} = \begin{bmatrix} \dfrac{\partial P}{\partial V}(\alpha_0, p) & \dfrac{\partial P}{\partial \vartheta}(\alpha_0, p) \\ \dfrac{\partial Q}{\partial V}(\alpha_0, p) & \dfrac{\partial Q}{\partial \vartheta}(\alpha_0, p) \end{bmatrix} \begin{bmatrix} \Delta V \\ \Delta \vartheta \end{bmatrix} = J(\alpha_0, p) \begin{bmatrix} \Delta V \\ \Delta \vartheta \end{bmatrix},$$

$$J(\alpha_0, p) = \begin{bmatrix} J_{VP}(\alpha_0, p) & J_{\vartheta P}(\alpha_0, p) \\ J_{VQ}(\alpha_0, p) & J_{\vartheta Q}(\alpha_0, p) \end{bmatrix}.$$

Sensitivity computing consists of finding the voltage dependence from the active power variation when setting $\Delta Q=0$ as well as from the reactive power when setting $\Delta P=0$. The results follow:

$$\Delta V = J_{RQ}(\alpha_0, p)^{-1}\Delta Q,$$
$$\Delta V = J_{RP}(\alpha_0, p)^{-1}\Delta P,$$

where the reduced Jacobian matrices $J_{RQ}(\alpha_0, p)$ and $J_{RP}(\alpha_0, p)$ are computed according to the equations:

$$J_{RQ}(\alpha_0, p) = J_{QV}(\alpha_0, p) - J_{Q\vartheta}(\alpha_0, p)J_{P\vartheta}(\alpha_0, p)^{-1}J_{PV}(\alpha_0, p),$$
$$J_{RP}(\alpha_0, p) = J_{PV}(\alpha_0, p) - J_{P\vartheta}(\alpha_0, p)J_{Q\vartheta}(\alpha_0, p)^{-1}J_{QV}(\alpha_0, p).$$

The diagonal coefficients of these matrices respectively represent the sensitivity coefficients of the dependence between ΔV and ΔQ (the first) and between ΔV and ΔP (the second) at each grid bus.

Injecting ΔQ, we find that the smaller the diagonal coefficients are, the greater system stability is. Conversely, the more the sensitivity coefficients grow, the lower the power system solidity/stability goes. The stability limit is reached when system sensitivity tends to infinity. Accordingly, static stability indices can be based on the values of the diagonal coefficients of the sensitivity matrices:

$$VSI_{SQ} = -\min\left\{\text{diag}(J_{RQ}(\alpha_0, p))\right\},$$
$$VSI_{SP} = -\min\left\{\text{diag}(J_{RP}(\alpha_0, p))\right\}.$$

Modal Analysis

From the diagonalisation of $J_{RQ}(\alpha_0, p)$ and $J_{RP}(\alpha_0, p)$:

$$J_{RQ}(\alpha_0, p) = T_Q\Lambda(J_{RQ})T_Q^{-1} \Rightarrow J_{RQ}^{-1}(\alpha_0, p) = T_Q\Lambda(J_{RQ})^{-1}T_Q^{-1},$$
$$J_{RP}(\alpha_0, p) = T_P\Lambda(J_{RP})T_P^{-1} \Rightarrow J_{RP}^{-1}(\alpha_0, p) = T_P\Lambda(J_{RP})^{-1}T_P^{-1}.$$

New variables are defined by decoupling each from the other:

$$\Delta V = J_{RQ}(\alpha_0, p)^{-1}\Delta Q = T_Q\Lambda(J_{RQ})^{-1}T_Q^{-1}\Delta Q \Rightarrow T_Q^{-1}\Delta V = \Lambda(J_{RQ})^{-1}T_Q^{-1}\Delta Q,$$
$$\Delta V = J_{RP}(\alpha_0, p)^{-1}\Delta P = T_P\Lambda(J_{RP})^{-1}T_P^{-1}\Delta P \Rightarrow T_P^{-1}\Delta V = \Lambda(J_{RP})^{-1}T_P^{-1}\Delta P.$$

In fact, from the Jacobian diagonal matrices it is possible to simply link each modal variation i to the injection of reactive or active power in the virtual "system nodes" by the transformation T^{-1} of the physical buses, thereby obtaining the so-called *modal voltages*: $T^{-1}\Delta V = \Delta v$. The use of diagonal matrices therefore allows decoupling and separately considering each singularity:

$$\Delta v_{Qi} = \frac{1}{\lambda_i} \Delta q_i,$$

$$\Delta v_{Pi} = \frac{1}{\lambda_i} \Delta p_i.$$

The smaller the absolute value of the eigenvalue λ_i the lower the related modal voltage value is. As the eigenvalue goes to zero, a small variation of the modal reactive power determines an infinite variation of the correspondent modal voltage. This implies voltage collapse.

Many variants starting from these basic considerations linked to the Jacobian matrix have been developed in the literature. While they are not the main issue of this chapter, more detail about them can be found in the referenced papers.

9.2.2 Basics of Off-line Indices Based on Load Margin

Another interesting category of static voltage stability indices is based on the load margin concept. The idea of load margin is very simple, being the sum of the possible load increments under the considered power system's operating condition up to the triggering of the voltage instability phenomenon.

There are different ways to compute load margins. Some consider the difference $\Delta P_i = P_{i\max} - P_{i\mathrm{act}}$, where $P_{i\max}$ is the maximum power allowed whereas $P_{i\mathrm{act}}$ is the operating power value at the ith bus.

$$VSI_{\lim} = \sum_{i=1}^{N} |\Delta P_i|,$$

$$VSI_{\lim} = \sum_{i=1}^{N} \sqrt{\Delta P_i^2}.$$

These easy-to-understand indices can provide very accurate results when detailed dynamic models of the process are used. The main impediment to these study limits is the consistently high computational effort owing to the large number of tests to be performed to bring about a correct understanding of the risk. In fact, because the voltage instability phenomenon is a prevailing local problem, results differ according to differences in localisation (system buses) and amount of load increase. Different load increase mappings correspond to different $VSI_{\lim}$ values. Therefore, these indices can be used in accordance with *dynamic security assessment* (DSA) criteria.

Starting from these basic indices, many other proposals have been made in the literature, always requiring a great computational effort. A simpler solution from the computational viewpoint is the so-called *local load margin*. This consists of computing for each bus the maximum local load that determines voltage instability, the loads of all other buses remaining fixed:

$$VSI_{\lim i} = \frac{P_{i\max} - P_{i\mathrm{act}}}{P_{i\max}}.$$

Even in this case, the study analysis requires a great many tests, which mainly refer to large power systems.

Load Margin and Sensitivity

Lastly, reference is made to a proposal of an iterative computing of load increase based on sensitivity coefficients, which takes into account system nonlinearity and the dynamic impact of the operating regulators [36].

Subsequent load flow computing at the $(k+1)$th iteration is based on the optimised load variation obtained as the minimum active load increment $\Delta C^{(k)}_{\min}$ at the kth step. $\Delta C^{(k)}_{\min}$ is related to the corresponding reactive load increment that, at the kth iteration, leads at least one controlling generator to its reactive power limit (OEL or UEL). Generators reaching their reactive power limits are switched from PV to PQ -type nodes.

The evaluation of $\Delta C^{(k)}_{\min}$ is based on the sensitivity coefficients $(\partial Q_{gi}/\partial C_{tot})^{(k)}$ related to the reactive power variation of the ith generator $\Delta Q^{(k)}_{gi}$ with respect to the overall active load increment of the system $\Delta C^{(k)}_{tot}$.

Sensitivity coefficients $(\partial Q_{gi}/\partial C_{tot})^{(k)}$ are computed by a linear combination of generator reactive power derivatives with respect to real and reactive power injection at the load nodes and real power variation at the generation nodes, taking into account load characteristics.

The total real load increment to be applied to each load bus according to the assumed evolution scenario is

$$\Delta C^{(k)}_{\min} = \min_{i=1,N^{(k)}_{pv}} \left\{ \frac{\left(Q^{(k)}_{gi\max} - Q^{(k)}_{gi}\right)}{(\partial Q_{gi} / \partial C_{tot})^{(k)}} \right\},$$

where $N^{(k)}_{pv}$ is the number of generators that at the kth iteration may be represented as V-P nodes because they have not yet arrived at their OELs.

$Q^{(k)}_{gi\max}$ and $Q^{(k)}_{gi}$ are the maximum and the current reactive power production levels, respectively, of the ith generator at the kth iteration. Obviously, the more $\Delta C^{(k)}_{\min}$ tends to zero, the narrower is distance from the voltage stability limit.

9.2.3 *Final Comment*

The synthetic and introductory § 9.2 on off-line voltage stability indices, far from being exhaustive, simply introduces the reader to a subject widely discussed in the literature. In this book, the main scope of Chap. 9 is to introduce the reader to real-time VSI as treated in § 9.3 and § 9.4.

9.3 Real-time PMU-based Voltage Instability Indicators

9.3.1 *Introduction*

On a timely recognition of voltage instability, worldwide interest in defining effective real-time voltage stability indicators has recently grown, under the assumption that very fast measurements of electrical system variables can be taken (e.g., by use of PMU [37] or phasor computing devices of comparable speed). By measuring local voltage and current phasors at an EHV bus, the voltage instability analysis can be performed via a consideration of the classic Thevenin equivalent of the remaining grid seen from that bus [38–41].

As is generally understood, instability is linked to the equality of the absolute value of the equivalent impedances (load and Thevenin), with this equality corresponding to maximal power transfer (refer to Chap. 8). We need to clarify that in general, of the two Thevenin equivalent impedances depicted in Fig. 9.1, whereas Z_L is physically measurable by load bus phasors, Z_{Th} is a true equivalent value. This means the phasors on the bus at the equivalent generator side cannot be provided by a PMU.

Unlike the computational point of view, very simple is the case where Z_{Th} is the physical line impedance between the two buses, both provided with PMUs. Usually, this is not the case, in practice; otherwise, so large an amount of PMUs would need to be installed in the transmission grid covering both sides of all main EHV lines. Realistically, a unique case of practical interest requiring a truly interesting analysis concerns the use of a single PMU phasor at a given bus, which allows us to deduce incoming voltage instability simply by taking local measurements. This objective is widely confirmed by a large variety of studies on this subject.

One of the first works linked to the objective [38] refers to the identification of E_{Th} and Z_{Th} based on a recursive least squares identification method applied to local voltage and current phasor measurements. Results are presented for P-constant loads as a sequence of power flow solutions resembling a continuation power flow.

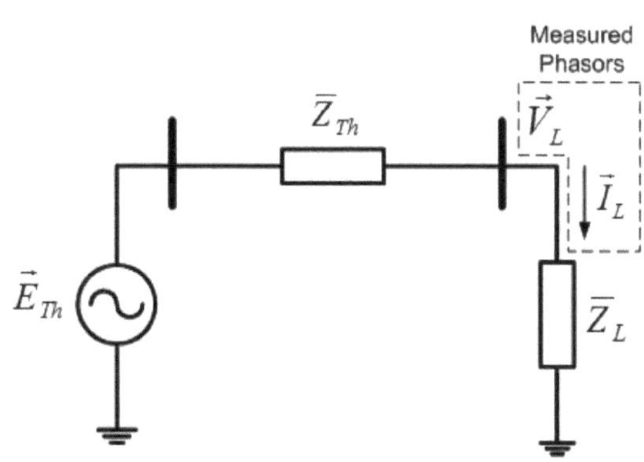

Fig. 9.1 Two-bus Thevenin equivalent circuit

References [41] and [42] differ in the identification method used, which is simplified to achieve a high-speed voltage instability risk evaluation. The work [41] uses the concept of *insensitivity* of the apparent power at the receiving end of the transmission line to infer voltage instability proximity, whereas [42] uses the same concept of Thevenin's equivalent and relies on Tellegen's theorem to identify the Thevenin parameters. In reference [39], the authors extend the previous analyses to ZIP loads and present a mechanism to include the Z-constant and I-constant portions of the load into the equivalent Thevenin impedance, allowing a conclusion that maximum loadability and voltage instability occur at the same point. They also propose monitoring of the status of the OELs of nearby generators and use the information to indicate voltage instability proximity.

In practice, the main critical aspects of the mentioned methodologies based on real-time measurements at a given bus are:

- Computing uncertainty of equivalent grid parameters depending on identification method and their high sensitivity to small changes of local measurements at a fast sampling rate;
- Computing time, which often does not allow sufficient speed for real-time applications;
- Significant performance differences of the real system with respect to the simple electrical model of the considered equivalent circuit when approaching the voltage instability condition;
- Unknown parameters of the ZIP load required in [39] when applied on-field;
- Absence of EHV "transit" buses from these analyses.

A "transit bus" is a bus that has no load directly connected to it (in the sense of a one-line diagram). It normally represents important EHV buses in bulk transmission corridors or nearby load centres having weak or no parallel paths.

Next, we propose how to overcome the above mentioned criticisms by defining a real-time indicator able to effectively support practical applications. The algorithm for fast-tracking the Thevenin parameters (voltage and reactance) is based simply (as required) on local voltage and current phasor measurements. Contrary to least squares-type identification methods, which generally need a large data window to suppress noise, the proposed algorithm has the good feature of being able to filter these high frequency oscillations without significantly delaying the identification process.

Some basic and fundamental considerations of voltage stability studies [3, 4] as already introduced in Chap. 8 are repeated here:

- The Thevenin equivalent must represent a detailed dynamic model. Therefore, the parameters of the Fig. 9.1 scheme change continuously and quickly while approaching maximum loadability;
- Local area OELs and OLTC transformer dynamics strongly impact bus maximum loadability and voltage instability [44];
- The mixed ZIP load is the more correct representation of real load characteristics;
- The maximum loadability point (the *V-P* "nose" curve tip) is different from the real voltage instability point, which in general occurs at lower voltage;
- The high speed of the *V-P* curve equilibrium point moves from the *V-P* nose curve tip to the first unstable point at lower voltage, confirming the practical

importance of nose tip identification to prevent voltage instability, as well as the need for a very fast identification procedure when the voltage instability limit is approached.

According to these basic considerations, the main characteristic of any real-time algorithm based on Thevenin's equivalent is its ability to identify equivalent parameters with higher precision, mainly in proximity to voltage instability, when OELs and OLTCs are operating and parameters are quickly changing. This faster-change dynamic also corresponds to relevant simultaneous and opposite-direction variations of load and Thevenin equivalent impedances. In other words, to be reliable and credible, and real-time VSI cannot be late or too conservative in identifying parameter values, or too precautionary with the co-related protection controls; rather, a VSI must be able to activate protection controls in a timely fashion (i.e., only when strictly needed), minimising the associated costs and in doing so, still be able preventing blackouts.

All other information from any real-time VSI coming far ahead of an actual instability event, even if it is useful, simply prompts an operator to pay more attention to process performance. No serious control effort is made unless the risk of an instability event is approaching 100 %. This is essential, since too much precaution compromises the system's efficient and economical operation. On the other hand, an overly conservative operation is unacceptable and costly to users, also.

A real-time algorithm should also be able to distinguish in a power system critical variations that determine voltage instability from those that, while large, still fall within normal operating conditions.

Lastly, as is obvious, any VSI, to be useful, is predictive, sending a timely alarm at the beginning of a voltage degradation process in a way that activates stabilising control. Therefore, real-time identification algorithms are not suited to discovering short-term voltage instability occurring at the instant of a large perturbation.

The algorithm proposed here was tested through time domain simulations performed in a large and realistic representation of the EHV Italian network. Results on both load and "transit" buses are presented.

9.3.2 Thevenin Equivalent Identification Algorithm

The identification problem covered here is basically a case of an underdetermined system, where a free variable is arbitrarily selected and the system is solved based on selection of the free variable. The free variable chosen is E_{Th}; then X_{Th} is calculated. The proposed theorem is based on the circuit shown in Fig. 9.1, establishing the direction in which the selection of the free variable should be updated (i.e., to a higher or lower value) for it to approach its correct physical value.

The mechanics of the theorem are based on the variation of load impedance and under the assumption that E_{Th} and X_{Th} are constant in the brief interval of their identification, from $i-1$ to i. Therefore, correct application of the theorem requires a very short sampling time to ensure its thesis is fulfilled. At the end of this section the reader will find a simple numerical example that clarifies the basic concept behind the theorem. Definitions of the theorem variables are presented first.

Definitions

$\Delta \bar{Z}_L$ Variation on load impedance

$\tilde{\Delta} X_{Th}^0$ Initial identification error on X_{Th}

$\Delta \vec{\bar{E}}_{Th}$ Complex error on phasor $\vec{\bar{E}}_{Th}$

$\tilde{\Delta} E_{Th}$ Given identification error on $E_{Th} = \left|\vec{E}_{Th}\right|$.

Theorem 9.1.
In a two-bus Thevenin equivalent circuit (Fig. 9.1) subject to load variation, the link between the estimation errors $\tilde{\Delta} E_{Th}$ and $\tilde{\Delta} X_{Th}$ of the Thevenin parameter modules is always given, under the following assumptions 9.2, according to Thesis 9.3:

Assumption 9.2

1. $\bar{Z}_{Th} = jX_{Th}$, *that is* $X_{Th} >> R_{Th}$*;*
2. X_{Th} *and* E_{Th}*, the correct values of the Thevenin parameters, be constant in the sampling interval;*
3. $\bar{Z}_L$ *is an inductive impedance;*
4. $\left|\bar{Z}_L\right| >> \left|\Delta \bar{Z}_L\right|$ *at each step of load variation.*

Thesis 9.3.
If load increases as

$$\Delta Z_L^i = \left|\bar{Z}_L^i\right| - \left|\bar{Z}_L^{i-1}\right| < 0, \quad \mathrm{i} = 1, 2, \ldots, n,$$

with $\tilde{\Delta} E_{Th} > 0$, then

$$0 \le \tilde{\Delta} X_{Th}^i \le \tilde{\Delta} X_{Th}^{i-1};$$

and with $\tilde{\Delta} E_{Th} < 0$, then

$$0 \ge \tilde{\Delta} X_{Th}^i \ge \tilde{\Delta} X_{Th}^{i-1}.$$

Or, analogously, if load reduces as

$$\Delta Z_L^i = \left|\bar{Z}_L^i\right| - \left|\bar{Z}_L^{i-1}\right| > 0, \quad \mathrm{i} = 1, 2, \ldots, n,$$

with $\tilde{\Delta} E_{Th} > 0$, then

$$\tilde{\Delta} X_{Th}^i \ge \tilde{\Delta} X_{Th}^{i-1} \ge 0;$$

and with $\tilde{\Delta} E_{Th} < 0$, then

$$\tilde{\Delta} X_{Th}^i \le \tilde{\Delta} X_{Th}^{i-1} \le 0.$$

Proof Referring to Figs. 9.1 and 9.2 (below), the circuit current measured in the load bus is associated with the last estimated system model, which generally includes parameter errors. On the other hand, the current being measured is coherent with the correct model (without errors on the parameters). Therefore, the following two equations are *consistent:*

$$\vec{\tilde{E}}_{Th} = \vec{E}_{Th} + \Delta\vec{\tilde{E}}_{Th} = \vec{E}_{Th} + \bar{\tilde{\Delta}}X_{Th}\overrightarrow{I_L},$$
$$\vec{E}_{Th} = \vec{V}_L + \bar{X}_{Th}\vec{I}_L.$$

Now, without loss of generality, $\vec{I}_L$ is considered to lie on the real axis. Accordingly:

$$\bar{\tilde{\Delta}}X_{Th}^{i-1} = j\tilde{\Delta}X_{Th},$$

$$\vec{\tilde{\Delta}}E_{Th} = j\tilde{\Delta}X_{Th} \times I_L = j\tilde{\Delta}E_{Th} \tag{9.1}$$

Equation (9.1) links the identification errors at any sampling time and shows the two errors have the same sign. Moreover, from the equality with the current I_L of the first two equations, it turns out:

$$j\tilde{\Delta}X_{Th} = \frac{j\tilde{\Delta}E_{Th}(jX_{Th} + \bar{Z}_L)}{\vec{E}_{Th}}. \tag{9.2}$$

We note that with the load held constant there is no way to identify the two errors.

Conversely, considering a load change during two subsequent sampling times $(i-1)$ and i and maintaining $\Delta\tilde{E}_{Th}$ constant between the two, it is possible to write:

$$j\tilde{\Delta}E_{Th} = j\tilde{\Delta}X_{Th}^{i-1}I_L^{i-1} \Rightarrow j\tilde{\Delta}X_{Th}^{i-1} = \frac{j\tilde{\Delta}E_{Th}}{I_L^{i-1}}, \tag{9.3}$$

$$j\tilde{\Delta}E_{Th} = j\tilde{\Delta}X_{Th}^{i}I_L^{i} \Rightarrow j\tilde{\Delta}X_{Th}^{i} = \frac{j\tilde{\Delta}E_{Th}}{I_L^{i}}. \tag{9.4}$$

Subtracting (9.3) from (9.4) yields

$$j\tilde{\Delta}X_{Th}^{i} - j\tilde{\Delta}X_{Th}^{i-1} = j\tilde{\Delta}E_{Th}\left(\frac{1}{I_L^{i}} - \frac{1}{I_L^{i-1}}\right), \tag{9.5}$$

$$j\left(\tilde{\Delta}X_{Th}^{i} - \tilde{\Delta}X_{Th}^{i-1}\right) = j\tilde{\Delta}E_{Th}\cdot\left(\frac{I_L^{i-1} - I_L^{i}}{I_L^{i-1}I_L^{i}}\right). \tag{9.6}$$

Equation (9.6) confirms the thesis:

In words, if E_{Th} is over-estimated, then X_{Th} is also over-estimated (9.1), and increasing the load changes $\tilde{\Delta}X_{Th}$, which assumes lower values. On the contrary, in case

E_{Th} is under-estimated, then X_{Th} is also under-estimated, and increasing the load changes $\tilde{\Delta}X_{Th}$, which assumes higher values. Analogous conclusions are inferred from the case of reducing the load.

We notice that the simplification of considering E_{Th} as fixed during load build-up resulted in an identification of X_{Th} as critically dependent on the value fixed for E_{Th} and able to achieve the correct value only at maximum power transfer (MPT): always, under normal operating conditions, $X_{Th} << \left|\bar{Z}_L\right|$; therefore, identification begins long before MPT approaches, with $X_{Th} << \left|\bar{Z}_L^0\right|$.

The following corollary is given to show that at MPT, the errors on both X_{Th} and E_{Th} cancel out, and a precise identification is possible.

Corollary 9.4.

If $X_{Th}^0 << \left|\bar{Z}_L^0\right|$ and $\Delta Z_L < 0$, that is $\left|\bar{Z}_L^i\right| - \left|\bar{Z}_L^{i-1}\right| < 0$, or if $X_{Th}^0 >> \left|\bar{Z}_L^0\right|$ and $\Delta Z_L > 0$, that is $\left|\bar{Z}_L^i\right| - \left|\bar{Z}_L^{i-1}\right| > 0$, then there will be a sampling interval i=n where: $X_{Th} + \tilde{\Delta}X_{Th}^n = \left|\bar{Z}_L^n\right|$. This condition corresponds to Thevenin's equivalent maximum loadability. On the other hand, $\left|\bar{Z}_L^n\right| = X_{Th}$ is the value of the load bus impedance at the maximum loadability point reached as if $\tilde{\Delta}X_{Th}^n = 0$.

Proof From Eq. (9.2), giving evidence of the load impedance change:

$$j\tilde{\Delta}X_{Th}^i = j\tilde{\Delta}E_{Th}\frac{jX_{Th} + \bar{Z}_L^{i-1} + \Delta\bar{Z}_L^i}{\vec{E}_{Th}}. \tag{9.7}$$

In Fig. 9.1, MPT is reached (with $\bar{Z}_{Th} = jX_{Th}$) when $\bar{Z}_L^{i-1} + \Delta\bar{Z}_L^i = jX_{Th}$.

At this limit condition, reachable under the given hypotheses when $i=n$, the following result is obtained from (9.7), which refers to the analysis with the estimation errors:

$$\lim_{i\to n} j\tilde{\Delta}X_{Th}^i = \left(\frac{j\tilde{\Delta}E_{Th}}{\vec{E}_{Th}}\right)\times 2jX_{Th}. \tag{9.8}$$

The corresponding circuit current is

$$\vec{I}_L^n = \frac{\vec{E}_{Th} + j\tilde{\Delta}E_{Th}}{2[jX_{Th} + (j\tilde{\Delta}E_{Th}/\vec{E}_{Th})jX_{Th}]} = \frac{\vec{E}_{Th}}{2jX_{Th}} = I_L^n. \tag{9.9}$$

This confirms the thesis that MPT happens when $\left|\bar{Z}_L^n\right| = X_{Th}$, even in the presence of estimation errors.

To sum up in words, the corollary states that at the sampling time n, $\tilde{\Delta}X_{Th}^{n}$ reaches a condition that when combined with $\tilde{\Delta}E_{Th}$ is equivalent to saying $\tilde{\Delta}E_{Th} = \tilde{\Delta}X_{Th}^{n} = 0$. Under this condition, we also verify that the MPT corresponds to $\left|\bar{Z}_L^i\right| = \left|jX_{Th}\right|$. Therefore, the approaching of MPT corresponds to cancellation of errors on E_{Th} and X_{Th}. At the nth sampling, load bus phasor measurements allow the correct estimation of X_{Th}. This also means E_{Th} cannot be maintained constant during the identification process; otherwise, parameter identification would be far from correct and the error last until MPT is reached.

Simple Numerical Example Showing Theorem used to Identify Thevenin's Equivalent Circuit

A very simple numerical example shows the basis of parameter adaptive identification. Referring back to Fig. 9.1, the correct values for the Thevenin voltage and impedance are assumed to be 20 V and 1 Ω, respectively. To explain the core feature of the proposed method, we suppose these values to be unknown. To actually estimate them we need to assume the values of E_{Th} and Z_{Th} to be constant between two subsequent measurements (this assumption is very reasonable considering a sampling rate of 20 ms). Starting the analysis with the load impedance equal to 9 Ω, the circuit current is 2 A and the voltage across the load impedance is of 18 V. In the next step, the load impedance decreases to a value equal to 8 Ω. Without changing the logic of our analysis, we could also assume that we had measured the load voltage and current and then calculated the load impedance. Now we separate the analysis in two ways—one for the case when we over-estimate the generator voltage value and one when we under-estimate it.

a. Over-estimation of E_{Th}—Table 9.1, when estimating E_{Th} = 21 V, shows the values of the circuit variables (columns 5 and 6). Note that when we over-estimate E_{Th} a decrease in Z_L is accompanied by a decrease in Z_{Th}.
b. Under-estimation of E_{Th}—Table 9.1, when estimating E_{Th} = 19 V, shows the values of the circuit variables (columns 7 and 8). Note that when under-estimating E_{Th}, a decrease in Z_L is accompanied by an increase in Z_{Th}.

In conclusion: when load impedance decreases, we can infer from this simple analysis that when variations of load impedance and Thevenin impedance are in the same direction, the value of the Thevenin voltage should be reduced; otherwise, it should be augmented.

Table 9.1 Circuit variables when over- and under-estimating E_{Th}

Step	Known variables			Estimated variables			
	Z_L [Ω]	I_L [A]	V_L [V]	E_{Th} [V]	Z_{Th} [Ω]	E_{Th} [V]	Z_{Th} [Ω]
1	9	2	18	21	1.5	19	0.5
2	8	2.22	17.76	21	1.46	19	0.56

A similar analysis could be done for the case when load impedance increases. In this case, the conclusions would be the opposite: when both impedance variations are in the same direction, the value of the Thevenin voltage should be augmented; otherwise, it should be reduced.

9.3.3 Description of Proposed Real-time Identification Algorithm

The objective we pursue is the computation of $\vec{E}_{Th}$ and $\overline{Z}_{Th}$ of the electrical circuit shown in Fig. 9.1, based on phasors $\vec{V}_L$ and $\vec{I}_L$ measured at the load bus.

From Kirchoff's law:

$$\vec{V}_L = \vec{E}_{Th} - \overline{Z}_{Th}\vec{I}_L \tag{9.10}$$

with $\overline{Z}_{Th} = R_{Th} + jX_{Th}$. Equation (9.10) shows that the considered problem has infinite solutions. In principle, two subsequent phasor measurements of the pair $\vec{V}_L$ and $\vec{I}_L$ can be used to compute $\overline{Z}_{Th}$ and $\vec{E}_{Th}$, under the assumption that they do not change during the time interval between the two subsequent measurements.

Accordingly, the corresponding matrix equation is at a high risk of being singular when the required very short sampling time interval is considered. Computations as such are very risky due to numerical difficulties, and the identification procedures based on this approach usually require time (a large data window) to converge. Only in the presence of significant system variations between two subsequent measurements is it possible to achieve an acceptable result. This happens when the system is collapsing; therefore, identification could arrive too late for preventive controls and special protections to be activated.

Because of this, we now propose a new algorithm to speed up the identification of the Thevenin equivalent parameters E_{Th} and Z_{Th}.

Identification Algorithm

Maximum power transfer to the load in the electrical circuit shown in Fig. 9.1 occurs when

$$\left|\overline{Z}_L\right| = \left|\overline{Z}_{Th}\right|, \tag{9.11}$$

with

$$\overline{Z}_L = Z_L \angle\theta = R_L + jX_L.$$

This circuit represents the entire network seen in an equivalent way by the considered bus. According to the phasor diagram in Fig. 9.2 the following relationship holds:

$$\vec{V}_\Delta = \overline{Z}_{Th} \times \vec{I}_L = R_{Th} I_L + jX_{Th} I_L, \tag{9.12}$$

$$\vec{E}_{Th} = \vec{V}_L + \vec{V}_\Delta, \tag{9.13}$$

with

$$\vec{E}_{Th} = E_{Th}\angle\beta,\ \vec{V}_L = V_L\angle\theta \quad \text{and} \quad \vec{I}_L = I_L\angle 0^\circ.$$

Separating (9.13) into real and imaginary parts yields

$$E_{Th}\cos\beta = R_{Th}I_L + V_L\cos\theta, \tag{9.14}$$

$$E_{Th}\sin\beta = X_{Th}I_L + V_L\sin\theta. \tag{9.15}$$

For the equivalent Thevenin impedance seen from an EHV bus we have $X_{Th} \gg R_{Th}$, and the assumption of $R_{Th} \approx 0$ is very reasonable. Therefore, an initial estimation for β is given by

$$\beta = \cos^{-1}\left(\frac{V_L\cos\theta}{E_{Th}}\right). \tag{9.16}$$

Since V_L and θ are measured quantities taken from the PMUs, the initial estimation of β still depends on E_{Th}. The admissible range for E_{Th} must be in agreement with electric circuit laws. Up to maximum power transfer (MPT) its minimum value ($E_{Th}^{\min}$) corresponds to load voltage, and its maximum value ($E_{Th}^{\max}$) corresponds to voltage when Z_L=X_{Th} (with R_{Th}=0). Under normal operating conditions the load impedance is much higher than the equivalent Thevenin impedance: a good starting estimation value for E_{Th} is the arithmetic average of its extreme values given by

$$E_{Th}^0 = \frac{E_{Th}^{\max} - E_{Th}^{\min}}{2}, \tag{9.17}$$

where $E_{Th}^{\min} = V_L$ and $E_{Th}^{\max} = V_L\cos\theta / \cos\beta$, with β obtained from: $\tan\beta = (Z_L I_L + V_L\sin\theta)/(V_L\cos\theta)$.

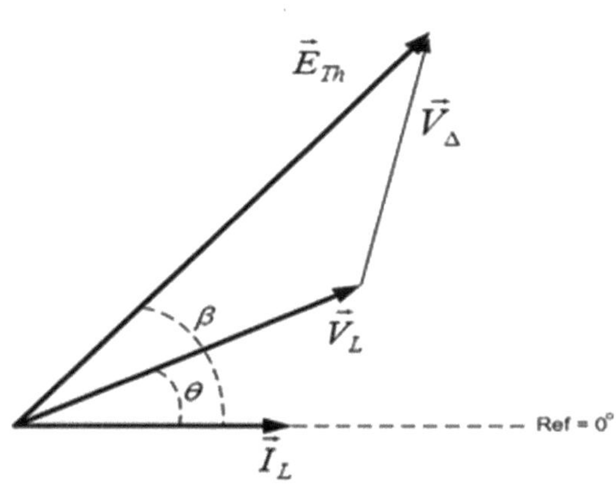

Fig. 9.2 Phasor diagram of the two-bus equivalent circuit

Even being inside its admissible range, E_{Th} as the free variable of the problem should be updated towards its correct physical value. § 9.3.2 introduces the considered theorem and shows a simple numerical example that elucidates the direction in which E_{Th} should be updated at each sampling time in order to speed up the X_{Th} identification convergence. The proposed algorithm assumes that E_{Th} and X_{Th} are constant in the brief $(i-1)$ to i time interval of their identification (according to the theorem), which requires a very short sampling time. In brief, the proposed adaptive algorithm will reduce E_{Th} when the variation of Z_L and the variation of the estimated X_{Th} are in the same direction; otherwise it will increase E_{Th}. This is evident from Eq. (9.7) with variable $\tilde{\Delta}E_{Th}$, i.e.:

$$\tilde{\Delta}\bar{X}_{Th}^{i} = \tilde{\Delta}E_{Th}^{i-1} \times \frac{\bar{X}_{Th} + \bar{Z}_{L}^{i-1} + \Delta\bar{Z}_{L}^{i}}{E_{Th}}. \tag{9.18}$$

The simple numerical example given in the 9.3.2 helps to further clarify this concept.

Knowing the direction E_{Th} should be updated at each sampling time, we now need to establish how large this variation should be. This quantity is calculated as follows:

$$\varepsilon_E = \min(\varepsilon_{\inf}, \varepsilon_{\sup}, \varepsilon_{\lim}) \tag{9.19}$$

with

$$\varepsilon_{\inf} = \left| E_{Th}^{i-1} - V_L^i \right|, \tag{9.20}$$

$$\varepsilon_{\sup} = \left| E_{Th}^{i-1} - E_{Th}^{\max(i)} \right|, \tag{9.21}$$

$$\varepsilon_{\lim} = \left| E_{Th}^{i-1} \times k \right|, \tag{9.22}$$

k being a pre-specified parameter chosen in such a way as to constrain the identification error within narrower bounds and i being the corresponding time step. Most of the time, $\varepsilon_{\lim}$ drives the identification process, so its specification has a major impact on the process. The quantities $\varepsilon_{\inf}$ and $\varepsilon_{\sup}$ are active only when the estimated E_{Th} is close to the edges of its feasible range. In the following section, the adaptive algorithm that tracks the correct value of E_{Th} to identify X_{Th} is presented. One possible variant of the algorithm is with respect to the calculation of E_{Th}.

Comment on the following identification procedure At Step 4, the calculation of β^i and X_{Th}^i can be made by using a moving average of E_{Th}, calculated over a window of appropriate size (m), instead of the instantaneous value of E_{Th}^i at iteration i. This variant has the advantage of filtering the identified variables, paying the price of a slower identification process. In summary, the proposed algorithm will identify the X_{Th} value as soon as the E_{Th} identification is reached by imposing an oscillation with small amplitude around their correct values with a frequency of half the sampling rate.

Algorithm to identify X_{Th}

Step 1. Estimate initial values for E^0_{Th} according to (9.17) and for β^0 according to Eq. (9.16), with E^0_{Th} already given by (9.17).

Step 2. Calculate X^0_{Th} from Eq. (9.15).

Step 3. Calculate E^i_{Th} according to the conditions:

If load impedance variation is negative, do

If $(X^{i*}_{Th} - X^{i-1}_{Th}) < 0$ then $E^i_{Th} = E^{i-1}_{Th} - \varepsilon_E$;

If $(X^{i*}_{Th} - X^{i-1}_{Th}) > 0$ then $E^i_{Th} = E^{i-1}_{Th} + \varepsilon_E$.

If load impedance variation is positive, do

If $(X^{i*}_{Th} - X^{i-1}_{Th}) < 0$ then $E^i_{Th} = E^{i-1}_{Th} + \varepsilon_E$;

If $(X^{i*}_{Th} - X^{i-1}_{Th}) > 0$ then $E^i_{Th} = E^{i-1}_{Th} - \varepsilon_E$.

If load impedance is constant,

$$E^i_{Th} = E^{i-1}_{Th}.$$

Step 4. Calculate β^i and X^i_{Th} from Eqs. (9.16) and (9.15), respectively.

Step 5. Increment *i* and go to Step 3.

Observation: X^{i*}_{Th} is an intermediate evaluation of X^i_{Th} that takes into account the present values of current and voltage phasor measurements and the previous values of E_{Th} and β.

9.3.4 *Sensitivity Analysis of the Identification Method*

Generally, the degree of success of identification methods relies on the parameter tuning. The above considered method has only one key parameter with major impact on the identification process: the parameter *k*. This section starts showing a sensitivity analysis on the parameter *k* having as an objective the selection of its proper value and admissibility range. The results are obtained from dynamic simulation tests performed on a large HV power system, considering a sampling rate of 20 ms.

Figures 9.3 and 9.4 show the estimated E_{Th} value for different choices of the parameter *k*, starting from a given initial value (E^0_{Th} = 290 kV). We can see that for large *k* the estimation of E_{Th} is faster (Fig. 9.3), but with oscillatory uncertainty, as evidenced by Fig. 9.4, where the saw-tooth periodicity is also linked to the stepping trend of the load increase. From the test experience, the identification algorithm shows good results when *k* is in the range of 0.01–0.1 % of nominal bus voltage.

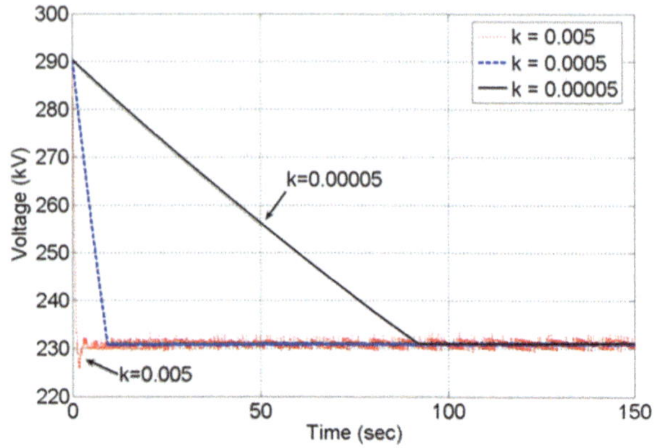

Fig. 9.3 Identification of E_{Th} as a function of k

One good practice is to adopt larger values of k during the initialisation phase (in the first seconds or minutes) of the identification process, then to switch g to a lower k value as soon as E_{Th} approaches a steady value. Figure 9.5 shows the corresponding final value of the estimated X_{Th} as a function of parameter k.

Figure 9.6 represents, at fixed k, the convergence of the estimated E_{Th} parameter by increasing the difference of the initial values of E_{Th}^{0} with respect to the correct estimation. Similar convergence curves are obtained for X_{Th}.

The results show a good performance of the proposed algorithm during the initial phase. $k=0.05\,\%$ was used in the results of Fig. 9.6. Figure 9.7 shows sensitivity with respect to filter parameter m. Figure 9.7a shows how the X_{Th} identification proceeds in the first seconds and is delayed while m increases (cases with $m=1$, 4, 10 and 100), in a period of time where a change in the identified value of X_{Th} is produced by load increase. Figure 9.7b shows X_{Th} up to the end of the simulation, when it approaches voltage instability due to load ramping increase. X_{Th} identification delays are evident as m gets larger. In principle, instantaneous identification would require m to equal 1. A good compromise is $m=4$.

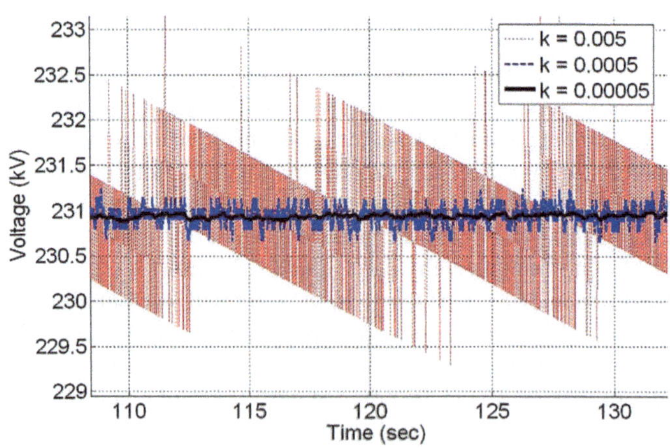

Fig. 9.4 Detail of E_{Th} identification as a function of k

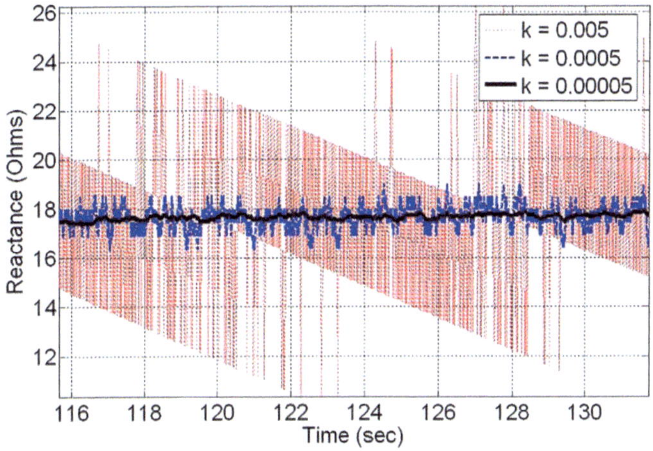

Fig. 9.5 Detail of X_{Th} identification as a function of k

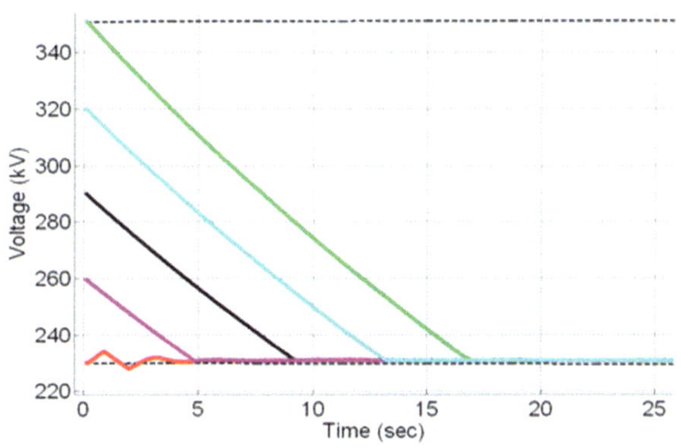

Fig. 9.6 Convergence of E_{Th} as a function of its initial value

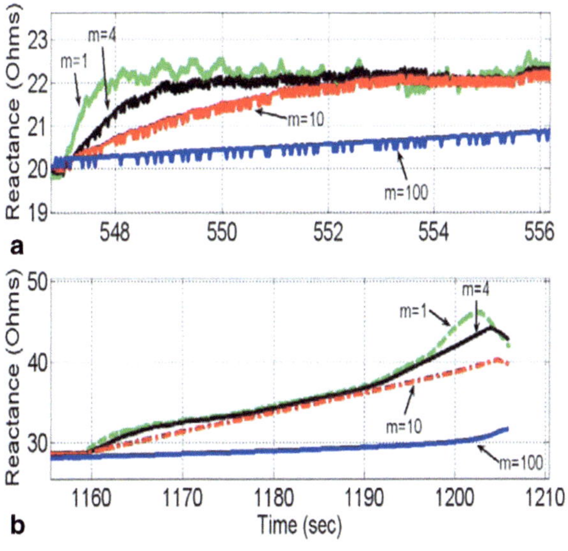

Fig. 9.7 Identification of X_{Th} as a function of m

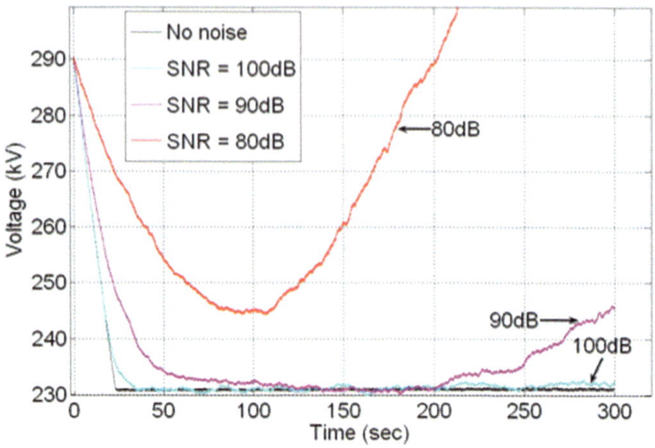

Fig. 9.8 E_{Th} convergence as a function of SNR without signal conditioning

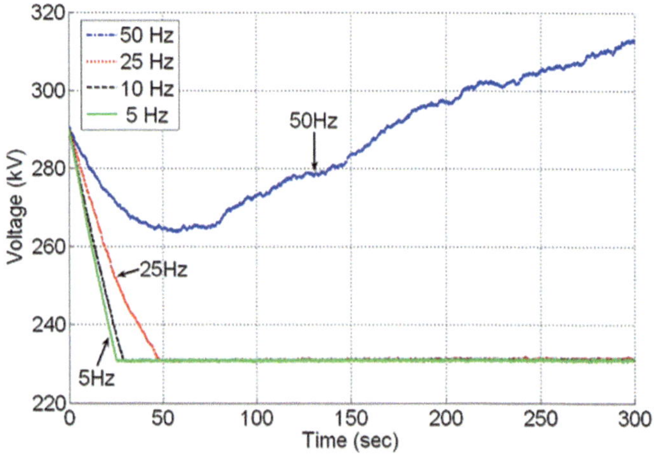

Fig. 9.9 Convergence of E_{Th} as a function of cut-off frequency of low-pass filter for the case SNR = 80 dB

Figure 9.8 shows the convergence of E_{Th} as a function of the signal-to-noise ratio (SNR) when white noise is added to the voltage and current phasor measurements. In the absence of a proper filter ($m = 1$), we observe in Fig. 9.8 that the noise increase with respect to phasor power causes the algorithm to drift from the correct value. Therefore, the required filter on the phasor signals must be checked not only on cut-off frequency tuning but also on SNR value effects. The maximum frequency of the white noise added to the signal in Fig. 9.8 coincides with the frequency of the phasor measurement sampling, i.e., 50 Hz. Figure 9.9 shows the identification of the equivalent. Thevenin voltage when SNR = 80 dB as a function of the cut-off frequency of a low-pass filter. We note that the performance of the proposed algorithm increases as low-pass filter cut-off frequency decreases. When the cut-off frequency

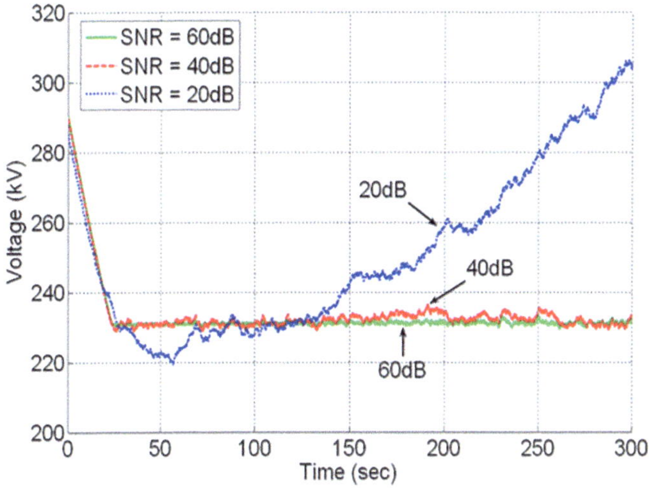

Fig. 9.10 Convergence of E_{Th} as a function of SNR with signal conditioning

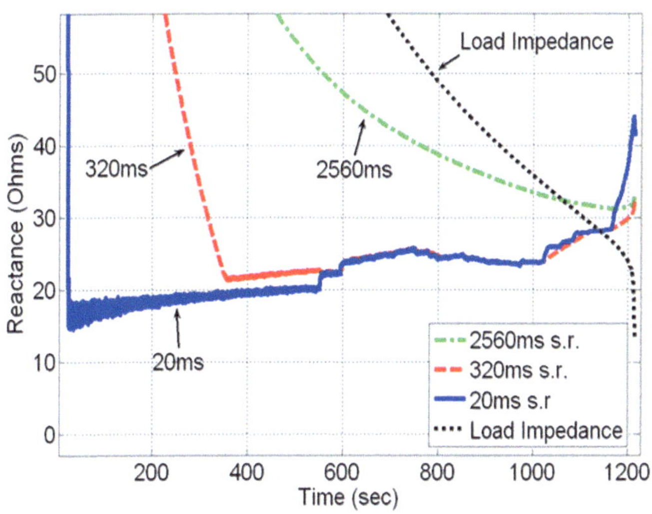

Fig. 9.11 Identification of X_{Th} as a function of the sampling rate (s.r.)

is half the measurement frequency, the algorithm is already able to properly track the correct parameters. The choice of filter cut-off frequency is not critical, as long as the system dynamics under analysis stay below 25 Hz (half the sampling-rate frequency). Thus, Fig. 9.10 proceeds with a sensitivity analysis of the identification algorithm, always starting from E_{Th}^{0} =290 kV, while varying SNR when phasor signals are filtered by a low-pass filter with a 5-Hz cut-off frequency.

The results show that the algorithm has an acceptable performance for SNR equal to or higher than 40 dB. Figure 9.11 closes the sensitivity analysis by showing the identification of X_{Th} as a function of the sampling rate (s.r.) of the phasor measurements. It can be seen that when the algorithm is utilised with an s.r. of 20 ms, X_{Th} is tracked with more precision, although it is more corrupted with high frequency noise.

With an s.r. of 2560 ms, the tracking of X_{Th} is sluggish, and it can be noted that before a steady value for X_{Th} is found, the system undergoes voltage instability. This is quite interesting and counterintuitive because it is difficult to say a priori that a sampling rate of a few seconds is not sufficiently small enough to track a problem that occurs in a timeframe of 1200s (for the selected initial value). The simulation in Fig. 9.11 is an incontrovertible example of the need for a high-frequency sampling rate.

The following are important conclusions of the sensitivity analysis on the proposed real-time algorithm:

- At the first on-field application, initial values of Thevenin's parameters obviously differ from the actual value; to speed-up the difference recovery, a few seconds are enough before the algorithm converges to the correct value, but with k greater than what it is during the subsequent, normal identification process.
- The signals used for real-time identification must be filtered with a cut-off frequency no greater than 25 Hz in the case of a data sampling of 20 ms. This also helps in reducing the SNR value, which cannot be greater than 40 dB;
- The identification speed should be high for a real-time application, and the 20-ms sampling rate provided by PMU is strongly recommended toward the end;
- The moving average on input measurements that support identification can use only a few samplings; otherwise, real-time is compromised;
- k value tuning, depending on specific field applications and characteristics of measurements used, will require a specific preliminary test during on-field installation of the identifier.
- In the case of remote identification, telecommunication of measurements used should be instantaneous. Because this is impossible, the solution must be local at the considered bus and possibly computed by the PMU installed on that bus. The identification result can be remotely transferred at high speed for centralised control/protection functions.

9.3.5 Algorithm Application to Dynamic Thevenin Equivalent

The identification algorithm is tested on an EHV bus (Bus#3) in the equivalent system shown in Fig. 9.12 and already introduced in Chap. 8. The system consists of a 370-MVA synchronous machine, a 20/400-kV step-up transformer, two parallel 400-kV/100-km transmission lines, a 400/132-kV step-down transformer and six parallel 132/20-kV distribution transformers with on-load tap changers (OLTC).

The complete network, equipment and dynamic data for this system are realistic and taken from the actual Italian system. The nominal load at Bus#5 is 160.0 MW (0.43 p.u.) and 0.0 MVAR. In all the cases, load increase rates are $\Delta P_L = 0.5$ MW/s and $\Delta Q_L = 0.5$ MVAR/s, according to (9.23).

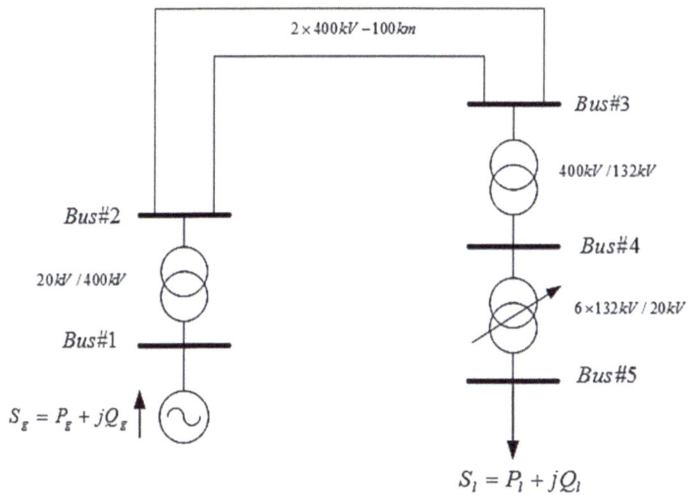

Fig. 9.12 One-line diagram of the single-machine test system

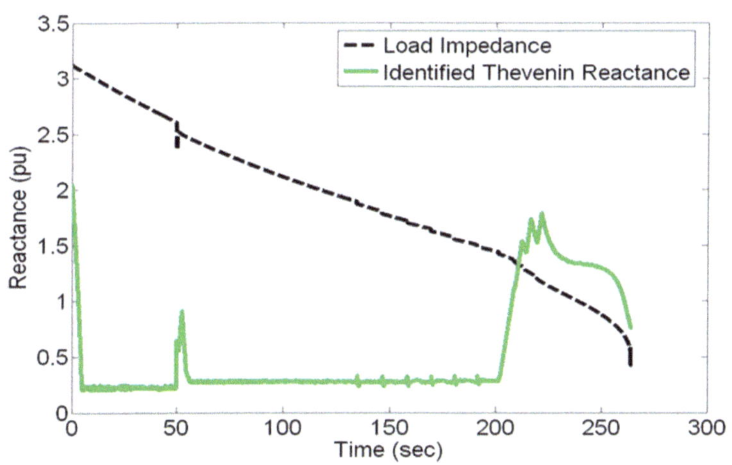

Fig. 9.13 Identification of X_{Th} as system load ramps up

$$P_L = (P_o + \Delta P_L) \times \left(\frac{V_L}{V_o}\right)^{\alpha}, Q_L = (Q_o + \Delta Q_L) \times \left(\frac{V_L}{V_o}\right)^{\beta}, \tag{9.23}$$

where P_o, Q_o and V_o are assumed to be nominal values. For active and reactive load dependency on voltage, the coefficients were set as $\alpha = 0.7$ and $\beta = 2.0$, respectively.

The test consists of ramping up the system load until there is an insurgence of voltage instability characterised by numerical convergence problems in the dynamic simulation program. In addition to the load increase, at 50 s one of the 400-kV transmission lines trips out. Due to the trip, a large transient (Fig. 9.13) in the identi-

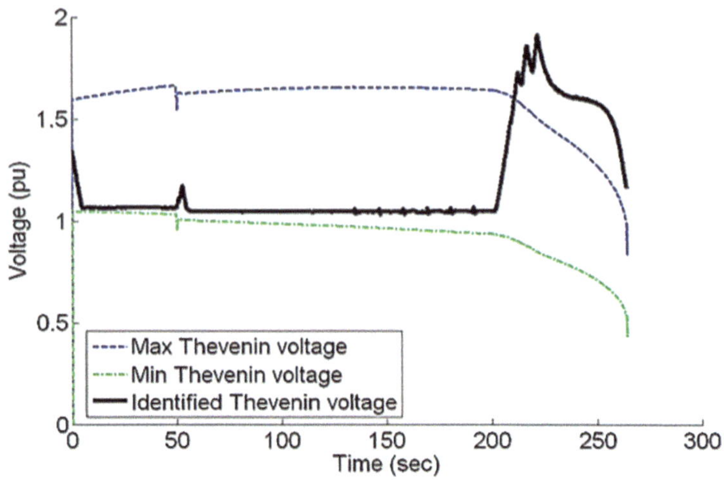

Fig. 9.14 Identification of E_{Th} as system load ramps up

fied Thevenin equivalent reactance is evident before reaching, after a few seconds, a new steady identified value. A little before 140 s the OLTC starts to actuate and at approximately 200 s the OEL is reached. The influence of the OLTC in the identification process is not critical. However, the influence of the OEL is of paramount importance, as can be seen in Fig. 9.13.

Both impedances become equal at maximum loadability point (MLP at 211 s), and the simulation runs up to approximately 264 s. The period between 211 and 264 s is critical, when the system if prone to voltage instability.

An additional test consisting of stopping the load increase, first at 242 s and then at 243 s, was performed. In the first case, the system remained at a stable equilibrium point, whereas in the second case the system lost stability. The test ensures the exact point of instability to be at 243 s. For practical purposes, the indicator gives in real time a very good assessment of the voltage instability risk. Figure 9.14 shows the corresponding identified equivalent Thevenin voltage together with its corresponding extreme values.

Performance Depending on System Dynamics

To show the identification algorithm's ability to distinguish the different closed-loop dynamic impacts on system instability, the four cases already defined and discussed in Chap. 8 are again compared from the identification algorithm point of view, with the results shown in that chapter:

- Case 1: OLTC and OEL in service;
- Case 2: OLTC out of service and OEL in service;
- Case 3: OLTC and OEL out of service;
- Case 4: Like Case 1 with one line tripped.

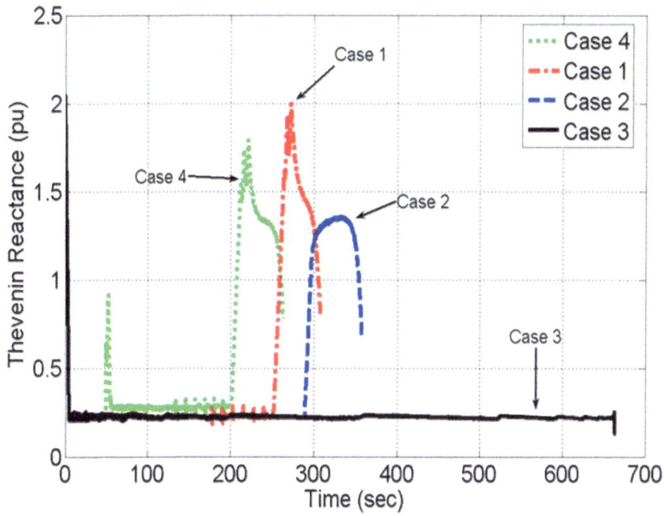

Fig. 9.15 Identification of X_{Th} for distinct consideration of close-loop controls

Figure 9.15 shows the identified X_{Th} for the four cases previously described. The following observations can be drawn from it:

i. Early voltage instability risk identification is inferred for Case 4 (dotted line). Closed-loop controls (OLTC and OEL) and transmission line tripping are responsible for voltage collapse anticipation;
ii. Inevitable large transient occurs in the identified X_{Th} in front of a system contingency (like the line tripping at 50 s). The fast recovery shows this could be a limitation of the proposed method only in the proximity of voltage instability, as a transient anticipation of the incoming event. Utilising a moving average of X_{Th}, consequently slowing down the identification process, can mitigate this transient. The cases in Figs. 9.13, 9.14 and 9.15 do not use a moving average;
iii. Correct increase of X_{Th} after line tripping;
iv. Sharp increase in X_{Th} when the OEL starts to operate in Cases 1, 2 and 4 (dash-dotted, dashed and dotted lines, respectively);
v. Anticipation of voltage instability by OLTC actuation (comparison between Cases 1 and 2) and its small influence on X_{Th};
vi. The correct invariability of X_{Th} in Case 3.

Overall, due to the real-time computing performance of the tested algorithm, each point of the represented traces is computed at the corresponding instant in the time scale; i.e., there is no delay between the system evolution and the corresponding algorithm computing output. Moreover, the proposed algorithm adequately identifies the equivalent Thevenin parameters, showing the relevant differences in terms of time and load values for each of the considered working conditions, as confirmed by the tests to be shown in the § 9.3.5.2 (Performance According to Load Characteristics).

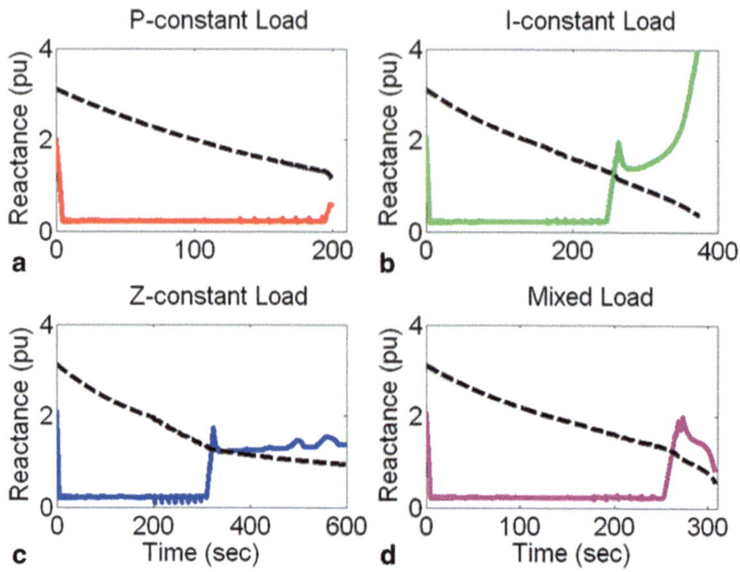

Fig. 9.16 Identification of X_{Th} as a function of load type

Performance According to Load Characteristic

The performance of the X_{Th} identification method is also tested for different load types, as shown in Fig. 9.16. The dashed line corresponds to load impedance. The power system is set in Case 1 and the load modelled as a static 100% P-constant (subplot (a)), 100% I-constant (subplot (b)), 100% Z-constant (subplot (c)) and mixed with $\alpha=0.7$, $\beta=20$ (subplot (d)).

Voltage instability occurs at the MLP (200 s) for the P-constant and after MLP (308 and 376 s) for mixed and I-constant load type, respectively. No voltage collapse occurs for the Z-constant load type after reaching the MLP (the simulation was deliberately stopped at 600 s). The ability of the method to distinguish among the different MLPs mainly driven by the start of OEL operation can be seen. For the mixed type (more realistic) and the I-constant loads, collapse occurs 46 and 119 s after MLP, respectively. In these last two cases, [45] and Chap. 8 show that, in the time period between MLP and voltage collapse points, the system is prone to voltage instability. Normally in such situation, system voltages are at a very low profile, and protective apparatuses are actuating.

9.3.6 Algorithm Application to the Italian 380/20-kV Network

The Italian system analysed contains the actual 380-kV and 220-kV networks. The system configuration has 2549 buses, 2258 transmission lines, 134 groups of thermal generators and 191 groups of hydro generators. Short- and long-term dynamic models are utilised in the time-domain simulations. The system load is approxi-

Fig. 9.17 Detail of the Firenze area showing the Poggio a Caiano 380-kV bus

mately 50 GW, represented as a static model with $\alpha=0.7$ and $\beta=2.0$ in (9.23). The system is under primary voltage and frequency control only.

Voltage and current phasors are measured with a sampling rate of 20 ms. The Thevenin parameter identification is updated every 20 ms, while the identified value is based on the sampled data of the last 80 ms ($m=4$).

The tests on the Italian system are performed at the Poggio a Caiano 380-kV load bus located at the Firenze area depicted in Fig. 9.17, and in the Baggio 380-kV transit bus located at the Milano area. The analysis performed in the Firenze area consists of increasing the local area load by 10 %/min, maintaining constant the power factor. Loads at Poggio a Caiano 380-kV and Casellina 220-kV buses were increased at a rate of 40 %/min due to the robust stability of the system operating point chosen.

The objective of such a load increase profile is to ensure the chosen buses be easily prone to voltage instability. Therefore, such a particular load increase is not chosen to find which of the system buses will be the first to recognise voltage instability and therefore to identify them as the weakest in terms of voltage instability so requiring proper local control to face the problem. Obviously, the algorithm could also be used for the latter objective analogous to the works reported in [40–42]. The load increase at the Milano area for the tests at the Baggio bus followed the same profile of 10 %/min described above [46].

The aim of the analysis performed at the Poggio a Caiano 380-kV bus, as in the single-machine equivalent system, is twofold: to show the impact of the system dynamics and of the local load characteristic in the performance of the proposed identification algorithm. Whereas, in considering the Baggio transit-bus the impact of system dynamics and surrounding load characteristic will be shown in § 9.3.6.2 (Baggio Transit Bus).

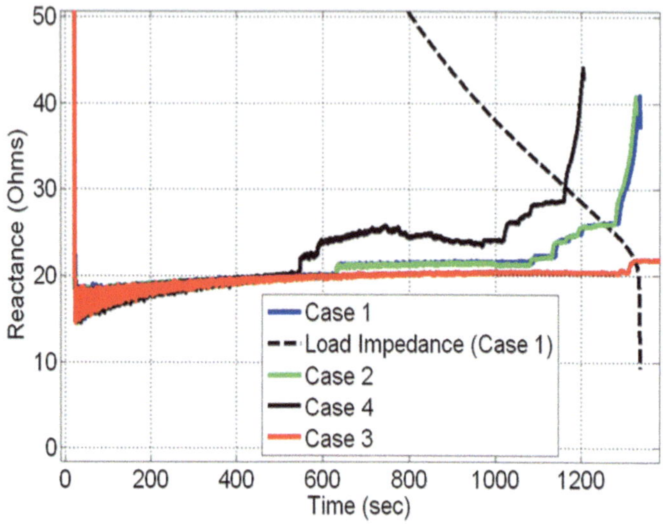

Fig. 9.18 Identification of X_{Th} for distinct consideration of close-loop controls

Poggio a Caiano Load Bus

Performance according to system dynamics

To elucidate in the Italian system the ability of the algorithm to identify the impact on the stability performance of different closed-loop dynamics, four cases are considered:

- Case 1: all OELs and only the Poggio a Caiano OLTCs in service;
- Case 2: all OELs in service and all OLTCs out of service;
- Case 3: all OELs and OLTCs out of service;
- Case 4: all OELs and OLTCs in service.

Cases 2 and 4 represent the more realistic situations, and Cases 1 and 3, though less realistic, are worth presenting for comparison purposes. Figure 9.18 gives evidence of the identified Thevenin reactance for all cases. It also shows the impedance load (dashed line) for Case 1 only. As one would expect, voltage instability occurs earlier in Case 4, as is pointed out well by the identification algorithm. The slight difference between Cases 1 and 2 is also well captured by the algorithm. In Case 3, X_{Th} is maintained fairly constant, since the OELs and OLTCs are out of service.

To further investigate the analysis done for the Poggio a Caiano bus, the results of two cases are detailed. Figures 9.19, 9.20, and 9.21 show simulation results for Case 1. Figures 9.22, 9.23, and 9.24 conversely refer to Case 4.

Figure 9.19 shows the identified X_{Th}. Figure 9.20 represents the evolution of the OEL indicator of six power stations electrically close to the Poggio a Caiano bus. When each OEL indicator reaches zero, the machine over-excitation limiter begins to work in closed loop. The OEL model is of the summing type with soft limiting,

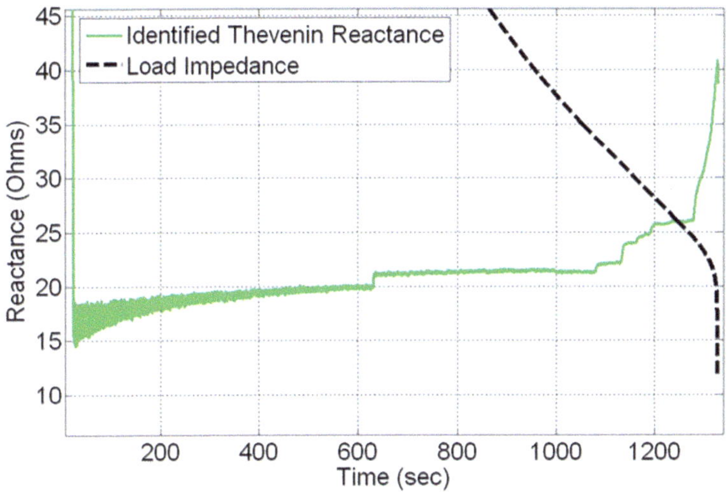

Fig. 9.19 Identification of X_{Th} in Case 1

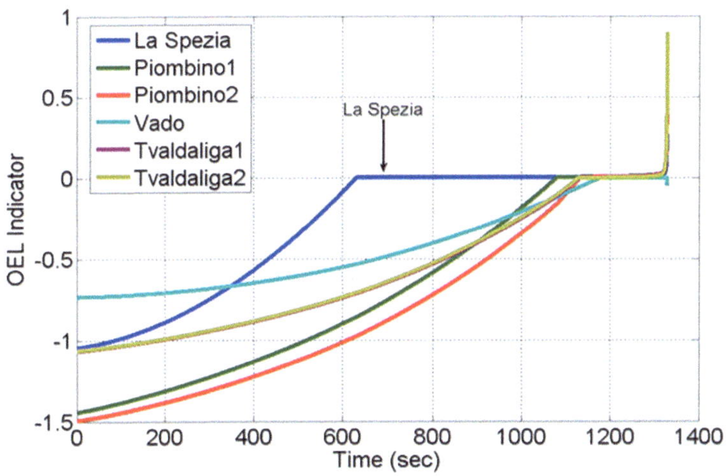

Fig. 9.20 OEL indicator of six electrically close groups of generators in Case 1

which retains the normal voltage regulator loop [47]. Figure 9.21 shows the tap position of the two OLTCs at the Casellina bus.

As can be seen, a continuous tap variation model is employed. This simplification is largely accepted in large power systems. OLTCs cease to operate when their tap position reaches the dimensionless value of 0.8 (lower tap limit).

Comparing the figures, we note that when the La Spezia OEL starts to operate (around 600 s) the identified Thevenin impedance value is visibly increased. This is explained by the fact that when the generator reaches its over-excitation limit the synchronous reactance becomes a part (and a big one) of the equivalent Thevenin

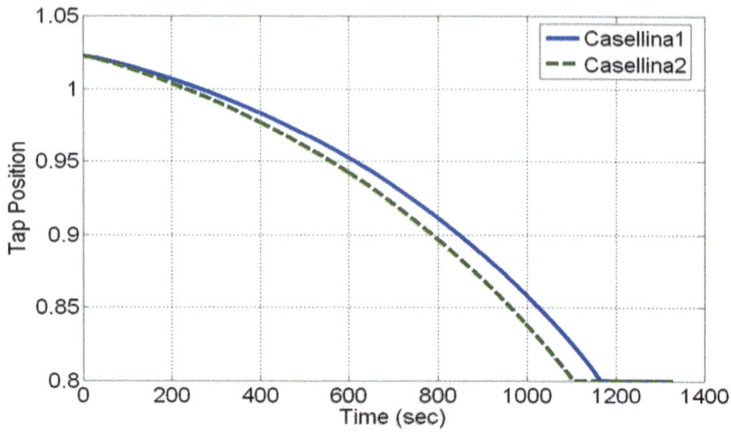

Fig. 9.21 Tap position of the two Casellina OLTCs in Case 1

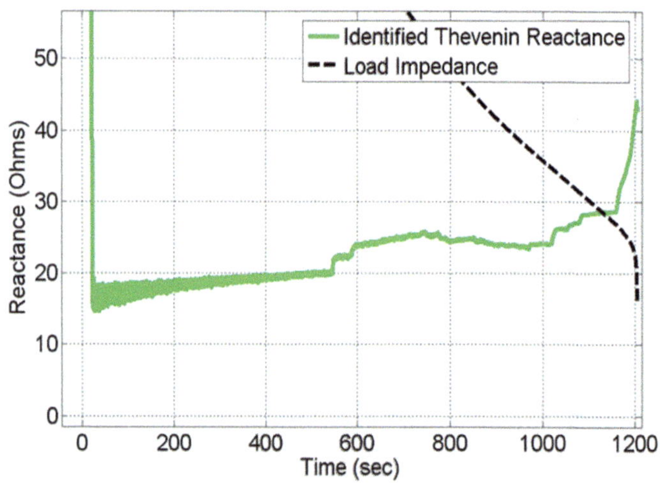

Fig. 9.22 Identification of X_{Th} in Case 4

impedance. The figures also show the fast variation of the Thevenin impedance (from 1100s to 1200s) when the system is approaching its MLP, with OLTCs ceasing to operate and the remaining nearby OELs starting to operate.

Similar conclusions can be drawn from the results obtained for Case 4. Figures 9.22, 9.23, and 9.24 respectively show the identified X_{Th}, the OEL indicator of the same six electrically close power stations and the tap position of certain OLTCs.

Performance According to Load Characteristic

Analogously to the analysis performed on the single-machine system, the performance of the proposed method is also tested in the Italian system for different load

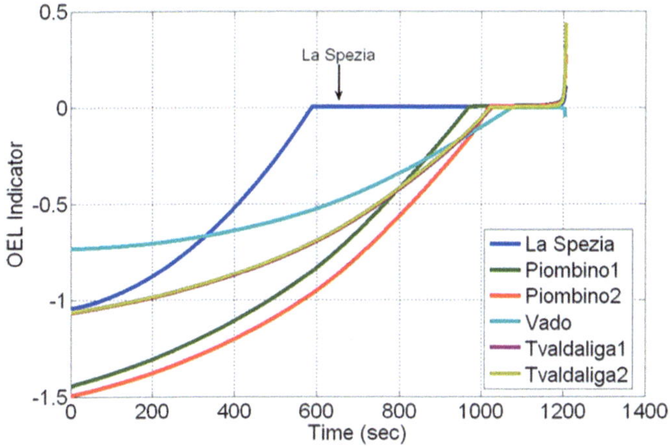

Fig. 9.23 OEL indicator of six electrically close power stations in Case 4

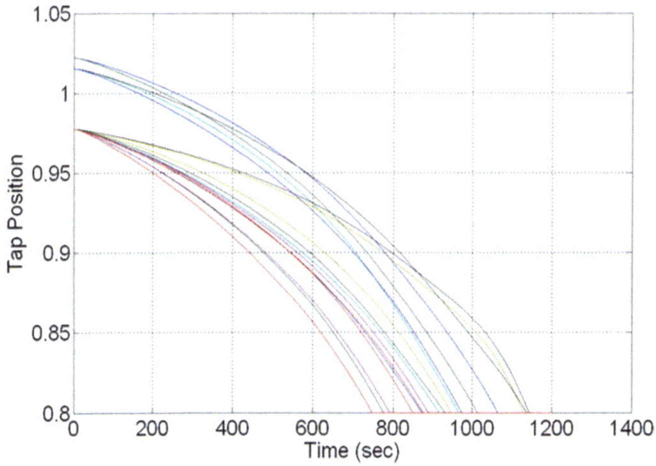

Fig. 9.24 Tap position of some OLTCs in Case 4

types. The power system is set in Case 4 and the loads modelled as a static 100% P-constant, 100% I-constant, 100% Z-constant, and mixed (original load type with $\alpha=0.7$ and $\beta=2.0$).

Figures 9.25 and 9.26 show the identified X_{Th} for the various load types and the simplest voltage instability risk indicator defined as the ratio of the identified Thevenin reactance (X_{Th}) and the load impedance (Z_L). In all cases, it is important to note that indicator slope changes when voltage instability is approaching. All the results match those of the equivalent model given in § 9.3.5.

Baggio Transit Bus

Not only for its practical interest, but also because it is the most burdensome for testing the identification procedure, the "transit" bus has been checked using the

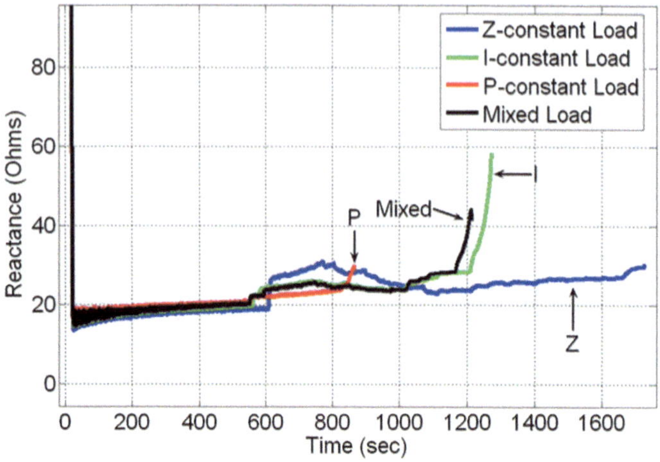

Fig. 9.25 Identified Thevenin reactance as a function of load type

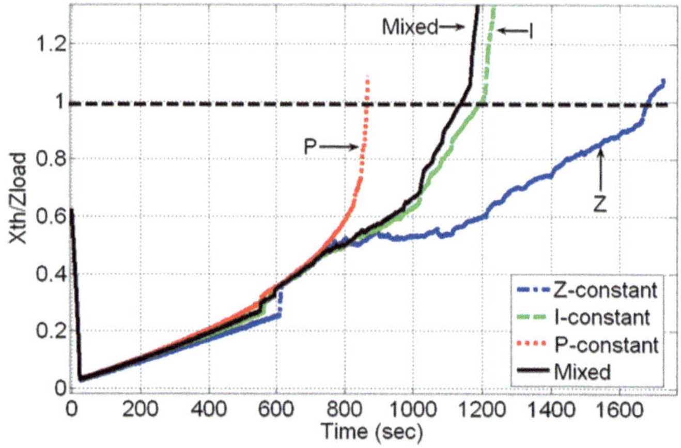

Fig. 9.26 Voltage instability indicator as a function of load type

proposed method. The transit bus can be modelled as a load bus according to the following assumptions: transmission lines with power leaving the bus represent the "equivalent load"; transmission lines with power entering the bus represent the Thevenin equivalent.

As mentioned before, the algorithm performance applied to the Baggio 380-kV transit bus provides distinct results dependent on different OLTC and OEL combinations. Three cases were defined as follows:

- Case 1: OELs and OLTCs in service;
- Case 2: OELs in service and OLTCs out of service;
- Case 3: OELs and OLTCs out of service.

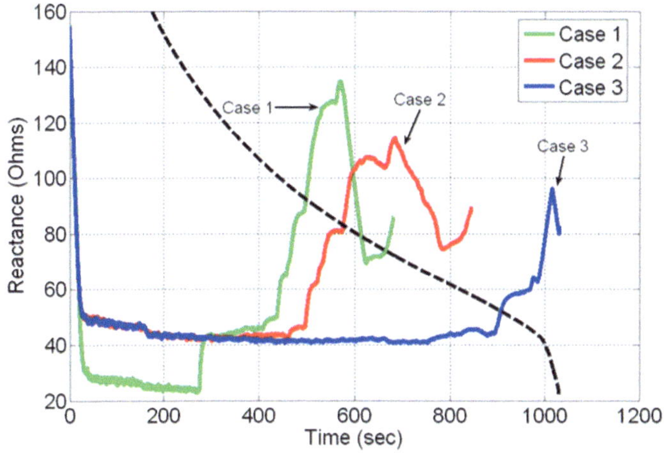

Fig. 9.27 Identification of X_{Th} for distinct consideration of close-loop controls

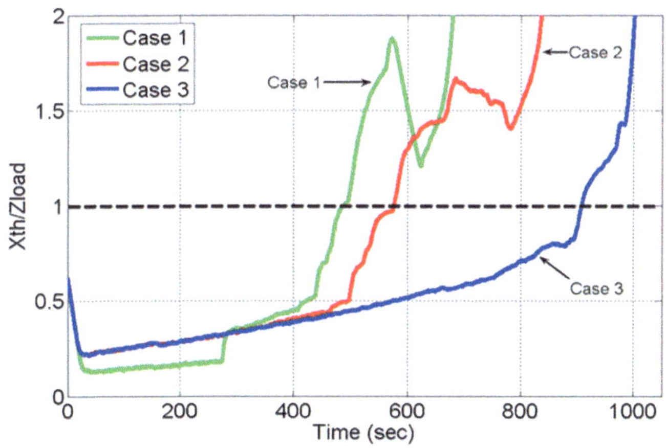

Fig. 9.28 Voltage instability indicator for distinct considerations of close-loop controls

Figure 9.27 shows both the identified Thevenin reactance for all cases and the impedance load (dashed line) for Case 3 only. The results qualitatively confirm those obtained in load buses like Poggio a Caiano and Bus#3 in Fig. 9.12, including the maintainability of X_{Th} fairly constant up to MLP in case the OELs and OLTCs are out of service. However, the point to highlight is the ability of the algorithm to clearly distinguish the differences with or without the control systems, even when considering a transit bus. The difference of X_{Th} in the first seconds, as seen in Case 1, is mainly due to OLTC operation. Figure 9.28 shows the voltage instability risk indicator corresponding to the three cases defined for the Baggio bus.

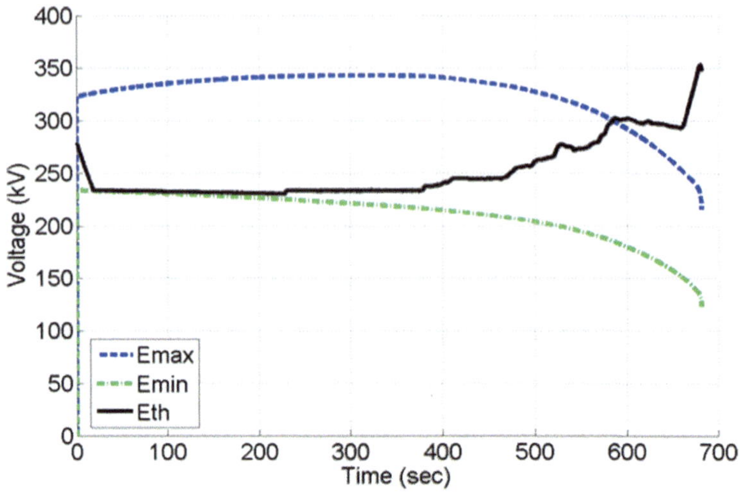

Fig. 9.29 Equivalent Thevenin voltage seen from Baggio bus

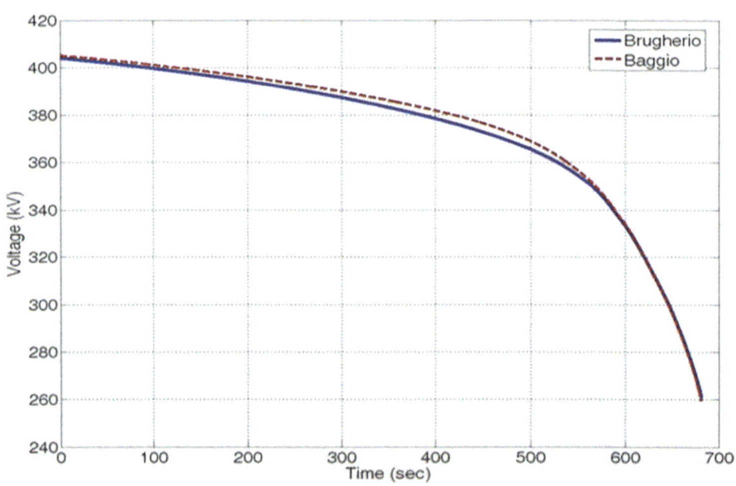

Fig. 9.30 Voltages at Brugherio load bus and Baggio "transit" bus

Figure 9.29 also shows for Case 2 the theoretical limit values for the equivalent Thevenin voltage in the allowed range ($E_{Th}^{\min}, E_{Th}^{\max}$). According to the simulation results, attention has to be paid after 400 s. It can be seen that before 400 s the indicator is reasonably flat, going to a steeper increase afterwards. Figure 9.30 shows voltage magnitudes of both the Brugherio (the load bus near the Baggio bus) and Baggio buses. At around 580 s (100 s before instability), the voltages at both buses are at 0.9 p.u. Thus, a voltage instability proximity that solely relies on voltage magnitudes may be too late. We should also note in Fig. 9.31 the instability

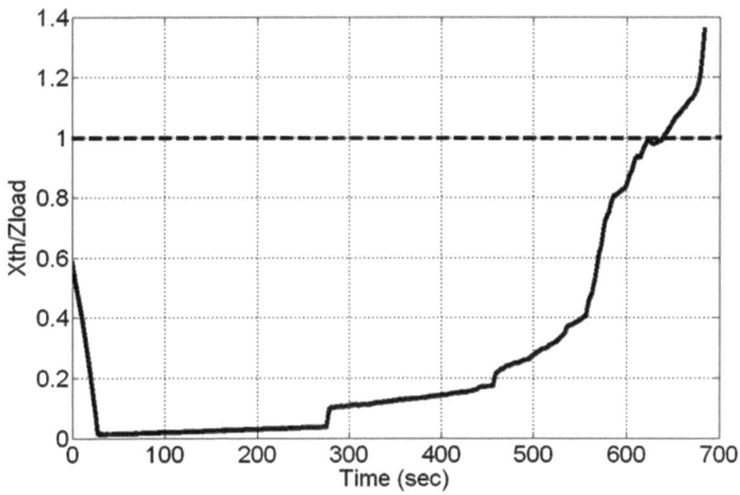

Fig. 9.31 The voltage instability risk indicator for Brugherio load bus

indicator based on the Brugherio bus data. The result clearly shows the instability is recognised with different timing by the system buses. In this case, according to the imposed load increase Baggio sees the instability tens of seconds before Brugherio.

9.4 Real-time Voltage Instability Indicators V-WAR–based

Premise A power system's precise dynamic simulations cannot be performed in real time due to uncertainties in network parameters and also because of the large amount of computing resources and time they would entail. A real-time VSI cannot be practically based on a detailed dynamic simulation.

A real-time index, being representative of the current power system situation, is necessarily different in terms of the conceptual and computational points of view from an index conceived for off-line forecasting studies, where consistent delays (many minutes or more) on result outputs are acceptable.

Next, a real-time dynamic indicator (first introduced by [48, 49]) of the proximity of the current network operating point to voltage instability is presented. It can be computed in a very simple way (a deterministic algorithm) and in a short time (a few seconds) by the on-field availability of an automatic real-time control system of the transmission network voltages (such as SVR and TVR).

This control solution (see Chap. 4) improves voltage quality, allowing a voltage profile across the power system with reduced voltage variations around the imposed values. Power system security is also enhanced by SVR control of generator reactive power reserves to be made available for emergency conditions, also achieving increased active power transfer capability. In addition, reduced total power system losses, by minimising reactive flows through better utilisation of reactive resources, improves the economics of the operation.

The basic idea of the following index proposal is to combine these quality, security and economy advantages, as provided by basic wide area regulation performance, with a further security enhancement achievable by using information provided by the SVR-TVR control system on incoming voltage instability. A real-time and on-line voltage stability index, for proper and effective grid voltage security assessment, easy to achieve and simple to use, is an awaited solution offered by advanced EHV network voltage control systems; such a VSI can be achieved through modern, very powerful, digital technology and the support of high level software tools.

9.4.1 The Real-time and On-line Index

In a power system scenario with SVR and TVR operating the transmission grid voltage support, further real-time information on the power system state and hierarchical regulation itself are available, being strongly representative of the reactive power reserves at one's disposal for real-time control of each pilot node.

Therefore, SVR and TVR impose through a composite regulation structure an effective coordination of reactive power resources (mainly those of the generators) in order to maintain a suitable network voltage profile by facing continuously changing reactive power demands as well as critical system perturbations. By doing so, they have the true view of the instantaneous control effort required to support the voltages at each SVR.

The provided description of voltage instability and collapse (Chap. 8) in the presence of SVR (Chap. 4) leads us to conclude that power system voltage degradation, following load increase, begins when the units of a grid area reach their over-excitation limits. Before this happens, the flat voltage imposed by SVR does not activate the OLTC operation that begins after the SVR saturation.

We also must remember the basic principle of SVR and TVR is a network subdivision into areas—a single area consisting of a number of buses having high electrical coupling to each other. Therefore, their voltages change in unison in front of local load variation or network perturbations, according to the trend of area pilot node voltage.

The SVR reactive power level $q_j(t)$ of the j-area represents instantaneously the control effort underway at the j-area and therefore the real-time reactive power load for j-area control units.

More precisely, the $q_j(t)$ value stands for the percentage of j-area unit reactive power with respect to their under- or over-excitation limits: in particular, when $q_j(t)$ reaches $+1$ the j-area voltage regulation is saturated because the operating points of all the j-area control units are fixed by their over-excitation limits.

With changing load, the pilot node voltage of a given grid area is therefore regulated through the SVR to the desired value unless all area control units reach their over-excitation limits. This approach to extreme operating conditions determines the achievement of area voltage instability limit.

According to this and also considering that voltage degradation usually takes a number of minutes to move from initial instability to irreversible collapse, it appears to be reasonable, simple and effective to compute directly inside the SVR a real-time and on-line indicator of the j-area proximity to voltage control saturation, mainly based on the actual value of area reactive power level $q_j(t)$.

As described in Chap. 8, the *V-P* curve nose tip representing maximum loadability point is also assumed as the reference extreme to recognise, in practice, after the nose tip is overcome, the approaching voltage instability.

The *V-P* nose tip (see Figs. 9.20 and 9.23) of a given SVR area is therefore achieved at contemporarily approaching OELs by area control generators. Therefore, the SVR area saturation event coincides with a high risk voltage instability in that area.

9.4.2 Voltage Stability Index Definition

With reference to the instantaneous reactive power level $q_j(t)$ of the j-area, the proposed proximity indicator $VSI_j(t)$ to the voltage instability limit is given by:

$$VSI_j(t) = q_j(t) + \rho \frac{\partial q_j(t)}{\partial t},$$

where:

- ρ is a suitable weight coefficient for introducing a derivative term with an useful lead effect;
- $-1 \le q_j(t) \le +1$, usually;
- $-1 \le q_j(t) \le +1 + \varepsilon_j(t)$, in case some j-area units can be transiently overloaded with respect to their OELs;
- $\varepsilon_j(t)$ is a parameter normally kept at 0, which under an RVR permit can take on positive growth values.

9.4.3 Voltage Stability Index Computation and Meaning

The above definition shows the real-time and on-line voltage stability index $VSI_j(t)$ is given by two terms:

- The *proportional term* is suddenly available by the RVR because it is related to variables already present and updated within a single time interval;

- Computation of the *derivative term* needs, conversely, a single or finite number of steps if as is often the case a suitable filtering action is conceived.

The voltage stability index $VSI_j(t)$ computation can therefore be carried out by the RVR within an updating time of a few seconds. The derivative term takes into account the $q_j(t)$ dynamics and expected short-term trend: the weight ρ allows an effective setting of such a "before-time" contribution, which is mainly useful as concerns the possible control action based on $VSI_j(t)$:

- The greater the positive value of the derivative term, the more $q_j(t)$ rapidly reaches its maximum limiting value;
- When the derivative term is negative, its contribution avoids the prolongation of useless control action because the actual network and load evolution will allow the SVR to recover distance from the limiting operation.

If the level $q_j(t)$ reaches the maximum value $+1$, in order to support the pilot node voltage to the optimal value defined by the TVR all the j-area control units begin to operate at their over-excitation limits. Under this condition if transient overloading is allowed, the over-excitation limits could be transiently exceeded.

Usually the RVR operates to avoid the operation of units at their over-excitation limits because this condition is normally strongly related to the tap-changer reverse action, which can anticipate the triggering of the voltage collapse mechanism. This is achieved by the control, in advance, of all possible reactive power resources (capacitor banks, reactors, synchronous or static compensators, etc.) installed in the j-area in such a way as to reduce the $q_j(t)$ operating value and therefore the control effort of the units. The RVR acts automatically on the reactive power resources under its control and moreover sends signals to the regional operator asking for manual switching of the remaining resources. When all these control actions concerning the j-area voltage are accomplished and area control units do not allow additional excitation overloading, then the difference $1-q_j(t)$ clearly represents the distance of the j-area from its saturation and, with a high probability and a few uncertainties here after resolved, from the voltage stability limit.

9.4.4 Crucial Role Played by Tertiary Voltage Regulation

The SVR carries out the optimal voltage plan defined and updated in real-time by the TVR according to both the forecasted and estimated operating plan and the real and measured grid working conditions. In other words, considering the j-area, the RVR carries out the set-point trend which is computed and updated with a slow dynamic (a few minutes) by the NVR on the basis of the optimal forecasting voltage plan and the actual network operating condition.

In practice, reactive power control margins will be strongly reduced by very critical grid situations when network voltage degradation becomes considerable. Under these conditions, the TVR attends to transmission network voltage degradation, progressively renouncing to the optimal short-term planned voltage profile (defined more for economic reasons than for security reasons). Therefore, the SVR will operate on its limits only when transmission network voltages are very low, notwithstanding all network reactive power resources being in operation for voltage support. Under these conditions, $VSI_j(t)$ becomes fully significant and reliable because operating limits are reached despite the reduced pilot node voltage plan and saturated reactive power control efforts.

In conclusion, $VSI_j(t)$ fully represents the *j*-area proximity to the voltage stability limit only when the TVR is operating; otherwise, it assumes the different meaning of *j*-area saturation when the set-point can be set by the operator through a calibrator or stored into the RVR according to forecasting studies. In both these cases, SVR of the *j*-area could reach the extreme operating condition with units at their limits only because the set-point value is inadequate for the present network situation. Without the TVR, voltage stability index $VSI_j(t)$ does not really represent the *j*-area distance from the voltage stability limit, but only the distance of the SVR from its operative limits.

9.4.5 Voltage Stability Index Control Function

The voltage stability index $VSI_j(t)$ represents useful real-time and on-line information for the regional dispatcher, who can draw from it when the given thresholds are exceeded in order to successfully recover the critical state of the area voltages. According to this area-alarm, an operator can decide to put in service the remaining reactive power resources of the *j*-area, carrying them out by remote controls or asking this be done by phone. Otherwise, if *j*-area high risk conditions of voltage instability are reached, the operator could manually shut down the *j*-area OLTCs.

Nevertheless, because $VSI_j(t)$ is a real-time, updated variable, it can be more effectively used for real-time automatic control actions (see Chap. 11).

9.4.6 Functional Performances

Figure 9.32 shows the generalised equivalent network "butterfly system" considered here for the analysis of real-time and on-line voltage stability index $VSI_j(t)$ performance.

This system pictured is a quite simple but quite realistic grid equivalent scheme (RVR apparatus could implement a similar, equivalent model aimed at operator RVR training and off-line checking), consisting of three areas (and three pilot nodes: 1, 4 and 7, respectively), each having station loads equipped with OLTCs and different sized units, most of which (Gsa, Gsb, Gsc) operate under the SVR and TVR and the others (Gpa, Gpb, Gpc) only under AVR, as depicted in Fig. 9.33.

The typical dynamic behaviour of the proposed area voltage instability indicators $VSI_j(t)$ can be shown through butterfly system performances obtained by simulating a progressive load charging in area A, with results observable in Fig. 9.34.

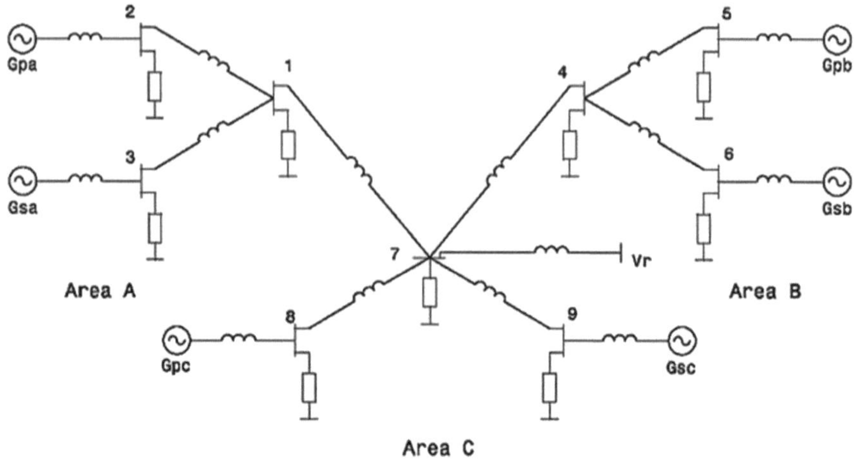

Fig. 9.32 The generalised network scheme considered for the $VSI_j(t)$ performance analysis

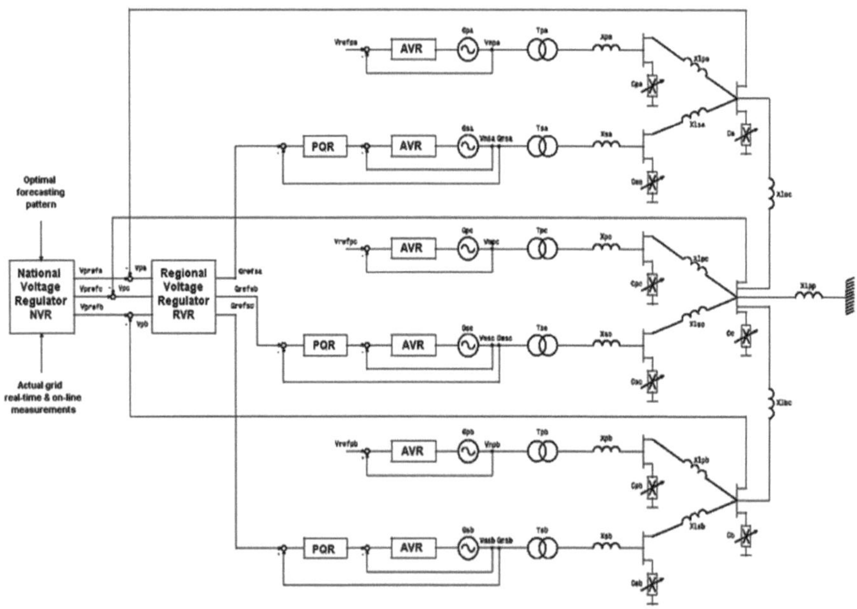

Fig. 9.33 The hierarchical control scheme considered for the $VSI_j(t)$ performance analysis

More precisely, the top three panels of Fig. 9.34 show in the first part of the transient the voltage instability of area A, following a local load increase which drags into collapse the areas C and B. The bottom three panels of Fig. 9.34 add information on voltage stability indices $VSI_j(t)$, which grow very rapidly as regards areas A (where the load increase takes place) and C (due to its electrical proximity to area

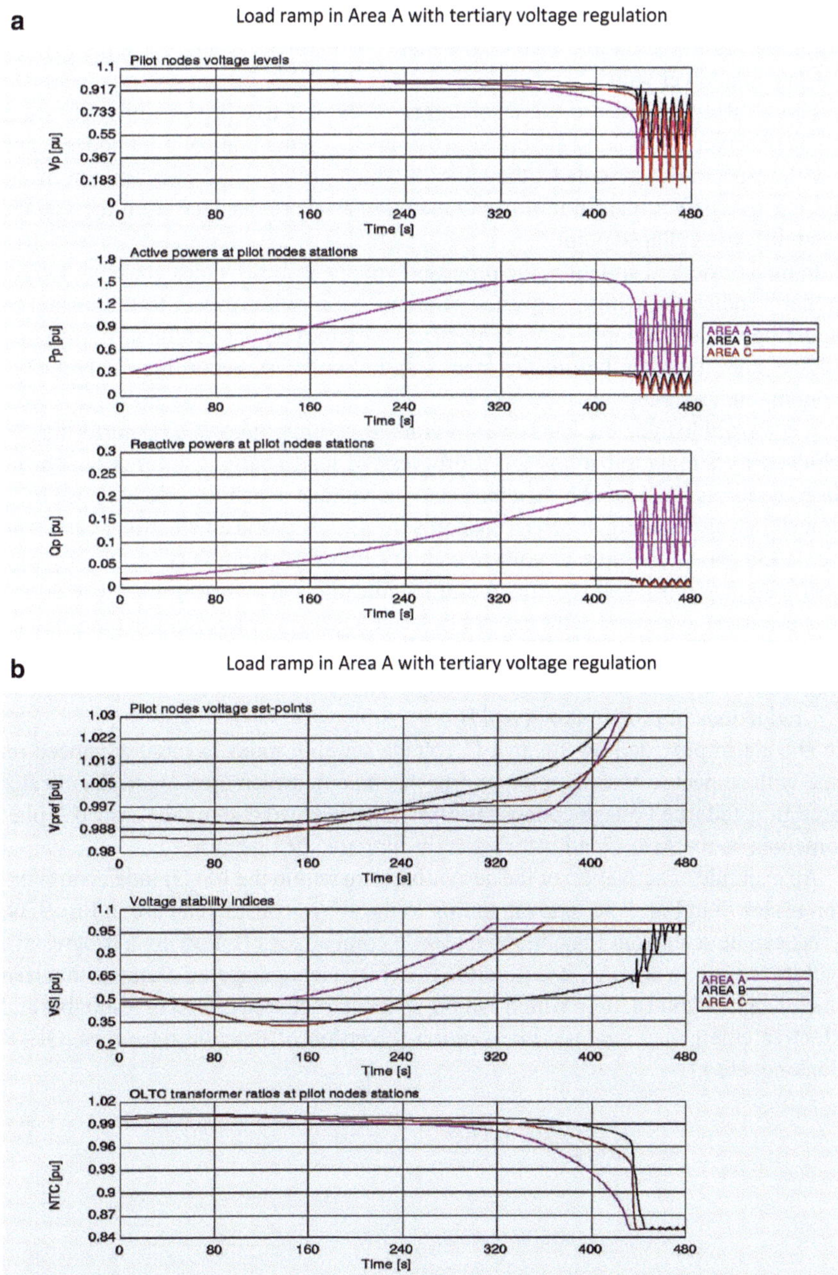

Fig. 9.34 $VSI_j(t)$ dynamic behaviour following Area A progressive load increase

A); instead, the area B index (due to both dynamic decoupling between the areas within the same region and electrical distance from areas A and C) remains quite constant at a lower value.

The TVR contributes to the discharging of the reactive power units' effort, as can be observed by looking at voltage set-point V_{pref} reductions of pilot nodes A and B with respect to pilot node C, the latter of which preserves a larger amount of the unit reactive power margin in order to maintain a secure distance from the voltage instability and collapse point.

In the considered situation, the proposed voltage stability index for area A gives an alarm indication with an advance of about 2 min with respect to the instant of triggering voltage collapse. Area B electrical variables do not exhibit any particular voltage stability crisis before the area A and C collapse, according to the corresponding index trend.

A comparable dynamic behaviour of voltage stability index $VSI_j(t)$ performance, achieved by simulation following a progressive load charging in area C, can be observed in Fig. 9.35a and b. In addition to the voltage instability and collapse transient shown again in Fig. 9.35a as concerns the more significant electrical variables, Fig. 9.35b shows the area C voltage stability index growing very rapidly (due to local load charging), and the area A and B indicators (in a very similar way due to their electrical symmetry with respect to area C) growing more slowly than area C. In the considered situation, the dynamic behaviour of the TVR is still evident, at the beginning of the transient, mainly observable in (b) in terms of the voltage set-point V_{pref} reductions of pilot nodes A and B.

The alarm provided by the area C voltage stability index is more advanced on time with respect to system collapse, but the delay between area A and B indicator saturation and true voltage collapse is drastically reduced due to the "cascade" phenomenon, so the Area C indicator strongly supports the operator.

An example of activation of the derivative term within the $VSI_j(t)$ index computation is shown in Fig. 9.36, corresponding to the above considered case in Fig. 9.34.

As can be seen from Figs. 9.34b and 9.36 comparisons (following a progressive load increasing in area A), it is possible to achieve an anticipated alarm with a lead time of a few minutes, time which can prove crucial for application of a suitable and effective emergency strategy intervention (insertion of reactive resources, OLTC blocking, etc.).

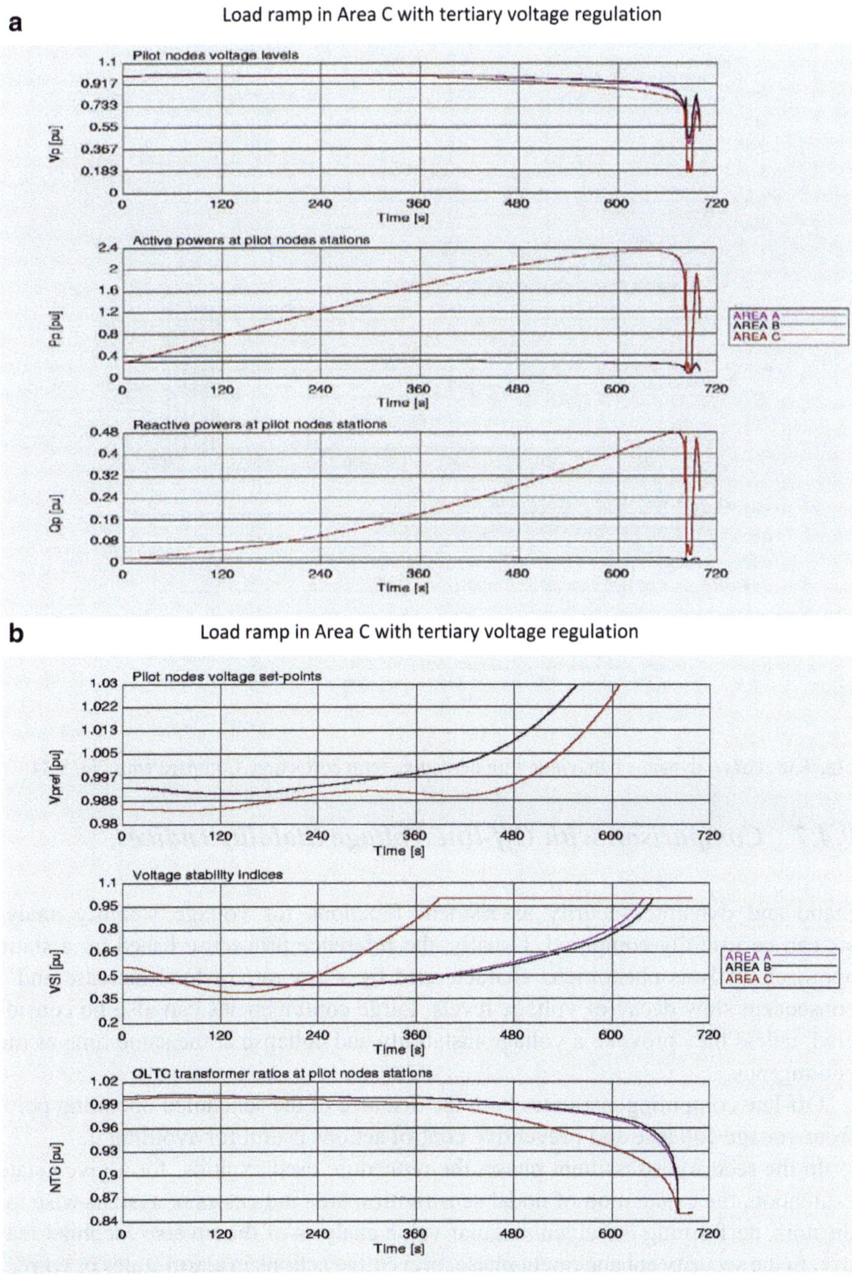

Fig. 9.35 $VSI_j(t)$ dynamic behaviour following Area C progressive load increase

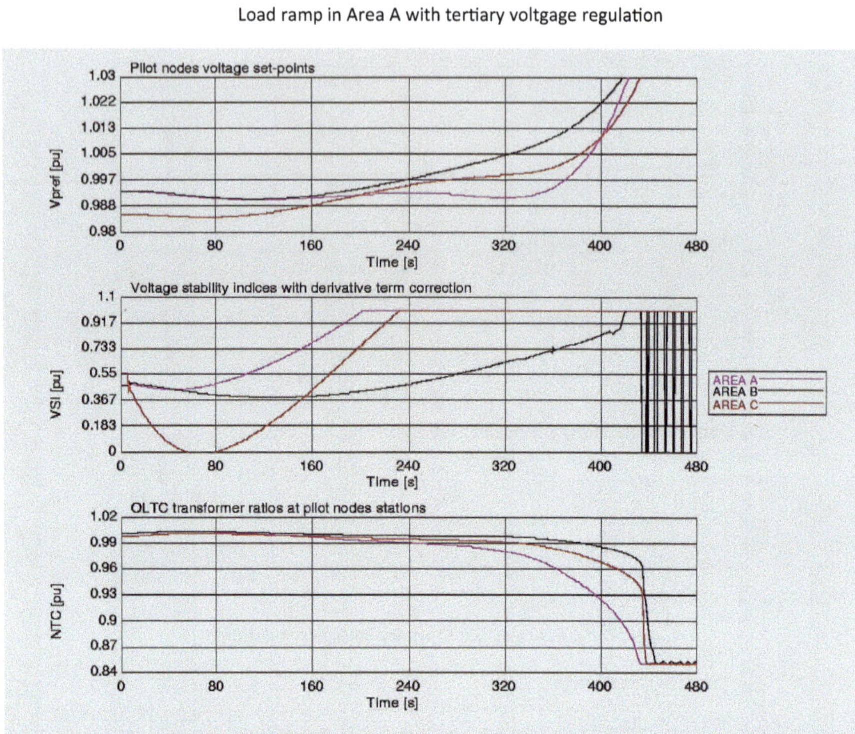

Fig. 9.36 $VSI_j(t)$ dynamic behaviour with derivative term correction. Compare with Fig. 9.34

9.4.7 Comparison with Off-line Voltage Stability Indices

Static and dynamic security assessment functions for voltage stability analysis can be usefully compared. Usually, the reference procedure based on a static approach analysis phenomena characterised by a low rate of load increase and a consequent slow decay of voltage levels. Large contingencies can also be considered, unless they provoke a voltage instability and collapse at the same time as the contingency.

Off-line computing evaluates both the distance of the scheduled operating point from voltage collapse and preventive control actions useful for avoiding it.

In the security assessment phase, the procedure used exploits, for a given state estimation, the calculation of nodal sensitivities, area indicators or system-wise indicators, performing the eigen/singular value analysis of the inverse Jacobian matrix. In the security enhancement phase, preventive actions in alarm states or corrective actions in emergency conditions are determined.

The sensitivities of the maximum singular value (second order information) are used as cost coefficients in a linear programming optimisation problem, which is solved for determining effective countermeasures.

Off-line procedure results can be used for an in-depth comparison and validation of the real-time and on-line index $VSI_j(t)$ fully described before, with reference

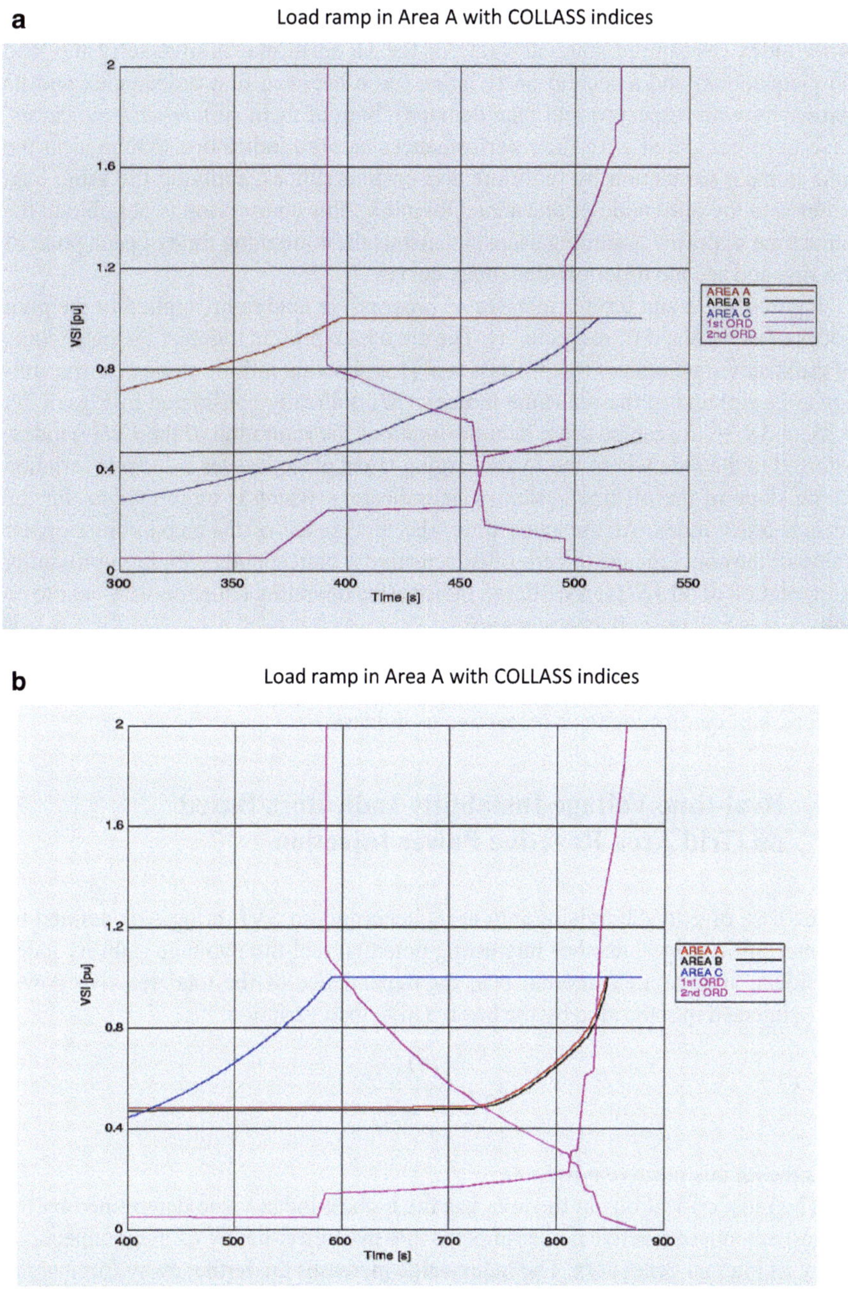

Fig. 9.37 $VSI_j(t)$ dynamic behaviour in comparison with off-line first and second order indices, Jacobian-based

to the generalised network in Fig. 9.32. More precisely, the behaviours of a first order index (maximum singular value of the Jacobian matrix inverse of the load flow equations) and a second order index (ratio between first order index and its sensitivity with respect to real load demand), both of them *system-wise* indicators, have been compared with their performances as *area* indicators, exhibited in the time domain simulation by real-time and on-line indices, applying the same load patterns to the pilot node of one area. Obviously, this comparison is possible in the same time scale, but assuming as instantaneous the computing time of each point of the first and second order off-line index curves.

Figures 9.37a and b refer to cases of progressive load ramp applied to the pilot nodes of areas A and C, respectively. The pictures represent together the behaviours of the steady state indices (the Collass tool [15] showing infinite slope near the voltage collapse) and of the real-time indices $VSI_j(t)$ already considered in Figs. 9.34, 9.35, and 9.36. As can be seen, in both situations the saturation of the $VSI_j(t)$ index, referred to the area where the load charging takes place, causes a sudden variation in the slope of the off-line system-wise indicators, which is more evident for the second order index. At the same time, the divergence of the maximum singular value or the convergence to zero of its sensitivity, corresponds with high proximity to saturation of all $VSI_j(t)$ area-based indices (the operating situation very near to an irreversible voltage collapse scenario).

In conclusion, the comparable results achieved through very different procedures validate the effectiveness and reliability of both on-line and off-line voltage stability indices, still confirming their respective usefulness.

9.5 Real-time Voltage Instability Indicators Based on Grid Area Reactive Power Injection

In the case of grid subdivision into areas according to SVR criteria or defined by country/utility edges, another interesting potential real-time voltage stability index considers, at each area internal bus, the dependence of the total reactive power Q_{in_tot} injected into the area by the bus reactive load variation:

$$I_{cr,i} = \frac{\partial Q_{in\_tot}}{\partial Q_i}.$$

Q_i is the *i*th bus reactive power.

The most critical bus in the area has the highest index value determined by the largest amount of reactive power injected into the area either by its interconnections or by its internal generators. The index value increases the farther away from which the reactive power comes and the higher the line losses are.

Because line currents also depend on the active powers, the above index has been improved by also considering the Q_{in_tot} link with the bus active power variation.

Based on a linearised analysis, the variation of total reactive power controlling the area (dQ_{in_tot}) can be linked with the bus's active and reactive load variations (dQ_i and dP_i) as follows:

$$dQ_{in\_tot} = \sum_{i=1}^{N} \frac{\partial Q_{in\_tot}}{\partial Q_i} \cdot dQ_i + \sum_{i=1}^{N} \frac{\partial Q_{in\_tot}}{\partial P_i} \cdot dP_i,$$

$$dQ_{in\_tot} = \sum_{i=1}^{N} \left(I_{crQ,i} \cdot dQ_i + I_{crP,i} \cdot dP_i \right),$$

with N representing the number of area load buses.

Two indices for the jth load bus are combined:

- Risk index linked with reactive power:

$$I_{crQ,i} = \frac{\partial Q_{in\_tot}}{\partial Q_i};$$

- Risk index linked with active power:

$$I_{crP,i} = \frac{\partial Q_{in\_tot}}{\partial P_i}.$$

Considering a constant power factor around operating point "0" and characterised at the ith bus by values of $P_{i,o}$, $Q_{i,o}$ and $\cos(\varphi_{i,o})$ ($A_{i,o}=P_{i,o}+jQ_{i,o}$), we can write:

$$dQ_{in\_tot} = \sum_{i=1}^{N} \left(I_{crQ,i} \cdot dQ_i + I_{crP,i} \cdot dP_i \right)$$

$$= \sum_{i=1}^{N} \left[I_{crQ,i} \cdot \left(\sin\left(\phi_{i,o}\right) \cdot dA_{i,o} \right) + I_{crP,i} \cdot \left(\cos\left(\phi_{i,o}\right) \cdot dA_{i,o} \right) \right]$$

$$= \sum_{i=1}^{N} \left[I_{crQ,i} \cdot \sin\left(\phi_{i,o}\right) + I_{crP,i} \cdot \cos\left(\phi_{i,o}\right) \right] \cdot dA_{i,o} = \sum_{i=1}^{N} I_{crPQ,i} \cdot dA_{i,o}.$$

The combined index $I_{crPQ,\,i}$ links active and reactive power risk indices by the $\sin(\varphi_{i,o})$ and $\cos(\varphi_{i,o})$ weights. Real-time computing of $I_{crQ,\,i}$ and $I_{crP,\,i}$ requires area computing based on PMU measurements/agents at area buses exchanging real-time data through a high-speed communication ring/hierarchy to compute and update at high speed the above indices. Through them it is possible to define an effective real-time area protection function as described at Chap. 11.

9.6 A Variety of Real-time Voltage Instability Indicators Based on Phasor Measurements Units Data

This section analyses and compares the performances of a number of proposed VSIs based on the § 9.3 real-time identification algorithm, elaborating grid bus local phasor measurements at a fast sampling rate. The main power of the proposed indicators lies in the real-time adaptive identification of Thevenin's voltage and impedance equivalents.

Algorithm effectiveness and robustness are guaranteed by a detailed sensitivity analysis and the comparison with off-line indices shown before; they are proven by the dynamic tests presented, with very clean and precise results, notwithstanding the continuous changing of the real system data (*50-Hz band*). This is thanks to the imposed identification high speed (a data-averaging based on the *last 80 ms* of sampling data).

After the reliability of Thevenin's equivalent identification algorithm is checked, the subsequent step consists in defining and testing a variety of real-time reliable indicators to be used in practice and being very simple, at low computation cost and mainly based on real-time identification results. They consider the distances between X_{Th}^{i} and $\left|\bar{Z}_{L}^{i}\right|$ or E_{Th}^{i} and $E_{Th}^{\max}$ (superior extreme), as well as the slope of the variables under identification to predict the approaching voltage instability.

The validity of and differences in the analysed indicators, and a consideration of the computing time they require, are checked at the EHV load and transit buses.

Risk reduction by combining more than one of the proposed indicators is analysed and tested, too. Important numerical results on the proposed real-time voltage instability risk indicators from the actual Italian EHV network are presented.

9.6.1 Real-time Indices Based on the Thevenin Equivalent Identification Method

The voltage instability identification algorithm described in § 9.3 allows the computation of a variety of real-time reliable indicators to be used in practice [45].

The simplest index, representing the ratio between X_{Th}^{i} and $\left|\bar{Z}_{L}^{i}\right|$, is here after named the *VSI-0* index. It indicates the maximum power transfer's correct instant (with an error on the order of a few tens of ms). Therefore, it is a useful reference for the other proposed indices. Analogously, the indicator *VSI-6* is another useful reference for comparison, being simply based on the voltage measurement at the considered EHV bus and not on the Thevenin's equivalent identification method.

The other proposed indicators VSI-1,…,VSI-5, in order to be useful in practice for power system control and protection, have to predict in real time and with high reliability the approaching of the *V-P* curve tip at the considered bus. Therefore, the correct indication of a high risk of voltage instability in advance of some or more seconds is their objective.

To test their correct performance and robustness, indicator parameters have been tuned on the basis of data coming from a very detailed dynamic simulation of the Italian transmission system in front of load increase and heavy unusual perturbations. Basically, all proposed indices have to be tuned on the threshold and the filtering. The threshold is the value the index function cannot overcome while the filter averages the *m* subsequent indicator values related to the *m* consecutive identification updates. The parameter *m* defines the size of the filter's "moving window" as

m times the sampling time. The results in § 9.6.2 refer to a sampling rate of 20 ms and $m=4$.

The proposed indices VSI-0, VSI-1, VSI-2 and VSI-3 are related to the Thevenin reactance identification, whereas VSI-4 and VSI-5 refer to the Thevenin voltage identification.

Index VSI-0

$$\frac{|\bar{X}_{Th}|}{|\bar{Z}_{\text{load}}|} \geq 1$$

This instantaneous index ($m=1$) is based at each step of sampling on the measurement and identification of Z_{load} and X_{Th}, respectively. VSI-0 indicates, as a reference, the instant of maximum power transfer.

Index VSI-1

$$\frac{|\bar{X}_{Th}|}{|\bar{Z}_{\text{load}}|} \geq 0.98$$

In the performed tests, index VSI-1 differs from VSI-0 simply for the threshold, diminished from 1 to 0.98, and for an averaging filter ($m=4$) to compute X_{Th}. Obviously, index *VSI-1* has as its objective to trigger voltage instability risk before VSI-0.

Index VSI-2

$$\bar{X}_{Th_i} + \rho_X \frac{d\bar{X}_{Th_i}}{dt} \Delta t > |Z_{\text{load}}|$$

Index VSI-2 makes use of the derivative term on X_{Th} to anticipate the instability limit. Gain ρ_X weights the derivative term contribution computed by considering a time interval Δt of "g" times the sampling interval. The filter on index computing is based on m steps. In the tests performed: $\rho_X=10$, $g=20$, $m=4$.

Index VSI-3

$$|\bar{Z}_{\text{load}}| - (\bar{X}_{Th_i} \sigma_{X_i}) \leq K_\sigma \sigma_{X\min}$$

Index VSI-3 makes use of the X_{Th} standard deviation σ_{Xi} to reduce the difference with respect to Z_{load}. The threshold is based on the minimum standard deviation to be achieved during the X_{Th} identification, weighted by the gain K_{σ}.

The filter on index computing is based on m steps. In the performed tests: $\sigma_{X\min} = 0.035,\ K_{\sigma} = 30,\ m=4$.

Index VSI-4

$$\left|\bar{E}_i\right| + \rho_E \frac{d\left|\bar{E}_i\right|}{dt} \Delta t > \bar{E}_{\max}$$

Index VSI-4 refers to the identification of Thevenin voltages $E_{Th}^i, E_{\max}^i$ and makes use of the derivative term on E_{Th}^i to anticipate the instability limit. Gain ρ_E weights the derivative term contribution computed by considering a time interval Δt of "g" times the sampling interval. The filter on the index computing is based on m steps. In the performed tests: $\rho_E = 10,\ g=20,\ m=4$.

Index VSI-5

$$\left|\bar{E}_{\max}\right| - \left|\bar{E}_i\right| < \varepsilon_{E_3} = 0.01\left|\bar{E}_{\max}\right|$$

Index VSI-5 refers to the identification of the Thevenin voltages $E_{Th}^i, E_{\max^i}$ and makes use of a threshold based on $E_{\max^i}$. The filter on index computing is based on m steps. In the performed tests, $m=4$.

Index VSI-6

$$\left|\bar{V}_i\right| + \rho_V \frac{d\left|\bar{V}_i\right|}{dt} \Delta t \le 0.85$$

Index VSI-6 refers to the measurement of Thevenin voltages V_i and makes use of the derivative to anticipate instability limit. The gain ρ_V amplifies the derivative term contribution, computed considering a time interval Δt of "g" times the sampling interval. The threshold cannot be higher than 0.85 in practice. The filter on index computing is based on m steps. In the performed tests: $\rho_V = 10,\ g=20,\ m=4$.

Table 9.2 Time in seconds to voltage instability given by each index

Bus	VSI-0	VSI-1	VSI-2	VSI-3	VSI-4	VSI-5	VSI-6
Brugherio	487.0	481.5	473.9	483.3	473.9	481.4	612.5
Poggio a Caiano	1138.5	1130.5	1117.1	1122.4	1117.1	1130.5	966.2

9.6.2 *Index Performance in Front of Load Increase*

This section shows the performance of the proposed indices applied to the EHV Italian electrical network. The data collected in computer simulation for the voltage and current phasors have a sampling rate of 20 ms. The Italian system analysed is the same considered in § 9.3.6, with dynamic models for the generator, AVR, OEL, governor, OLTC and loads represented as a static model with the voltage dependence of active and reactive power having exponents $\alpha=0.7$ and $\beta=2.0$, respectively.

Two sets of tests are performed at the load buses: one at the Brugherio 380-kV bus in Milano area and the other at the Poggio a Caiano 380-kV bus in Firenze area. These buses are the ones already considered for the identification algorithm tests.

Application to Brugherio and Poggio a Caiano Buses

The index test considers load increasing at the considered bus while the power factor is maintained as constant. At the neighbouring buses, load grows at a given rate, lower than at the considered load bus. The objective in testing the proposed indices is to check how they would behave as voltage instability point is reached at the bus most prone to instability. The load increase is consistent due to the power system's robust stability at the considered operating condition.

Table 9.2 summarises the performance of the proposed indices in terms of the time (in seconds), where a voltage instability alarm would be triggered.

Figures 9.38a–f show at the Brugherio bus simulated results obtained for each of the indices defined in § 9.6.1; at Poggio a Caiano bus, Figs. 9.39a–f show analogous results.

The results obtained for the Brugherio and Poggio a Caiano buses indicate similar performance in the indices derived from the identified Thevenin parameters, i.e., from VSI-0 to VSI-5. A different behaviour is observed from Index VSI-6, which is based on bus voltage magnitude. Figures 9.40a and b show the simulation of Index VSI-6 for the Brugherio and Poggio a Caiano buses, respectively.

In Figs. 9.38–9.40 and Table 9.2, one can see the anticipated time given by those indices that use derivative terms (shaded columns in Table 9.2). This places an important role on the weight of the derivative term. As an example, Fig. 9.41 shows the influence of the derivative term ρ_X in the index VSI-2 for the Brugherio bus. The plots are taken for values of $\rho_X=1$, 10 and 50. The larger the ρ_X, the wider the curves are. As expected, the anticipation effect due to the derivative term can be clearly

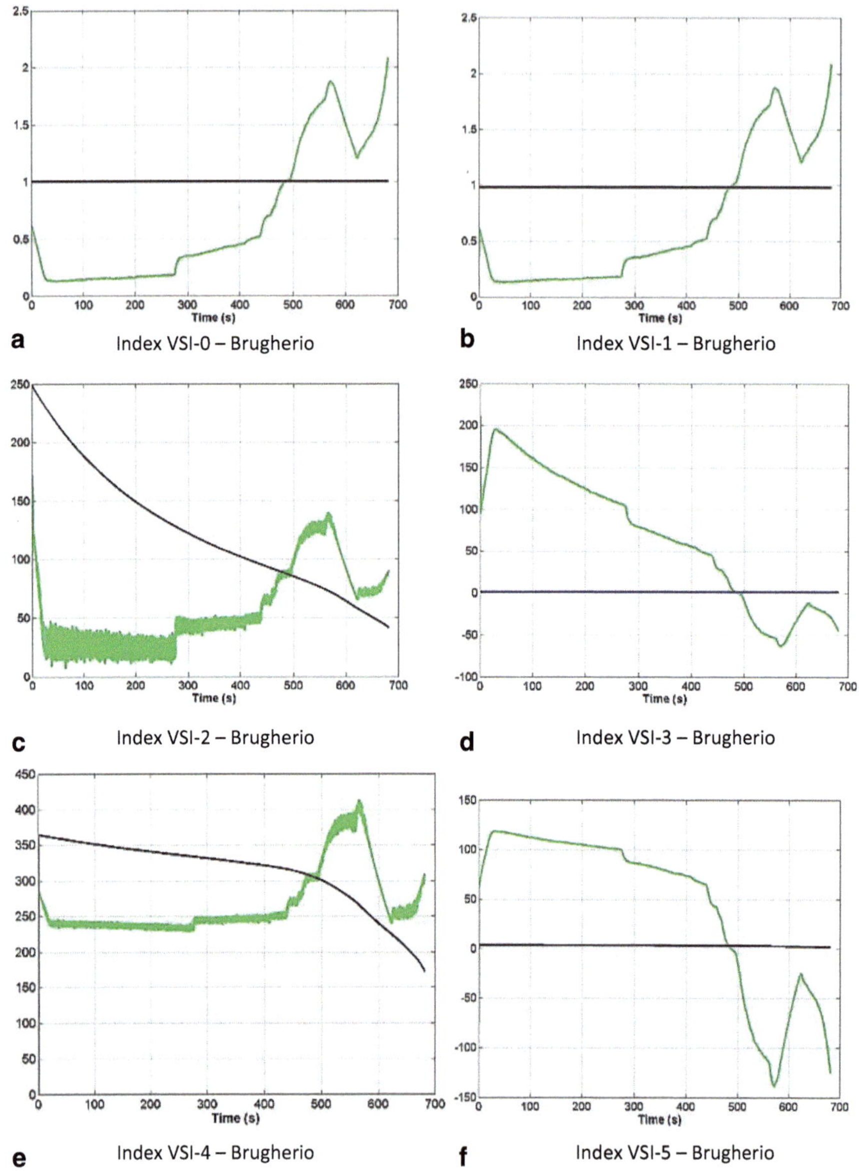

a Index VSI-0 – Brugherio

b Index VSI-1 – Brugherio

c Index VSI-2 – Brugherio

d Index VSI-3 – Brugherio

e Index VSI-4 – Brugherio

f Index VSI-5 – Brugherio

Fig. 9.38 VSI performance at Brugherio bus

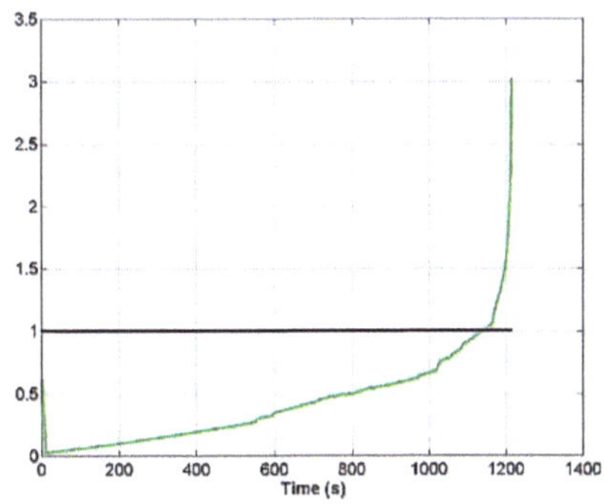

(g= a) Index VSI-0 – Poggio a Caiano

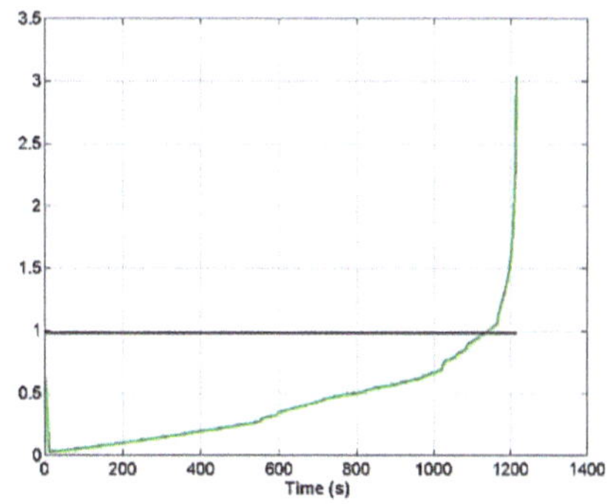

(h = b)Index VSI-1 – Poggio a Caiano

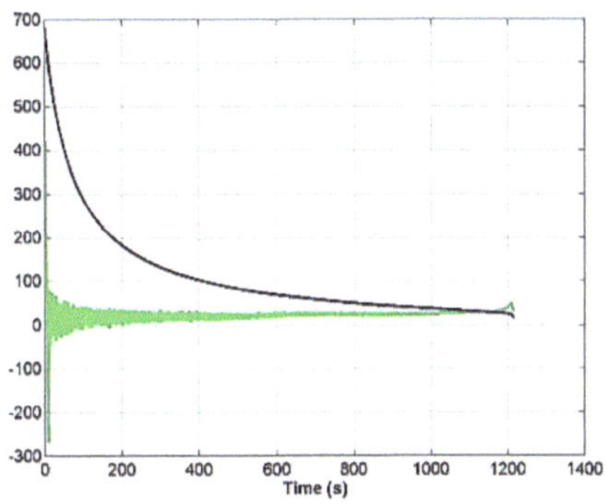

Index VSI-2 – Poggio a Caiano

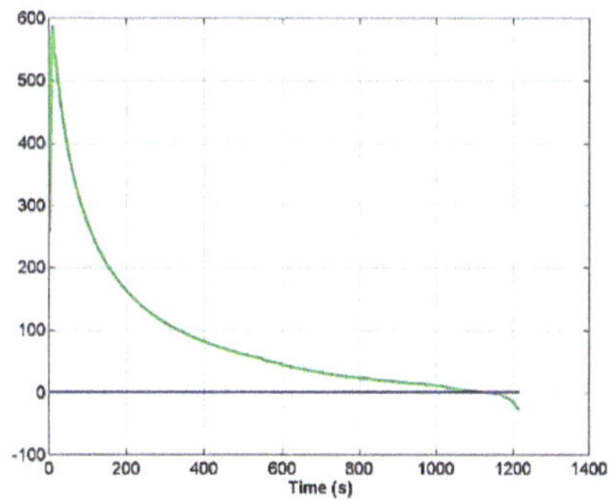

Index VSI-3 – Poggio a Caiano

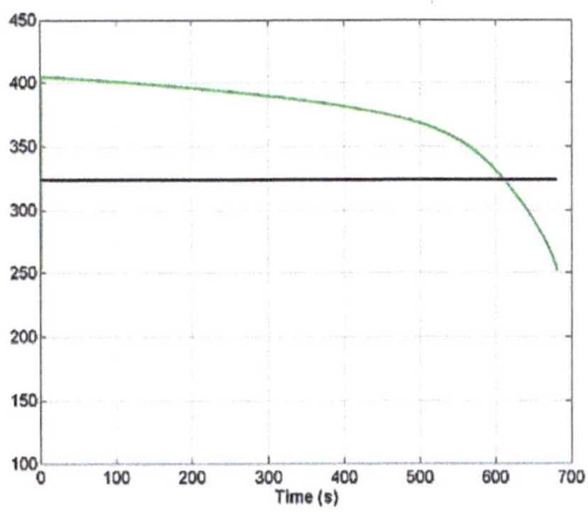

Index VSI-4 – Poggio a Caiano

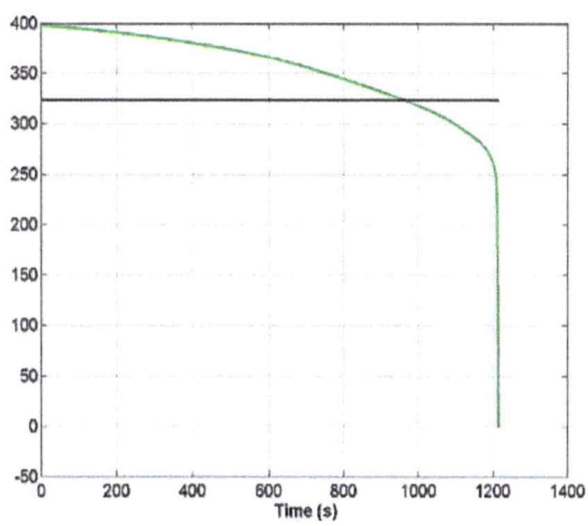

Index VSI-5 – Poggio a Caiano

Fig. 9.39 VSI performance at Poggio a Caiano bus

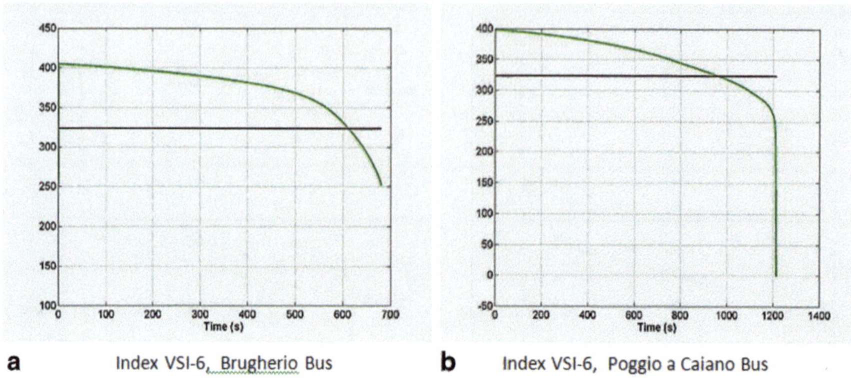

Fig. 9.40 **a** Index VSI-6, Brugherio bus; **b** Index VSI-6, Poggio a Caiano bus

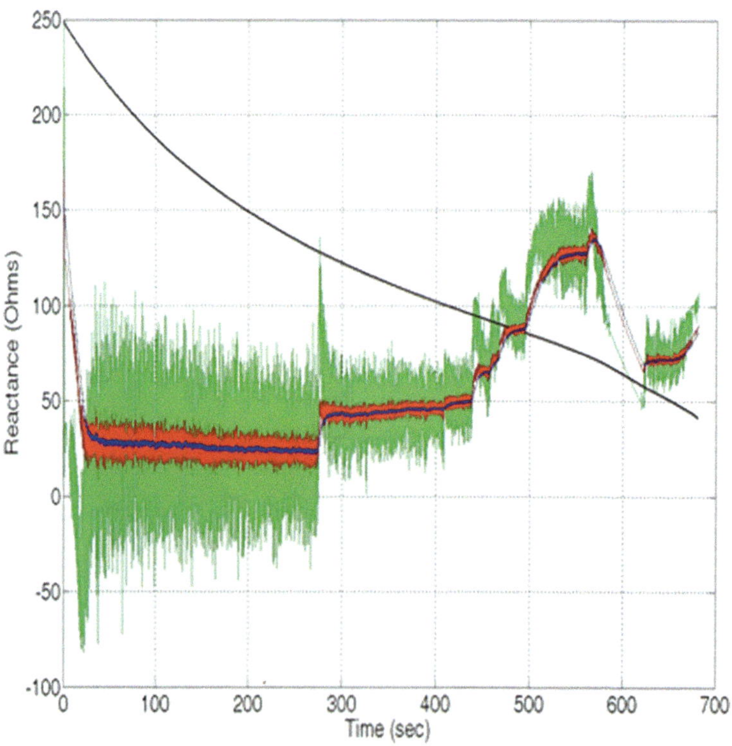

Fig. 9.41 Influence of the weight on the derivative term

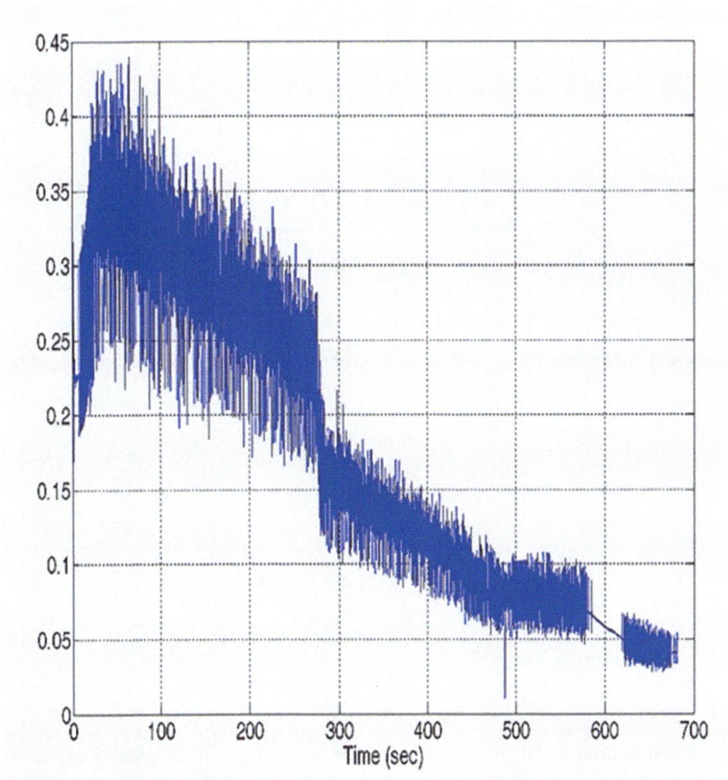

Fig. 9.42 Standard deviation for VSI-3

seen. The times the impedances become equal are 485, 474 and 439 s for $\rho_X = 1$, 10 and 50, respectively.

An important term for Index VSI-3 is the standard deviation computed in a sliding window of the identified Thevenin reactance. Figure 9.42 shows the standard deviation σ_{Xi} used in VSI-3. From simulated plots like this that it is possible to establish a priori the term $\sigma_{X\min}$.

It is worth noting the decrease in standard deviation as system approaches instability.

9.6.3 Index Performance in Front of Large Perturbations

The robustness assessment of the proposed indices is done by simulations in front of large perturbations widely affecting the identification process. Two sets of tests are shown—one considering a saw-tooth load variation and the other considering a step load variation. Tests consider the Poggio a Caiano load bus (Figs. 9.43, 9.44).

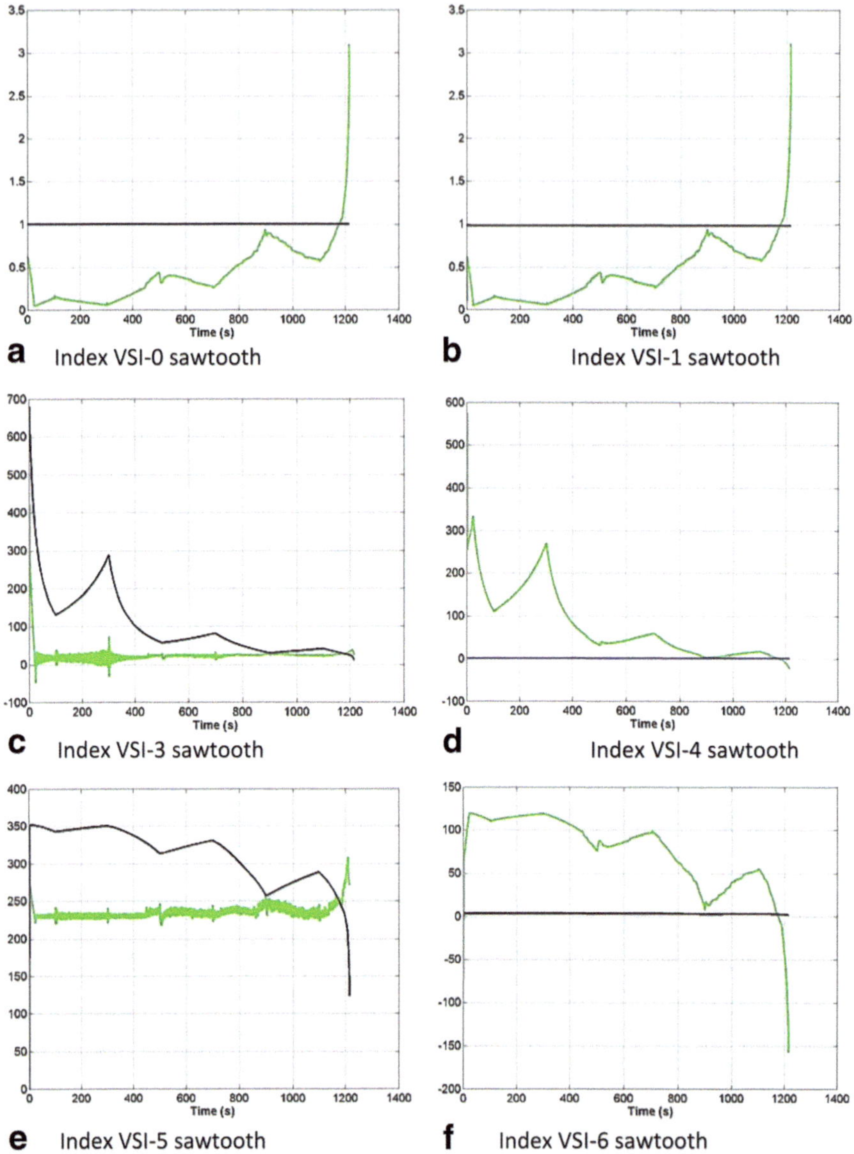

Fig. 9.43 VSI performance at Poggio a Caiano bus, saw-tooth test

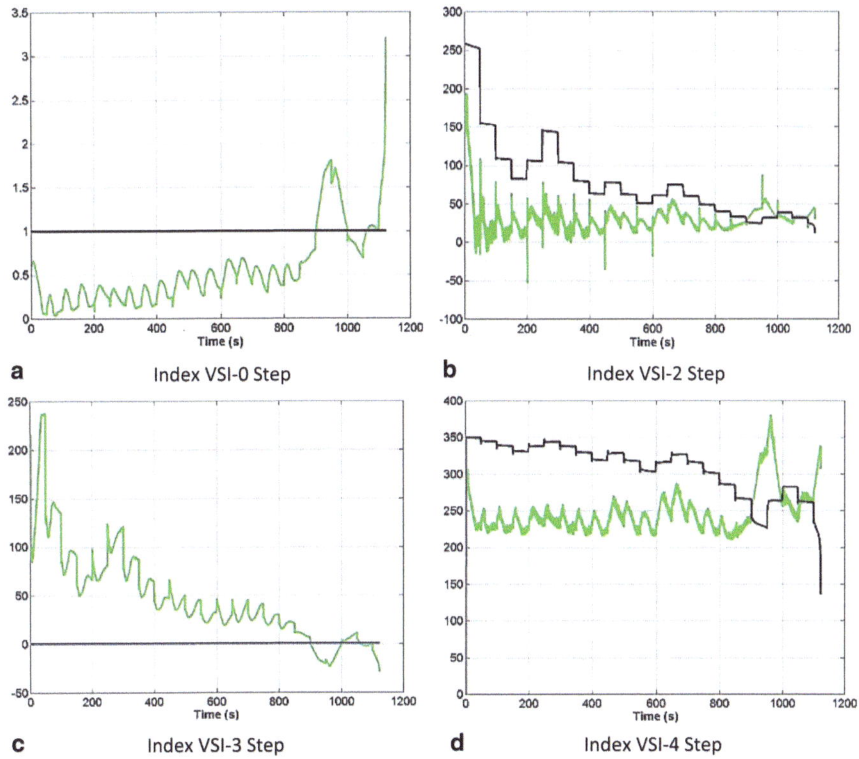

Fig. 9.44 VSI performance at Poggio a Caiano bus, stepping test

Table 9.3 Time in seconds to voltage instability given by each index

Load variation	VSI-0	VSI-1	VSI-2	VSI-3	VSI-4	VSI-5	VSI 6
Saw-toothed	1176.3	1173.7	1172.4	1171.4	1172.8	1173.7	848.3
Large steps	900.0	900.0	850.0	900.0	900.0	900.1	–[a]

[a] Reaches the threshold since the first load step variation

Table 9.3 summarises the performance of the proposed indices in terms of the time (s) where a voltage instability alarm would trigger.

Once again the different behaviour of VSI-6 compared to the other indices can be seen. Thus, we remark that relying only on bus voltage magnitude to anticipate voltage instability or collapse is not a very reliable strategy to adopt.

9.7 Final Remarks

The clear scope of the chapter is summarised below:

1. First, it distinguishes off-line voltage instability indicators from real-time ones, giving evidence to their relevant difference vis-à-vis practical contribution. Off-line indices are used for (planning, dispatching) studies, different than real-time indices, which are able to support moment-by-moment the power system operator and mainly contribute to the increase of power system security by their use in automatic control and protections.
2. The voltage instability phenomenon varies according to the operating closed-loop and real-time controls in the power system and it depends as well on real on-field load characteristics, which change for many reasons throughout time; fixing their model amounts to a limiting assumption. The different results coming from both operating control configuration and load model characteristics as well were adequately detailed.
3. Evidence is also given to the consistent difficulty of defining and developing a reliable and effective "real-time" identification of incoming voltage instability, not achievable by classic identification methodologies nor by on-line detailed process modelling. They are characterised by too-slow elaboration and computing, apart from the incompatible delay on state estimation upgrading they require.
4. Real-time voltage instability identification can come solely from an unconventional and computationally simple algorithm based upon on-field, real-time fast measurements provided by PMUs (short sampling time (ms) is the key to a correct real-time identification of the Thevenin equivalent) or coming from a real-time, wide-area voltage control system.

The objective of VSIs when operating on-line and in real time is to correctly anticipate in a short period the approaching of voltage instability. The achievement of this objective will surely allow relevant improvement in power system control and protection with the aim of increasing system stability and security. The proposed identification algorithm and voltage instability indicators underwent an in-depth check by dynamic tests performed on the Italian large power system through a detailed simulation model and data acquisition at the sampling rate of 20 ms. The effectiveness of the identification algorithm was confirmed by the very satisfactory results of the proposed indicators with minimum differences from one to another in recognising and indicating the risk of instability. Indicators including derivative terms can help one anticipate incoming risk, but the performed test in the presence of large perturbations showed the need to contain the gains of the derivative terms to avoid untimely/false alarms. The performed tests refer to a heavy load increased according to a continuous ramp, but also to saw-toothed and step variations, thereby checking the robustness of the proposed indicators in front of sign-changing in the load variation slope up to $\pm\infty$ in the case of stepping variation.

Results are very clean and precise, notwithstanding the continuous changing of the real system data (50-Hz band) and high speed ($m=4 \Leftrightarrow 80$ ms) imposed on identification. Reliable real-time indicators described can be effectively used in real transmission systems, being very simple and having a low computational cost. Their use in a wide-area multi-bus with more than one PMU is considered in Chap. 11.

References

1. CIGRE-TF 38-02-10 (1993) Modelling of voltage collapse including dynamic phenomena. CIGRE Brochure 75, Summary in Electra, pp 71–77
2. IEEE/PES, Power System Stability Subcommittee Voltage stability assessment—Procedures and guides. Special Publication, SP101 PSS
3. Van Cutsem T, Vournas C (1998) Voltage stability of electric power systems. Kluwer, Norwell
4. Taylor CW (1994) Power system voltage stability. McGraw-Hill, New York
5. Abe S, Fukunaga Y, Isono A, Kondo B (1982) Power system voltage stability. IEEE Trans Power Syst 101(10):3830–3840
6. Ajjarapu V, Lee B (1998) Bibliography on voltage stability. IEEE Trans Power Syst PWRS-13(1):115–125
7. Sauer PW, Pai MA (1990) Power system steady-state stability and the load flow Jacobian. IEEE Trans Power Syst 5(4):1374–1383
8. Lee BH, Lee KY (1991) A study on voltage collapse mechanism in electric power systems. IEEE Trans Power Syst 6(3):966–974
9. Van Cutsem T, Moisse C, Mailhot R (1999) Determination of secure operating limits with respect to voltage collapse. IEEE Trans Power Syst 14(1):327–335
10. CIGRE-TF38-02-11 (1994) Indices predicting voltage collapse including dynamic phenomena. CIGRE Report
11. Chow JC, Fischl R, Yan H (1990) On the evaluation of voltage collapse criteria. IEEE Trans Power Syst 5(2):612–620
12. Cañizares CA, de Souza ACZ, Quintana VH (1996) Comparison of performance indices for detection of proximity to voltage collapse. IEEE Trans Power Syst 11(3):1441–1450
13. Marannino P, Bresesti P, Delfanti M, Montagna M (1994) Voltage collapse proximity indicator for very short-term security assessment. Seminar on Bulk Power System Voltage Phenomena III–Voltage Stability and Security, Davos
14. Berizzi A, Bresesti P, Marannino P, Granelli GP, Montagna M (1996) System-area operating margin assessment and security enhancement against voltage collapse. IEEE Trans Power Syst 11:1451–1462
15. Berizzi A, Finazzi P, Dosi D, Marannino P, Corsi S (1998) First and second order methods for voltage collapse assessment and security enhancement. IEEE Trans Power Syst 13(2):543–551
16. Berizzi A, Zeng YG, Marannino P, Vaccarini A, Scarpellini PA (1998) Voltage collapse assessment in an on-line DSA procedure. Bulk Power System Dynamics & Control, IREP-IV, Santorini
17. Gao B, Morison GK, Kundur P (1992) Voltage stability evaluation using modal analysis. IEEE Trans Power Syst 7(4):1529–1542
18. Seydel R (1994) Practical bifurcation and stability analysis from equilibrium to chaos. Springer, New York
19. Overbye TJ, De Marco CL (1990) An energy-based security measure for assessing vulnerability to power system collapse. IEEE Trans Power Syst 5(2):419–427
20. Prapost KL, Loparo KA (1994) An energy function method for determining voltage collapse during a power system transient. IEEE Trans Circ Syst 41(10):635–651
21. Bompard E, Carpaneto E, Chicco G, Napoli R (1996) A dynamic interpretation of the load flow Jacobian singularity for voltage stability analysis. Int J Elec Power Ener Syst 18(6):385–395
22. Cañizares CA (1995) Conditions for saddle node bifurcations in AC/DC power systems. Int J Elec Power Ener Syst 17(1):61–68
23. Cañizares CA, Alvarado FL (1993) Point of collapse and continuation methods for large AC/DC systems. IEEE Trans Power Syst 8(1):1–8
24. Ajjarapu V, Feng Z, Long B (1999) Identification of voltage collapse through direct equilibrium tracing. IEEE Trans Power Syst 15(1):342–349
25. Ajjarapu V, Christy C (1992) The continuation power flow: a tool for steady state voltage stability analysis. IEEE Trans Power Syst 7(1):416–423

26. Haurer JF (1991) Application of Prony analysis to the determination of modal content and equivalent models for measured power system response. IEEE Trans Power Syst 6(3):1062–1068
27. Cañizares C, Ghasemi H, Reeve J (2004) Prediction of instability points using system identification. Bulk Power System Dynamics & Control, VI-IREP, Cortina D'Ampezzo
28. Van Cutsem T, Mailhot R (1997) Validation of a fast voltage stability analysis method on the Hydro-Québec system. IEEE Trans Power Syst 12(1):282–292
29. Begovic MM, Phadke AG (1989) Voltage stability assessment through measurement of a reduced state vector. IEEE/PES Summer Meeting, Long Beach
30. Savulescu SC, Oatts ML, Pruitt JG, Williamson F, Adapa R (1993) Fast steady-state stability assessment for real-time and operations planning. IEEE Trans Power Syst 8(4):1557–1569
31. Lamaitre C, Paul JP, Tesseron JM, Harmad Y, Zhao YS (1989) An indicator of the risk of voltage profile instability for real-time control application. IEEE/PES Summer Meeting, Long Beach
32. Ingelsson B, Lindström PO, Karlsson D, Runvik G, Sjödin JO (1997) Wide-area protection against voltage collapse. IEEE Comp Appl Power 10:30–36
33. La Scala M Trovato M Torelli F (1996) A neural network-based method for voltage security monitoring. IEEE Trans Power Syst 11(3):151–163
34. Doraiswami R, Appan RG, Liu W, Vijayakumar B (1995) An intelligent scheme to monitor the status of a power system. Elec Power Syst Res 33:151–163
35. Corsi S, Pozzi M, Marannino P, Zanellini F, Merlo M, Dell'Olio G (2001) Evaluation of load margins with respect to voltage collapse in presence of secondary and tertiary voltage regulation. Bulk Power System Dynamics & Control, IREP-V, Onomichi
36. Phadke AG (1993) Synchronised phasor measurements in power systems. IEEE Comp Appl Power 6(2):10–15
37. Vu K, Begovic MM, Novosel D, Saha MM (1999) Use of local measurements to estimate voltage-stability margin. IEEE Trans Power Syst 14(3):1029–1035
38. Milosevic B, Begovic M (2003) Voltage-stability protection and control using a wide-area network of phasor measurements. IEEE Trans Power Syst 18(1):121–127
39. Zima M, Larsson M, Korba P, Rehtanz C, Andersson G (2005) Design aspects for wide-area monitoring and control systems. Proc IEEE 93(5):980–996
40. Verbic G, Gubina F (2004) A new concept of voltage-collapse protection based on local phasors. IEEE Trans Power Deliv 19(2):576–581
41. Smon I, Verbic G, Gubina F (2006) Local voltage-stability index using Tellegen's theorem. IEEE Trans Power Syst 21(3):1267–1275
42. Corsi S., Pozzi M (2006) A real-time EHV bus-bar indicator of local voltage instability. IFAC Symposium on Power Plants and Power Systems Control, Calgary
43. Corsi S, Taranto GN (2007) Voltage instability: the different shapes of the "nose". Bulk Power Systems Dynamics & Control, IREP-VII Conference, Charleston
44. Corsi S, Taranto GN (2008) A real-time voltage instability identification algorithm based on local phasor measurements. IEEE Trans Power App Syst 23(3):1271–1279
45. Corsi S, Taranto GN, Guerra L (2008) New real-time voltage stability indicators based on phasor measurement unit. CIGRE Conference, Paris
46. IEEE Task Force on Excitation Limiters (1995) Recommended models for overexcitation limiting devices. IEEE Trans Ener Conver 10(4):706–713
47. Arcidiacono V, Corsi S (1994) A real-time voltage stability index for bulk system with secondary voltage regulation. Seminar on Bulk Power System Voltage Phenomena III–Voltage Stability and Security, Davos
48. Corsi S, Arcidiacono V, Bazzi U, Chinnici R, Mocenigo M, Moreschini G (1996) The regional voltage regulator for Enel's dispatchers. CIGRE Conference Session Group 39

Chapter 10
Voltage Control on Distribution Smart Grids

This chapter deals with control structures and functionalities contributing to distribution grid smartness in the presence of distributed generators (DG).

Primary cabin (PC) voltage control of a distribution grid that hosts the DG is analysed, assuming a control scheme that includes transformers (TR) with on-load tap changer (OLTC), but also generators and flexible AC transmission systems (FACTS), allowing their reactive power remote control. Therefore, more than one contemporary closed-loop control operates on the same variable (voltage) or on the strictly linked variable (reactive power). These combined efforts require proper coordination among operating control loops that generally have different dynamic performances. Promising proposals of alternative coordinated PC controls, including reactive power flow between HV and MV bus bar control, are presented. They are aimed to a great extent to achieve effective automatic regulation via simple and practical solutions.

The considered control schemes take into account different PC operating conditions, including the islanded network covered by a proper functional variant. The PC regulator considered here is also provided with automation functionalities that operate timely and proper switching of capacitor banks and shunt reactors when these are remotely controllable.

10.1 Introduction

A new primary cabin control and protecting system, mostly oriented to the case of the "active" distribution grid with operating distributed generators, has as its main objective the provision of adequate and complete support to automatic control of voltage, frequency and active/reactive power flows of the MV grid. Moreover, PC control can contribute to the back-up feeding of a neighbouring islanded PC grid.

To this end, the control system considered here necessarily must be a peripheral part of a more complex real-time control structure automatically coordinated by the distribution dispatching centre (DDC). More precisely, a distribution grid hierarchical control system is required for each DDC, having decentralised specific

S. Corsi, *Voltage Control and Protection in Electrical Power Systems,*
Advances in Industrial Control, DOI 10.1007/978-1-4471-6636-8_10

functionalities at the PC level while coordinating more PC controls by centralised functions at the DDC's distribution management system (DMS).

The introduced hierarchy will likely assign high speed and peripheral functions to an intelligent PC control system while it manages and coordinates slower control functions at a higher centralised regulation level.

10.1.1 Generalities

Generality of PC Voltage Regulation

1. In principle, each generator of the PC MV network automatically regulates generator edge voltage, thereby indirectly supporting voltage on its line-interconnecting bus. We cannot exclude the case of generators able to parallel the grid and operate until their MV buses are voltage-sustained by other generators. Both kinds of generators can be found, and their potential contribution to PC voltage control obviously differs;
2. The candidates for PC voltage regulation are HV/MV TR bus bars, traditionally the MV bus. In this case, the new PC voltage regulator could control all the local reactive power resources aimed at maintaining the desired voltage value on the MV bus. Operating the TR OLTC and realistically assuming its voltage stepping control to be active on the regulated MV bus bar, an adequate coordination and decoupling between the two parallel voltage control loops is required.
3. The impact that the new PC voltage regulation and its dynamics have on local MV protections must be properly assessed as well as its interaction with the neighbouring PC.
4. At the DDC, a centralised regulator aimed at defining, in real-time, voltage set-point references for each controlled PC is required. This DDC-DMS control function must take into account each PC MV grid structure and resource, electrical connections among controlled PCs, HV side voltages in the supervised MV area and reactive power flows at HV/MV interconnections. Moreover, any possible control requests on the HV voltage support, coming to the DDC by transmission network TSO, must be taken into account and processed.

Generality of PC Frequency Regulation

5. An exceptional case of when PC frequency regulation is required follows an instance of PC MV grid islanding until a direct connection to the HV bus bar or an indirect connection through the MV neighbouring grid is restored. The objective of PC frequency regulation is that an MV island continues to operate through control of local generators. Generator tripping is to be avoided during an islanding fault; in fact, generators are needed for their contribution to island local load balancing.

6. This control functionality requires that PC generators be controllable by a PC frequency regulator, which must be able to quickly recognise islanding passage, sufficient island controllability after the passage and containment of frequency and power variation within an acceptable range for both generators and local loads.
7. The impact of the PC frequency regulator on the local protections requires a proper dynamic design and assessment, including the possible dynamic interaction with the surrounding electrically coupled PC when it is part of the island.

Generality of PC Power Flow Regulation at the HV Bus Bar

This innovative control functionality consists of regulating in real time the active and reactive power flows between the PC HV and MV bus bars. Accordingly, a PC production market by DG resources at MV to sustain the HV grid is made possible. More precisely:

8. Each PC generation area will be controlled by the local PC power flow regulator by taking into account the available control margins, the allowed MV/HV line overloading and the operation bounds imposed by the DDC.
9. The DDC, having authority on a given PC, should manage its power flow exchange with the HV grid according to all the linked aspects by also including energy metering and computing and certifying the supply provided by each generator.
10. The DDC dispatcher defines set-point references of active and reactive power flows of each PC as well as their real-time updates by taking into account: production availability of each controlled PC; the present/actual (at any moment) PC internal load value; existing contractual obligations and operating limits; and the requirements coming from the TSO, without neglecting to take into account the operation and security needs of the MV grid.
11. The impact of power flow control at the HV bus bar and its control dynamics on PC protection must be thoroughly assessed, including the dynamic interaction with the neighbouring PC under different operating conditions.

Generality of PC Back-Up Feeding by Neighbouring MV Network

After losing its HV connection and feeding, a PC MV network can, in principle, be re-fed by a neighbouring MV grid. This new feeding can be manually or automatically operated by the PC control. The connection is enabled by a local/remote compatibility check of the lines' overload bounds as well as a check of the operating scheme of the feeding PC MV bus bar and its MV/HV interconnection control.

10.1.2 Chapter Objective

The "smartness" of a distribution grid derives from the availability of its control and automation functions, like those briefly introduced above. These functions may differ in some aspects because of possible variations in the PC electrical scheme, the type and size of distributed generators and differences in controllability.

This chapter deals only with distribution network voltage and reactive power regulation and possible alternative schemes [1–19] by orienting choices of these schemes as much as possible towards simplification of control complexity and proposing cost-effective solutions.

On control schemes, the de facto solution proposed here is a PC control philosophy that does not end with evaluation of the principle but rather with identification of more feasible and effective control solutions to put in practice, according to MV grid characteristics and operation needs. To this end, we should ask whether there is any reason for—or practical interest in—PC HV/MV interconnection reactive power flow regulation. Moreover, we should ask, "Are PC situations with large, continuous reactive power flow towards the HV bus bar plausible and of the most common sort?" If not, we should clarify whether cases of real interest host a limited number of generators at each PC MV grid so local reactive power resources are limited and their HV voltage support unrealistic. In this case their proper and appreciable use should be concentrated more on MV voltage support.

Lastly, we must ask, "Is it not true that the best contribution to transmission network voltage support (including local load voltage support) consists of providing achievable control of PC voltage through local control resources without asking it support the transmission grid?" Therefore, the preliminary and preferred way to address the choice of PC control solution is to identify more precisely viable and meaningful scenarios, ones that would be of interest to the MV ancillary services market and which would provide a solid hope of success and utility.

10.2 Generalities on Medium Voltage Grid and Primary Cabin Schemes

The next figures represent the basic medium voltage grid and primary cabin unified schemes of Italy. A distribution grid that is subject to high DG penetration is assumed. Such a grid (Fig. 10.1) has a radial structure, with electrical lines characterised by an approximately unitary ratio between the resistance and reactance values. Therefore, power distribution to the loads penalises line feeder voltage values. With generators located on the one trunk, power flow is altered both in magnitude and direction, causing changes in voltage profiles. Such variations are significantly greater as ratio of DG to load increases.

In principle, among all the DG resources, synchronous generators contribute to sustaining local voltages in the same way as do generators interfacing the grid by means of static converters. Commonly, the DG is largely enslaved to industrial production processes; therefore it is more closely tied to possible changes of the active power produced, used to control voltages. As is obvious, the hypothesis assumed here is that active power flows cannot be modified by PC centralised voltage control unless the protection functionalities require it.

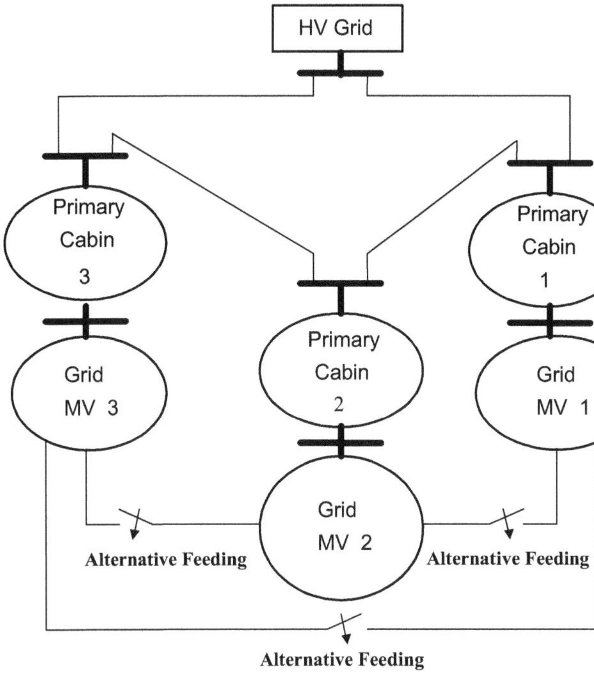

Fig. 10.1 Basic scheme of medium voltage grid

MV grids can be interconnected or electrically separated. Through their primary cabins, MV grids are connected to the HV grid. In the case of islanding, the MV grid is isolated from the HV and the other local MV grids, as well. The possibility of feeding an isolated MV grid by a neighbouring MV grid is also considered.

A PC operating scheme usually shows one HV bus bar and two separated MV bus bars (see Fig. 10.2). Along feeder lines the other nodes are of loads, compensators and generators. The choice of the voltage-controlled bus bars strongly impacts control scheme complexity, reliability, cost and DG management.

Fig. 10.2 Primary cabin unified scheme in Italy grid

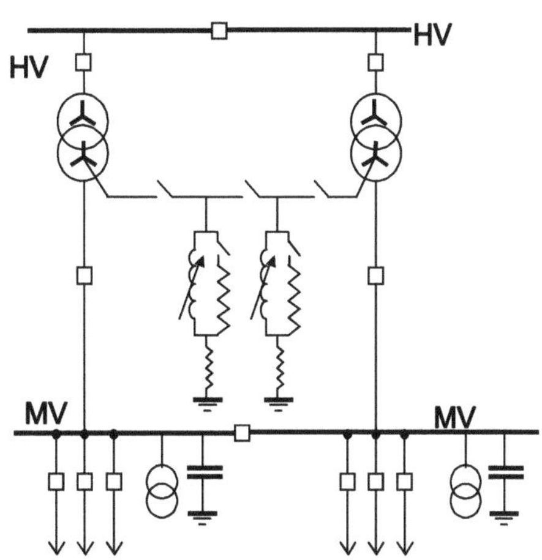

In agreement with the present proposal, all generators involved must be remotely controlled with continuous adjustment of their primary AVR voltage set-point values, as required by the *PC voltage regulator* (PCVR), which will impose on the local MV or HV bus bars the voltage set-point value sent by the DDC.

10.3 Generalities of Primary Cabin Voltage Control

We must keep in mind that PC MV bus bar voltages assume values very dependent on nearby transmission grid voltage controls. This involves the PC regulator's continuous and considerable effort, which may in some cases achieve high frequency variation and saturating conditions. Therefore, notwithstanding the need for this regulation to be quick and adequately sized to ensure effective controllability, sometimes it might not be up to the task.

It is clear from the above that PC voltage needs to be controlled by effective coordination between a regional transmission dispatcher (TRD) and a distribution dispatch centre (DDC) competent with regard to the PC considered. An in-depth analysis of such coordination is not done here, but the point is a relevant one vis-à-vis grid operation problems and data exchange among grid control centres.

We have also said that the PCVR will be expected to maintain, with relatively fast dynamics, PC bus bar voltage values (remotely set DDC) by facing changes in the local load, in the voltage variation coming from the transmission grid, and then upgrades of reactive power flow exchanged with the HV bus bar, as well as changes in the local provision of reactive power due to a modified operating point or to thermal problems or limitations imposed on each generator.

Primary cabin voltage regulation will also be responsible for maintaining adequate voltage on the HV/MV bus bars, even in front of large local perturbations. In addition to continuous generator control, it will also operate discontinuous switching of local capacitor banks and shunt reactors (for a significant amount of reactive power control), limiting the number of such operations to the minimum extent necessary and only in extreme conditions at approaching control generator saturation.

PCVR ultimately will operate an effective and timely coordination of local resources, taking into account the dynamic characteristics of each, minimising and sharing the control effort among them and limiting switching manoeuvres to the minimum amount necessary, thereby increasing the life of the power factor correction components. Doubts persist as to which control scheme best meets these clear objectives.

To define a PCVR control scheme proposal, it is first important that we clarify the answers to the following questions:

- Is it better to directly regulate MV voltage at the cabin bus bars or at the buses along MV lines?
- How does increasing the complexity of the voltage control structure improve its effectiveness?

- How is reactive power control among generators that operate on the same PC properly shared?
- How are PCVR and OLTC controls coordinated, with both operating on the same PC bus bar?

Considering the first two questions and the practicability of possible alternative solutions for a distribution grid, the following guidelines are apparent:

- The PCVR should pass from classical discrete and slow transformer tap control, completely dependent on HV side voltage robustness supported by the transmission grid with autonomous resources, to fast automatic regulation of PC bus bar voltage based on local reactive power resource control (i.e., the generators of the distribution grid). Furthermore, on the one hand, restricting voltage regulation to PC bus bars simplifies and significantly reduces the complexity of the control solution for adjusting the voltages of all load and generation buses; on the other hand it does not limit load bus control performance, as will be made clear.
- MV voltage regulation focuses on the objective of the quality of voltage ancillary service provided to the load and then to MV customers. In this case, the contribution to support the local HV network is less, valuable and therefore it is difficult for the ancillary service provided to be seen and remunerated by the TSO. Because, as seen, a PC has more than one separated MV bar, the voltage regulation of each individual bus bar via its own control resources requires a (PC) multiple voltage control system (one control for each MV bus bar). Therefore, many feeder regulators will operate in parallel, requiring a specific configuration and management of the voltage regulation itself, depending on the state of the switches connecting the MV bars.
- Voltage regulation at the HV side shifts the objective to HV grid support, always ensuring a good voltage quality at the MV side. In this case, the ancillary service offered can be more easily recognised and remunerated by the TSO, and its effectiveness is to be considered comparable, if not superior, to that of reactive power flow control towards HV. Moreover, in this case, normally only one voltage needs to be regulated: the PC HV bus bars are normally exercised at a closed disconnector. This greatly simplifies the structure and complexity of the control system with respect to MV side voltage regulation.

It should be added that HV side control moves the problem of dynamic coupling between control loops from inside to outside the primary cabin, among PCs electrically coupled at the HV side. In the presence of this regulation mostly provided by PCVR, the OLTC greatly reduces the number of manoeuvres, with obvious advantages to the life of the transformer tap changer.

On the recognition of the service provided, each controlled generator may be remunerated according to its "capability", made available to the control and hours of operation under automatic coordination by the PCVR. Minor, and to be ignored in terms of economic recognition, is the amount of reactive power delivered or absorbed by each generator.

From these preliminary comments on PC voltage control, with MV grid hosting the distributed generators, some important points can be fixed:

- Classic OLTC tap control is a poor solution. A new, multifunction, control solution is desired;
- The DDC should "in principle" update local PC voltage set-points in real time by taking into account HV side voltage values, HV/MV reactive power flows and the commercial *V, Q* bounds.
- Attributing to tap manoeuvres objectives other than local voltage support is the way to address possible advances in PC voltage control solutions.

To add more detail: OLTC-tap manual or automatic stepping control of transformer MV value, maintained inside a confidence band around a set-point voltage value, constitutes a conventional that:

- Requires solid voltage at transformer HV side in order to effectively operate;
- Is affected by too-slow dynamics with respect to real grid voltage variation;
- Contributes to an increased risk of voltage instability that could require the locking of the OLTC transformers.

On the contrary, automatic and continuous reactive power control of rotating and static generators is novel PC voltage control, which assumes that:

- All rotating synchronous and asynchronous generators as well as synchronous compensators contribute to distribution grid voltage support;
- An analogous contribution comes from static compensators (SVC, STATCOM, and UPFC) and generators that are grid-connected through static converters provided with reactive power control;
- Compensating equipment like switchable capacitor banks (SCB) and shunt reactors (SR) also contribute, even if with lower dynamics.

Therefore, considering automatic closed-loop control of these resources, as is the case with PC secondary voltage regulation (PC-SVR), it is possible to speedily regulate PC MV (HV) buses to their desired voltages and increase power system security.

On the remaining two questions and the practicability of possible alternative solutions for distribution grid voltage control, the PCVR should:

- Operate an effective and timely coordination of PC resources by taking their dynamics into account and properly sharing the control effort among them;
- Minimise overall PC control effort as well as the number of switching manoeuvres to increase the life of PC compensating equipment.

About coordination between the PCVR and OLTC:

- Because the two are distinct closed-loop automatic regulators, unstable and conflicting controls of the same voltage is to be avoided;
- Taking into account their different dynamics, the PCVR would have to regulate PC voltages, while the OLTC would guarantee the transformer ratio allowing the full use of over- and under-excitation resources of the PC generators, as well as avoid machinery voltage limits being overcome.
- The OLTC could alternatively control reactive power exchange with the HV grid, limiting it inside a confidence band.

10.4 PCVR Basic Control Schemes

The PCVR proposed here is a continuous, centralised, closed-loop control system, aimed at regulating the voltage of the PC HV bus bar or the MV bus bars at each independent PC MV bus.

PCVR-HV is a unique voltage regulator, controlling in real time all PC generators and static compensators as a PC-SVR would do. The control structure is very simple in this case. The PCVR-HV mainly sustains the local HV side voltage, always guaranteeing acceptable values at the MV side. The service offered to HV side can be easily recognised by the PC and remunerated by the transmission grid dispatcher.

A PCVR-MV consists of parallel independent voltage regulators of PC MV buses, each controlling an independent MV bus bar with its feeder generators. This is a more complex control structure and entails more complex logics than PCVR-HV, and possible dynamic interactions among parallel controls have to be avoided. Lastly, a PCVR-MV mainly sustains local MV side voltages.

We need to point out that MV feeders with generators have a more solid voltage profile than those without, due to the local AVRs. Also, nearby load buses gain a voltage benefit from local AVR control. Furthermore, line buses with generators under PCVR control show fewer higher/lower voltages than corresponding PC MV bus bars, due to their reactive power delivery/absorption.

The voltage profile along each feeder line (Fig. 10.3) can therefore be obtained by the automatic value limitation of each reactive power control signal $Qref_i$, aimed to maintain a generator's local MV voltage inside a proper confidence band. Recognising that this simplifying feeder voltage objective is sufficient, we can easily achieve it by the PCVR through line bus voltage measurements and PC generator remote control.

The PCVR-MV may also control the transformer tap in a coordinated way, as described next.

10.4.1 OLTC Operation in Presence of PCVR

Characterised by the well-known discrete stepping slow control, the main task of an OLTC is to adjust the transformation ratio VHV/VMV, always guaranteeing the full use of generators inside their over- and under-excitation limits, and to maintain the machinery between minimum and maximum voltage ties. Therefore, the OLTC is very useful to the correct working of the PCVR when its main objective is the balancing of MV voltage deviations for external reasons (HV grid voltage steps) or for internal ones (large load variations).

This OLTC support can, in principle, be required by the PCVR:

- Either in the case of sufficient autonomy by the PC MV grid on local voltage control, able to maintain at about zero the reactive power exchange between MV and HV bus bars;

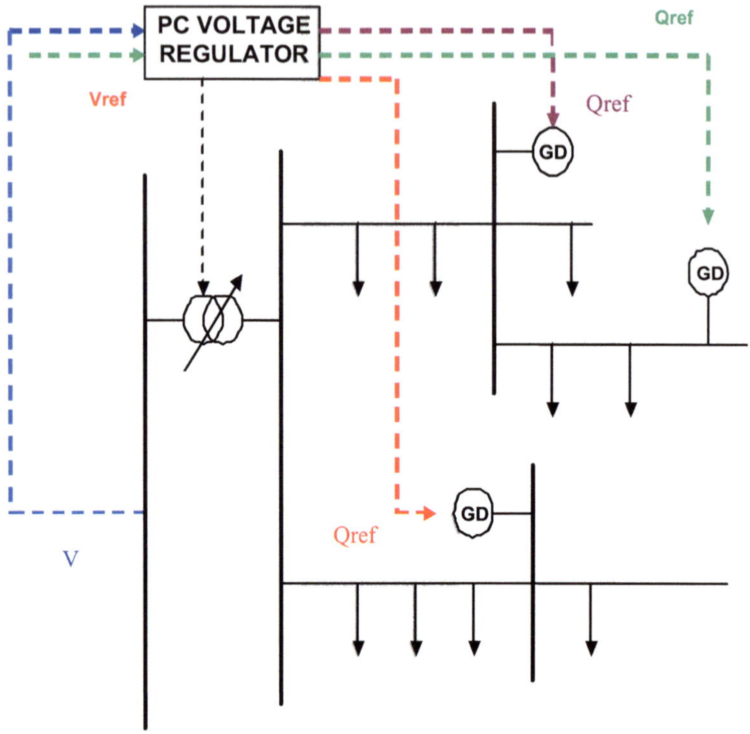

Fig. 10.3 Primary cabin voltage regulation scheme

- Or with constant reactive power flow exchange defined by MV and HV operators' reciprocal support agreement;
- Or, in the general case, with a variable reactive power flow allowed by a reciprocal MV-HV dynamic support agreement.

The objective of adapting OLTC control is to guarantee, under different operating conditions, full generator use via frequent recovery of operating conditions in the middle of their normal operating fields; this is a complex, not-easily managed, tap control. On the other hand, the PCVR, characterised by faster dynamics than the OLTC, speedily operates and first seeks to regulate MV bus bar voltage, so charging the overall control effort on the local generators until saturation is reached. In this way, the OLTC remains almost unused and becomes active only after the PCVR's supervening saturation.

From what is said above, it clearly appears that the two voltage controls must observe each other, and that OLTC will be called to operate first to bring PCVR out of saturation and then to maintain the generator margin of controllability at safe values. This functional requirement is not optimised from the classic OLTC voltage control loop, which will rarely work in the presence of the faster PCVR. Therefore, it will be necessary to subject the tap control to the PCVR's functional state, maintaining unchanged the characteristics of OLTC open-loop dynamics and controlling the UP and DOWN commands as follows:

- Achieving containment of the generator control effort when too high;
- With lesser priority, maintaining reactive power flow exchange with the HV grid as consistently as possible with contractual arrangements associated with the "voltage ancillary service" either provided by the HV grid or supporting the HV grid via the PC MV generators.

Obviously, in a PC grid emergency, the PCVR will quickly become saturated and the OLTC will be controlled to slowly recover the MV voltage, in case HV support is still consistent. OLTC control acts either in an open loop by the PCVR or in a closed loop by itself.

10.4.2 Islanded Grid Voltage Regulation

Islanded grid voltage regulation is very simple, being fully entrusted to the PCVR and therefore to the internal PC grid reactive power resources. The control scenarios defined above and linked to local reactive power resources are applied in this case.

10.4.3 Automatic Voltage Regulation of HV or MV PC Bus Bars

The proposed control system is characterised by a structure of a centralised type that regulates the voltage of the PC HV bus bar or of the PC MV individual bus bars. The structure, which is apparently the same in both cases, must be detailed, providing two substantially different solutions.

Because a homogeneous distribution of generators within a PC is not guaranteed, some MV feeders may have a few generators while other feeders can host many control generators. Therefore, a centralised system that regulates HV bus bar voltage also supports voltages along MV feeders which lack generators; this is a better situation than what happens in the case of MV separate bus bar controls whose mutual support in voltage regulation is less effective.

On an MV feeder, buses with control generators will have, as mentioned, voltages slightly higher or lower than those of respective MV PC bus bars. In any event, MV bus bars without generators will have the greatest voltage variation, even if variation is definitely less than when PCVR is absent.

With this type of centralised PC control, the maintenance—within normal operating range—of generator voltage values along feeder lines can be obtained either locally on the generator control interface or by centralised PC control. In the first case, the PCVR interface at the generator side not only calculates in real time the generator's reactive power set-point values, but it restricts these references in delivery or absorption for high or low voltages, respectively. In the second case, the interface at the generator level is simplified and the calculations and limits as set out above are transferred to the centralised PC control.

The architecture of the control system proposed is represented in Fig. 10.3; it is a centralised control of the HV bus bar aimed to maintain voltage V to reference V_{ref} through specific reactive power controls Q_{ref} sent to each of the feeder generators.

The figure may also show a single PC MV bus bar, therefore also lending itself to describing the voltage regulation of such a bus by simply shifting measurement V from the HV to the MV bus bar. It therefore consists of a PC centralised voltage control of each MV bus bar to which at least one generator is connected. Each control signal Q_{ref} will avoid, on local generation buses, voltages that lie outside a properly defined confidence band. Thus, it is assumed that all MV generators considered are provided with primary reactive power control.

Moreover, as was already introduced, OLTC transformation ratio optimisation is occasionally required by the PCVR basically to increase the degree of generator controllability.

Feeder bus voltage measurements are useful for real-time monitoring of when the normal operation confidence range is exceeded and then intervening in the generator control variable $Q_{\text{ref}i}$ for the necessary voltage limitation.

The electrical connection reclosure between PC MV bus bars does not require special intervention in the case of PCVR HV side voltage regulation, whereas transformer paralleling at the MV side involves PCVR management of the MV bars' OLTCs with a unique coordinating control (done with the already existing functionality or still easy to implement on the PC control).

Instead, in the case of MV bus bar voltage regulation, the reclosing of the electrical connection determines the transition of the two parallel PCVR MV voltage regulators from the "master-master" to the "master-slave" configuration.

In PC "island" operation, we distinguish the PC grid separateness from the HV transmission grid or the islanding of the single MV feeder from its transformer.

In the first case and in the presence of HV bus voltage regulation, this control continues to operate as if the interconnection to the HV grid were still active. The only difference is the lack of exchange of reactive power with the transmission network. If this flow were a relevant input before islanding, then the island voltage regulation would go rapidly into saturation.

In the same situation but with MV bus bar voltage regulation, similar considerations apply in a scenario of multiple voltage regulators in parallel, one for each MV bus bar. In the second case, that of a single MV feeder to separate, automatic mutual support between MV feeders does not operate, and the MV island must be able to independently maintain an acceptable feeder voltage level through its own control resources. Therefore, MV bus bar voltage regulation continues to operate as if the operating interconnection to the transformer were active. The only difference is the lack of reactive power exchange with the transmission grid and the other MV feeders.

If these flows were relevant inputs before islanding, then the MV island voltage regulation would go rapidly into saturation. Conversely, if the operating control before the islanding occurrence were on the HV bus voltage, this control would continue to operate without the support of the islanded feeder resources. Instead, islanded MV generators continue to operate under local primary voltage control only, unless they are aggregated to an MV island voltage regulator, which would be activated for this purpose and last until the resources return to the HV control, at which time the MV feeder would again be put in parallel to the PC transformer.

10.4.4 Block Diagrams of PCVR Control Functions

The PCVR control signal that is sent to each generator is the set-point of the generator reactive power control loop, characterised by a first order dynamic with a dominant time constant on the order of a few seconds.

This reference, calculated by the PCVR separately for each controlled generator, takes the reactive power over- and under-excitation limits into account, which obviously change depending on power system operating conditions. Such a power limit is called *DFIG*—Q_{lim} in the Fig. 10.4 block diagram, for the case of the wind DFIG (double fed induction generator); the case of the synchronous generator is named *Synchronous*—Q_{lim} in the figure.

Limiting values are computed in real time by the PCVR according to specific fitting functions for each generator, which are mainly based on real, experimental "capability" measurements of the generators themselves.

The Q_{ref} request normally changes within the field of controllability defined by the Q_{lim} but can undergo further limitation in case the generator bus bar exceeds the fixed voltage confidence band (V_{lim}). Also, the limitation imposed by V_{lim} requires a specific control scheme, shown in Fig. 10.5.

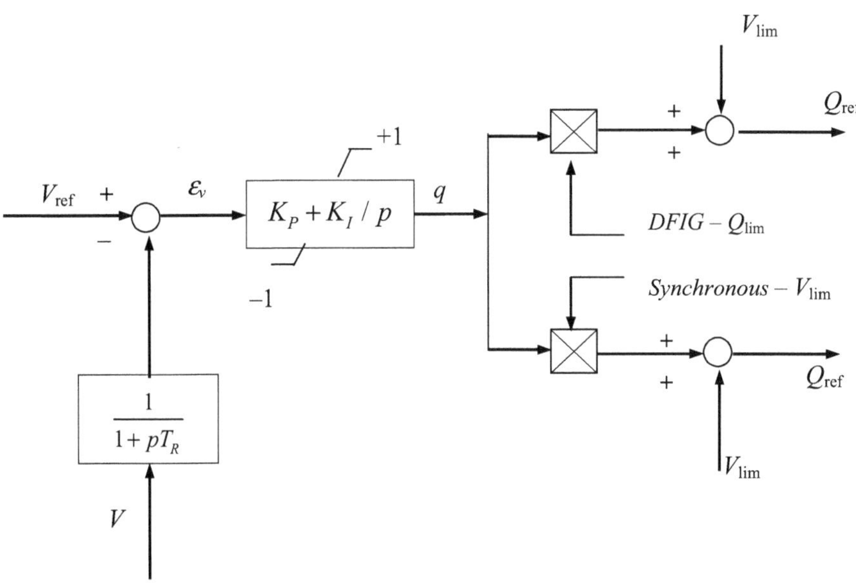

Fig. 10.4 Block diagram of primary cabin voltage regulator: PCVR

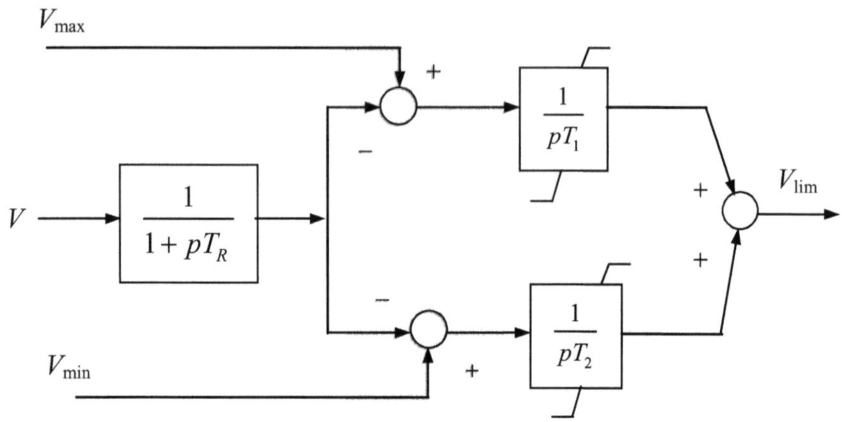

Fig. 10.5 Q_{ref} limiting loops maintaining generator voltages inside $V_{max} - V_{min}$ field

The regulator in Fig. 10.4 can be used for voltage regulation of both PC HV and MV bus bars, depending on the point of voltage measurement *V acquisition* and correspondingly on the controlled generators (there are as many outputs Q_{ref} as there are generators controlled by the PC voltage regulator).

The voltage regulator of proportional-integral type automatically updates its output q as a function of the difference between the reference voltage (V_{ref}) and the current value of this voltage (V). At steady state this difference vanishes.

A filter operates on the voltage measurement, to be calibrated according to need, always ensuring a voltage control loop dominant dynamic on the order of tens of seconds.

A useful but probably rarely needed additional function, one which ensures a voltage profile along the line within limits V_{min} and V_{max}, is obtained by integrating, in the control, a correction factor dependent on overcoming these limits so as to reduce the Q_{ref} requested if V_{max} is exceeded, or to increase it if V_{min} is violated (Fig. 10.5).

The block diagrams represented in Figs. 10.4 and 10.5 show the following:

1. The main PI control law defines the output as a function of the difference $V_{ref} - V$. The input V is measured and filtered. The PI regulator output provides control level q with a positive sign if the reactive power is supplied by the controlled generators and vice versa with a negative sign if reactive power has to be absorbed. Output variables sent toward power plants represent reference values of reactive power. Their values are achieved by multiplying the PI output q by the actual limit values of the generator reactive powers.
   The dynamics of the voltage regulating loop considered must be faster than the OLTC voltage control.
2. The q range of negative values fixes the absorbed reactive power from 0 to 100 %, while the q positive range fixes the values of the delivered reactive power also from 0 to 100 %. This means the PI control law defines instant by instant the control effort percentage of each generator with respect to its operating limits.

As mentioned, these limits can be identified out of line with the appropriate "fitting" of data from the experimental measurements. In fact, calibration declared by manufacturers could provide values that do not correspond to actual plant situations.

3. The Fig. 10.5 control scheme contains two limiting loops of integral type becoming active at the overcoming of the admitted limits (V_{max}, V_{min}). These cycles, which overlap the Fig. 10.4 control scheme, must be characterised by slow dynamics with respect to the main voltage loop.
   Because the active limiting cycle acts in a contrasting way to the main integral regulating loop, abnormal behaviour does not result if more generators participate in the voltage regulation. When instead only one controlled generator is operating under the PCVR, the limit will push the regulator into saturation, from which it will be released only upon generator recovery from those limits. In other words, only one of the two cycles will operate, and that which limits will be the dominant one.
4. The PCVR will change the transformation ratio, as described through the Fig. 10.3 scheme, aimed to dynamically increase the PCVR control margin, therefore the PCVR interacts with the automatic classic OLTC slow control, which in this case is operated in open loop. The new OLTC task obviously reduces the tap range in controlling the bus bar voltage variation. This not-strictly-necessary-but-useful improvement is based on the monitoring and maintenance of the time until generator saturation by activation of the tap up/down command when q values approach the $-1/+1$ level.

10.5 Automatic Reactive Power Flow Regulation on the PC HV Bus Bar

This section proposes a possible solution for a control system that regulates the reactive power flow exchanged by the PC with the HV upright of the step-up transformer, in the presence of MV distributed generators.

The proposed *primary cabin reactive-power regulator* (PCQR) regulates the flow of reactive power sent to the HV/MV transformer from each MV line with distributed generation. The voltage on the MV side is instead classically controlled by varying the discrete transformation ratio of the tap. This system consists then of two parts:

- An MV voltage regulator using tap control of HV/MV transformation ratio;
- A reactive power regulator of flow between HV/MV transformer sides, by controlling reactive power of each feeder generator.

The two controls are necessarily conditioned, as the reactive power in the feeders also depends on voltage values at both feeder and PC bus bars. In addition, the required reactive power to the generators has to suffer restrictions which are necessary to avoid a situation of local voltages exceeding normal operating range.

The control system proposed for the PCQR is characterised by a decentralised type of structure regulating the flow of reactive power on a single MV feeder. This structure is required mainly because of possible differences in the DG generator location on the various PC feeder lines. In other words, because a homogeneous distribution of PC generators is not guaranteed, each MV line may have no, a few or many generators. Therefore, a centralised regulation controlling all the assets of PC generation with a single variable is likely to create imbalances, especially on MV node voltages.

This observation leads us to propose a control that regulates reactive power flow on each feeder connected to the HV/MV transformer, provided with at least one generator, and taking into account the range of MV voltage values allowed. The feeder generators control the feeder reactive power flow on their own.

To ensure an acceptable bus voltage profile, each feeder line must be observed independently and autonomously managed by intervention in its reactive power flow control to maintain feeder bus bar voltages inside the allowed band.

The architecture of the proposed control system is represented in Fig. 10.6.

The main feeder generators are subservient to the respective control signal values ("*Q*ref *X*-feeder *i*" in Fig. 10.6), updated in real time. These already take into account reactive power limitations on production/absorption of each generator and its voltage constraints.

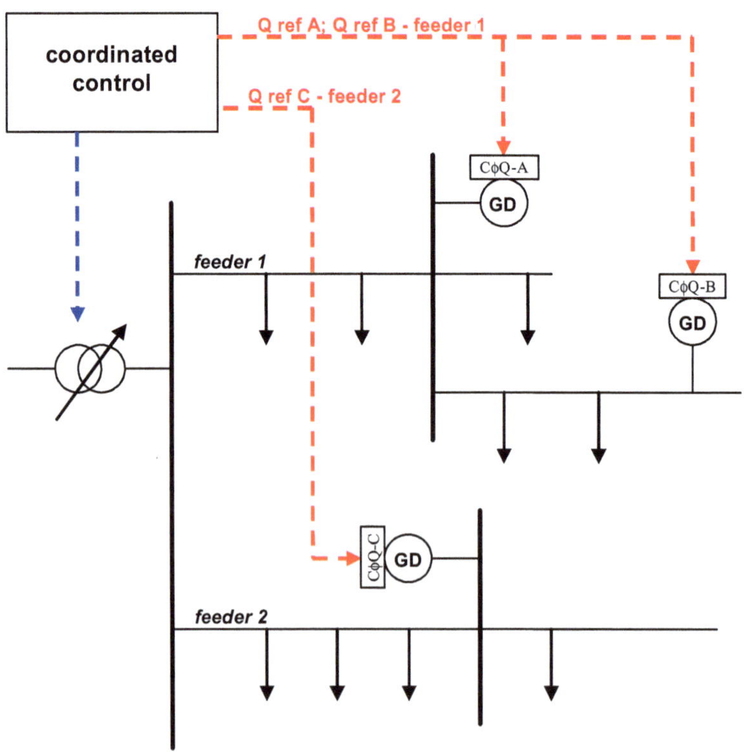

Fig. 10.6 Single-wire scheme of coordinated control of reactive power flow and voltage

In this respect, it is useful to refer to the Fig. 10.4 control scheme, substituting V_{ref} with $|Q_{flowref\,i}|$ and V with $|Q_{flow\,i}|$. Obviously, the direction of the power flow must be taken into account at the control logic management stage, as is clarified in what follows. In this way, we obtain the control scheme of the individual "feeder *i*" power flow. Figure 10.4 also makes use of the voltage limitation scheme in Fig. 10.5.

The sum of the $Q_{flowrefi}$ (all with the same sign, of course) represents the reference of reactive power exchanged with the PC HV bar: $Q_{flowref}$. Primary cabin-coordinated control defines the value of the $Q_{flowrefi}$ update according to value of $Q_{flowref}$, the operating and controllable feeder generators and the Q_{lim} capabilities made available to the control.

Therefore, like the PCVR, a PCQR requires that all control generators be provided with primary reactive power control.

We also note that coordinated control commands the HV/MV transformer OLTC to keep the MV bus bar near the nominal voltage. Obviously, OLTC discrete control of approximately 30 steps has much slower dynamics than reactive power flow control.

Figure 10.7 shows in greater detail PC centralised controller data exchange to and from the generators, including the data related to tap control. This type of OLTC control acts by taking into account the MV bar voltage measurement and possibly the optimisation curves that identify, for a given active and reactive power flow through the transformer, the OLTC control voltage reference that best marries allowed voltage operating values along the feeders connected to the transformer.

Upright voltage measurements are important for keeping track of when the normal operation voltage band is exceeded in real time, and therefore for the PCQR's automatic intervention on control variable Q_{refx} for necessary feeder limiting control on generator *x*.

10.6 Analysis of PCVR and PCQR Control Logics and Results

Next we justify the simplicity and effectiveness of the proposed HV or MV PC bus bar voltage regulation, based on reactive power control of the feeder generators. To this end, reference is made to Fig. 10.8, which shows a simplified PC electrical scheme, represented by an equivalent feeder. Analysis results based on this equivalent scheme can be easily extrapolated to the multi-feeder case.

PCVR-coordinated control regulates, in a closed-loop, the V_{MV} or V_{HV} bus bar voltage by controlling the reactive power of the feeder generators by a reactive power level: *Levq* represents the percentage, with respect to capability limits, that each generator in the feeder makes available.

The inequality $1 \geq Levq \geq 0$ represents the p.u. of the $Q^{+}_{lim\,i}$ (the generator *i* over-excitation reactive power limit), whereas $-1 \leq Levq < 0$ represents the p.u. of the $Q^{-}_{lim\,i}$ (G*i* under-excitation reactive power limit).

In the case of a generator grid-connected by inverters, the $Q^{\pm}_{lim\,i}$ represent the maximum deliverable ($Q^{+}_{lim\,i}$) and absorbable ($Q^{-}_{lim\,i}$) reactive powers by the *i*th invert-

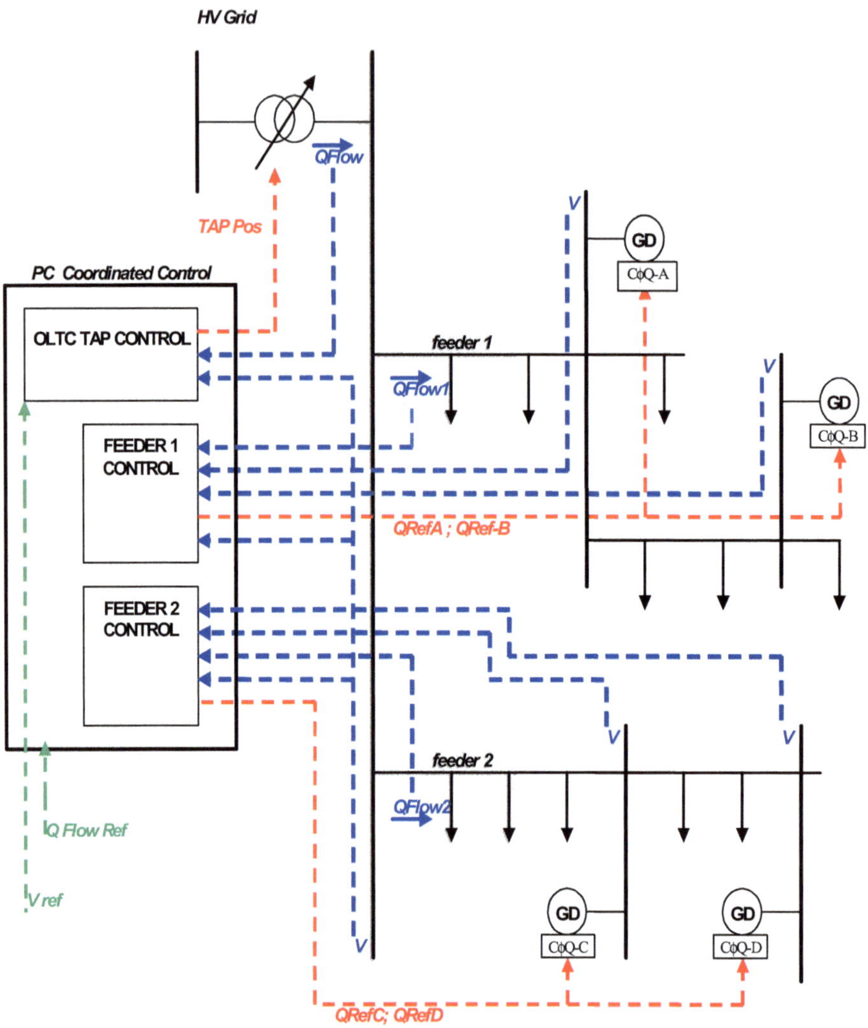

Fig. 10.7 Single-wire scheme of reactive power flow coordinated control, with details of the exchanged signals

Fig. 10.8 Scheme of the primary cabin equivalent scheme

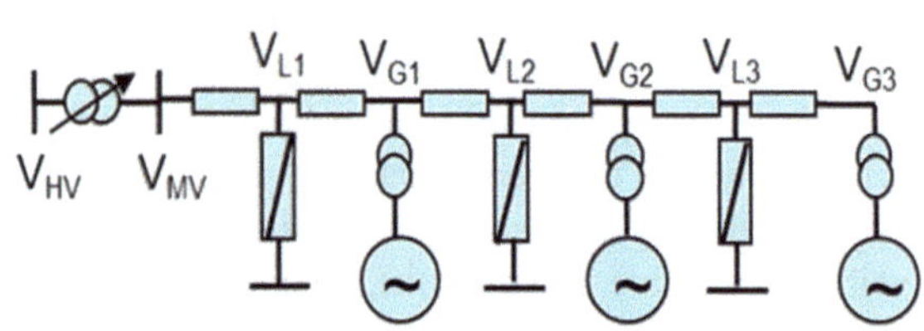

er. Other operating limits $V^{\pm}_{\text{lim}i}$, already introduced, fix the band ($V^{-}_{\text{lim}i} \leq V_{Gi} \leq V^{+}_{\text{lim}i}$; usually, $V_N - 5\% \leq V_{Gi} \leq V_N + 5\%$), limiting generator voltage variation.

Therefore, for each generator:

$$Q^{-}_{\text{lim}i} \leq Q_{Gi} \leq Q^{+}_{\text{lim}i},$$

$$V^{-}_{\text{lim}i} \leq V_{Gi} \leq V^{+}_{\text{lim}i}.$$

Taking into account these generator limits, the impact of the proposed PC voltage control is described next, showing its effect under different operating conditions as well as the OLTC-tap parallel control contribution to its correct functionality.

We note that all the feeder generators are charged at the same percentage with respect to their capability limits. Therefore, all feeder generators reach their over- or under-excitation saturation at the same time. A different case is that of the limiting voltage being reached at load and generator buses (usually $\pm 5\%$). Voltage-limiting can be activated at any feeder's single bus according to load variation and generator operating points.

Reactive power flow obviously impacts bus voltage profiles, and generator reactive power control can determine relevant feeder voltage variations with respect to operating conditions under primary voltage control alone. The following figures indicate PCVR voltage control characteristics.

Through the PCVR, a voltage set-point value V_{ref} is imposed on the PC MV grid not only according to the day-before forecasting plan but also taking into account the transformer reactive power flow and its sign. PCVR must also monitor feeder bus distance from $V^{\pm}_{\text{lim}i}$ values as well as generator approach to $Q^{\pm}_{\text{lim}i}$ limits, in order to maintain with continuity PCVR voltage regulation far from the onset of saturation.

With the reactive power flow that exits from the feeder (Q_{exit}), V_{ref} can be reduced, thereby lowering the Q value and reducing feeder generator contribution to PC voltage support. Under this operating condition, feeder voltages are higher than V_{ref}, in any event, and asking for a reduced V_{ref} value corresponds to two possible objectives:

- Lowering the Q_{exit};
- Forcing the generators to recover controllability by leaving $Q^{+}_{\text{lim}i}$ or $V^{+}_{\text{lim}i}$.

V_{ref} and therefore Q_{exit} can be increased until the generators reach their $Q^{+}_{\text{lim}i}$ or $V^{+}_{\text{lim}i}$.

With reactive power flow entering the feeder (Q_{enter}), the imposed V_{ref} can be maintained until the generators reach their $Q^{+}_{\text{lim}i}$. Conversely, V_{ref} can be increased in case the feeder generators have enough of a control margin to also compensate the lowered reactive power flow Q imported from the HV grid. Under this operating condition, feeder voltages are generally lower than V_{ref}. Exceptionally, the generators located at the feeder's extreme end could reach voltage values higher than V_{ref}. Lowering V_{ref} determines the lowering of generator contribution to PC voltage support while increasing the imported Q flow: the final result is a general lowering of feeder voltages with PC V_{MV} maintained at the V_{ref} value.

Generally speaking, PCVR controls all feeder generators up to their limits, thereby achieving the objective of PC closed-loop voltage regulation by maintaining feeder voltages inside a defined confidence band. Reactive power flow from or to the HV grid can be modified by changing the V_{ref} value. Lowering V_{ref} increases Q_{enter} or reduces Q_{exit}. Obviously, the PCVR will also control compensating equipment located in the feeder by infrequently switching them on/off, doing so only to recover generator controllability or to keep feeder voltages from overcoming confidence band thresholds. This simple, additional control is not discussed in what follows; it is assumed to operate after the PCVR control analysed here has already worked up to saturation.

Referring to Fig. 10.8, voltage profile is analysed and represented along the feeder for different operating conditions by describing the control logics required by the PCVR and PCQR.

10.6.1 Case of Reactive Power Flow Entering Feeder by HV Bus Bar

A feeder's normal operating condition, with feeder generators dominantly supporting feeder node voltages, is represented in Fig. 10.9, where the V_{MV} voltage is regulated at the V_{ref} set-point value up to when generators reach their capability limits. After that, feeder voltage lowering will determine the increase of reactive power input Q from HV.

Figure 10.10 shows the voltage profile difference with respect to Fig. 10.9 in the case of some generators reaching their Q^{+}_{lim}. The entering reactive power flow increases while the generators are performing at maximum effort to sustain the lowered feeder voltage profile.

Voltage lowering can also result in $V_{MV} < V_{\mathrm{ref}}$, with no possibility of recovering the voltage difference, due to voltage control saturation (i.e., with generators at

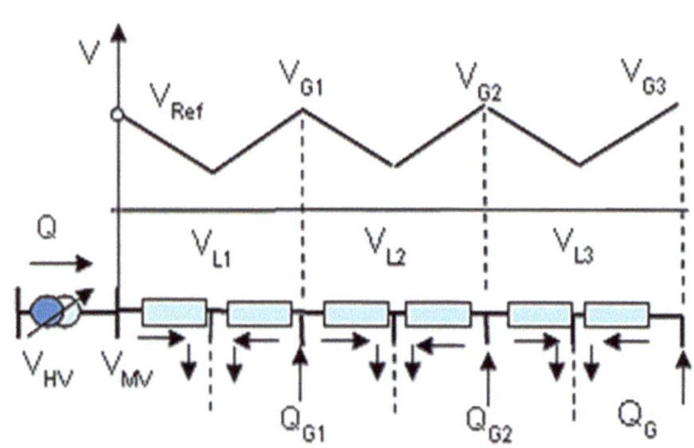

Fig. 10.9 Feeder voltage profile mostly supported by local reactive powers through regulating V_{MV}

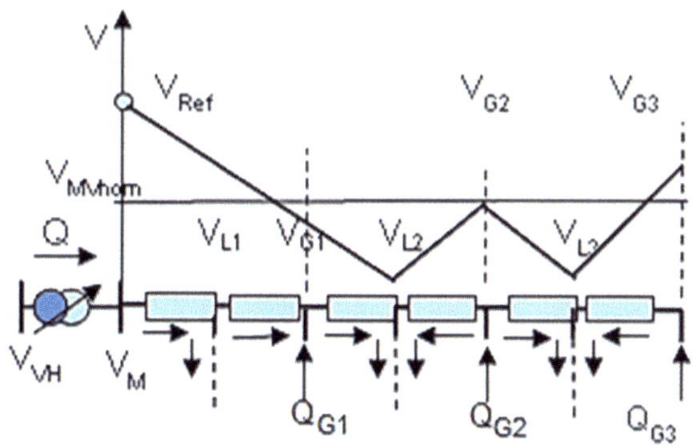

Fig. 10.10 Feeder voltage profile with generators at Q^+_{lim} and $V_{MV} < V_{\mathrm{ref}}$

Q^+_{lim} as in Fig. 10.10). At this operating condition, the one improvement possible can be obtained by increasing V_{MV} through PC transformer tap-changing operated by PC-CC (coordinated control), as is later described.

Always considering the entering reactive power flow from the HV grid case, other possible feeder operating conditions are described in Figs. 10.11, 10.12 and 10.13.

Figure 10.11 shows the case in which feeder voltage regulation operates to achieve $V_{MV} = V_{\mathrm{ref}}$ with controllable generators:

$$Q^-_{\mathrm{lim}i} \le Q_{Gi} \le Q^+_{\mathrm{lim}i},$$

$$V^-_{\mathrm{lim}i} \le V_{Gi} \le V^+_{\mathrm{lim}i}.$$

The same figure can be used to describe the following cases:

a. $V_{\mathrm{G3}} = V^+_{\mathrm{lim3}}$: In this case, PC-CC holds the reactive power delivery increase required of G_3 while allowing any reduction. At the same time, further reactive power delivery increase can be obtained by G_1 and G_2 following V_{ref} or the feeder load increase.
b. $V_{\mathrm{L1}} \le V_{\mathrm{L1min}}$: In this case, PC-CC has to increase the V_{ref} value in a way that recovers V_{L1}, and this is successfully achieved as long as feeder voltage control does not reach its saturation with all the control generators at Q^+_{lim} or V^+_{lim}. After this, OLTC-tap control has to operate as described later, with $V_{MV} < V_{\mathrm{ref}}$.
c. Figure 10.12 shows, the same as Fig. 10.11, the case of a feeder voltage regulation with $V_{MV} = V_{\mathrm{ref}}$ and all generators under control. The analysis already described for Fig. 10.11, when operating condition exceeds allowed confidence bands (points (a) and (b)), is the same for (a) as for (b) when V_{L2} is considered instead of V_{L1}, and again $V_{MV} < V_{\mathrm{ref}}$.

Figure 10.13 shows the last example of reactive power entering into the feeder having a heavy load at its extreme end. Analogous to the Figs. 10.11 and 10.12 cases, Fig. 10.13 represents a normal operating condition under PCVR control, unless:

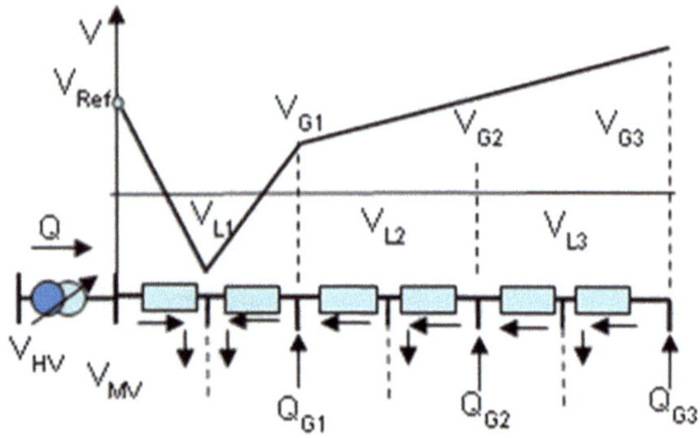

Fig. 10.11 Voltage and reactive power flows in a feeder with heavy load near PC

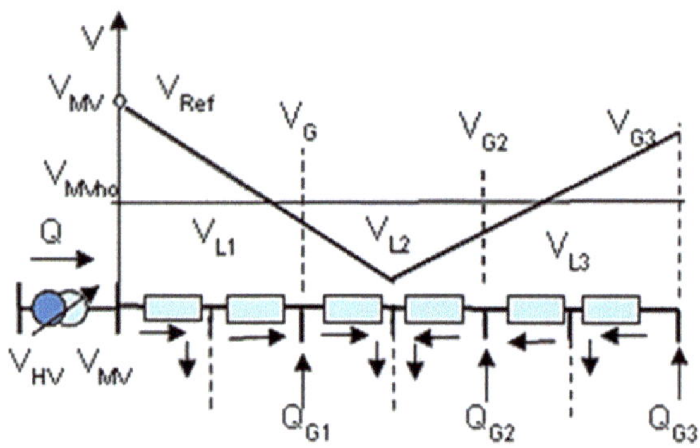

Fig. 10.12 Voltage and reactive power flows in a feeder with heavy load in middle of feeder

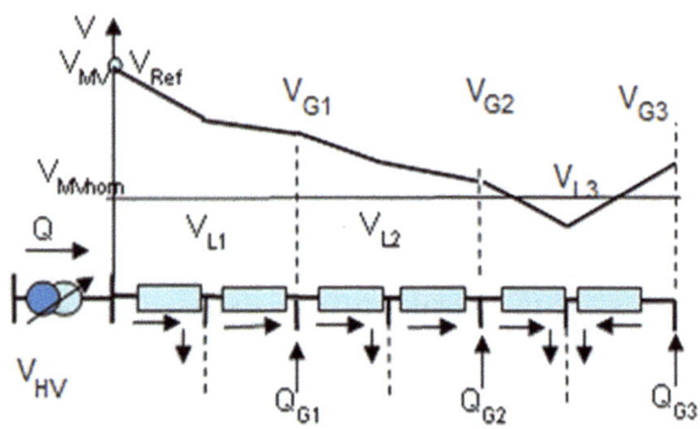

Fig. 10.13 Voltage and reactive power flows in a feeder with heavy load at extreme end of feeder

d. $V_{G1} = V^{+}_{\lim 1}$: Same comment as in a) case, substituting G_3 with G_1.
e. $V_{L3} \le V_{L3\min}$: Same comment as in b) case, substituting V_{L1} with V_{L3}.
f. $V_{L1} \ge V_{L1\max}$: In this case, the V_{ref} value is too high and the PC-CC has to operate a V_{ref} reduction, also taking into account the consequent voltage reduction at V_{L3}.

In all the cases shown in Figs. 10.11, 10.12 and 10.13, the V_{ref} increase will result in the reduction of the entering reactive power flow from the HV bus bar and a large voltage support by the feeder generators. With entering reactive power, voltages in the feeder are generally lower than V_{ref}, with the possible exception at the extreme feeder-end generation bus (i.e., V_{G3}).

10.6.2 *Case of Reactive Power Flow Sent by Feeder into PC HV Bus Bar*

The normal operating condition of a feeder with generators totally supporting the local voltages is represented in Fig. 10.14.

The V_{MV} voltage is regulated at V_{ref} through a control of the generators' reactive powers that totally supports feeder load voltages. V_{MV} voltage regulation can be maintained at the V_{ref} set-point value up to when the generators reach their capability limits. After that point, decreasing feeder voltage will result in the reduction of the reactive power flow leaving the feeder, and $V_{MV} < V_{ref}$.

Voltage lowering can result in $V_{MV} < V_{ref}$, as in Fig. 10.15, with no possibility of recovering the voltage difference, due to control saturation (generators at $Q^{+}_{\lim i}$ or $V_{G3} = V^{+}_{\lim 3}$).

At this operating condition the one improvement possible can be obtained by increasing V_{MV} through PC transformer tap-changing, as operated by PC-CC (de-

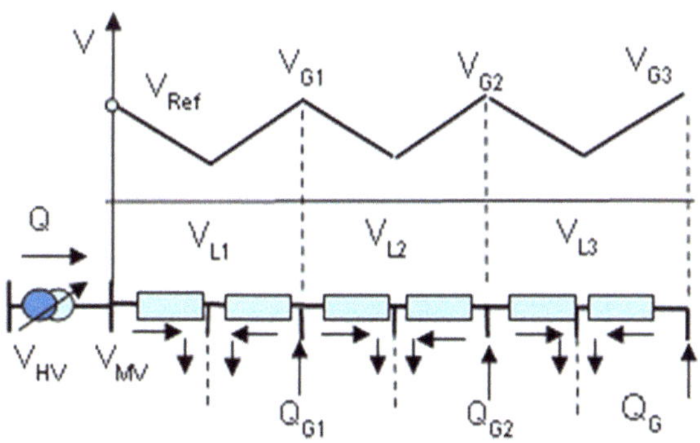

Fig. 10.14 Feeder voltage profile mostly supported by local reactive power flows through regulating V_{MV}

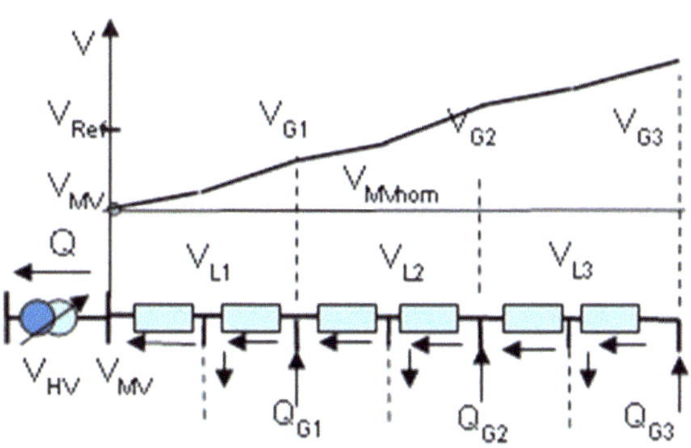

Fig. 10.15 Feeder voltage profile with local generators at $Q^{+}_{\lim i}$ or $V_{G3} = V^{+}_{\lim 3}$ and $V_{MV} < V_{ref}$

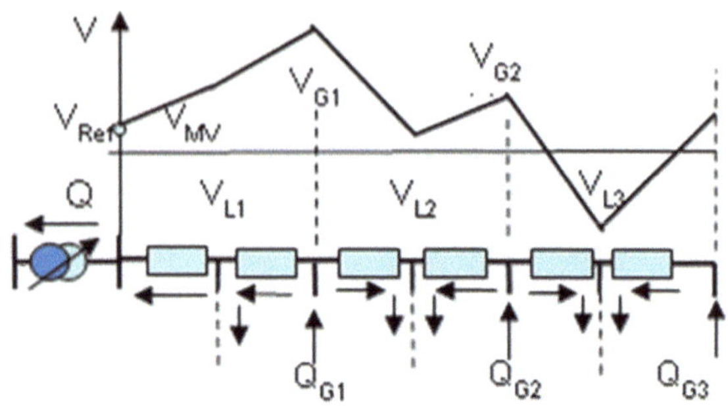

Fig. 10.16 Feeder voltage profile with local generator $V_{G1} = V^{+}_{\lim 1}$, local load $V_{L3} \leq V^{-}_{\lim 3}$ and $V_{MV} = V_{ref}$

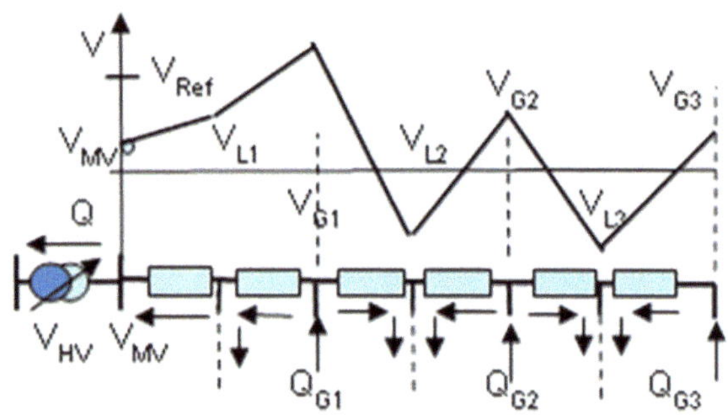

Fig. 10.17 Feeder voltage profile with local generator $V_{G1} = V^{+}_{\lim 1}$; local load $V_{L3} \leq V^{-}_{\lim 3}$ and $V_{MV} < V_{ref}$

scribed later). Continuing our consideration of the case of exiting reactive power flow from the feeder towards the HV bus bar, other possible operating conditions are described in Figs. 10.16 and 10.17 related to saturated V_{MV} control.

Figure 10.16 may represent the cases $V_{L3} \leq V^{-}_{\lim 3}$ and/or $V_{G1} = V^{+}_{\lim 1}$:

a. $V_{L3} \leq V^{-}_{\lim 3}$: In this case, the PC-CC requires V_{ref} to increase in a way that recovers V_{L3}. This is achieved as long as the feeder voltage control doesn't reach saturation, with all control generators at $Q^{+}_{\lim}$ or $V^{+}_{\lim}$. After this, OLTC control has to operate as described in the next section, with $V_{MV} < V_{ref}$.
b. $V_{G1} = V^{+}_{\lim 1}$: In this case, the PC-CC holds the reactive power delivery increase provided by G_1, while allowing any possible reduction. At this time, a further reactive power delivery increase can be obtained by following the V_{ref} increase or the feeder load increase.

Analogous to Fig. 10.16, the case represented in Fig. 10.17 requires similar controls, such as those described at points f) and g), until $V_{MV} = V_{ref}$. When $V_{MV} < V_{ref}$, voltage control is saturated; the way out is OLTC control required by PC-CC, described next.

10.6.3 OLTC Tap Control by PC-CC Operating as PCVR

Operating conditions requiring an HV/MV transformation ratio control by the OLTC are strictly linked to PCVR voltage regulation saturation.

As already seen, voltage regulation reaches saturation when all feeder generators approach their operating limits. More precisely, referring to the case of:

$$1, 2, 3,\ldots,i,\ldots,n \text{ generators,}$$

$$1, 2, 3,\ldots,j,\ldots,m \text{ loads,}$$

- *OLTC-Control-A* operates as follows:
  - If $V_{MV} < V_{\mathrm{ref}}$ and all feeder generators reach their $Q^{+}_{\mathrm{lim}i}$ or $V^{+}_{\mathrm{lim}i}$, for each i: Then the V_{HV}/V_{MV} transformation ratio has to be reduced up to recovery $V_{MV} \geq V_{\mathrm{ref}}$; $V_{\mathrm{load}\,j} > V^{-}_{\mathrm{load}\,j}$ for each j.

Otherwise, V_{ref} has to be reduced, gaining again PCVR controllability.

- Conversely, *OLTC-Control-B* operates as follows:
  - If $V_{MV} > V_{\mathrm{ref}}$, and all feeder generators reach their $Q^{-}_{\mathrm{lim}i}$ or $V^{-}_{\mathrm{lim}i}$ for each i, then transformation ratio V_{HV}/V_{MV} must be increased up to recovery $V_{MV} \leq V_{\mathrm{ref}}$; $V_{\mathrm{load}\,j} < V^{+}_{\mathrm{load}\,j}$ for each j. Otherwise, V_{ref} has to be increased, regaining PCVR controllability.

OLTC-Control-A increases the V_{MV} value in a way that better sustains feeder voltages and allows the generators to exit from their superior limits Q^{+}_{lim} or V^{+}_{lim}. The result of *OLTC-Control-A* is an increase in reactive power flow from the HV bus bar to the feeder or a reduction of the reactive power flow towards the HV bus bar.

On the contrary, *OLTC-Control-B* reduces the V_{MV} value in a way that reduces feeder generator operation in under-excitation, so limiting their reactive power absorption and allowing them to exit from their inferior limits Q^{-}_{lim} or V^{-}_{lim}.

The result of *OLTC-Control-B* is a reduction of reactive power flow from the HV bus bar to the feeder or an increase of the opposite Q flow towards the HV bus bar.

Next are some examples of the described OLTC control allowing PCVR voltage regulation by exiting its saturation and gaining controllability.

OLTC-Control-A must be activated in some of the PCVR cases considered previously and reviewed here:

- In Figs. 10.11 and 10.12, the (a) and (b) cases after generator saturation is reached and $V_{MV} < V_{\mathrm{ref}}$;
- In Fig. 10.13, the (c) and (d) cases after generator saturation is reached and $V_{MV} < V_{\mathrm{ref}}$;
- In Fig. 10.15, the case after generator saturation is reached and $V_{MV} < V_{\mathrm{ref}}$;
- In Figs. 10.16, 10.17, the (f) and (g) cases after generator saturation is reached and $V_{MV} < V_{\mathrm{ref}}$.

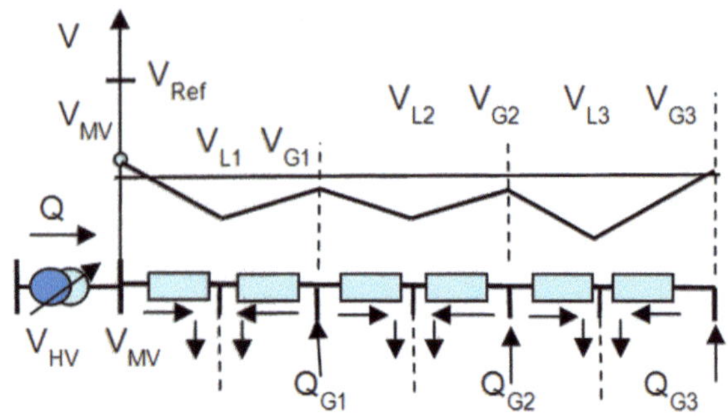

Fig. 10.18 Feeder low voltage profile with PCVR saturated, Q entering, $V_{MV} < V_{ref}$, with generators at Q^{+}_{lim}

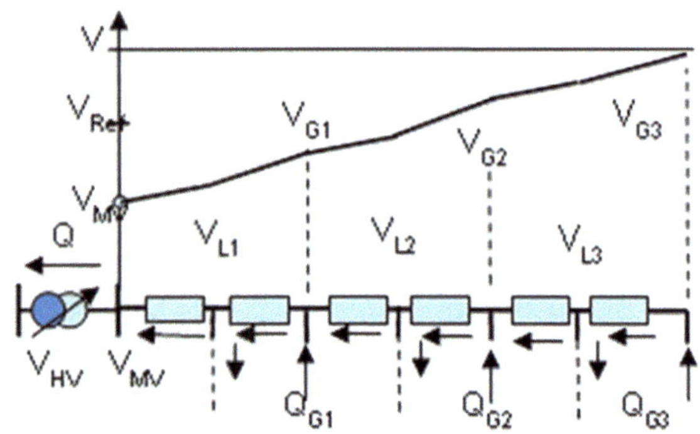

Fig. 10.19 Feeder low voltage profile with PCVR saturated, Q exiting, $V_{MV} < V_{ref}$, with generators at Q^{+}_{lim}

Low Voltage in the Feeder with $V_{MV} < V_{ref}$

The Fig. 10.18 case requires *OLTC-Control-A* to reduce the V_{HV}/V_{MV} transformation ratio in a way that increases V_{MV} by the support of the transmission grid. This tap control increases feeder voltages and reactive power entering into the feeder by the HV bus bar.

Upon reaching operating condition $V_{MV} \geq V_{ref}$, generators leave their Q^{+}_{lim} limits, and PCVR control starts operating again with a higher voltage profile in the feeder. The Fig. 10.19 case requires, as did the Fig. 10.18 case, that *OLTC-Control-A* determines V_{MV} increase by transmission grid support. This is the case of a weak HV grid, and tap *Control A* reduces the voltage support towards the HV bus bar, but improves as long as possible the feeder voltage profile by higher values. As before, with $V_{MV} \geq V_{ref}$ the PCVR departs from control saturation, and the new higher voltage profile in the feeder can be again controlled.

High Voltage in the Feeder with $V_{MV} < V_{ref}$

Another case requiring tap *Control A*, with reactive power flow injected into the HV bus bar and PCVR saturation, is represented in Fig. 10.20. With $V_{Gi} = V^{+}_{lim\,i}$ for each i, reducing V_{HV}/V_{MV} increases V_{MV}, thereby reducing generator reactive power delivery towards the HV bus bar.

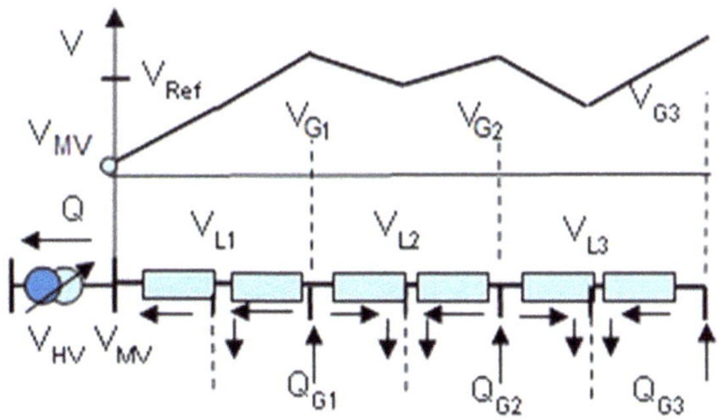

Fig. 10.20 Feeder high voltage profile with PCVR saturated, Q exiting, $V_{MV} < V_{ref}$, with generators at V_{lim}^{+}

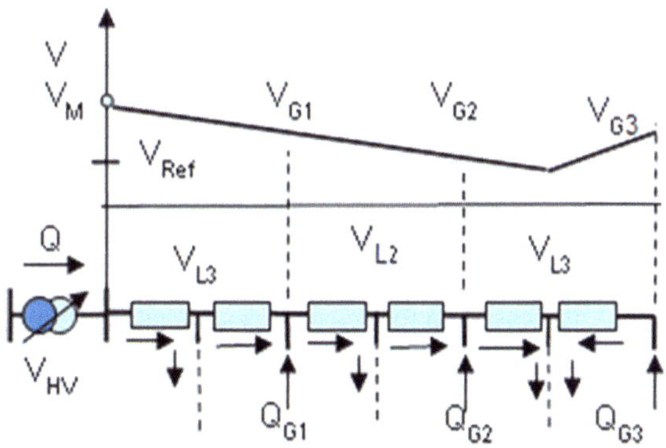

Fig. 10.21 Feeder high voltage profile with saturated PCVR, Q entering, $V_{MV} > V_{ref}$, with generators at Q_{lim}^{-}

Tap *Control A* determination of $V_{MV} \geq V_{ref}$ allows PCVR generators to leave V_{lim}^{+}, returning to normal operating condition.

The Fig. 10.21 case requires *OLTC-Control-B* to increase the transformation ratio V_{HV}/V_{MV} in a way that reduces V_{MV} by increasing the electrical distance from the HV grid. This tap control reduces feeder voltages and reactive power entering the feeder by the HV bus bar.

Reaching the operating condition $V_{MV} \leq V_{ref}$, generators then depart from their Q_{lim}^{-} limits, and PCVR control again starts to operate with a lowered voltage profile of the feeder.

10.6.4 OLTC Control by PC-CC During PCQR Operation

Operating conditions requiring infrequent change of transformation ratio HV/MV happen only when reactive power flow regulation by the PCQR reaches limiting thresholds.

As already seen, the saturation conditions of the Q flow regulation happen when all feeder generators reach their operating limits. More precisely, referring to the case of:

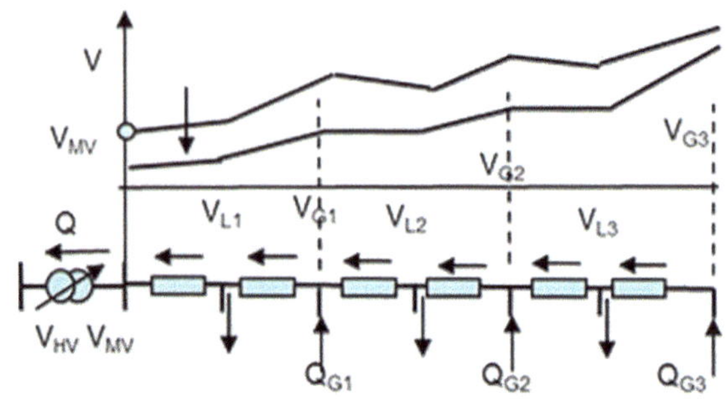

Fig. 10.22 Feeder voltage profile and reactive power flows with local generators at Q^{+}_{lim} or V^{+}_{lim} and $Q_{\text{exit}} < Q_{\text{ref}}$

$$1, 2, 3, \ldots, i, \ldots, n \text{ generators,}$$

$$1, 2, 3, \ldots, j, \ldots, m \text{ loads,}$$

- *OLTC-Control-C* operates as follows:
  - If $Q_{\text{exit}} < Q_{\text{ref}}$ (Fig. 10.22) or $Q_{\text{enter}} > Q_{\text{ref}}$ and all the feeder generators reach their $Q^{+}_{\text{lim}i}$ or $V^{+}_{\text{lim}i}$ for each i, then transformation ratio V_{HV}/V_{MV} must be increased up to recovery as soon as possible.

$$Q_{exit} \geq Q_{ref} \quad or \quad Q_{enter} \leq Q_{ref},$$

$$V_{\text{load}\,j} > V^{-}_{\text{load}\,j} \text{ for each } j.$$

Otherwise, Q_{ref} must be reduced (to Q_{exit}) or increased (to Q_{enter}), regaining the PCQR controllability.

- Conversely, *OLTC-Control-D* operates as follows:
  - If $Q_{\text{enter}} < Q_{\text{ref}}$ (Fig. 10.23) or $Q_{\text{exit}} > Q_{\text{ref}}$ and all feeder generators reach their $Q^{+}_{\text{lim}i}$ or $V^{+}_{\text{lim}i}$ for each i, or $Q_{\text{exit}} > Q_{\text{ref}}$, and all feeder generators reach their $Q^{-}_{\text{lim}i}$ or $V^{-}_{\text{lim}i}$ for each i, then transformation ratio V_{HV}/V_{MV} must be reduced (increase Q_{enter}/reduce Q_{exit}), always maintaining the PCQR into saturation until:

$$Q_{enter} \geq Q_{ref} \quad or \quad Q_{exit} < Q_{ref},$$

$$V_{\text{load}\,j} > V^{-}_{\text{load}\,j} \quad \text{for each } j.$$

Otherwise, Q_{ref} has to be reduced (Q_{enter}) or increased (Q_{exit}), regaining the PCQR controllability.

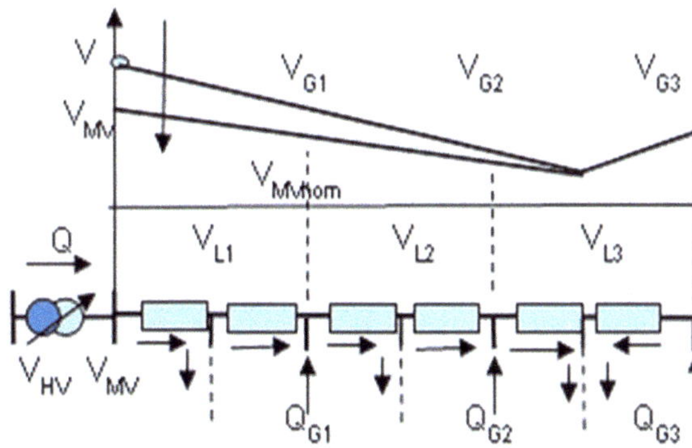

Fig. 10.23 Feeder voltage profile and reactive power flows with local generators at $Q^{+}_{\lim}$ or $V^{+}_{\lim}$ and $Q_{enter} > Q_{ref}$

Increasing V_{HV}/V_{MV}, V_{MV} is lowered while Q_{exit} is increased until $Q_{exit} \geq Q_{ref}$. With $Q_{exit} > Q_{ref}$ and all generators at $Q^{-}_{\lim}$ or $V^{-}_{\lim}$, reducing V_{HV}/V_{MV} increases V_{MV} until $Q_{exit} \leq Q_{ref}$.

Increasing V_{HV}/V_{MV}, the V_{MV} and Q_{enter} are lowered until $Q_{ref} \geq Q_{enter}$. Instead, with $Q_{ref} \geq Q_{enter}$ and all generators at $Q^{-}_{\lim i}$ or $V^{-}_{\lim i}$, reducing V_{HV}/V_{MV} increases V_{MV} together with Q_{enter} until $Q_{ref} \leq Q_{enter}$.

10.7 Conclusions

Evidence is given of the practical feasibility of the distribution grid voltage control in the presence of sparse generators. The PCVR and PCQR proposed are based on reactive power control of local generators. A classic slow OLTC control based on HV side voltage support is substituted by a fast control of the local reactive power resources (PCVR) to manage as autonomously as possible the local voltage and the reactive power flow linked with the transmission grid.

With PCVR/PCQR, the voltage/reactive power ancillary service can be easily operated in keeping with economic transactions.

The simplest and easiest to operate PCVR considers HV side voltage control; that is, PCVR-HV. Less simple but more suited to local customer voltage support is PCVR-MV.

With PCVR, the OLTC changes its task by achieving the main objective of maintaining local generator controllability as much as possible. But with PCQR, the OLTC controls MV voltage to maintain controllability of the required reactive power flow exchange with the HV side.

The local distribution dispatching centre should manage the local PCVR or PCQR (voltage or reactive power set-points) according to PC MV/HV voltage and reactive power exchange planning and dispatching. This should be in agreement with national dispatcher needs.

During PCVR and PCQR operation, MV feeders are maintained inside a voltage confidence band.

The proposed solutions are apparently very simple control systems, ones that are not as critical, complex or expensive as those having distributed controllers along the MV feeders. In comparing the two solutions, we cannot ignore the implication of reactive power flow regulation for the market management of this resource, with its high fragmentation of modest contributions, all to be measured with precision so their economic advantages may be recognised. Furthermore, reactive power flow control towards the HV bus is not simple to manage, as it is subject to continuous variations, with the obvious, attendant compliance issues of rigid supply contracts.

To close, it should also be noted that the two control proposals require all important generators of a given PC to be involved in voltage or reactive power regulation, to avoid conflicting and compensating effects on the control action by the remaining outside-control generators.

References

1. IEEE Standards Association: IEEE Standard 1547–2003 (2003) IEEE Standard for interconnecting distributed resources with electric power system
2. Hadjsaid N, Canard J-F, Dumas F (1999) Dispersed generation impact on distribution networks. IEEE Comp Appl Power 12(2):22–28
3. Hiscock L, Hiscock N, Kennedy A (2007) Advanced voltage control for network with distributed generation. Paper 0148, 19th international conference on electricity distribution, Vienna, May 2007
4. Choi J-H, Kim J-C (2001) Advanced voltage regulation method at the power distributed systems interconnected with dispersed storage and generation systems. IEEE Trans Power Syst 16(2):329–334
5. Viawan FA, Karlsson D (2008) Coordinated voltage and reactive power control in the presence of distributed generation. IEEE/PES general meeting-conversion and delivery of electrical energy in the 21st century, Pittsburgh, July 2008
6. Lu F-C, Hsu Y-Y (1997) Fuzzy dynamic programming approach to reactive power/voltage control in a distribution substation. IEEE Trans Power Syst 12(2)
7. Bompard E, Carpaneto E, Chicco G, Napoli R (1999) Voltage control in radial systems with dispersed generation. International conference on electric power engineering, IEEE PowerTech Conference, Bucharest, Romania, 1999
8. Tlusty J (2005) Management of the voltage quality in the distribution system within dispersed generation sources. 18th International conference on electricity distribution, CIRED, Turin, June 2005
9. Bonhomme A, Cortinas D, Boulanger F, Fraisse J-L (2001) A new voltage control system to facilitate the connection of dispersed generation to distribution networks. CIRED, Conference Publication No. 482 IEE 2001, June 2001
10. Vovos PN, Kiprakis AE et al (2007) Centralized and distributed voltage control: impact in distributed generation penetration. IEEE Trans Power Syst 22(1):437–483
11. Sansawatt T (2010) Integrating distributed generation using decentralized voltage regulation. IEEE/PES General Meeting, Minneapolis
12. Di Fazio AR Fusco G Russo M (2012) Decentralised voltage regulation in smart grids using reactive power from renewable DG. IEEE International Energy Conference and Exhibition (ENERGYCON) pp 580–586

13. Delgado N, Costa N, Maia B (2012) Control architectures to perform voltage regulation on low voltage networks using DG. Integration of renewables into the distribution grid, CIRED Workshop, May 2012
14. Vaccaro A, Zobaa AF (2013) Voltage regulation in active networks by distributed and cooperative meta-heuristic optimizers. Electr Power Syst Res 99:9–17
15. Chen TH, Wang MS, Yang NC (2007) Impact of distributed generation on voltage regulation by ULTC transformer using various existing methods. WSEAS International Conference on PS, Beijing, China, 2007
16. El-Khattam W, Salama MMA (2002) Impact of distributed generation on voltage profile in deregulated distribution system
17. Tanaka K, Oshiro M, Toma S, Yona A, Senjyu T, Funabashi T, Kim C-H (2010) Decentralized control of voltage in distribution systems by distributed generators. IET Proc Gener Transm Distrib 4(11):1251–1260
18. Baran ME, El-Markabi IM (2007) A mulitagent-based dispatching scheme for distributed generators for voltage support on distribution feeders. IEEE Trans Power Syst 22(1):52–59
19. Le ADT, Muttaqui KM, Negnevitsky M, Ledwich G (2007) Response coordination of distributed generation and tap changers for voltage support. Australasian Universities Power Engineering Conference, December 2007

Chapter 11
Wide Area Voltage Protection

The main objectives of EHV wide area voltage protection (V-WAP) are to face voltage instability and to increase power system security. Some innovative and very promising V-WAP control solutions are presented here.

V-WAP control solutions are largely based on the voltage instability indices presented in Chap. 9, operating OLTC blocking and area load-shedding according to more or less simple control logic.

The unique capability and strength of the WAP solution we consider first is due mainly to its effective coordination with the modern wide area voltage regulation (V-WAR) system, where secondary and tertiary voltage regulations operate according to their place in the hierarchy. Evidence is given to the practical feasibility and simplicity in defining and developing a wide area voltage protecting solution, one that is very effective in drastically reducing up to eliminating the risk of voltage collapse. The main simplification comes by way of an already existing and adequate subdivision of the power system into *areas* given by the operating SVR, which also fixes the areas of the protection scheme intervention and provides for each of them, in real time, protecting control efforts aimed at voltage support. On this basis, the area voltage stability index already proposed in Chap. 9 in § 9.4 is used in a timely fashion to avoid or block area voltage instability according to a simple and incontrovertible control logic.

The second WAP control solution is also based on power system subdivision into *regions* and on a real-time index based on the dependence of the total reactive power absorbed by the region (partially contained by the internal generators and coming through the interconnections with the surrounding grids), due to the internal active and reactive power changes in each of the region load buses (see § 9.5).

The index value increases with the distance traveled by the reactive power as well as the line loss amount. Therefore, the index also considers the relationship that total reactive power absorbed by a region has with the active power variation inside the region.

On this basis, the variation of the total reactive power controlling the region is linked with each bus's active and reactive load changes and used by a proper logic to safely protect the region from incoming voltage instability.

S. Corsi, *Voltage Control and Protection in Electrical Power Systems*,
Advances in Industrial Control, DOI 10.1007/978-1-4471-6636-8_11

We also consider the combination of the two WAP indices when both indices refer to the same grid subdivision into areas, as a WAP-protecting redundancy.

Simulation results of a dynamic analysis related to the North Italy power system show the effectiveness of the proposed protection schemes which minimise the load shedding operation to achieve the continuity and stability of the controlled power system in front of a ramping load increase.

In addition, we also look at and analyse the wide area protection required to operate alone, without the presence of any automatic voltage control system in the area under protection. Under this operating condition, an alternative protection functionality is considered for comparison purposes and to observe the achieved simulation results; even if less effective with respect to the case in the presence of V-WAR, it demonstrates the possibility for a significant increase in system security.

The third V-WAP control solution considered here is independent of the area/region concept and makes use of real-time voltage instability risk indicators based on a reliable identification, through bus local phasor measurements at a fast sampling rate (i.e., using a phasor measurement unit (PMU), of the Thevenin equivalent seen at any instant by each relevant grid bus (see § 9.3). The prediction criterion is largely based on on-line and true real-time recognition of the system operation approaching the nose tip on the real *V-P* curve seen by a local PMU at each considered bus.

We consider the use of a single bus voltage instability indicator in the case of the wide area multi-bus system, through elaboration and coordination of information coming from the PMUs installed in the main system buses. The objective is to recognise higher risk zones in the wide area and to reinforce the prediction of overall wide area voltage collapse risk.

The proposed VSI based on PMU use (see § 9.3) is one of the few options available until now that allows true real-time operation of a V-WAP protecting solution.

A further WAP solution described here, whose real-time performance can be stated under a given hypotheses and simplification of the power system model used and Jacobian computing, may also require the support of on-line (or computed) data by the operating conditions of generator OELs and transformer OLTCs. Thus, these are centralised protecting solutions whose limitations are discussed.

Lastly, we point out that voltage instability prediction as used by WAP can either be anticipated or delayed by tuning the speed of the identification algorithm and the proposed indicator derivative terms, as well. Examples of instability prediction as managed by WAP are also shown.

The proposed WAP algorithms' performance and related predictive indicators are presented through a detailed simulation analysis of different power system grids.

11.1 Introduction

In considering transmission network voltage control we have to remember the distinction, already given, between V-WAR and V-WAP:

- V-WAR: Continuous- or discrete-time control actions devoted to sustaining network voltages at values around the normal operating conditions, facing load

changes and contingencies. These actions, up until now widely described as "regulating controls", are aimed to stabilise and maintain in operation the overall system, feeding all loads with continuity;
- V-WAP: Extreme step control actions that are continuously ready to act but rarely operate on the real system and only during extreme conditions when the risk of losing the system or part of the system is high. These "protecting controls" are aimed to confine the incoming problem and to minimise the part of the system to be lost and/or the part of the load to be shed.

Historically, the two different controls have been analysed, designed, applied, operated and maintained by different specialists usually not coordinated with each other; they are the control and protection specialists. Moreover, the already developed protections are mostly independent controls dedicated to power system components and lines. Only recently has the subject of wide area protection (WAP) or system protection scheme (SPS) been growing in consideration, after the recent widespread blackouts which have occurred in many parts of the world and the increased need for new protecting solutions able to improve overall power system reliability and continuity of power system operation.

On the contrary, some "regulating control" solutions operating at utility, regional or national levels already exist in part; and in part some are under development or being considered for field applications. Load frequency control schemes (LFC) of centralised, decentralised and hierarchical types, as well as secondary (SVR) and tertiary (TVR) voltage regulation or coordinated voltage/reactive powers regulation, are examples of these applications, and they already operate on large power systems.

Looking to the future, the objective to increase power system security, reliability, operation quality and management economy is increasingly taken into account. From this perspective, system wide area regulating controls (WAR) and the system wide area protection solutions (WAP) represent the most concrete and feasible ways to achieve the required objectives. WAR and WAP have to operate together, in a coordinated way, according to the different roles they play, reciprocally taking into account the other's existence and operating state, including limitations, failures, saturation, out-of-service, etc. They can help each other recognise the phenomena in progress and make decisions for their specific control actions [6–18, 23].

Obviously, it is very different for a protection system to operate in the presence or absence of a regulating system over the same grid area. Its control decisions, in fact, should be strongly related to and coordinated with those of the regulating system, also taking into account its operating conditions.

A new idea is taking hold in which regulating and protecting systems exchange information and support each other.

This coordination is, in general, not simple to define, and it is strongly dependent on the characteristics, functionality and performances of the two WAP and WAR schemes.

In principle, the schemes should be designed in concert for a correct separation of their roles and synchronisation of their interventions, as well, but mainly for the most effective optimisation of each of their achievable contributions and reciprocal support. Such a dynamic coordination and functional synergy between regulating and protection wide area schemes is, in general, a complex problem that differs from one case to another. Nevertheless, a lucky circumstance does exist in which a strong simplification and a wide harmony between the two real-time schemes is easily achievable: this is the case of a wide area voltage stability protection combined with a transmission network voltage control based on a hierarchical real-time control structure realising SVR and TVR voltage regulations [7–9, 14, 19, 20]. The might of this solution is unique, as we will see.

The development of WAP without WAR is in principle more complex and less effective. On this concern, phasor measurement unit (PMU) technology offers a new and powerful opportunity for innovative and effective system protections [11, 12, 21, 24, 25, 27, 29, 30]. Accurate, very fast, time-tagged phasor data related to the main buses of a wide area network can in fact be used by area WAR or WAP systems to better recognise incoming phenomena and to appropriately select the corresponding most powerful control action(s) and to operate in a timely way the correct control where needed. This is in accordance with advanced, intelligent and adaptive functionality that must be defined for both the WAR and WAP systems.

Another kind of promising voltage stability indicator to be used for wide area real-time WAP requires a burdensome and complex regional monitoring system based on computing the derivative/variation terms of the dependence of reactive power absorption on a grid region with respect to variations of internal load and generation. There are two critical aspects related to this solution:

1. Noise on a control signal due to derivative term computing: the high speed required by wide area voltage instability protection justifies use of the derivative term, which can determine uncertainty and delay on protection controls.
2. A too-detailed and costly wide area monitoring system that collects a large amount of real-time data, and that properly manages the data's partial unavailability and failures.

Last to be mentioned is the effort based on achieving a voltage wide area protection V-WAP based on the classic power system modelling and Jacobian computing, starting from state estimation fast update combined with information on generator OEL and transformer OLTC operation up to saturation [1, 2, 4, 5, 13, 15–17, 22, 26, 28].

Computing effort is generally high when state estimation is required: real-time computing is not guaranteed, and state estimation updating at high speed is not always available on time. Moreover, this represents a centralised solution requiring a large amount of data, as transferred by telecommunications.

11.2 Area Voltage Protection Based on SVR-TVR and Real-Time Indicators

Premise In recent years many indices and criteria for power system voltage stability have been studied and proposed by the literature (Chap. 9). The general objective has been to give a measure of the margin between point of operation and instability limit, thereby providing an early warning of a potential critical situation in the grid.

Here reference is made to § 9.4 real-time VSI index-based on the real-time trend of the V-WAR control variables of a hierarchical structure, largely introduced in this book as a combination of secondary (SVR) and tertiary (TVR) voltage regulations.

It is useful to briefly review the basic concepts of the SVR control system structure and performance:

- The reactive power of the generating units is the principle resource; it is already available on-field, and it is low-cost and simple to control for network voltage support;
- The voltage control system considers the grid's dominant buses only (a small amount), thus allowing a sub-optimal but feasible and reliable control solution;
- Joint-buses to the dominant bus (pilot node), those having high electrical coupling to it, form a "control area" with voltages close to each other;
- The control structure—based on the subdivision of the grid into control areas—automatically and, as much as possible, independently regulates each area pilot node voltage;
- The area control resource encompasses essentially the reactive powers of the largest units in the area (control plants), which mainly influence local pilot node voltage.

The basic idea of TVR comes from the need to increase a system's operating security and efficiency through centralised coordination of the decentralised SVR structure:

- Pilot node voltage set-points must be adequately updated and coordinated with dynamics slower than SVR, considering the real condition of the overall grid and avoiding pointless and conflicting inter-area control efforts;
- Pilot node voltage set-points can be computed and updated in real time considering the global control system structure and its real-time measurements;
- Pilot nodes voltage set-points must be optimised to minimise grid losses while still preserving voltage control margin.

The SVR reactive power level $q_j(t)$ of the j-area represents instantaneously the control effort underway at the j-area and, therefore, the real-time reactive power load for the j-area control units. More precisely, $q_j(t)$ value stands for the percentage of j-area unit reactive power with respect to under- or over-excitation limits: in particular, when $q_j(t)$ reaches $+1$, j-area voltage regulation is saturated because the operating points of all j-area control generators are fixed by their over-excitation limits. Changing the load, the pilot node voltage of a given grid area is therefore regulated, through the SVR, to the desired value, unless all the area control units reach their over-excitation limits. Under TVR, approaching this extreme operating condition determines achievement of area voltage instability limit.

§ 9.4 refers to a VSI based on SVR and TVR which appears reasonable, simple and effective, to be computed directly inside the SVR.

A real-time, on-line indicator of j-area proximity to voltage instability, one that is based mainly on the actual value of the area reactive power level $q_j(t)$ when TVR is operating, appears to be proper and effective for power system protection. Intuitively, it can in fact give a clear answer to the following questions:

1. How is the best localisation and amount of load to be shed selected?
2. How is a timely, securely reliable high voltage instability risk and its localisation recognised?
3. How are *wide area voltage regulation* and *wide area voltage protection* coordinated so they both operate on the same network wide area?
4. How is untimely conflicting control between a *wide area protection system* and a *real-time regulating system of transmission network voltages* avoided?

From here, we briefly comment on the real-time and on-line indicator of the j-area proximity to voltage instability.

11.2.1 Basics of Real-time SVR-TVR $VSI_j(t)$ Index Computing

With SVR and TVR: The already proposed (see § 9.4) proximity indicator $VSI_j(t)$ to voltage instability is

$$VSI_j(t) = q_j(t) + \rho \left.\frac{\partial q_j(t)}{\partial t}\right|_{\Delta t} \Delta t,$$

where ρ is a suitable weight coefficient for introducing a derivative term with a useful lead effect; $-1 \le q_j(t) \le +1$; Δt is the time interval used to filter derivative term samplings: a moving average filter of duration Δt then operates on the computing of the derivative term.

In Practice:

$$VSI_j(t) = q_j(t) + \rho \frac{1}{m} \left(\sum_{\tau=t}^{t-(m-1)\delta} \frac{\partial q_j(\tau)}{\partial \tau} \right) \Delta t$$

$\Delta t = t-(m-1)\delta$; δ = sampling time interval;
m = number of sampling intervals used by derivative term filter

Without TVR: The voltage stability index $VSI_j(t)$ does not really represent the j-area distance from the voltage stability limit, but only distance of the SVR from its operating limits.

$VSI_j(t)$ is a real-time variable, updated with a few seconds delay with respect to the actual j-area operating state. Therefore, it can be effectively used for real-time automatic protecting controls, provided the TVR is operating.

11.2.2 Basics of Real-time V-WAR and V-WAP Coordination

The dynamic coordination and functional synergy between regulating and protection wide area schemes is, in general, a complex and therefore often neglected problem. Nevertheless, the circumstance of SVR and TVR operating on a power system offers a strong simplification and an easily achievable and extensive harmony between the two V-WAR and V-WAP real-time control schemes.

Figure 11.1 introduces the combined operation on a wide area power system of overlapped and coordinated V-WAR and V-WAP controls.

Use of $VSI_j(t)$ Indicator by V-WAR

As the proximity indicator $VSI_j(t)$ to voltage instability can be used by a V-WAR (based on SVR and TVR), it has already been introduced and is summarised here:

- When $VSI_j(t)$ exceeds the first threshold value corresponding to the j-area consistent voltage control effort and then to a small voltage control margin, the proposed V-WAR control logic will order the automatic switching of the j-area capacitor banks, reactors and update of SVC set-points according to an appropriate priority. On-line checking of the switching fatigue of controlled reactive power equipment will restrict automatically their number of commutations according to design prescriptions and network voltage criticism. Switching timing will be related also to the value of the $VSI_j(t)$ derivative term and its dynamics.
- If, after all j-area reactive power resources are used up, a local network working condition with small reactive power control margins still persists, $VSI_j(t)$ could then reach a second critical threshold which allows the SVR to modify automatically voltage set-points or transformer ratio of j-area OLTCs in order to obtain a reduction of the load seen by the local transmission network.
- After these regulating control "first barriers" are taken care of, the $VSI_j(t)$ basic logic could indicate when a considered very critical threshold is overcome, or when all j-area control units reach and permanently reside at the upper terminal voltage limit, or when the pilot-node voltage reduction by TVR under the lowest allowable value must be inhibited. In any of the above mentioned extreme working conditions, the control logic will automatically shut down the j-area OLTCs.

After this last control is operated by SVR in the j-area, all possible automatic regulating controls at the SVR's disposal for supporting the j-area pilot node voltage will have been exhausted. Besides, the SVR will have drawn the operator's attention by monitoring the j-area voltage value and the stability index value and asking for pos-

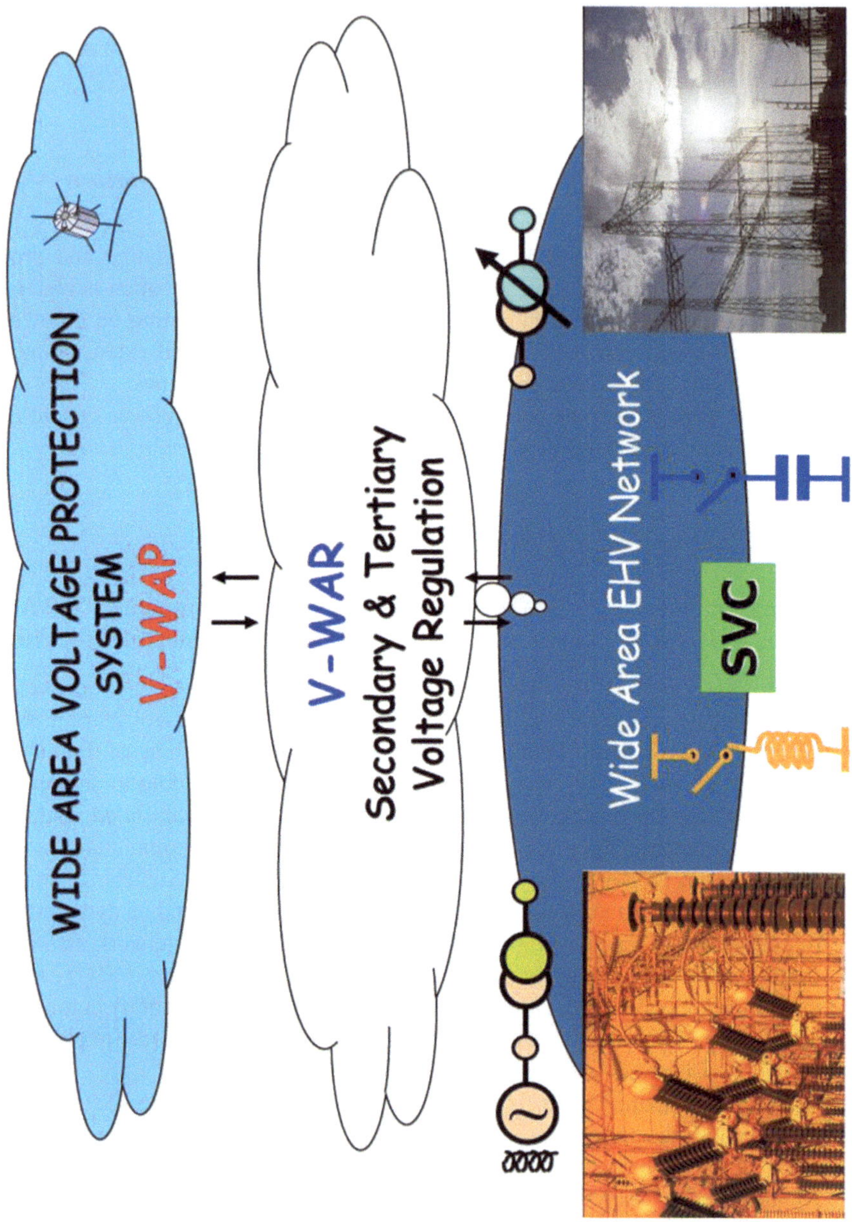

Fig. 11.1 V-WAR and V-WAP combined and coordinated controls of a wide area power system

sible manual extreme controls. At this advanced stage, wherein with the control system reaches saturation condition, the *j*-area WAP should enter into operation, coming to the aid of the *j*-area WAR (which cannot contribute more) and substituting for it.

The criteria through which WAP will operate from now on are described in the following section. When critical operating conditions no longer obtain, the $VSI_j(t)$ value will decrease progressively up to authorising the restoration of the loads shed by WAP, while the SVR will release the *j*-area OLTCs and reduce the reactive power resources that are operating.

11.2.3 Wide Area Voltage Stability Protection Philosophy Based on SVR-TVR $VSI_j(t)$

From the above considerations and due to TVR control, the SVR will operate up to its limits, reached only when transmission network voltages are low, notwithstanding all network reactive power resources are in operation for voltage support.

Under these conditions, $VSI_j(t)$ becomes fully significant and reliable for protection control because the *j*-area operating limits are reached despite the reduced pilot node voltage plan (by TVR) and saturated reactive power control efforts. Therefore, a wide area voltage protection system (V-WAP) can be simply defined hereafter only in those cases when the SVR-TVR (representing the V-WAR control system) is operating in the same power system [7–9, 14, 19, 20].

Use of $VSI_j(t)$ Indicator by the V-WAP

As said, after the *j*-area pilot node voltage regulation has exhausted its task, V-WAP becomes active according to the following guidelines:

- The V-WAP structure simply follows the SVR areas and the possible changes at their edges. This is the first, very important, exchange of information between V-WAR and V-WAP for their alignment in terms of area edges.
- The *j*-area V-WAP, adequately informed of SVR operating conditions, will leave the corresponding V-WAR the task of regulating voltages and maintaining stability in the *j*-area, until the SVR correctly works (information coming from SVR auto-diagnostics and V-WAP monitoring).
- The V-WAP system will therefore be authorised to operate only when it is able to significantly take the place of the V-WAR control, which happens when the two following conditions are verified:
  - The *j*-area control system has reached its own saturation limits, after operating all its available continuous controls on the generators and discontinuous controls on the other reactive power resources, as well as on the transformer OLTCs.
  - The area real-time voltage instability index confirms a high instability risk in the area.

At that time, V-WAP will enter into operation in the j-area and command the following control actions:

- Confine the part of the power system which is the first cause of the voltage instability (whenever this part has been automatically recognised inside the *j*-area);
- Shed local loads in the percentage and with the frequency required by real-time needs. The sole objective is the removal of a serious voltage instability risk in the *j*-area under its control.

This relevant intervention is decided simply on the basis of the reduced margins of the *j*-area control system with respect to its saturation. Therefore, a unique area protective decision it is proposed here, guided by the vicissitudes of the corresponding (same area) control system. In this sense, the proposed protective solution is not conventional and appears extremely simple because it is largely based on the measurements and on the state of operation of the *j*-area V-WAR.

V-WAP protective controls have to command the following actions, according to Fig. 11.2:

1. If locally available, the paralleling of hot-running reserves for reactive power support, already alerted at the time the V-WAR regulator reached its saturation. The time delay required by the reactive reserve to be injected into the network should be coordinated with the $VSI_j(t)$ index by anticipating the threshold-activating V-WAP. This first protective action, even if expensive, does not determine the cuts of the feeding of the power system area users.
2. Progressive reduction of the local *j*-area loads, starting (if possible) from the prevailing inductive load, when they are known. Priority is given to confining the first cause of voltage lowering whenever the *j*-area nodes having very low voltages and high reactive power absorption are recognisable: the protecting control opens the lines feeding those loads.
3. After the selection of the loads to be shed according to given characteristics or connected to selected buses, and always maintaining a continuous monitoring on the persistence of the phenomenon underway, V-WAP progressively sheds local area loads by following possible defined priorities. Shedding speed will also be dependent on the $VSI_j(t)$ index trend and will be reduced to zero only when this trend changes slope, as well as when local voltages again assume normal values. Protective action will stop only when the SVR and TVR have again reached their normal operating conditions, outside their saturation limits.

Without SVR-TVR, all considerations fall again into the conventional case, where V-WAP operates alone because V-WAR is not operating, with the exception of the PVR. In this case V-WAP protective controls are the same as before but are usually simply linked to low voltage thresholds.

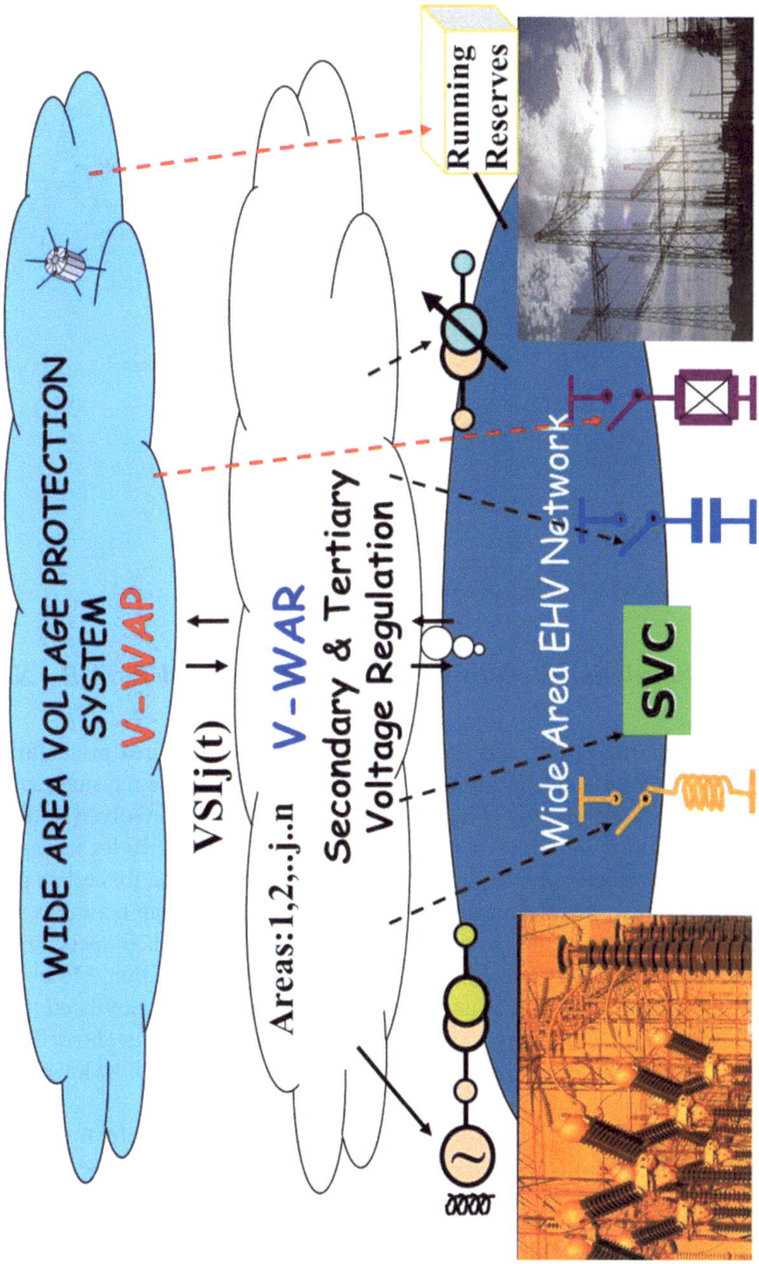

Fig. 11.2 Combined data exchange with SVR-TVR and coordinated protecting controls by V-WAP in the wide area power system

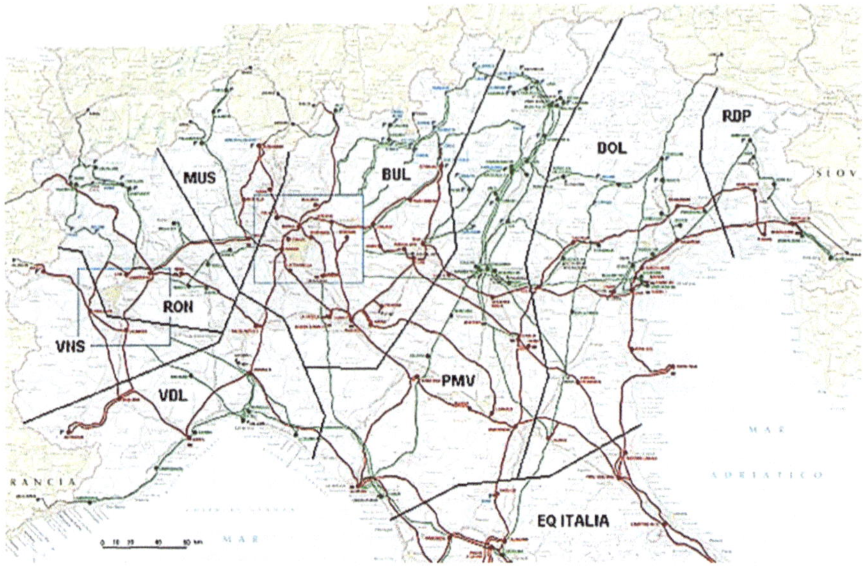

Fig. 11.3 Structure of the North Italy transmission system

11.2.4 Simulation Results of V-WAP Based on SVR-TVR $VSI_j(t)$

To demonstrate the power and limits of the proposed wide area protection controls, simulation tests are performed on the Italian transmission system, and particularly in the North Region (Fig. 11.3), which was strongly involved in the 2003 blackout. The power system dynamic model adopted is simplified for its aspects of low impact on voltage stability phenomena: equivalents are used for certain production and load areas, mainly for the South Italy and the European networks. On the contrary, detailed models are adopted for the generators and their speed and AVR controls, including the dynamic limits in over- and under-excitation. Moreover, the on-load tap changers of grid transformers and their controls are simulated in detail, as is load dependence on voltage, which is exponential: at voltages above 0.85 p.u., active and reactive power exponents are 0.5 and 0.8, respectively. At lower voltage values, load decreases are greater.

As far as SVR and TVR control systems are concerned, these refer to the original Italian solution: a detailed model of SVR is used, corresponding to the on-field application. Regarding TVR, its optimising real-time algorithm is simplified, primarily guaranteeing a control margin to the SVR when its saturation is approached. This functionality operates until pilot nodes voltages are maintained above 0.9 p.u.

Load shedding based on the $VSI_j(t)$ index is controlled according to the protective criteria above defined, using adequate filtering for the computation of $VSI_j(t)$. The shedding of 5 % of the j-area nominal load is executed when $VSI_j(t)$ reaches a value of 0.99 p.u. and the pilot node voltage is lower than 0.95 p.u. If the need for load shedding persists, the subsequent 5 % step in the same area is operated 30 s after the previous one.

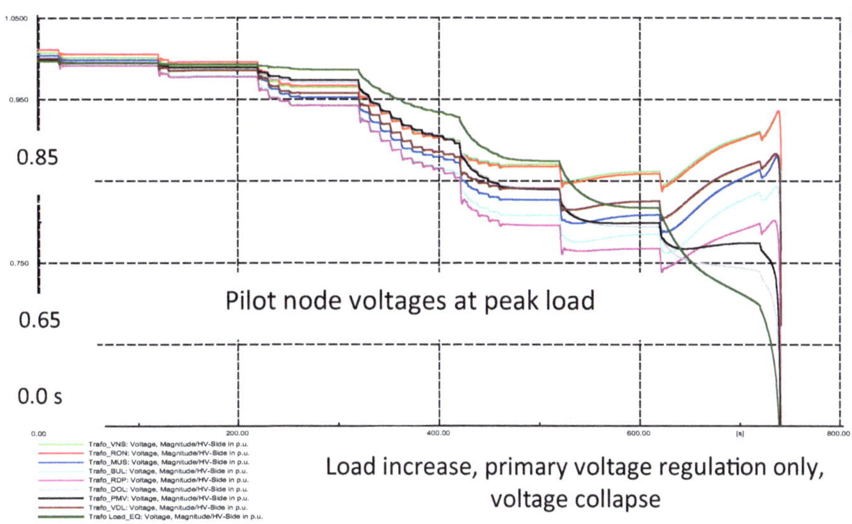

Fig. 11.4 Case of power system with PVR alone. Voltage transients of the main EHV buses at peak load, without V-WAP

Load shedding without SVR, based on voltage thresholds, also considers "potential" pilot nodes and operates 5 % of local load shedding when the voltage value below the 0.82 p.u. threshold lasts longer than 3 s.

The tests performed refer to the same percentage of load increase throughout the system. The load increases every 100 s, starting from 20 s, with the first two steps each at 2 % of nominal load and subsequent steps at 1 % each, up to when it reaches the last increase at 820 s. This large load increase, due to the power system's very stable original operating condition, allows us to a significantly and reliably check differences between six selected cases:

Comments: Without SVR and V-WAP, voltages collapse at 620 s (see Fig. 11.4). We note that the static *V-P* curve "nose" cannot be properly traced from these transients due to the dynamics introduced by the operating control loops. At the second step, generators reach their over-excitation limits and the subsequent transients are characterised by alternating interventions of generator limits and transformer OLTCs. Voltages in some areas reach very low values (some values lower than 0.85 p.u.) before growing again, due to the greater load reduction at low voltage. Voltage collapse, as a combination of the processes underway when it becomes irreversible, is clearly evidenced by a further load step increase, pushing voltages down.

Collapse is avoided with SVR and OLTC blocking based on $VSI_j(t)$ (Fig. 11.5), with voltages around 0.9 p.u. at 600 s and with no load shedding necessary. Coming to Fig. 11.6, the very simple classic protection based on voltage threshold criteria is shown for comparisons with the other cases. Notwithstanding that voltage goes below 0.85 p.u., voltage collapse (Fig. 11.3) is avoided and load shedding amounts to 7.64 %.

The transients in Fig. 11.7 show the overcoming of voltage lowering and instability problems, notwithstanding a load increase lasting up to 820 s, with voltages con-

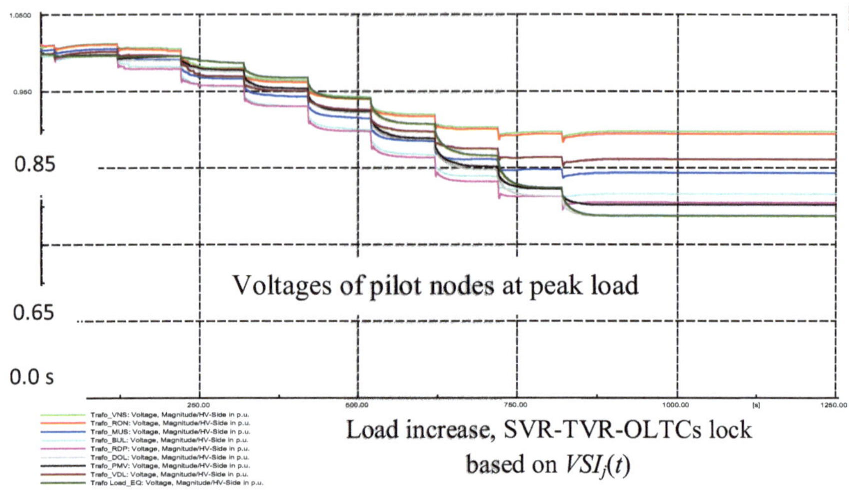

Fig. 11.5 Case with SVR-TVR and OLTC lock via $VSI_j(t)$. Pilot node transients at peak load, no load shedding

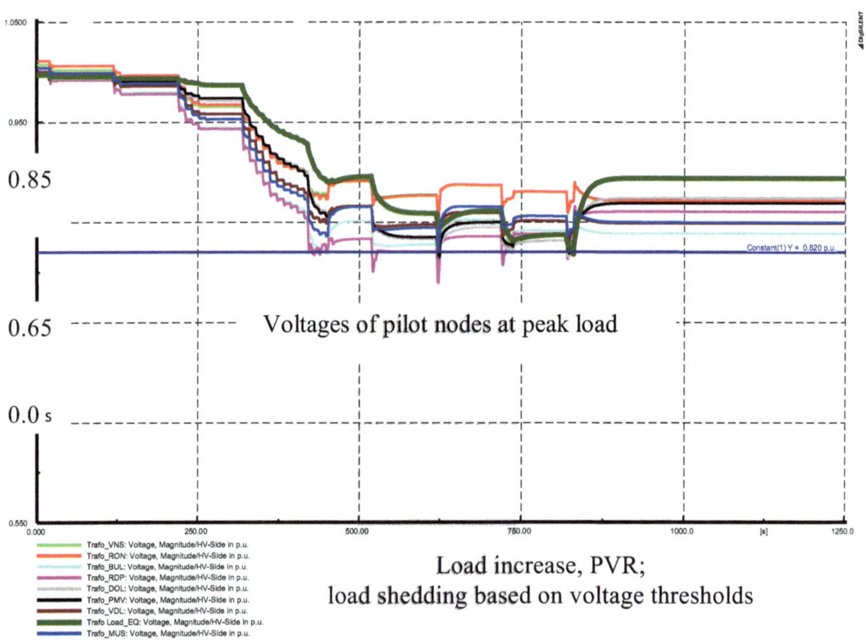

Fig. 11.6 Case with PVR alone. EHV bus voltage transients at peak load, with V-threshold load shedding

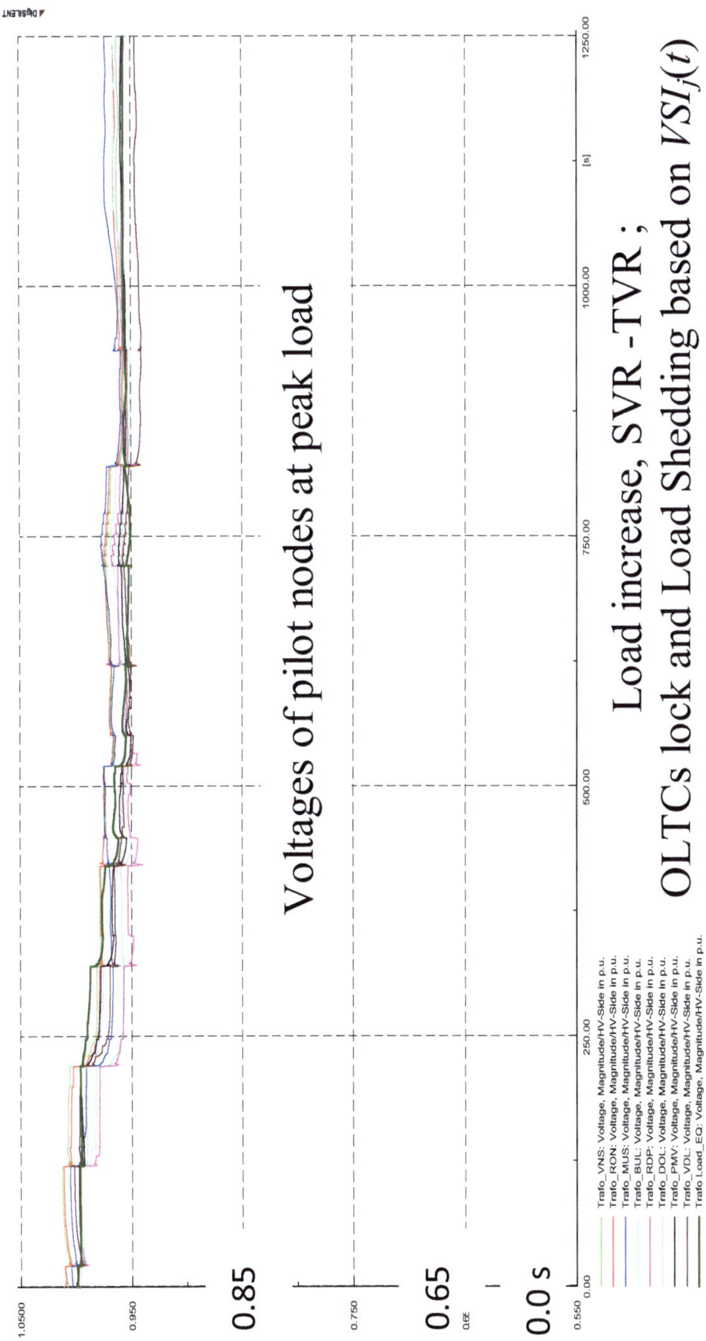

Fig. 11.7 Case with SVR-TVR. OLTC lock and load shedding $VSI_j(t)$-based

trolled around 0.96 p.u. The effects of the SVR are more evident after the first load step, even if it works up to the end. The TVR mainly operates along subsequent load steps. The first load shedding (LS) correctly operates at 321 s (after the fourth load increase) in the area with the lower voltage, being that its SVR is near saturation from the beginning. Before and after the fifth load step increase, the second LS (at 405 s) and the third LS (at 425 s) happen in the other two areas due to their voltages ramping down. In short, LS occurs in only a few areas, doing so in from two up to four steps: the very small LS amount is thanks to timely control based on the measurement of the voltage regulating system effort and to the area selection where the automatic voltage control is unable to overcome its saturation.

Those parts of the transients far from load steps and LS are characterised by alternating intervention of generator limits and OLTCs. Voltage changes operated by TVR—reduction after load increase and recovery after LS—are also shown by the transients in Fig. 11.7.

Figures 11.8 and 11.9, respectively, show the V-WAP control on OLTC locking/unlocking and load shedding steps in the North Italy SVR areas during the Fig. 11.7 traces, determined by the use of SVR-TVR $VSI_j(t)$ indices.

With line distance protections, even in front of possible separations the voltage profile is maintained very high, showing a relevant stability result when operating the $VSI_j(t)$-based V-WAP [8, 9]. This performance is also confirmed at low load operating conditions or in case of load variation localised in a given part/area of the system. With distance protections, Fig. 11.10 replaces Fig. 11.4, showing the separation between North and South Italy, with the South unstable. Analogously, Fig. 11.11 replaces Fig. 11.7, showing a relevant stability result, notwithstanding the separation.

The ability of WAP control based on VSI index is clearly evidenced by the analysis of a situation similar to that related to the 2003 blackout [6, 9, 10], where stability improvement was remarkable. Figures. 11.12 and 11.14, linked to a low load case and the untimely operation of distance protections, replaces Figs. 11.4 and 11.7. Figure 11.12 shows, under PVR alone and while load is increasing, a cascade of events after the Lavorgo and Bulciago line openings. Figure 11.13 shows an effective way to reverse the Fig. 11.12 instability by a voltage threshold V-WAP protecting system (shedding: 1892 MW, 6.90 % of nominal). The ability of a V-WAP control based on a VSI index is clearly evidenced by Fig. 11.14, where the stability improvement is remarkable (load shedding 1750 MW, 6.39 % of the nominal).

11.3 Area Voltage Protection Based on Reactive Power Inflow Real-time Voltage Stability Indicator

Premise The V-WAP control solution here presented is still based on the power system subdivision into areas and considers the total real-time reactive power inflow at each area that comes from the surrounding areas/grids interconnections. Obviously,

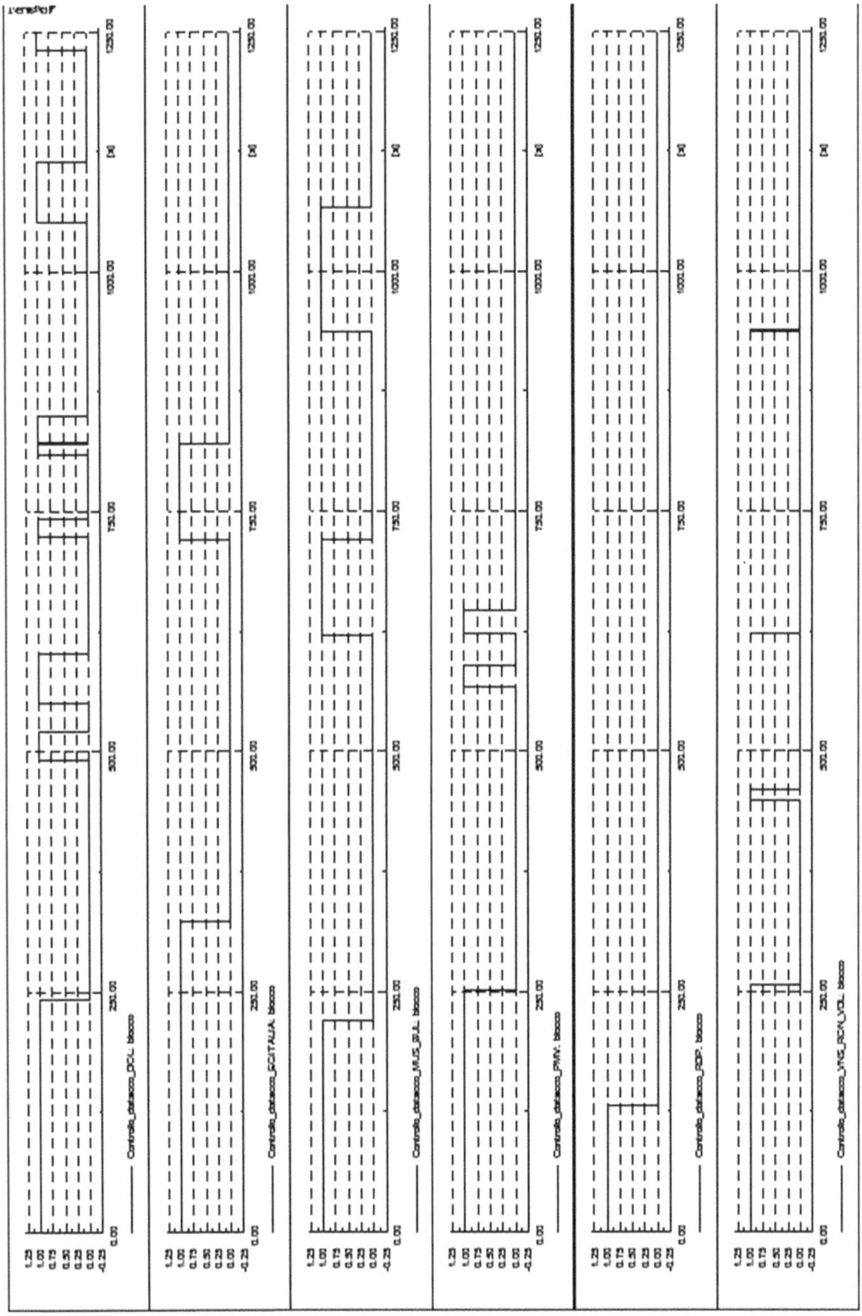

Fig. 11.8 Load increase, $VSI_j(t)$-based WAP-SVR-TVR: OLTC lock-unlock

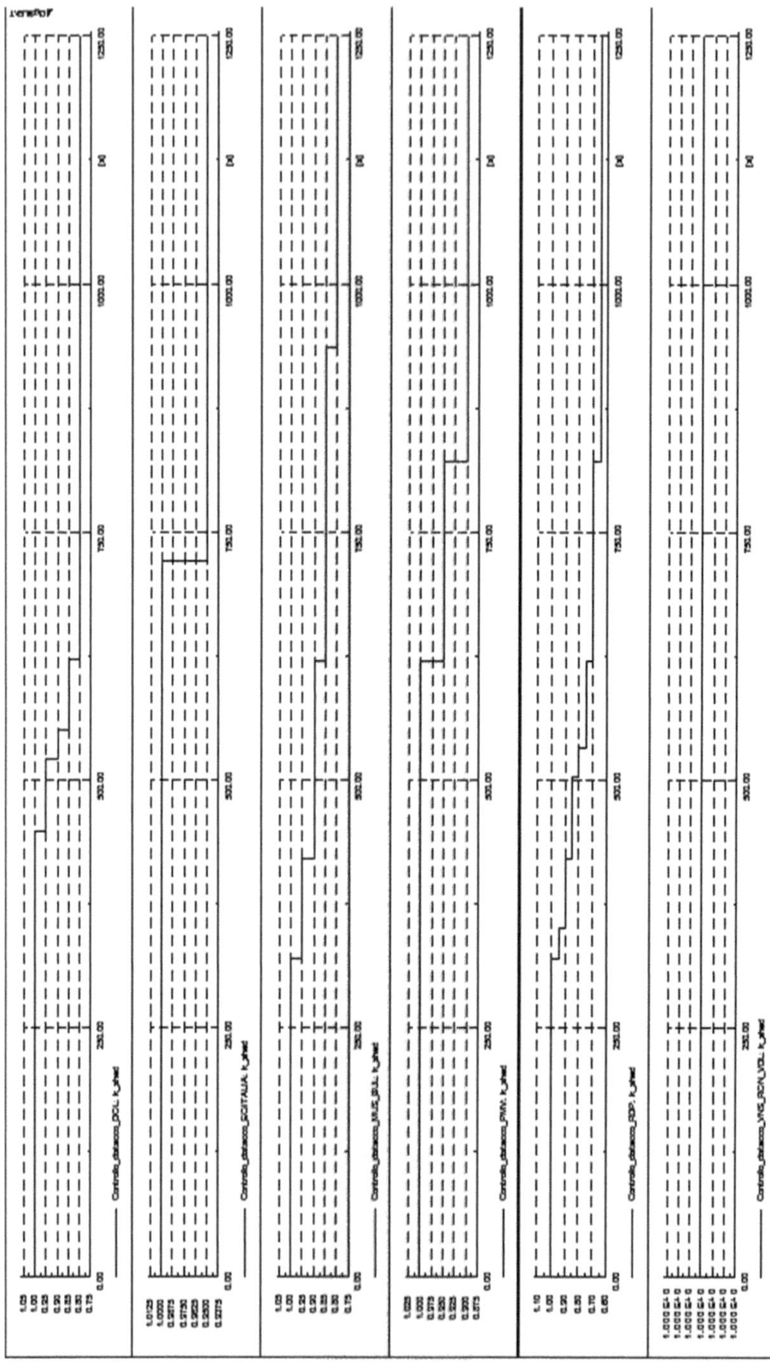

Fig. 11.9 Load increase, $VSI_j(t)$-based WAP-SVR-TVR: load shedding step controls in six SVR areas, based on $VSI_j(t)$

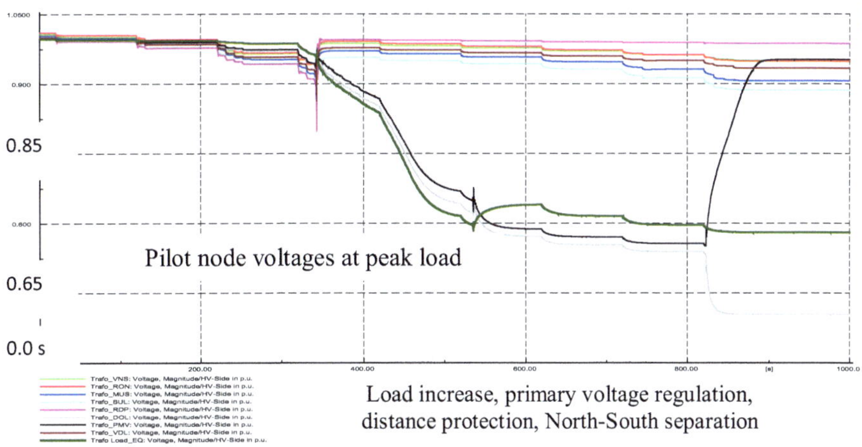

Fig. 11.10 Load increase, primary voltage regulation, distance protection, North–South separation

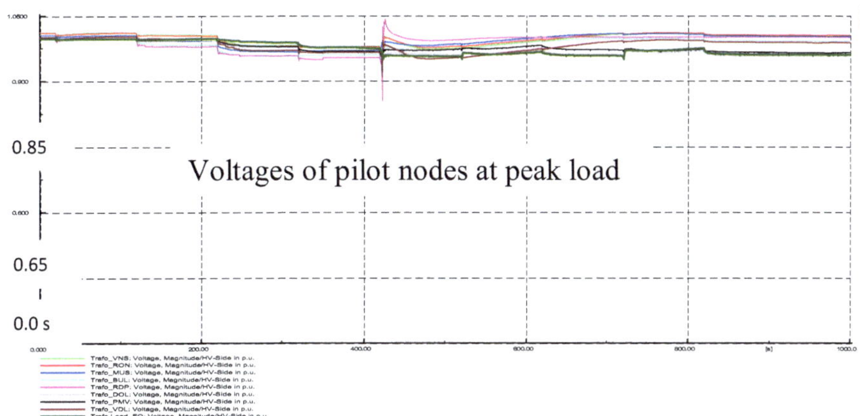

Fig. 11.11 Load increase, SVR-TVR-OLTC block, distance protections, North–South separation, load shedding based on $VSI_j(t)$

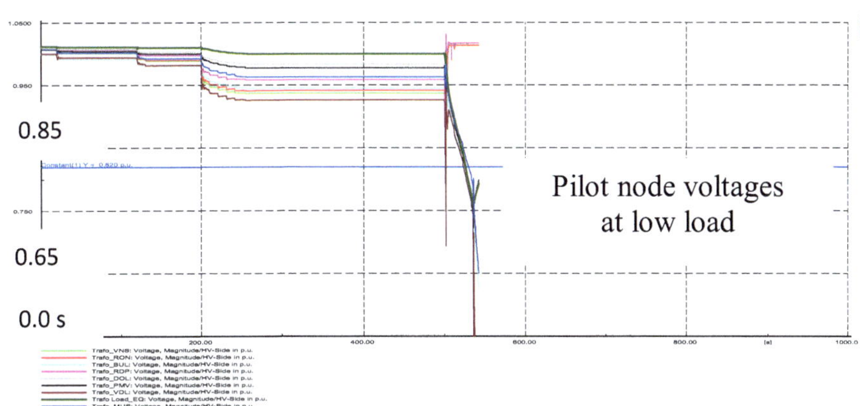

Fig. 11.12 North Italy main 400-kV buses with PVR alone and line protections: "distance and maximum current protections". Lavorgo (200 s) and Bulciago (500 s) lines opening. Grid separation

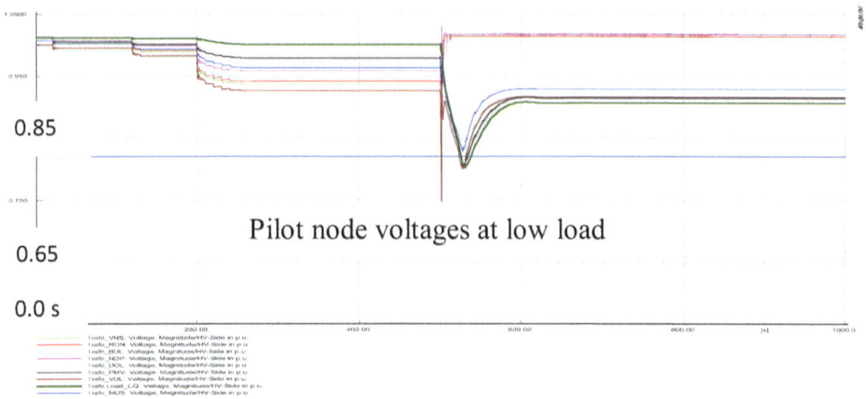

Fig. 11.13 Like Fig. 11.18 but including load shedding based on voltage thresholds. Grid separation

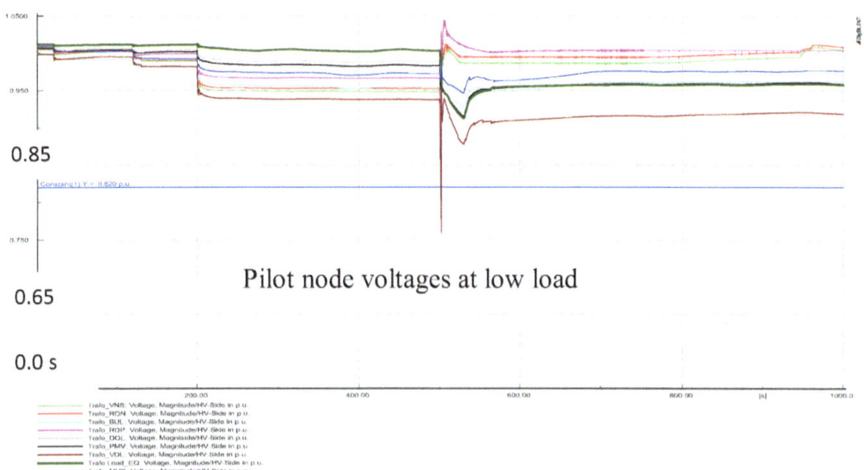

Fig. 11.14 North Italy main 400-kV buses. Loads increase with PVR-SVR-TVR and line protections: "distance and maximum current protections". Load shedding based on $VSI_j(t)$. Grid separation

the inflow is partially contained by internal generators, compensating equipment and line reactive power production. More precisely, the considered V-WAP algorithm makes use of a real-time VSI index (see § 9.5) based on the dependence of the area reactive power inflow on internal active and reactive load changes at each area bus.

As already seen, this area index increases in value the farther the reactive power comes and the more internal line losses increase. On this basis, the variation of the total reactive power inflow into the grid area is used by proper V-WAP logic to safely protect the area from incoming voltage instability.

11.3.1 Basics of Real-time $VSI_i(t)$ Index Linked to V-WAP Referring to a Power System Area-i:

In the case of an area based on the SVR subdivision or defined by the country/utility edges, the here-considered voltage stability index mainly refers to the dependence of total reactive power Q_{in_tot} inflow into the area by each internal bus reactive load variation:

$$I_{cr,i} = \frac{\partial Q_{in\_tot}}{\partial Q_i},$$

where Q_i is the reactive power absorbed by the ith bus.

The most critical bus in the area has the highest index value determined by the largest amount of the reactive power either injected into the area by its interconnections or its internal generators. This preliminary index value tends to increase the farther away the reactive power comes, as well as the higher the line losses are.

Because line currents also depend on active powers, the above index should be improved by also considering the Q_{in_tot} link with bus active power variation.

Based on a linearised analysis, the variation of the total reactive power controlling the area (dQ_{in_tot}) from outside can be linked with the bus active and reactive load variations dQ_i, dP_i, as already seen in § 9.5:

$$\begin{aligned} dQ_{in\_tot} &= \sum_{i=1}^{N} \frac{\partial Q_{in\_tot}}{\partial Q_i} \cdot dQ_i + \sum_{i=1}^{N} \frac{\partial Q_{in\_tot}}{\partial P_i} \cdot dP_i \\ &= \sum_{i=1}^{N} (I_{crQ,i} \cdot dQ_i + I_{crP,i} \cdot dP_i), \end{aligned}$$

with N representing the number of area load buses.

Therefore the following two indices for the ith load bus are combined:

- Risk index linked with reactive power: $I_{crQ,i} = \dfrac{\partial Q_{in\_tot}}{\partial Q_i}$;
- Risk index linked with active power: $I_{crP,i} = \dfrac{\partial Q_{in\_tot}}{\partial P_i}$.

Considering a constant power factor around operating point "0" characterised, at the $i$thbus, by the $P_{i,o}$, $Q_{i,o}$ and $\cos(\varphi_{i,o})$ values, it is possible to write:

$$\begin{aligned} dQ_{in\_tot} &= \sum_{i=1}^{N} (I_{crQ,i} \cdot dQ_i + I_{crP,i} \cdot dP_i) \\ &= \sum_{i=1}^{N} [I_{crQ,i} \cdot (\sin(\phi_{i,o}) \cdot dA_{i,o}) + I_{crP,i} \cdot (\cos(\phi_{i,o}) \cdot dA_{i,o})] \\ &= \sum_{i=1}^{N} [I_{crQ,i} \cdot \sin(\phi_{i,o}) + I_{crP,i} \cdot \cos(\phi_{i,o})] \cdot dA_{i,o} = \sum_{i=1}^{N} I_{crPQ,i} \cdot dA_{i,o}. \end{aligned}$$

The combined index $I_{crPQ,i}$ links the active power and reactive power risk indices by the $\sin(\varphi_{i,o})$ and $\cos(\varphi_{i,o})$ weights.

The real-time computing of $I_{crQ,i}$ and $I_{crP,i}$ requires a burdensome and complex effort such as is usually required by derivative terms, and through it is possible to define an effective real-time area protection function in case of a lack of both filtering uncertainty and too lengthy computing of a reliable VSI index value.

11.3.2 Wide Area Voltage Stability Protection Philosophy Based on $dQ_{in\_tot}(t)$ Indicator

In case the area is one of those defined by SVR, local voltages are linked to each other as well as to local generators; therefore, all area buses are involved in local voltage/reactive power variations. On the contrary, when we consider a large power system like an area, a local voltage problem can be seen by only a part of the area. This difference significantly impacts indicator reliability, as we show in the following.

To compute the amount of load to be shed, an area security degree is defined which considers the difference between the maximum reactive power that can be injected into the area $Q_{\mathrm{in\_tot\_max}}(t)$ and the real value $Q_{\mathrm{in\_tot}}(t)$:

$$MQ(t) = Q_{\mathrm{in\_tot\_max}}(t) - Q_{\mathrm{in\_tot}}(t),$$

$$mq(t) = \frac{MQ(t)}{Q_{\mathrm{in\_tot}}(t)}.$$

With regard to the generators, their maximum reactive power is fixed, in each instant, by the over-excitation limits. As far the interconnecting lines are concerned, the maximum reactive power transfer to the area by each can be deduced from the Thevenin equivalent seen by the line from the edge of the protected area.

V-WAP control logics operate the area OLTC lock and load shedding according to both the real-time $mq(t)$ and the combined index $I_{crPQ,i}$ values.

Load Shedding

Load shedding logic acts according to the information from the margin of reactive power injected into the network, $mq(t)$, and by the nodes' criticality indicators. The $mq(t)$ value defines the time and the amount of load to be shed in the area. When $mq(t)$ goes below the allowed confidence band (see Fig. 11.15), having smq_{sup} and smq_{inf} thresholds, for a time lasting more than ΔT_{shed} from the previous shedding command, the new load shed amount is $\Delta A_{\mathrm{car\_tot}}(t)$:

$$\Delta A_{\mathrm{car\_tot}}(t) = k_{\mathrm{shed}} \cdot \left(smq_{\mathrm{sup}} - mq(t)\right) \cdot Q_{\mathrm{in\_tot}}(t).$$

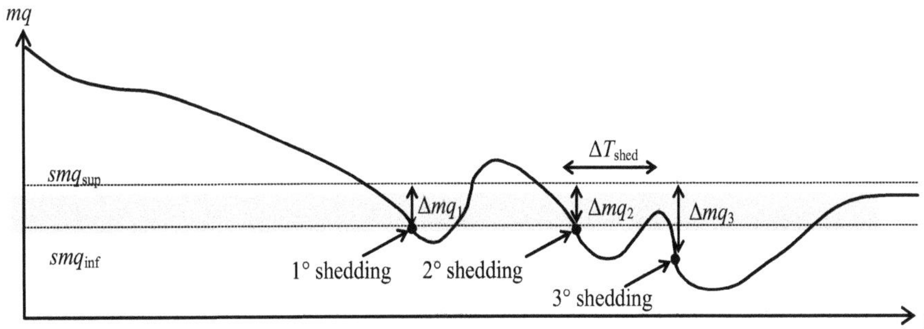

Fig. 11.15 Control logic, $mq(t)$-based, to define timing and amount of load to be shed

ΔT_{shed} is required to wait for the effect of the previous operated load shedding; $Q_{\text{in\_tot}}$ is the measured reactive power injected into the area grid. By this logic, the load to be shed increases when $mq(t)$ decreases and vice-versa. Figure 11.15 gives an example of the described logics in front of progressive load increases.

Referring to Fig. 11.15, at the beginning $mq(t)$ progressively decreases down to the lower threshold smq_{inf}. This determines the first load shedding followed by a margin recovery. Increasing the load, smq_{inf} is again reached with a consequent new load shedding. Because of the more critical conditions, the protecting action is less effective than before and again the margin suddenly falls. This time the new load shedding is operated after the waiting, in the interval ΔT_{shed}. The third control operates a greater load shedding than before due to the higher distance from smq_{sup}. Having defined the timing and amount of load to be shed, the index $I_{crPQ,i}$ is used to determine the way $\Delta A_{\text{car\_tot}}$ has to be shared among the load buses in the area, according to the bus combined indices $I_{crPQ,i}$:

Load Shedding Criteria:

- The load to be shed at the *i*th bus must be below a given percentage of the nominal apparent bus load;
- The load shedding at the *i*th bus is allowed only when its voltage is below a given threshold: $V_{\text{shed},i}$.

Once the maximum load to be shed at each *i*th bus, $\Delta A_{\text{car—max}i}$, is settled, the total apparent power to be shed, $\Delta A_{\text{car\_tot}}$, is shared among the buses, with the highest criticism defined by the index $I_{crPQ,i}$, subject to the $\Delta A\text{car\_max},_{i}$ bond. In this way the injected reactive power is minimised with the consequent maximisation of the $mq(t)$ margin. The $dQ_{\text{in\_tot}}(t)$ indicator is used to represent the area criticism.

In the presence of SVR this V-WAP protection logic will enter into operation the same way as described in § 11.2, where the SVR-TVR $VSI_j(t)$ indicator (*j*-area) was considered. Moreover, before load shedding, it will operate in the order the OLTC lock, following a logic based on $mq(t)$ like the one represented in Fig. 11.15, but with greater thresholds (larger band), followed by the paralleling of hot running reserves for reactive power support. More precisely, OLTC locking is shared among the buses, with the highest criticism defined by the index $I_{crPQ,i}$, while the paralleling of hot running reserves should be as much as possible near those buses.

11.3.3 Simulation Results of V-WAP Based on $dQ_{in\_tot}(t)$

Considering risk index $dQ_{\text{in\_tot}}(t)$:

- OLTC blocking is based on the following control parameters:
  - $smq_{\text{block\_inf}} = 3.5\,\%$,
  - $smq_{\text{block\_sup}} = 4.5\,\%$,
  - $k_{\text{block}} = 0.1$.
- *Load shedding* is based on the following control parameters:
  - $smq_{\text{inf}} = 1.75\,\%$,
  - $smq_{\text{sup}} = 2.25\,\%$,
  - $k_{\text{shed}} = 6.0$,
  - $\Delta T_{\text{shed}} = 30$ s,
  - $V_{\text{shed}} = 0.88$ p.u.,
  - $\Delta A_{\text{car\_max}} = 5\,\%$.

The performed tests refer to the same percentage of load increase in the entire Italian power system, shown in Fig. 11.2. The load increases every 100 s, starting from 20 s, with the steps each amounting to 2% of nominal load, up to reaching the last increase at 820 s. This large load increase allows the differences to be checked to significantly among the six selected cases:

- Figure 11.16: With PVR alone, OLTC lock and load shedding $dQ_{\text{in\_tot}}(t)$-based. Voltage transients of buses named as the pilot nodes in § 11.2;
- Figure 11.20: With SVR-TVR, OLTC lock and load shedding based on $VSI_j(t)$ and $I_{crPQ,I}$ indicators. Voltage transients of buses named pilot nodes in § 11.2.

Figure 11.16 shows the effects of V-WAP control based on risk indices $I_{crP,i}$ and $I_{crPQ,i}$. Indices $I_{crP,i}$, related to bus-*i* active power, assume values between 6.0 and 28.9, whereas $I_{crQ,i}$, dependent on bus-*i* reactive power, range between 5.1 and 16.6; both are therefore very important in defining critical conditions. In this case, load shedding amounts to 6.54 % while voltage differences with respect to the Fig. 11.16 case, based on voltage thresholds, are not very relevant.

The voltage transients in Fig. 11.20 are very similar to those in Fig. 11.17, even when the load V-WAP function is related to the $VSI_j(t)$ and $I_{crPQ,i}$ combined indices. In this case, risk index $I_{crPQ,i}$ assumes lower values than before, in Fig. 11.16, mainly because of:

- The TVR, which maintains a control margin in the generators;
- The pilot node higher voltage values at the pilot nodes.

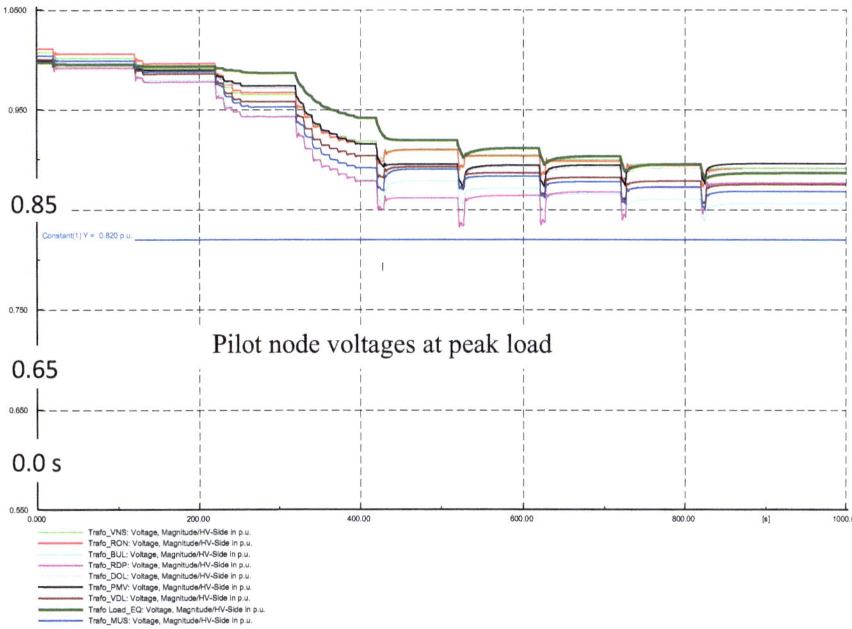

Fig. 11.16 Load increase, PVR alone; OLTC lock and load shedding based on $I_{crPQ,i}$

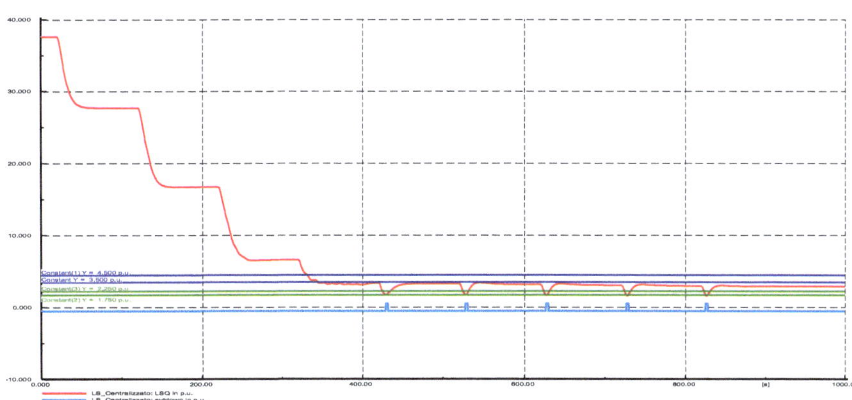

Fig. 11.17 Trend of $mq(t)$. Case of load increase, PVR alone; OLTC lock and load shedding based on $I_{crPQ,i}$

Even with voltage transients being very similar to Fig. 11.7, load shedding in this case is reduced from 11.26 to 10.37 %, due to the highest capillary/effective load selection inside the area operated in accordance with risk index $I_{crPQ,i}$. On the contrary, the complexity of $I_{crPQ,i}$ index computing is very high in comparison to SVR-TVR $VSI_j(t)$ alone.

With distance protection, even in front of possible separation, voltage profiles are maintained to be very high, showing a relevant stability result when $VSI_j(t)$-based V-WAP operates. This performance is also confirmed at low-load operating conditions or in the case of load variation localised in a given part/area of the system. Conversely, $I_{crPQ,i}$-based V-WAP control confirms its utility only when SVR is operating or when the $I_{crPQ,i}$ and $VSI_j(t)$ indices are combined to define the V-WAP strategy.

V-WAP control based only on $I_{crPQ,i}$ becomes less robust in terms of voltage quality/stability mainly in front of protection operations and localised load changes (these results are not shown here due to limited space). The capability of WAP control based on $VSI_j(t)$ indices is clearly evidenced by [8, 9, 14], where the stability improvement is remarkable, again in front of a situation similar to that related to the 2003 blackout in Italy [10].

Figure 11.17 shows the $mq(t)$ trends when a power system is under PVR alone. Increasing the load, the $mq(t)$ indicator progressively goes down to reach thresholds that activate OLTC locking commands (Fig. 11.18) and load shedding according to the sequence shown in Fig. 11.19.

Figure 11.21 shows the entering/exiting of OEL operation by generators in each SVR area when a power system is under SVR-TVR control. Increasing the load, the OEL operation is entered, while OLTC locking commands contribute to exclude OEL from their operation (Fig. 11.22), and load shedding is according to the sequence shown in Fig. 11.23.

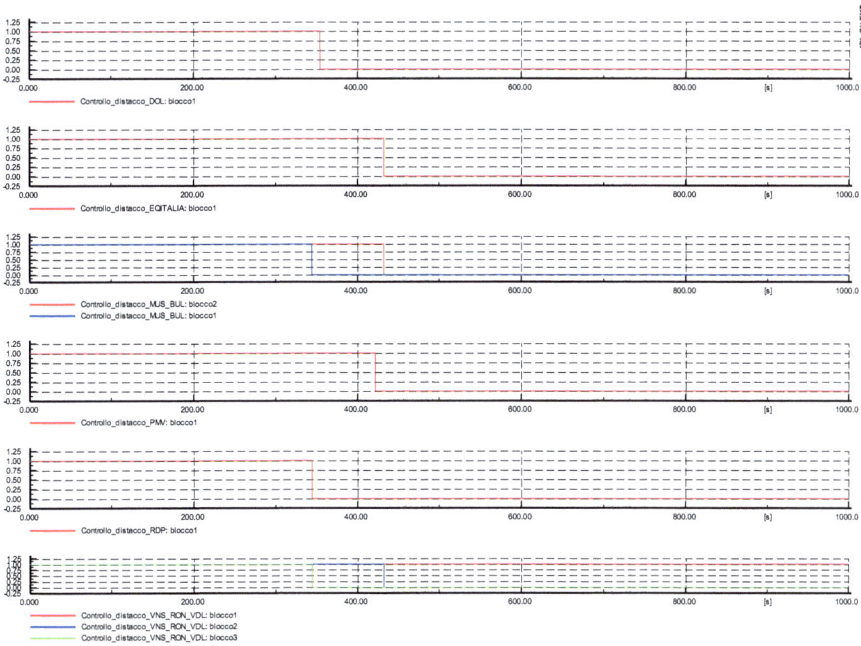

Fig. 11.18 OLTC lock commands. Load increase, PVR alone; OLTC lock and load shedding based on $I_{crPQ,i}$

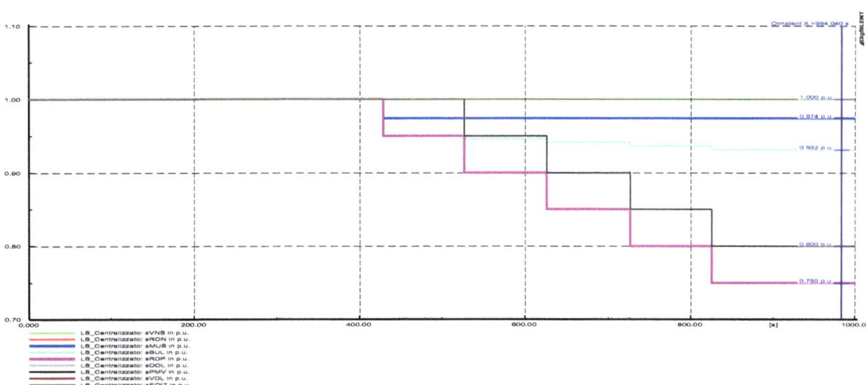

Fig. 11.19 Load shedding command traces: load increase, PVR alone; OLTC lock and load shedding based on $I_{crPQ,i}$

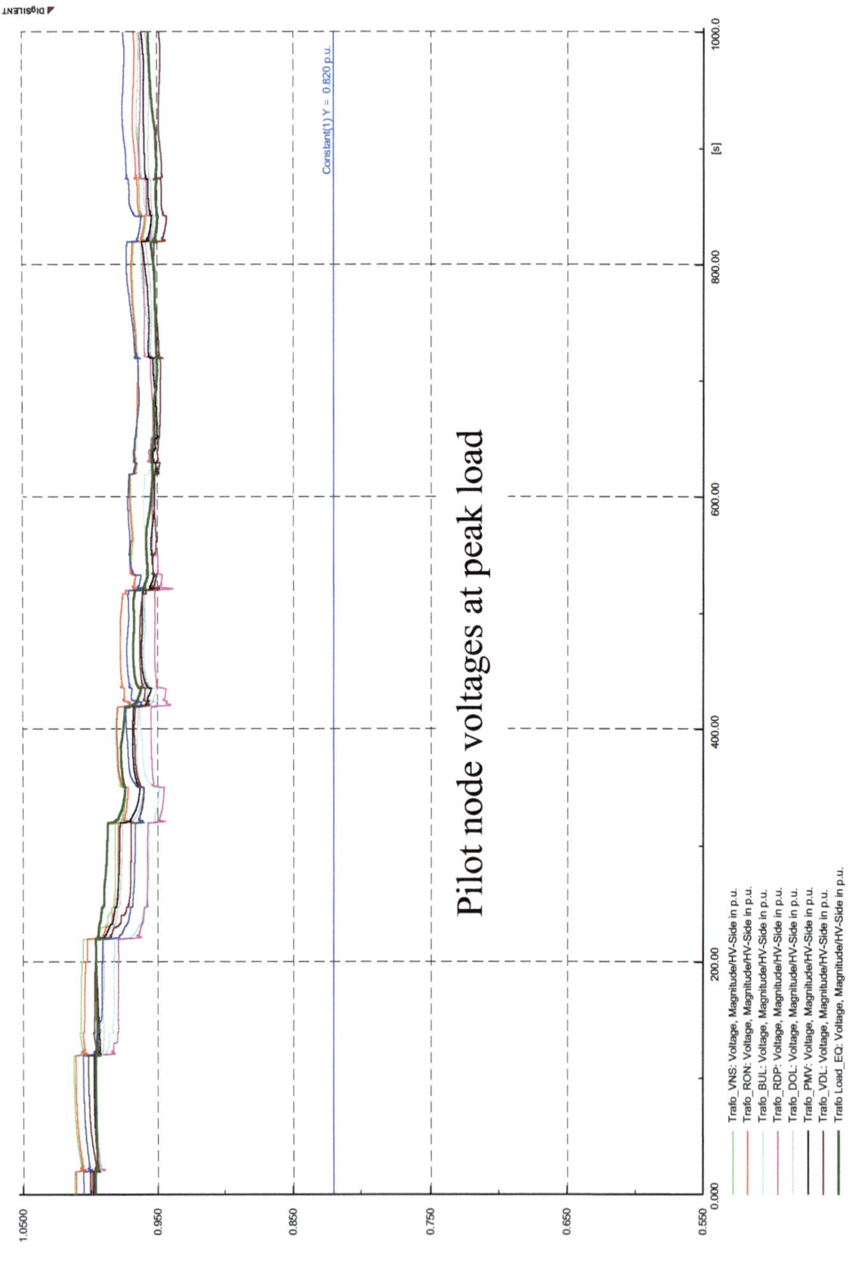

Fig. 11.20 Load increase, SVR-TVR; OLTC lock and load shedding based on SVR-TVR $VSI_j(t)$ and $I_{crPQ,i}$

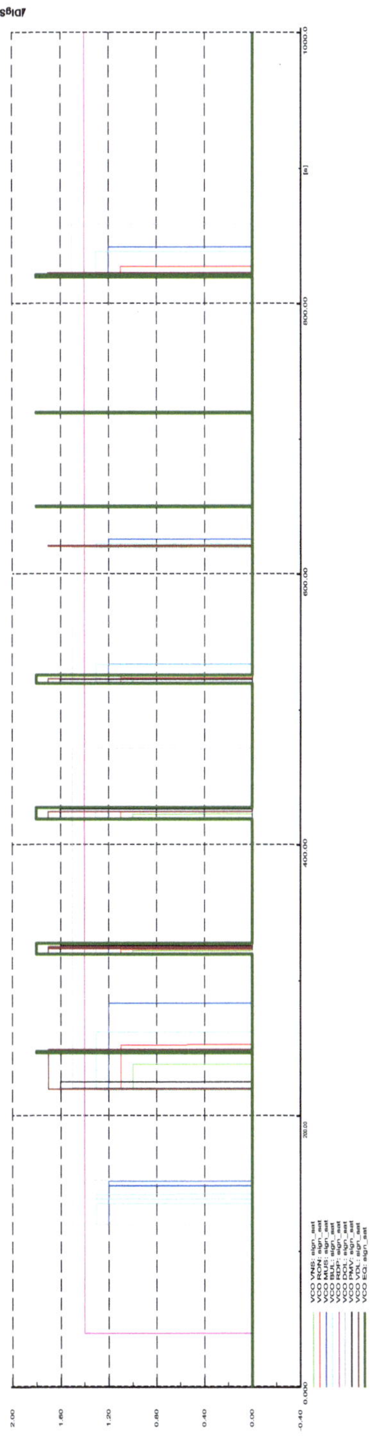

Fig. 11.21 OEL intervention traces. Case of load increase, SVR-TVR; OLTC lock and load shedding based on SVR-TVR-$VSI_j(t)$ and $I_{crPQ,i}$

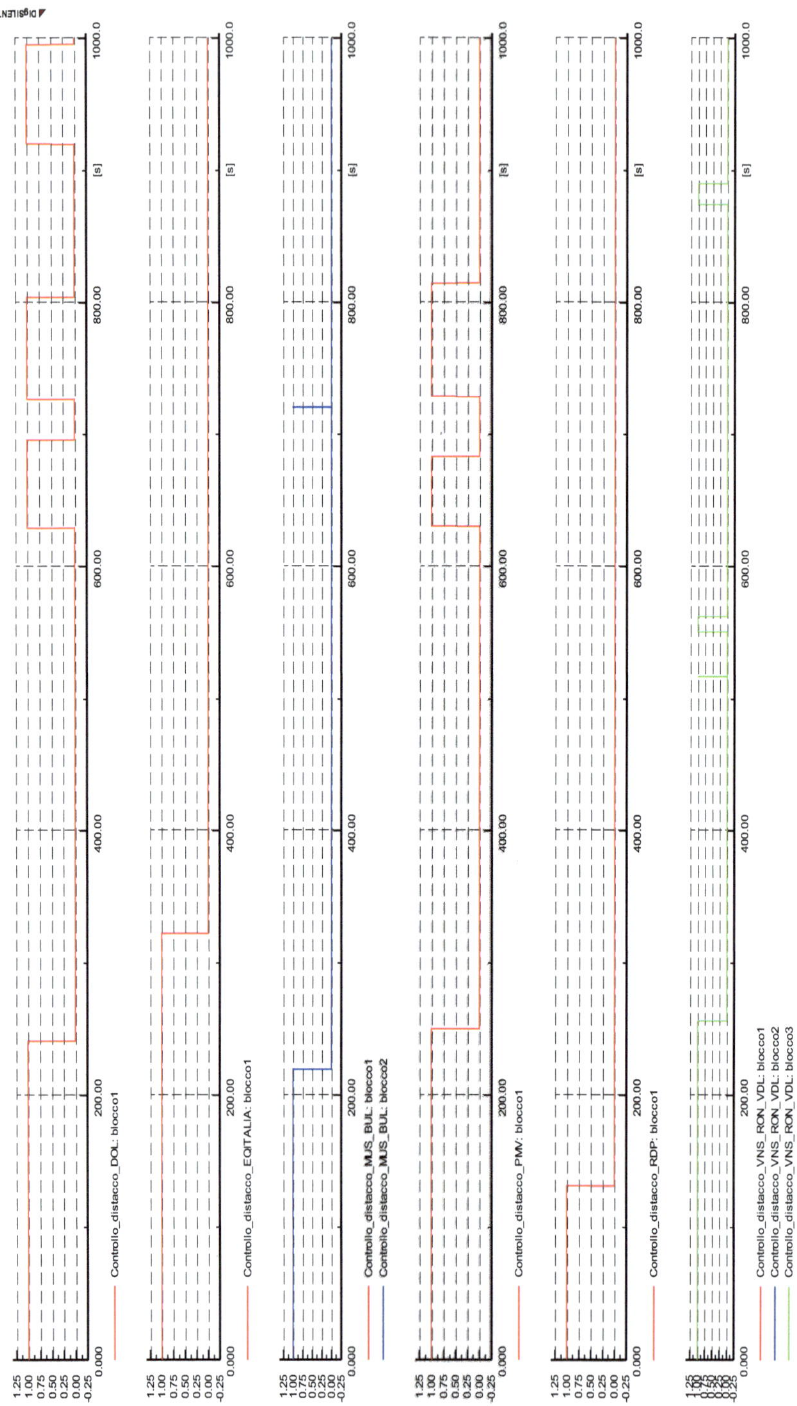

Fig. 11.22 OLTC lock/unlock area command traces. Case of load increase, SVR-TVR in service; Commands of OLTC lock and load shedding based on SVR-TVR VSI_j t) and $I_{crPQ,i}$

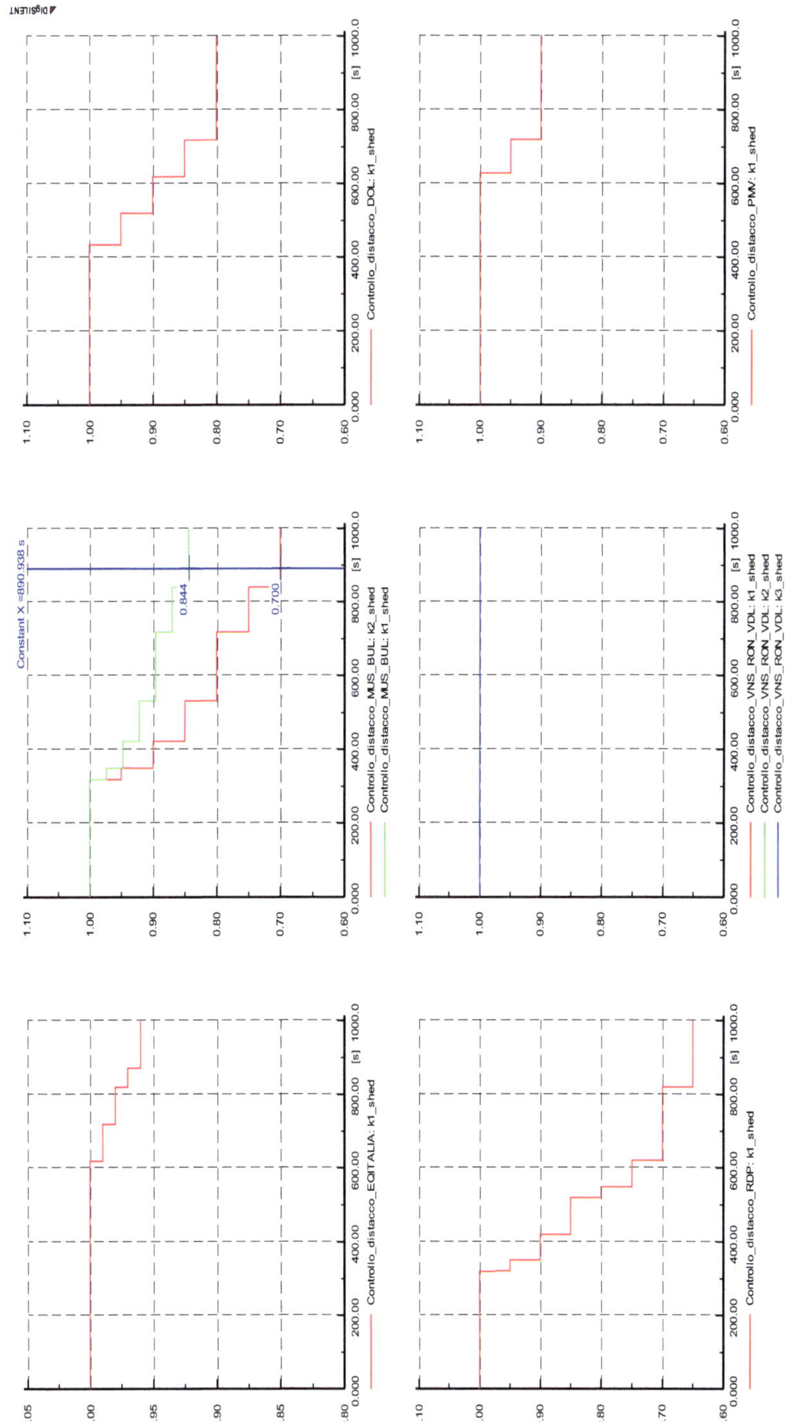

Fig. 11.23 Load shedding: area command traces. Case of load increase, SVR-TVR in service; Commands of OLTC lock and load shedding based on SVR-TVR $VSI_j(t)$ and I_{crPQ}

11.4 Area Voltage Protection Based on PMU and Related Real-time Voltage Stability Indicator

Premise The V-WAP instability analysis performed by considering the classic Thevenin equivalent seen by a given grid bus [11, 12, 21, 24, 25, 27–30]. Based on measuring local voltage and current phasors at the EHV bus considered, a real-time voltage instability indicator can be computed by identifying (the updating interval is between 20 and 80 ms) the parameters of the Thevenin equivalent analysed (refer to § 9.3).

As is generally understood, instability is widely linked to a condition of equality between the absolute value of two equivalent impedances—load and Thevenin impedances—with this equality corresponding to the maximal power transfer. Concerning this, some basic and fundamental considerations should be made clear [25]:

(i) The first point is that the Thevenin equivalent must represent a detailed dynamic model. Therefore, its parameters change continuously and quickly while approaching maximum loadability.
(ii) The second refers to local area generator OELs and transformer OLTCs, whose dynamics strongly impact bus maximum loadability and voltage instability [11].
(iii) The third refers to the P-constant load analysis, which is not adequate for a correct identification in real systems, where a mixed ZIP load should be adopted.
(iv) The fourth point distinguishes maximum loadability point (*V-P* nose curve) from real instability point, which in general occurs at very low voltage.
(v) The last basic consideration refers to the high speed with which the *V-P* curve's equilibrium point moves from the *V-P* curve nose tip to the first unstable point at lower voltage. This fact confirms the practical importance of nose tip identification to prevent voltage instability, as well as the need for a very fast procedure to identify the approach of the voltage instability limit.

On this basis, the proposed VSI in [12], described in § 9.3, is capable of identifying in real-time the Thevenin equivalent, with higher precision in proximity of the voltage instability where the OELs and OLTCs operate. This faster-change dynamic also corresponds to relevant simultaneous and opposite-direction variations of load and Thevenin equivalent impedances.

A V-WAP system based on this method requires more than one installation of PMU on the main HV buses in the considered wide area, jointly with real-time identifiers of local bus voltage instability risk indicators. Based on wide area indices values a proper and timely load shedding can be progressively ordered around the bus with the highest index value.

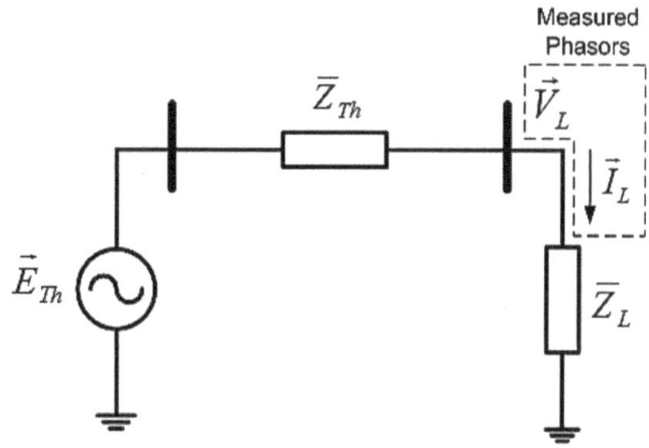

Fig. 11.24 Two-bus Thevenin equivalent circuit

11.4.1 Basics of Real-time VSI-PMU(t) Index Linked to V-WAP

Synthetic Description of Proposed Algorithm

The maximum power transfer to the load in the electric circuit shown in Fig. 11.24 occurs when

$$\left|\bar{Z}_L\right| = \left|\bar{Z}_{Th}\right|, \text{ with } \bar{Z}_L = Z_L \angle\theta = R_L + jX_L.$$

This circuit represents the entire network "seen" from the considered bus in an equivalent way. Moreover, at the transmission grid, $X_{Th} \gg R_{Th}$, and the assumption of $R_{Th} \approx 0$ is very reasonable.

The admissible range for E_{Th} must be in agreement with electric circuit laws. Up to the maximum power transfer (MPT), its minimum value ($E_{Th}^{\min}$) corresponds to the load voltage, while its maximum value ($E_{Th}^{\max}$) corresponds to the voltage when $Z_L = X_{Th}$ (with $R_{Th} = 0$).

Under normal operating conditions the load impedance is much higher than the equivalent Thevenin impedance.

Even being inside its admissible range, E_{Th} as the free variable of the problem should be updated towards its correct physical value (see § 9.3 and [12], where the E_{Th} updating direction is determined at each sampling time, according to a demonstrated theorem, in order to speed up the X_{Th} identification convergence). The proposed algorithm assumes that E_{Th} and X_{Th} are constant in the brief "i—(i—1)" interval of their identification, which requires a very short sampling time. This is allowed by the electrical measured quantities provided by the local PMU.

In brief, the proposed adaptive algorithm will reduce E_{Th} when both variations of Z_L and estimated X_{Th} have opposite direction; otherwise E_{Th} will increase. This is evident from the following (provided by [12], see § 9.3.2, Eq. (9.7)):

$$\tilde{\Delta}\bar{X}^{i}_{Th} = \tilde{\Delta}E^{i-1}_{Th} \cdot \frac{(\bar{X}_{Th} + \bar{Z}^{i-1}_{L} + \Delta\bar{Z}^{i}_{L})}{E_{Th}} = \tilde{\Delta}\bar{X}^{i}_{Th-1} + \frac{\tilde{\Delta}E^{i-1}_{Th}(\Delta\bar{Z}^{i}_{L})}{E_{Th}}.$$

After knowing the direction E_{Th} should be updated at each sampling time, we must establish how large this variation should be. This quantity is calculated as already shown in § 9.3:

$$\varepsilon_E = \min(\varepsilon_{\inf}, \varepsilon_{\sup}, \varepsilon_{\lim})$$

with

$$\varepsilon_{\inf} = \left|E_{Th}^{i-1} - V_L^i\right|,$$

$$\varepsilon_{\sup} = \left|E_{Th}^{i-1} - E_{Th}^{\max(i)}\right|,$$

$$\varepsilon_{\lim} = \left|E_{Th}^{i-1} \times k\right|,$$

with k a pre-specified parameter chosen to constrain the identification error within narrower bounds, and i the corresponding time step.

Most of the time, $\varepsilon_{\lim}$ drives the identification process, so its specification has a major impact on the process. The quantities $\varepsilon_{\inf}$ and $\varepsilon_{\sup}$ are active only when the estimated E_{Th} is close to the edges of its feasible range. Based on the above, the adaptive algorithm tracks the correct value of E_{Th} to identify X_{Th}. This is pursued by using local load real time values of V_L, I_L and angle θ between the two phasors, to calculate at each instant i the corresponding X_{Th}^i value by using a moving average of E_{Th} over a window of appropriate size (m), instead of the instantaneous value of E_{Th}^i at iteration i. This has the advantage of filtering the identified variables, paying the price of a slower identification process.

In summary, the proposed algorithm will identify the X_{Th} value as soon as identification of E_{Th} is reached by imposing an oscillation with small amplitude around their correct values and with half the sampling rate frequency. When $|Z_L| = |X_{Th}|$, the *V-P* curve nose tip has been reached and real-time indicators, based on this result, can be used.

All the performed tests and simulations assume a sampling rate of 20 ms, with identification updated every 20 ms. The identified value is based on the sampled data of the last $m \times 20$ ms (for tests on the Italian system $m=4$). The parameter k was set within the range of 0.01–0.1 % of nominal voltage.

Tests have been satisfactorily performed at both the load bus and "transit" bus (no local load bus with equivalent load computing based on the exiting currents).

The identification algorithm works correctly with high reliability until the local load at the considered HV bus is continuously changing. Therefore, the algorithm must hold as soon as the local load variation is below a given slope. A minimum threshold of 0.5 %/min can be assumed as an impassable limit for the identification.

Figure 11.25 shows two curves for the identification of X_{Th} as a function of the rate of change for the Poggio a Caiano bus load in the Italian grid. From 0 to 200 s, the load rate of change is 40 %/min (76 MW/min) in both curves. From 200 to 600 s, Curve 1 and Curve 2 correspond to a load rate of change of 20 %/min and 0.5 %/min, respectively. The algorithm is capable of continuing to correctly identify X_{Th} when the load ramp changes from 40 to 20 %/min. However, it starts to have difficulties when the rate of change is 0.5 %/min.

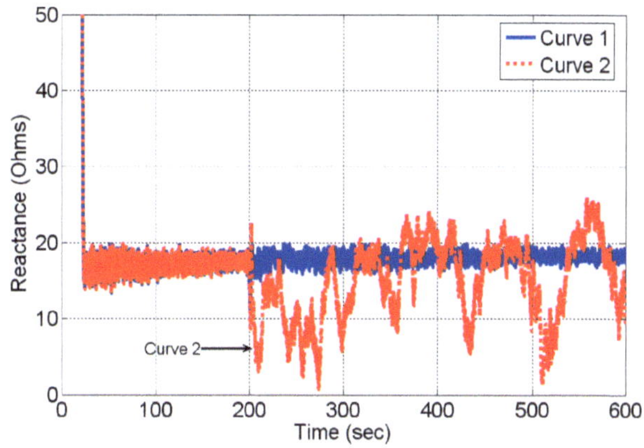

Fig. 11.25 X_{Th} identification as a function of the load rate of change

From the tests performed, it can be concluded that the algorithm is able to adequately and robustly identify the correct value of X_{Th} for a slope variation higher than 1 %/min with respect to the initial load (this corresponds to a few kW/s).

When the load variation is less than this value, there is a need to stop the identification algorithm. Inside this deadband the value of X_{Th} to be considered is that achieved by the last sampling time with a correct identification.

This transient identification stopping by a given bus does not mean a V-WAP limitation or uncertainty on system security is attributed to that bus. In fact, the process does not reduce its voltage stability margin with respect to invariant local loads. Other grid buses provided with PMU and with variant local loads will continue their local VSI index updating.

11.4.2 Wide Area Voltage Stability Protection Philosophy Based on VSI-PMU(t)

Voltage Instability Risk Prediction

The results of both Chap. 8 and § 9.3 give evidence to the fact that the nose tip in the dynamic *V-P* curve is reached before the irreversible voltage instability begins. Therefore, nose tip recognition is helpful information that anticipates the incoming voltage instability determined by the system load variation underway and the operating controls on the generators, transformers and compensating equipment.

What is less simple to predict in power systems is the speed the operating point has, at any *V-P* curve seen by each bus, when moving from the tip to the irreversible instability point. This in fact depends on the load variation trend, robustness (short circuit power) of the bus, perturbation consistency and its electrical distance from the system buses with PMU, and so on. Therefore, in case the nose tip identification is considered too-late information or if one wishes to check other possibilities for tuning the identification speed, we show the proposed algorithm that allows an easy tuning of crossing time between the identified and the real local load reactance, therefore anticipating or delaying the nose tip recognition with respect to its correct time.

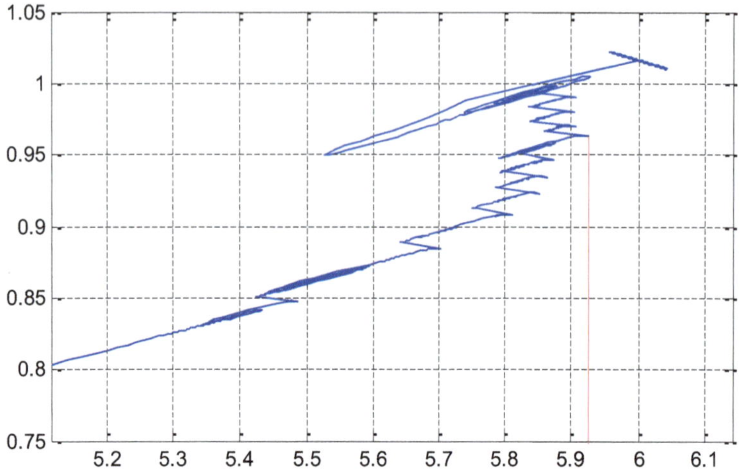

Fig. 11.26 Dynamic *V-P* nose curve at Bus 1041 in Nordic 32 power system following 4032–4034 line opening

Referring to § 11.4.1, it can be seen that identification speed depends on the parameter $\varepsilon_E = \min(\varepsilon_{\inf}, \varepsilon_{\sup}, \varepsilon_{\lim})$ and mostly on the *k* coefficient of $\varepsilon_{\lim}$. Therefore, *k* tuning simply allows changing the slope of the Thevenin reactance identification curve around its correct value: that is, the curve determining the crossing point of Thevenin's and load reactance at the *V-P* curve nose tip.

Prediction Modulation by Tuning Control Parameter of Identification Algorithm

As said, the prediction can be anticipated or delayed by simply changing the value of a control parameter of the WAP algorithm, as is shown now for the case of Nordic 32 power system, considering the opening at $t=270$ s of an important North–South line between Buses 4032 and 4044, transferring approximately 750 MW (see Fig. 11.29). The contingency determines voltage instability followed by system collapse. By analysing the phenomenon at HV Bus 1041, one of the buses more sensitive to the incoming instability risk, the following results are obtained:

The dynamic *V-P* curve nose tip in Fig. 11.26 is characterised by a sawtooth shape due to the system's OLTC and OEL operation after the contingency.

The red line in the previous plot indicates the maximum line power MLP after the contingency. This occurs at time $t=336.1$ s. The *k* parameter value used in the identification algorithm that makes crossing of load and Thevenin's impedances occur at 336.1 s, with load bus voltage >0.95 p.u., is $k=0.00035$. The achieved result is shown in Fig. 11.27.

The anticipation of the algorithm crossing, and therefore of the alarm switching alerting the operator, can be simply achieved by a small increase of the control parameter *k*. With $k=0.001$, the prediction of the nose tip approaching is anticipated at about 20 s, as shown in Fig. 11.28.

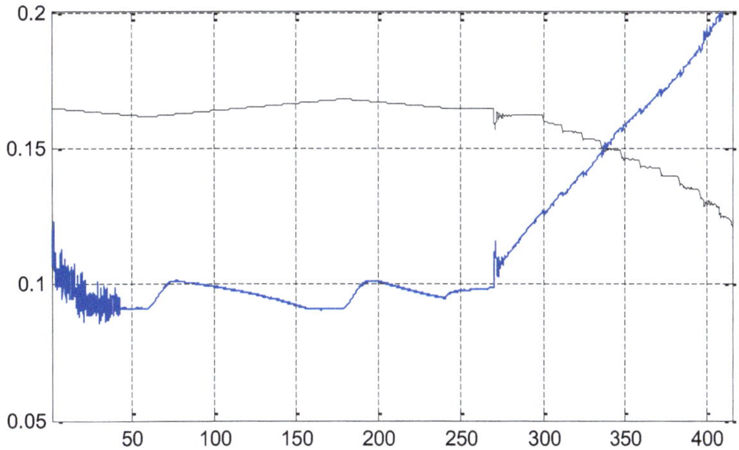

Fig. 11.27 Load and Thevenin impedances crossing at 336.1 s (k=0.00035)

11.4.3 V-WAP Based on VSI-PMU(t) Simulation Results

Wide Area VSI Warning

Being the proposed WAP algorithm based on the computing of the local voltage instability risk through the PMU high speed measurements, each bus with PMU is able to check the instability risk seen by its grid node, as already seen in § 9.3.

In principle the voltage degradation problem differs from one grid point to another as well as the maximum line loadability seen by each bus. Therefore the grid will show different local maximum loadability risks values when comparing his main

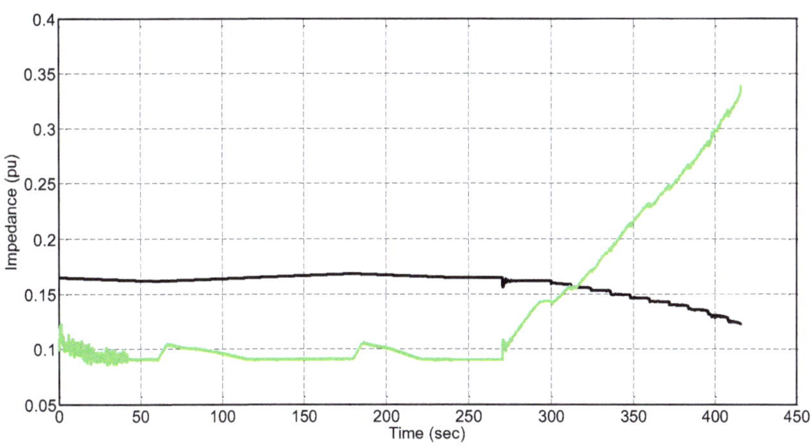

Fig. 11.28 Load and Thevenin impedances crossing at 315 s (k=0.001)

buses information. Now the point is how to correctly use the single risk information coming from each bus and all together at power system level to correctly recognise and distinguish the voltage instability high risk limited to a power system region or pertinent the overall wide area system.

Usually the voltage instability phenomenon begins in a given part of the power system and can be faced locally by avoiding his propagation through local controls and protections. Conversely, when the problem is not confined and a process of cascading events begins, in that case the voltage instability risk refers to the overall system. Lastly the system instability can also begin just after a very heavy contingency to which a more or less fast unstable transient follows.

Many scenarios could therefore be possible: The most common will be the case of localised voltage problem in a given part of the grid, with the risk of local blackout mainly due to untimely protection interventions or local contingency.

The VSI information from the PMU will therefore help to recognise the region edges of the underway voltage lowering risk and the loading increase in given tie lines. Without any proper and timely control based on the regional VSI information, the phenomenon could expand and the number of useful local VSI information increases. This puts in evidence a possible dynamic process progressively changing the powerfulness of the area VSI indicator, at the beginning mostly based on very local information at a given bus or small set of bus and progressively integrating the information coming from an enlarged set of PMU related to a system region up to a wide area. Obviously the more the phenomenon expands the higher is the voltage instability risk for the overall system up to determining the general voltage collapse.

As described, all these mentioned increasing risk phases can be monitored in real time by a "PMU network" linking the main system buses, with the aim to exchange and elaborate the full power system information based on the local VSI real-time indicators. This decentralised or centralised monitoring and computing is a real strong help for significantly improve the power system security and operation:

1. Validating the VSI local information by taking into account of the grid topology and the electrical distance among the buses with data;
2. Monitoring in real-time of the incoming risk and the nodes with higher risk;
3. Monitoring in real-time of the incoming risk speed and its expansion trend linked to cascading events;
4. Supporting in real-time and where properly needed the operator's manual controls on reactive power reserves, OLTCs, international tie-lines powers exchange, loads shedding.
5. Automatic voltage and stability controls based on real-time VSIs and their intelligent elaboration.

To be mentioned that each system perturbation, including protection intervention, load shedding, generator trip, line opening as well as the OLTCs and OELs interventions and saturations, change the shape of the *V-P* curve nose and the speed the operating point runs along the curve, with possible jumps due to large contingency. This instantaneously moves the system from stability to instability and vice-versa. In other words, the more the risk increases the higher is the speed toward instability and collapse. This reason clearly allows an understanding of the need for predictive

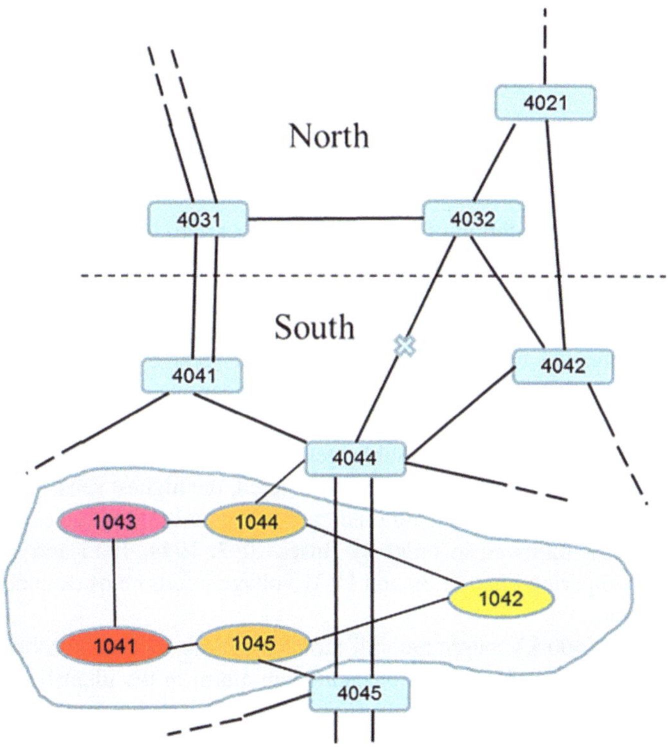

Fig. 11.29 Nordic 32 grid: Mapping by coloured buses of the area at a higher risk of VI, at 315 s: 4 buses at 400 kV and 1 bus at 130 kV (those buses mainly involved)

real-time information on the incoming voltage instability risk, thus giving an opportunity to operators and to the control system to face on time the phenomena underway and to minimise their negative impact.

Adding some details on wide area VSI, the proposed local algorithm can be applied as follows, under the hypothesis of a proper selection of system buses with PMU able to adequately monitor the entire power system and assuming a centralised, high speed elaboration of bus information [13]:

a. Selection of buses with higher instability risk recognised by the local PMU analysis;
b. Mapping voltage instability risk on the grid topology, recognising its extension and propagation speed (Fig. 11.29);
c. Elaborating "local VSIs" into "region-wide area VSI" by selecting the buses with the highest local VSI values, weighting each of them with the local short circuit power normalised by the system's maximum power. The instantaneous sampling and sum of these values gives a reliable region/area VSI at that instant;
d. Checking VSI trend from a region toward a wide area and its value in comparison with high risk slope thresholds as well as maximum allowed limit. Thresholds are computed off-line by system simulations of high risk operating conditions.

Table 11.1 VI recognition and span from the line 4032–4044 outage, Nordic 32

Voltage instability (VI)	Bus 1041	Bus 1042	Bus 1043	Bus 1044	Bus 1045
VI recognition time (s)	315	340	320	325	325
Span VI recognition after outage (s)	45	70	50	55	55

The practical implementation of this proposal appears to be very simple and does not require high speed telecommunication from the remote to the central control system, due to local PMU computing of local VSI. A sampling rate of 1 s should suffice for classic control actions based on the above mentioned region/area VSI. Conversely, referring to WAP protection control, local bus VSI, updated every 20 ms, can also be used for local decentralised protections.

The example of predictive area VSI for the Nordic 32 grid after contingency, described in § 11.4.2, is based on available local VSI of system buses around Bus 1041 (the bus which has, for the contingency considered, the highest sensitivity to VI).

Figure 11.29 gives the following predictive area results: Bus 1041 has the highest VI risk (red), followed in order by Buses 1043, 1044, 1045 and 1042, as in Table 11.1. Comparing Figs. 11.30 and 11.31 voltage trends helps us understand the Fig. 11.29 result.

Although the 400-kV buses are still far from a high voltage instability risk, at 130-kV in the south PMU indications sound an alarm in the identified area, with Bus 1041 being closest to instability (Fig. 11.29). This voltage instability can be stopped by V-WAP intervention, summarised here:

- The VSI real-time trend from a region toward a wide area, and its value in comparison with high risk slope thresholds and a maximum allowed limit, can be used to stop VI, up to reversing the risk of the map enlarging to the point of overlapping the topology map. This objective can be pursued by activating WAP control, VSI-based, internally to the dynamic critical region by locking the OLTC and progressively shedding loads there according to the limiting slope and overcoming thresholds.

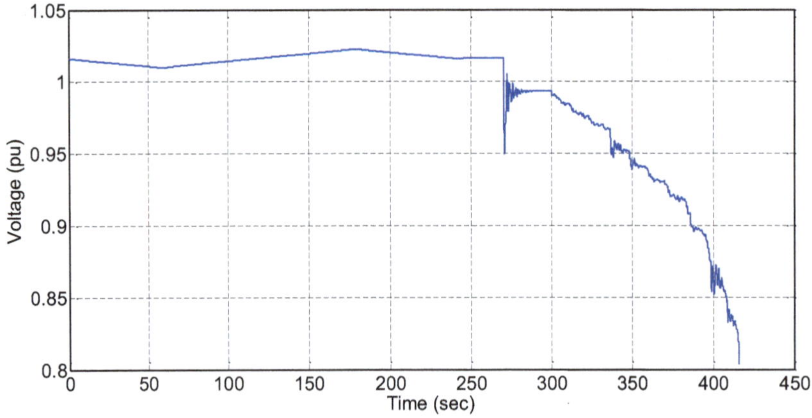

Fig. 11.30 Nordic 32 grid: Voltage at Bus 1041, contingency described in § 11.4

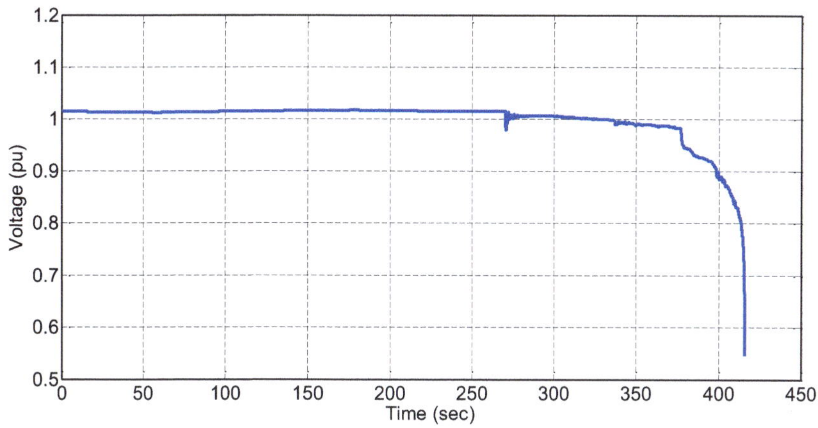

Fig. 11.31 Nordic 32 grid: Voltage at Bus 1042, contingency described in § 11.4.2

11.5 Area Voltage Protection Based on System Jacobian Computing Combined with OEL and OLTC Real-time Information

Voltage instability identification algorithms based on computing and real-time updating of a large power system mathematical model are potential candidates for supporting wide area voltage protecting systems. There is a wide variety of proposals concerning this, mostly based on the computing of the Jacobians, sensitivity matrices and eigenvalues [1–5, 13, 15–17, 22, 26]. Moreover, the mathematical models used are of a static or dynamic nature, more or less simplified in specific aspects in order to minimise computing effort.

One critical aspect these methods must consider is their necessary link to power system state estimation updating. More precisely, for a power system V-WAP to be effective and safe it must be frequently updated according to a real system's timely state estimation. These high speed dynamics assure V-WAP will operate its severe protecting controls on load shedding, or on line opening, or on switching-on cold or hot reactive power reserves in a way that is coherent and adequate with the power system's present operating condition. Protection decisions based on a power system's actual operating condition might be too late if they are made by simply referring to a load flow model that was pertinent minutes earlier. Such protecting controls could be too delayed to produce transient instability recovery or inadequate in terms of proper control effort to stop the instability underway.

This crucial aspect of a voltage instability controls often relegates the considered voltage instability protecting methods to academic analyses or to system studies, far from being widely used in real power systems as a secure and reliable V-WAP under all the possible operating conditions, mainly after severe contingencies.

Moreover, the algorithms under consideration often require heavy computing and are time consuming for many reasons:

1. *Large mathematical model elaboration.* In principle the power system model should be of a dynamic nature, including generator AVR and OEL and transformer OLTC control, as well. Furthermore, the loads model used has a relevant impact on the result.
2. *Algorithms computing derivative terms or matrix inversions.* All the derivative terms computed by the model require proper filtering if they are to be reliable, while large matrix inversion is time consuming.

These and other related considerations on the use of static models or eigenvalue computing make any hopes unlikely, now as in the future, that the algorithms considered will be used for real-time V-WAP, notwithstanding the many attempts to find a solution this way.

Section 9.2 provides information and references on some of these wide-area voltage instability indicators [1–5, 13, 15–17, 22, 26, 28], whose computing time is generally too long with respect to the considered instability dynamics (mainly from the time a real instability is about to begin up until the system collapse).

Both sensitivity analysis [1, 2, 4, 5, 26] and eigenvalue or singular value analysis of various Jacobian matrices [17, 22] are also frequently used in voltage stability analysis. Initially intended to provide preventive security indices [3], these linearisation-based techniques may prove useful in off-line study and recognition of the unstable system behaviour. They have been comparatively investigated in [22], where eigenvalues were computed in selected snapshots of the evolution of an unstable system, and in [26] and [4], where sensitivity analysis is coupled to a simplified time-domain simulation.

Methods that do not rely on a power system dynamic model require, in any event, a reliable model, especially for loads in emergency conditions. One of these [15] is aimed to devise a wide-area criterion for detecting long-term voltage instability; it does not try to anticipate load response, but rather it anticipates generator limitations recognised by internal computation on a power system mathematical model based on PMUs that provide synchronised bus voltage phasors. A system equilibrium model is fitted to each snapshot.

The theoretical background of this approach, one of sensitivity analysis, is extended to tracking eigenvalue movement around a maximum load power point. The sensitivities of reactive power generation to reactive power loads are considered. It is their change in sign that makes up the proposed criterion.

It is useful that we comment on the wide area voltage instability recognition method based on a power system model, in agreement with the introduction of the authors in [16], in which they refer to the adopted simplification and assumptions, they say:

- It is not clear how to reconcile the model used with measured system evolution in case discrepancies are caused, for instance, by events not accounted for in the model.
- In principle, the computations proposed could rely on bus voltages provided by a standard state estimator processing SCADA measurements. However, SCADA data are not collected and state estimators are not run at the rate considered (and

hence some proposed filtering would not be possible). Furthermore, standard (nonlinear) state estimators may encounter convergence problems in degraded system conditions.

- The data collected by remote terminal units (RTUs) suffer from a time skew that could make the proposed computations unreliable in the presence of significant transients.
- This proposal does not consider the important problems of PMU placement, communication infrastructure, measurement pre-processing by a state estimator, etc. Instead, it simply assumes that PMUs provide synchronised bus voltage phasors.
- Circuit breaker statuses are supposed to be provided by the same equipment or by the SCADA system.
- The algebraic model from which the Jacobian is derived assumes the short-term dynamics to be at equilibrium.

Therefore, this interesting contribution to VI analysis is a clear view point of the unsuitability of model-based wide area-VSI algorithms for on-field, real-time V-WAP applications.

A different case is that of generator AVR-OELs and transformer OLTC states and measurements that are directly acquired from the field without significant delay and used for the V-WAP objective. Here there are no safe theories on the way this information is to be properly used, but the widely shared belief is that their dynamics impact heavily on voltage instability. More precisely, OEL and OLTC operation significantly determines voltage instability, but VI indicators based on them could excessively anticipate or delay the actual instability time, therefore proving too precautionary and untimely when used for V-WAP applications. On the other side, waiting for all OEL and OLTC to operate in the considered wide area could make it too late to save a process.

Various attempts to use such real-time information [28], apart from the difficulty/complexity of data collection and timely exchange, often appear anticipatory, approximate and uncertain; They also depend on the situation analysed and the electrical distance of generators and transformers with active OLTC from the bus/buses most affected by voltage instability. To be reliable and ready for V-WAP applications, these measurement-based solutions require a consolidated theory and evidence of on-field positive results.

11.6 Conclusions

We presented a V-WAP protecting solution, one that is quite simple and effective at eliminating the risk of voltage instability in a given network area under SVR and TVR control. The main simplifications derive from the fact that any network area where WAP must operate is already dynamically defined by an operating SVR which, as we know, is a type which is decentralised over a number of power system areas.

We showed that in the presence of TVR the proposed real-time VSI index not only assists SVR in a control action but, above high risk thresholds, also becomes the reference for V-WAP intervention. This is the second, very important simplification.

The third simplification comes from ascertaining that, in the presence of TVR and SVR, V-WAP can operate corretly in a timely and selective, precise way by only considering measurements, voltage stability index, logic states, etc. that come from V-WAR; it does this by a very simple processing and computing of these data. When SVR is absent, or in case of PVR alone, the proposed WAP protection based on the voltage thresholds of the main bus (or on the thresholds of the pilot nodes, if they are known) can be easily used for timely load shedding with acceptable improvement of stability.

Simulation results confirm the strength of the proposed coordination between V-WAR and V-WAP in accordance with the "area concept", congenial to both SVR and WAP schemes. It should be noted that the proposed WAP, which is VSI-based, gains advantages by recognising a dangerous situation at relatively high voltages. This kind of operated control minimises load shedding due to timely reciprocal coordination and support between WAR and WAP: A compromise exists between voltage values to be sustained, when they are greater than a minimum threshold, and amount of load to be shed to avoid voltage collapse. In addition, using the risk index improves selection, inside an area, of the more critical load buses.

Though effective, a V-WAP solution based on the region/area concept and on risk indices based on reactive power inflow variation in the area considered, with PVR alone, is less effective than the previous solution (i.e., V-WAR/WAP coordination). In any event, it requires complex and dedicated data measuring, acquisition and elaboration with derivative terms that could compromise the required real-time.

A solution based only on reactive power inflow risk index fails in some cases.

A different V-WAP proposal makes use of a voltage instability algorithm to recognise in real time the approach of the loadability limit in the equivalent system seen by an EHV bus provided with PMU.

The *V-P* curve nose tip also indicating a high voltage instability risk and its recognition by real-time identification is an alternative to be used to predict the incoming instability that, if not stopped, could determine the system collapse. The prediction of the incoming nose tip can be also tuned by simply changing the control parameter value of the proposed identification algorithm and related VSI indicators, to achieve the anticipated or delayed result.

The bus VSI, locally computed by use of PMU phasor fast measurements, provides the information risk recognised by one bus. By collecting in real time, at a centralised level, the VSI information of the other main buses in the power system, it is possible to map the phenomenon extension underway, its expansion trend from one point in a region up to a wide area. Moreover, it is also possible to associate to the overloading expanding area an integrated VSI based on the risk values coming from area buses, each weighted by the local short circuit power. Therefore, a system operator and the system controls can be significantly helped by the *local VSI* and also by the *area VSI* to localise and control the risk of voltage instability and to prevent the system collapse in a timely fashion.

The effectiveness of the proposal is confirmed by dynamic tests performed on different power systems through detailed dynamic simulation models. The results are very clean and precise, notwithstanding the imposed continuous changing of the real system data (50-Hz band) and the high speed identification (updating every 20 ms).

Distinguishing off-line from real-time voltage instability indicators, we offered evidence of V-WAP performances based on their relevant differences vis-à-vis a practical contribution.

Not-real-time indices are to be used for (planning, dispatching) studies only, different from the case for true real-time indices, which are able to support moment by moment the power system operator and which mainly contribute to the increase of power system security by in automatic control and protection.

Attempts are frequently made to improve off-line VSIs by using them in real-time V-WAP by studying and assessing alternative solutions, even if this has seen a small hope of success. Their computing time is generally too long with respect to the instability dynamics considered (mainly from the time a real instability is about to begin up to the system collapse).

The attempt that combines a simplified computing of the power system Jacobian and sensitivity matrices with information on the generators (OEL operation) and transformers (OLTC operation up to saturation) has been analysed, even though the slow updating of power system state estimation represents a limitation that has been difficult to overcome.

References

1. Ajjarapu V (2006) Computational techniques for voltage stability assessment and control. Springer, New York
2. Begovic M, Phadke AG (1992) Control of voltage stability using sensitivity analysis. IEEE Trans Power App Syst 7(1):114–123
3. Cañizares C (ed) (2002) Voltage stability assessment: concepts, practices, and tools. Special Publication. IEEE Power Engineering Society, Piscataway
4. Capitanescu F, Van Cutsem T (2005) Unified sensitivity analysis of unstable or low voltages caused by load increases or contingencies. IEEE Trans Power App Syst 20(1):321–329
5. Carpentier J, Girard R, Scano E (1984) Voltage collapse proximity indicators computed from an optimal power flow. In: Proc. 1984 8th Power Syst Comput Conf (PSCC), Helsinki, Finland, pp 671–678
6. CIGRE-TF38-02-11 (1994) Indices predicting voltage collapse including dynamic phenomena. CIGRE Report
7. Corsi S (2005) Wide are voltage regulation and protection: when their co-ordination is simple. Proceedings of Power Tech 05, St. Petersburg, Russia, June 2005
8. Corsi S (2009) Wide area voltage regulation and protection. Proceedings of power tech–09, Bucharest, Romania, June/July 2009
9. Corsi S (2009) Wide area voltage protection. IET Proc Gener Transm Distrib 4(10):1164–1179

10. Corsi S, Sabelli C (2004) General blackout in Italy: Sunday September 28, 2003, h. 03:28:00. Proceedings of IEEE-PES General Meeting, PSSS Panel on Recent Blackouts, Denver
11. Corsi S, Taranto GN (2007) Voltage instability: the different shapes of the "nose". Bulk Power System Dynamics & Control, IREP-VII, Charleston, South Carolina, August 2007
12. Corsi S, Taranto GN (2008) A real-time voltage instability identification algorithm based on local phasor measurements. IEEE Trans Power App Syst 23(3):1271–1279
13. Corsi S, Taranto GN (2012) Voltage instability alarm by real-time predictive indicators, Proceedings IEEE—General Meeting, San Diego, August 2012
14. Corsi S, Cappai G, Valadè I (2006) Wide area voltage protection. CIGRE Conference, Paris
15. Glavic M, Van Cutsem T (2009) Wide-area detection of voltage instability from synchronized phasor measurements. Part I: Principle. IEEE Trans Power App Syst 24(3)
16. Glavic M, Van Cutsem T (2009) Wide-area detection of voltage instability from synchronized phasor measurements. Part II: Simulation results. IEEE Trans Power App Syst 24(3)
17. Löf P-A, Smed T, Andersson G, Hill DJ (1992) Fast calculation of a voltage stability index. IEEE Trans Power App Syst 7:154–164
18. Madani V, Novosel D, Apostolov A, Corsi S (2004) Innovative solutions for preventing wide area disturbance propagation. Bulk Power System Dynamics and Control, IREP-VI. Cortina d'Ampezzo, Italy, August 2004
19. Marannino P, Zanellini F, Merlo M, Corsi S, Pozzi M, Dell'Olio G (2001) Evaluation of load margin with respect to voltage collapse in presence of secondary and tertiary voltage regulation. (Symposium) Bulk Power System Dynamics & Control, IREP-V, Onomichi, Japan, August 2001
20. Marannino P, Zanellini F, Berizzi A, Medina D, Merlo M, Pozzi M (2002) Steady state and dynamic approaches for the evaluation of the loadability margins in the presence of the secondary voltage regulation. MedPower Conference, Athens, November 2002
21. Milosevic B, Begovic M (2003) Voltage-stability protection and control using a wide-area network of phasor measurements. IEEE Trans Power App Syst 18(1):121–127
22. Morison GK, Gao B, Kundur P (1993) Voltage stability analysis using static and dynamic approaches. IEEE Trans Power App Syst 8(3):1159–1171
23. Rehtanz C (2006) Wide area monitoring and control for transmission capability enhancement. CIGRE Technical Brochure. Working Group C4:601
24. Smon I, Verbic G, Gubina F (2006) Local voltage-stability index using Tellegen's theorem. IEEE Trans Power App Syst 21(3):1267–1275
25. Taylor CW (1994) Power system voltage stability. McGraw-Hill, New York
26. Van Cutsem T (1995) An approach to corrective control of voltage instability using simulation and sensitivity. IEEE Trans Power App Syst 10(2):616–622
27. Verbic G, Gubina F (2004) A new concept of voltage-collapse protection based on local phasors. IEEE Trans Power Deliv 19(2):576–581
28. Vournas C, Van Cutsem T (2008) Local identification of voltage emergency situation. IEEE Trans Power Syst 23(3):1239–1248
29. Vu K, Begovic MM, Novosel D, Saha MM (1999) Use of local measurements to estimate voltage-stability margin. IEEE Trans Power Syst 14(3):1029–1035
30. Zima M, Larsson M, Korba P, Rehtanz C, Andersson G (2005) Design aspects for wide-area monitoring and control systems. Proc IEEE 93(5):980–996

11.6 Conclusions

Appendix

Appendix A

Synchronous Machine Ideal Model

The three-phase electrical machine can be schematised [A.1] with three windings distributed along the stator circumference at 120° apart from each other. The fourth winding determining the magnetic field is solid with the rotor. The three stator windings and the field winding are respectively named "*a*", "*b*", "*c*" and "*f*".

Two reference axes, solid with the rotor, are called the "direct" ("*d*") and "quadrature" ("*q*") axes. The angle θ is between the direct axis and the phase "*a*" axis.

- The direct (or polar) *d*-axis is coincident with the *f*-axis, that is, with the magnetic field produced by the rotor.
- The quadrature (or interpolar) *q*-axis is 90° ahead of *d*.

Under the hypotheses to overlook the damping windings, magnetic saturations and iron losses, the instantaneous voltage at each winding can be computed as follows:

$$V = RI + \frac{d\Psi}{dt},$$

where *R, I* and Ψ are the winding resistance, current and flux linkage. Solving this equation requires knowledge of the auto and mutual inductances and their dependence on angle $\theta(t)$.

With $\Omega(t)$ being the rotor angular speed (nominal value Ω_N),

$$\vartheta(t) = \int_0^t \Omega(\gamma)d\gamma + \vartheta_0.$$

S. Corsi, *Voltage Control and Protection in Electrical Power Systems*,
Advances in Industrial Control, DOI 10.1007/978-1-4471-6636-8

The Park transform allows us to rewrite the equations referring to the *a*-, *b*-, and *c*-axes by taking as new references the *d*- and *q*-axes solid with the rotor.

From here, the Park transform result is shown as:

$$\begin{aligned} V_d &= -\Omega\Psi_q + \frac{d\Psi_d}{dt} + RI_d, \\ V_q &= \Omega\Psi_d + \frac{d\Psi_q}{dt} + RI_q, \\ V_f &= \frac{d\Psi_f}{dt} + R_f I_f, \end{aligned} \tag{A.1}$$

where I_d and I_q are positive if they determine a demagnetising effect (the synchronous maching operates as a generator with exiting currents).

Overlooking armature winding resistence R and representing the operator d/dt by "p", the "per unit" equations become

$$\begin{aligned} V_d &= -\frac{\Omega}{\Omega_N}\Psi_q + \frac{p}{\Omega_N}\Psi_d, \\ V_q &= \frac{\Omega}{\Omega_N}\Psi_d + \frac{p}{\Omega_N}\Psi_q. \end{aligned}$$

The terms p/Ω_N consider only flux variations; therefore, they are null at steady state and numerically negligible during transients because they are divided by Ω_N.

As a consequence, the generator terminal voltage amplitude V_m substantially depends on the magnetic flux and the rotor speed:

$$V_m = \sqrt{V_d^2 + V_q^2}.$$

The exciter fixes the flux while the term Ω/Ω_N significantly contributes to voltage value. Under normal operating conditions, $\Omega/\Omega_N \approx 1$; therefore, in these cases the exciter acquires more roles in imposing the V value:

$$\begin{aligned} V_d &\approx -\Psi_q, \\ V_q &\approx \Psi_d. \end{aligned}$$

Returning to Eq. (A.1) and considering the Ψ_d, Ψ_q, Ψ_f dependence from the auto and mutual inductances, eliminating Ψ_f by substitution results in the following:

$$\Psi_d = A(s)V_f - L_d(s)I_d,$$
$$\Psi_q = -L_q(s)I_q,$$
$$I_f = B(s)V_f + sA(s)I_d,$$

where transfer functions $A(s)$, $B(s)$, $L_d(s)$, $L_q(s)$ depend on machine-building characteristics.

The per-unit model becomes

$$\Psi_d = a(s)V_f - L_d(s)I_d,$$
$$\Psi_q = -L_q(s)I_q,$$
$$I_f = b(s)V_f + \frac{\left(A_N/P_f^*\right)s}{\Omega_N a(s)I_d},$$

where A_N is alternator apparent nominal power; P_f^* is exciting power value determining nominal stator voltage at no-load, and operating point on the air-gap linear characteristic. Moreover:

- $a(0)=b(0)=1$;
- $L_d(0)=X_d$ (*direct axis synchronous reactance*);
- $L_q(0)=X_q$ (*quadrature axis synchronous reactance*).

Starting from this introductory and simplified model, the reader can now intuitively understand its possible extension to less simplified representations.

General, Linearised Model

A general alternator model, also valid when including the damping circuits on the *d*- and *q*-axes, is here after shown:

$$\Psi_d = \frac{1+sT_a}{(1+sT'_{d0})(1+sT''_{d0})}V_f - X_d\frac{(1+sT'_d)(1+sT''_d)}{(1+sT'_{d0})(1+sT''_{d0})}I_d = V_q,$$
$$\Psi_q = -X_q\frac{1+sT'_q}{1+sT'_{q0}}I_q = -V_d, \tag{A.2}$$
$$X'_d = X_d\frac{T'_d}{T'_{d0}},\ X'_q = X_q\frac{T'_q}{T'_{q0}},\ X''_d = X_d\frac{T''_d}{T''_{d0}},\ X''_q = X_q\frac{T''_q}{T''_{q0}}.$$

These equations also show the alternator subtransient constants on both the direct and quadrature axes. They take into account the damping windings and massive rotor effects. On the direct axis the impact is negligeble due to the dominant effect of the field winding, whereas on the quadrature axis the damping winding represents the sole dynamic effect.

Neglecting the T''_d and T''_{d0} subtransient time constants, the (A.2) become

$$V_q = \frac{1}{\left(1+sT'_{d0}\right)}V_f - X_d\frac{\left(1+sT'_d\right)}{\left(1+sT'_{d0}\right)}I_d = \frac{V_f-\left(X_d-X'_d\right)I_d}{\left(1+sT'_{d0}\right)} - X'_dI_d,$$

$$V_d = \frac{\left(X_q-X'_q\right)I_q}{\left(1+sT'_{q0}\right)} + X'_qI_q. \tag{A.3}$$

Because V_d and V_q are the components on the d- and q-axes of the V_m vector (having constant amplitude during a sinusoidal regime) solid with the rotor, then V_a, V_b, V_c represent the V_m vector projections with respect to the a, b, c phase axes, respectively.

Analogously, for i_m and Ψ. This means:

$$\begin{aligned}
\overline{v_m} &= V_d + jV_q,\\
\overline{i_m} &= I_d + jI_q,\\
\overline{A} &= \overline{v_m i_m}^{*} = \left(V_d + jV_q\right)\left(I_d - jI_q\right) = P + jQ,\\
P &= \left(V_dI_d + V_qI_q\right) = \text{activepower},\\
Q &= \left(V_dI_d - V_qI_q\right) = \text{reactivepower}.
\end{aligned}$$

Typical values of the synchronous machine are given in Table A.1:

Generator Operating on a Large Power System

We refer to the case represented in Fig. A.1:

The scheme shows a generator connected through a transformer and an equivalent line to an "infinite bus" characterised by infinite ($S_{SC}=\infty$) apparent power and constant voltage V_R. This schematisation is corrently used to analyse the synchronous machine dynamics. In evidence is the generator voltage control loop that operates in parallel to the turbine speed regulator.

The overal reactance connecting the generator to the grid is obtained combining the transformer reactance X_T with the equivalent line reactance X_L: $X_e=X_T+X_L$.

The synchronous machine Eq. (A.3) represent the generator model, Park transformed. The electrical connection between the alternator and the infinite bus require new equations, here after introduced.

Table A.1 Synchronous machine typical parameter values

Axis	Parameter	Turbo alternator	Hydraulic generator
Direct (d)	X_d	1.9 p-u.	1.1 p-u.
	X_d'	0.3 p.u.	0.35 p.u.
	X_d''	0.25 p.u.	0.33 p.u.
	T_{d0}'	7–10 s	6 s
	T_d'	1.1 s	1.9 s
	T_{d0}''	0.01 s	0.08 s
	T_d''	0.008 s	0.07 s
Quadrature (q)	X_q	1.7 p.u.	0.7 p.u.
	X_q'	0.25 p.u.	0.33 p.u.
	T_{q0}'	0.25 s	0.15 s
	T_q'	0.04 s	0.07 s

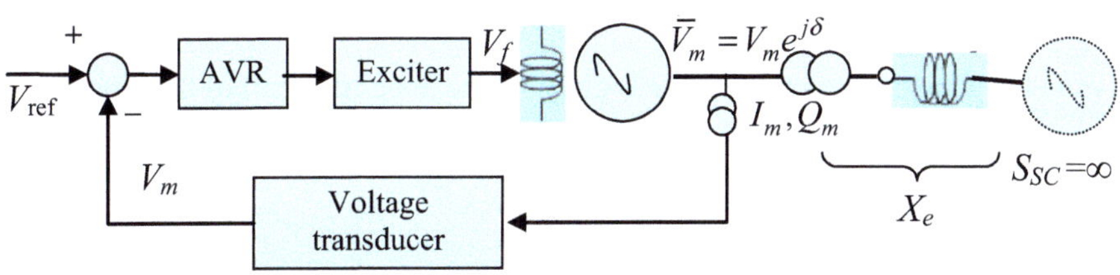

Fig. A.1 Generator connected to a prevailing grid (infinite bus) with AVR in closed loop

The angle difference between the $\overline{v_R}$ phasor (solid to the infinite bus rotor: ($\delta_R = 0.0$) and the q-axis of the generator rotor is called δ (Fig. A.2).

The electrical link between the two generators is described by the following equations:

Fig. A.2 Phasor diagram of the infinite bus rotor with respect to the d- and q- axes of the alternator rotor

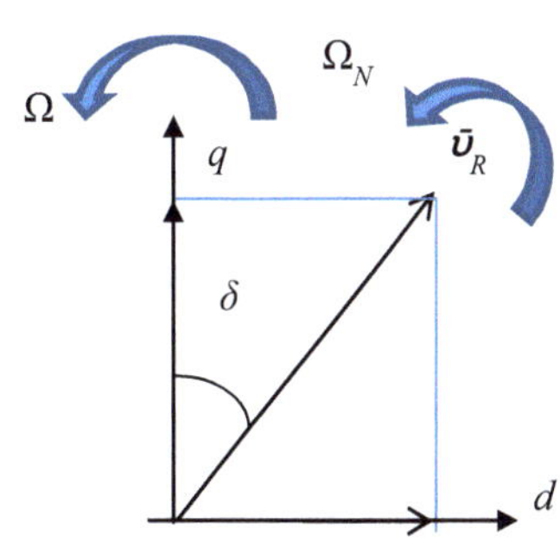

$$\overline{v_m} = \overline{v_R} + jX_e\overline{i_m},$$
$$\overline{i_m} = \frac{\overline{v_m} - \overline{v_R}}{jX_e} = j\frac{\overline{v_R} - \overline{v_m}}{X_e} = \frac{V_q - V_R\cos\delta + j(V_R\sin\delta - V_d)}{X_e}.$$

Therefore,

$$I_d = \frac{V_q - V_R\cos\delta}{X_e}, I_q = \frac{V_R\sin\delta - V_d}{X_e}.$$

The generator voltage amplitude is $V_m = \sqrt{V_d^2 + V_q^2}$, whereas V_f is provided by the voltage regulator AVR having $\mu(p)$ as the control function.

The generator electrical power is given by the real part of the product:

$$P_e = \Re_e\left\{\overline{v_m}\overline{i_m^*}\right\} = V_d I_d + V_q I_q,$$

while the generator reactive power is given by the imaginary part of the product:

$$Q_e = \Im_m\left\{\overline{v_m}\overline{i_m^*}\right\} = V_q I_d - V_d I_q.$$

All the above equations related to the link with the equivalent grid provide a non-linear model.

Mechanical Equations

Moving to the mechanical part of the process, where P_m represents the motor power from the turbine, the link between the rotor speed and the accelerating power is

$$\frac{\Omega}{\Omega_N} = \frac{1}{pT_m}(P_m - P_e),$$
$$T_m = \frac{J\Omega_{MN}^2}{A_N},$$
$$\delta = \frac{1}{p}(\Omega - \Omega_N), \tag{A.4}$$

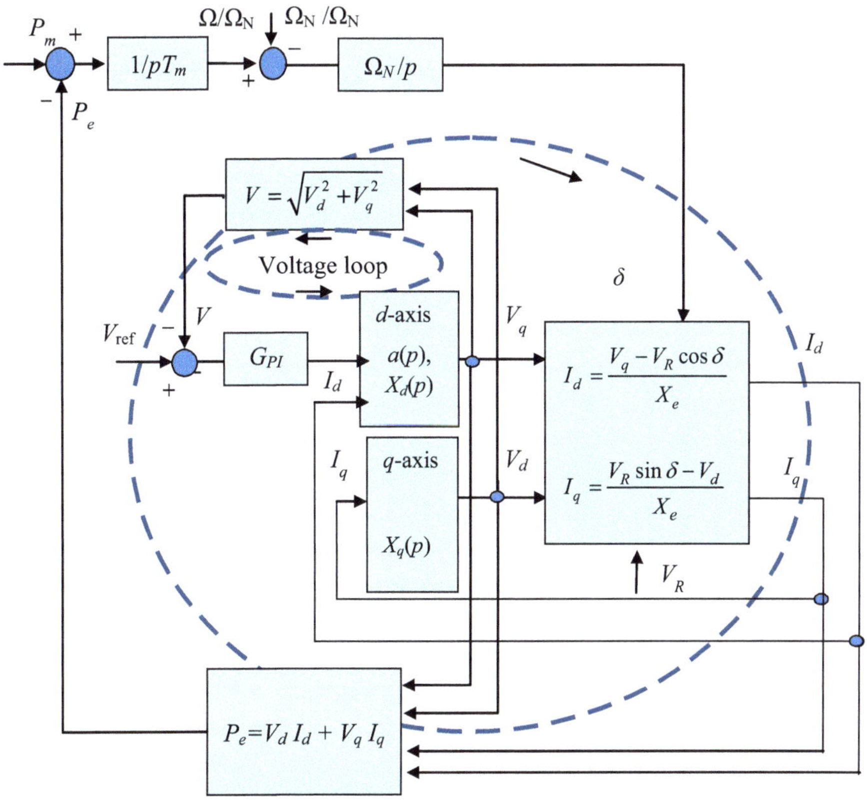

Fig. A.3 Block representation of generator model in Fig. A.1

where:

T_m = at no-load, generator starting time with the nominal mechanical torque;
J = moment of inertia of rotating masses (generator rotor + turbine shaft);
Ω_{MN} = mechanical nominal angular speed while Ω_N is electrical nominal angular speed;
δ = angle given by integral of speed difference between generator and prevailing grid rotating at Ω_N.

The above electrical and mechanical equations together describe the block diagram in Fig. A.3. Examining the scheme, the system alternator-grid puts in evidence two control loops:

- The voltage loop operating through the voltage regulator;
- The electromechanical nature loop, with dynamics determined by the above mechanical equations characterised by two series integrators.

As can be seen, these two control loops interact with one another.

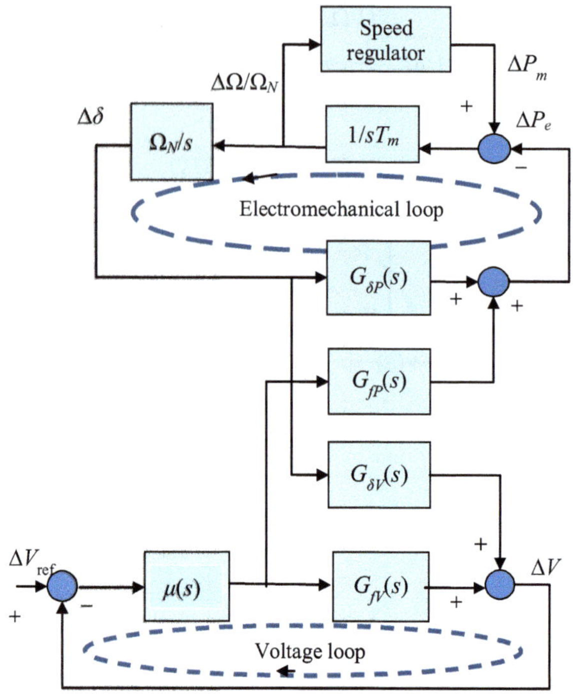

Fig. A.4 Block diagram of the linear model of the generator in Fig. A.1

The Linearised Model

The considered nonlinear system, linearised by imposing small variations around an operating point, and therefore representable with transfer functions by substituting "p" for the complex variable "s", is provided by Fig. A.4:

In fact, from § A.1, the alternator-grid linearised model is given by Eq. (A.3):

$$V_q = \frac{V_f}{1+sT'_{d0}} - X_d \frac{1+sT'_d}{1+sT'_{d0}} I_d,$$
$$V_d = X_q \frac{1+sT'_q}{1+sT'_{q0}} I_q. \tag{A.5}$$

In the field of interest, high frequency phenomena are analysed; therefore, Eq. (A.5) can be further approximated as follows:

$$V_q \cong \frac{V_f}{sT'_{d0}} - X'_d I_d$$
$$V_d \cong X'_q I_q \tag{A.6}$$

Moreover, because $X'_d \approx X'_q \approx X_i$, (A.6) becomes

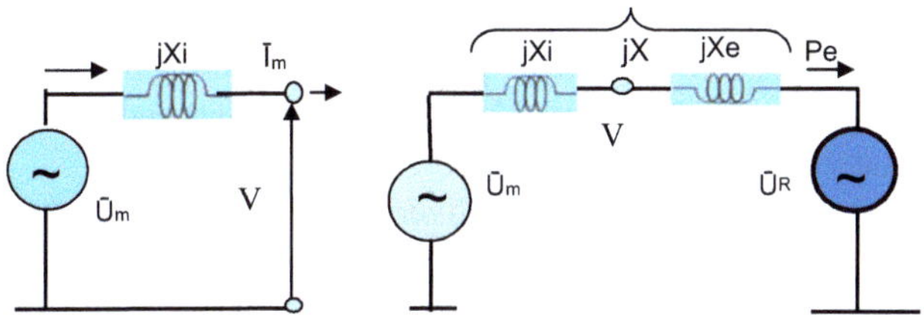

Fig. A.5 Equivalent scheme of the alternator (*left*) grid interconnected (*right*)

$$\begin{aligned} V_q &\cong \frac{V_f}{sT'_{d0}} - X_i I_d = e'_q - X_i I_d \\ V_d &\cong X_i I_q; \end{aligned} \tag{A.7}$$

Therefore,

$$\overline{v_m} = V_d + jV_q = je'_q - jX_i\left(I_d + jI_q\right) = \overline{e_i} - jX_i\overline{i_m}.$$

This equation corresponds to Fig. A.5, representing the equivalent scheme of the alternator within the frequency field of interest.

As said, equations that model the electrical link between the two system buses indicate the nonlinear relationships:

$$\overline{i_m} = \frac{\overline{v_m} - \overline{v_R}}{jX_e} = \frac{\overline{e_i} - \overline{v_R}}{j\left(X_i + X_e\right)} = \frac{je'_q - \left(V_R \sin\delta + jV_R \cos\delta\right)}{j\left(X_i + X_e\right)},$$

$$I_d + jI_q = \frac{e'_q - V_R \cos\delta}{X_i + X_e} + j\frac{V_R \sin\delta}{X_i + X_e} = \frac{V_q - V_R \cos\delta}{X_e} + j\frac{V_R \sin\delta - V_d}{X_e}.$$

Moreover, at the V_R bus:

$$P + jQ = \overline{v}_R \overline{i_m}^*,$$

$$\overline{e_i} = je'_q = e'_q e^{j\delta}, \qquad \overline{v_R} = jV_R = V_R e^{j0} = V_R,$$

$$P + jQ = V_R j\frac{e'_q e^{-j\delta} - V_R}{X_t + X_e},$$

$$P = \frac{e'_q V_R \sin\delta}{X}, \qquad Q = \frac{e'_q V_R \cos\delta - V_R^2}{X}.$$

Without line losses, the active power equation is the same for both the extreme buses, while the expression of reactive power only refers to the Q entering into the infinite bus. In fact, the Q delivered by the generator internal bus differs by the amount Im^2X due to the reactive power absorbed by the line.

In general, referring to Fig. A.5, the delivered power and generator voltage are given by nonlinear equations:

$$P = f\left(e'_q, V_R, \delta\right), \qquad V_m = g\left(e'_q, V_R, \delta\right).$$

These nonlinear equations link the electromechanical and voltage loops.

Figure A.4 shows the linearised links obtained by differentiating the above P_e and V_m equations. They are:

$$\Delta P = K\Delta\delta + h\Delta e'_q, \qquad \Delta V_m = h_1\Delta\delta + h_2\Delta e'_q.$$

with

$$K = \left[\frac{\partial P}{\Delta\delta}\right]_{P_e = P_e^0}, \qquad h = \left[\frac{\partial P}{\Delta e'_q}\right]_{P_e = P_e^0},$$

$$h_1 = \left[\frac{\partial V_m}{\Delta\delta}\right]_{V_m = V_m^0}, \qquad h_2 = \left[\frac{\partial V_m}{\Delta e'_q}\right]_{V_m = V_m^0}.$$

From these results it is evident that linearisation based on sensitivity gives proportional coefficients and not transfer functions, as was preliminarily introduced in the Fig. A.4 links.

Therefore, based on the approximations used, the linear model of the system is represented in Fig. A.6, which includes the speed regulator and additional stabilising feedback, discussed in Chap. 3.

The scheme clearly evidences the two loops' interaction due to the h and h_1 blocks:

- A variation in δ determines a ΔV_m and, through the voltage regulator, a ΔV_f;
- A variation in e'_q determines a ΔP and, through the electromechanical loop, a $\Delta\delta$.

The electromechanical and voltage loops are coupled, unless parameters h and h_1 are zero. This happens if $\delta^0 = 0$ (that is, $P_e^0 = 0$) because under this condition,

$$i_d^0 = i^0, \qquad V_q^0 = V_m^0.$$

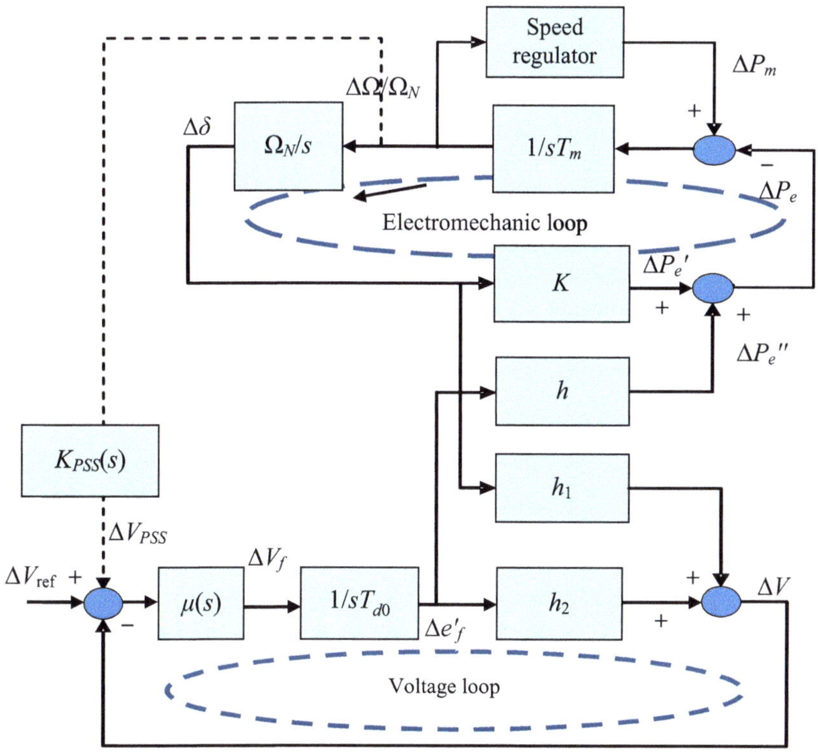

Fig. A.6 Block diagram of linear model of power station in Fig. A.1

Therefore,

$$V_d^0 = 0, \qquad i_q^0 = 0,$$

$$h = \left[\frac{\partial P}{\Delta e_q'}\right]_{P_e = P_e^0} = \frac{V_R \sin\left(\delta^0\right)}{X} = 0.$$

Moreover,

$$V_m = V_q = a(s)V_f - X_d(s)I_d,$$
$$\Delta V_m = \Delta V_q = a(s)\Delta V_f - X_d(s)\Delta I_d,$$
$$\Delta I_d = \frac{\Delta V_q - V_R \sin\left(\delta^0\right)\Delta\delta}{X_e}.$$

ΔV_m being independent from $\Delta\delta$ determines that $h_1 = 0$.

To sum up, with $\delta^0 = 0$:

- Active power is zero;
- ΔP_e is independent from ΔV_f, and $h = 0$;
- ΔV_m is independent from $\Delta\delta$ and $h_1 = 0$.

Conversely, with $\Delta P_e^0 \neq 0$, the two control loops interact unless the voltage control loop is open. In fact, under manual voltage control, V_f is a constant value, therefore it is independent from $\Delta\delta$.

Analogously, with constant V_f, no contribution comes to P_e from the voltage loop.

Reference

A.1. Kimbark EW (1956) Power system stability, vol 3. Wiley, New York

Index

S. Corsi, *Voltage Control and Protection in Electrical Power Systems*,
Advances in Industrial Control, DOI 10.1007/978-1-4471-6636-8

MIX
Papier aus verantwortungsvollen Quellen
Paper from responsible sources
FSC® C105338
FSC
www.fsc.org

If you have any concerns about our products,
you can contact us on
ProductSafety@springernature.com

In case Publisher is established outside the EU,
the EU authorized representative is:
Springer Nature Customer Service Center GmbH
Europaplatz 3, 69115 Heidelberg, Germany

Printed by Libri Plureos GmbH
in Hamburg, Germany